B

Neuroembryology

The Selected Papers

B

Viktor Hamburger

Viktor Hamburger

Neuroembryology

The Selected Papers

Introduction by Ronald W. Oppenheim

1990

Birkhäuser
Boston Basel Berlin

Viktor Hamburger
Department of Biology
Washington University
St. Louis, MO 63130

Library of Congress Cataloging-in-Publication Data
Hamburger, Viktor, 1900–
    Neuroembryology: the selected papers of Viktor Hamburger.
      p. cm,
    Includes bibliographical references.
    ISBN 0-8176-3459-2
    1. Developmental neurology. I. Title.
    [DNLM: 1. Nervous System—embryology—collected works. WL 101
H199n]
    QP363.5.H36 1990
    591.3'34—dc20
    DNLM/DLC
    for Library of Congress   89-18247

ISBN 0-8176-3459-2
ISBN 3-7643-3459-2

Printed and bound by Edwards Brothers, Incorporated, Ann Arbor, Michigan.
Printed in the U.S.A.

9 8 7 6 5 4 3 2 1

Contents

---

* The numbers in brackets following the titles of the papers are from the Bibliography of Viktor
Hamburger (pages xv–xxii in this volume).

Contents

Introduction
Viktor Hamburger: Pioneer Neuroembryologist, Teacher, Colleague, and Friend

Ronald W. Oppenheim

The publication of collected works of famous and influential novelists and poets is a commonplace occurrence, often bringing together the well-known and the obscure, the famous and the infamous, and the first rate and the second (or worse) rate efforts of writers. One goal of publishing the collected works of literary figures is to pay homage to a meritorious individual's lifetime efforts. Another goal, I believe, is to draw attention to forgotten or previously unknown masterpieces and thereby attempt to forestall the tendency of many of us to ignore all but the latest publications in a field. In contrast to literature, science, with few exceptions (e.g., Darwin, Freud), seldom honors its leaders, living or dead, by the publication of collected works or even selected compilations of their writings. Although "Festschriften" are commonplace in science, and perhaps to some extent subserve the same goals as literary collected works, it is noteworthy that they seldom include the voice of the honoree. This difference between literature and science is, in my view, largely owing to the regrettable tendency of most scientists to believe the old adage that "old news is no news" (or old science—and "old" scientists—is science not worth knowing about). Consequently, it is propitious that the present compilation of some of the writings of one of this century's foremost neuroembryologists, Viktor Hamburger, be published at this time.

As is elegantly and thoughtfully described by Viktor in some of the essays in this collection, neuroembryology began around the turn of the present century, largely through the pioneer efforts of Wilhelm His, S. Ramón y Cajal, and Ross Harrison. Accordingly, although it is not possible to assign a specific time or date to its foundation, the centenary of the establishment of neuroembryology is close at hand and I can think of no more fitting event to celebrate 100 years of neuroembryology than the present set of papers by Viktor Hamburger. Born in 1900 in Silesia, then a province of Germany, Viktor has been at the center of neuroembryology for more than 60 years. During that time he has made important contributions to most of the major issues in neuronal development, and in some areas he has been the acknowledged leader for half a century. Having begun his research on amphibian embryos in Hans Spemann's lab in Freiburg in the early 1920's, he later switched to his beloved chick embryo while working as a research fellow in Frank Lillies' laboratory in Chicago in the 1930's. In the ensuing 50 years, he deviated only once from the chick embryo model, to examine the development of motility in the rat fetus in an attempt to determine the correctness of his views regarding vertebrate behavioral ontogeny (Narayanan, et al., 1971; and see below).

Although it isn't evident from the papers reprinted here, it was largely through his efforts that the chick embryo was established as the species of choice for most neuro-embryological investigations. His early research papers on the chick, together with his *Manual of Experimental Embryology* (1942) and the later morphological stage series of the chick embryo (1951) done with H. Hamilton, placed the chick on the embryological map. The stage series has been honored as one of the most frequently cited papers in biology.

Because the present set of papers is composed solely of reviews and essays, one regrettably misses the logic and elegance of experimental design and interpretation of results that characterize Viktor's more empirical papers. Nonetheless, in all of his writings (even his earliest papers written in English) one is struck by the simple and succinct, yet elegant and effective, English prose. As one of his students, and an English-speaking native as well, I still strive to attain some semblance of his admirable writing style.

Although all scientists conduct research within some kind of conceptual framework, it is evident in many of his publications that Viktor eschews unbridled speculation and grand theorizing. It would be an exaggeration, but not a very great one, to characterize his approach in the words of the TV detective, Sergeant Friday, who admonished victims with "only the facts madam, just the facts." For instance, in a letter to a Dutch historian of science who was attempting to pin down his theoretical approach to behavioral development, Viktor responded," I have noted repeatedly that ethologists, behaviorists and comparative psychologists are much more inclined to discuss theories and specula-tions and form hardened dogmas which they then try to prove by experiments. And if the experiments do not come out right they make the most absurd hypotheses to fit the data. These people do not seem to understand that other people, like experimental embryologists, can be motivated by the fascination of *phenomena*, entirely without preconceived ideas. They are in love with the living embryo, whereas many psychologists are in love with theories and speculations." However, once one has some inkling of the phenomena then, as he has put it, "one is guided by some frame of reference that may be called a working hypothesis, but also by hunches, preconceptions, and even strong personal preferences that may be rooted in very deep strata of the personality" (1988, p. 33).

Before making a few statements about the specific papers contained in this collection, I feel compelled to comment briefly on Viktor Hamburger the man. What one often misses in such collected works, is some measure of the individual, outside of his professional contributions. To all of his friends and colleagues, Viktor is known as a warm, witty, self-effacing, and altogether decent person who, unlike many famous scientists, has never unduly promoted his own accomplishments. In addition to his interests in science and biology, Viktor also has wide ranging but deep interests in several other areas that go well beyond that of the dilettante. Art, literature, poetry, politics, philosophy, music and modern dance are just some of the topics on which he can converse with great enthusiasm and expertise. Furthermore, the remarks quoted above about "love of the embryo" serve to reveal still another facet of a complex personality, that of the naturalist, the devotee of animals, plants, and the environment (especially the mountains). As he has explained: "At age 14 I was exploring the countryside around the small German town in which I grew up. I collected plants, animals, fossils and brought home frog and salamander eggs and watched them develop and metamorphose. There was never any doubt in my mind that I would become a naturalist. Years later I made the conscious choice to study Zoology" (Hamburger, 1989; reprinted here). In short, Viktor fits perfectly the definition of the endearing Yiddish term "ein mensch," a bona fide human being. I know that I speak for all of his friends and colleagues when I say that I consider it a special privilege to know him. Perhaps his long-time friend Tom Hall said it best: "With Viktor, one cannot simply have an affair of the mind, it ultimately becomes an affair of the heart."

The selective reviews and essays Viktor has chosen for inclusion here cover a wide range of topics, from neuroembryology and psychobiology to philosophy and history of science, again reflecting the interests of an active mind fascinated by diverse phenomena. In the first set of papers on Developmental Neurobiology (Section I), one can trace the development of ideas about most of the central questions of neuronal ontogeny in this century. In the first of these, originally published in German in *Naturwissenschaften* in 1927, Viktor established the framework that guided most of his research over the next 60 years (the only missing ingredient was the problem of behavioral development in the embryo, a subject that occupied Viktor and his students for over 10 years in the 1960's and 70's). Here, already in 1927, one finds the description of three major research areas that one has come to identify closely with Viktor Hamburger: the influence of innervation on limb development; the guidance of nerves to their targets; and the effects of the growing limb on innervating nerve centers, including cell death and survival. Although Viktor has made many important and original contributions to all of these problems, it was, in my opinion, his efforts on the latter that will stand as his single most influential legacy to neuroembryology. Following the pioneering efforts of M. Shorey and S. Detwiler, Viktor began experiments, first with amphibians and later with the chick embryo, that led eventually to the discovery of the nerve growth factor (NGF). Despite some claims to the contrary (Levi-Montalcini, 1988), it is obvious from the record that he played a central role in the discovery of NGF (Hamburger, 1989; Purves and Sanes, 1987). Consequently, it was a great disappointment to his friends and colleagues in neuroembryology when he was not included in the 1986 award of the Nobel prize to R. Levi-Montalcini and S. Cohen. In this respect, he is in good company. Ross Harrison also failed to share in the 1935 Nobel prize with Spemann for his fundamental contributions to experimental embryology. According to the records of the Nobel Committee, "opinions diverged, and in view of the rather limited value of the method (*tissue culture*) and the age of the discovery, an award could not be recommended" (1962, p. 259). As Viktor has noted, "What was actually of limited value was the judgement of the Committee and not Harrison's achievements" (1980, p. 611). Some things never change! Harrison's failure to win the Nobel prize has in no way lessened his impact on embryology and I'm certain that posterity will be equally kind to Hamburger.

As mentioned above, beginning in the early 1960's Viktor's research interests took a new and rather unexpected turn. At the age of 60, when most of us are contemplating retirement, and while still chairman of the Zoology Department at Washington University with a heavy teaching load, he initiated a seminar series of studies on the ontogeny of behavior in the chick embryo. Despite the fact that plans for the study of the embryology of behavior were not explicitly included in his 1927 article in *Naturwissenschaften* (see above), the decision to study behavior may, nevertheless, have had its seeds sown at about the same time. As Viktor has related, "Shortly before I came to this country (*1932*) I had laid out plans for experiments on bird behavior—and I might have joined the camp of ethologists, if I had stayed in Germany." Although many of the pioneer neuroembryologists in this century (e.g., Coghill, Detwiler, Weiss, Sperry, Windle, and even Harrison) included behavioral development as an important topic for study by the neuroembryologist (Oppenheim, 1982), by 1960 such studies had fallen into disfavor and the field was largely moribund. Viktor's studies and ideas on this topic, especially as summarized in his 1963 review in the *Quarterly Review of Biology*, rekindled an interest in this field and initiated experimental studies on vertebrate and invertebrate neurobehavioral development that continue to the present time. A fundamental influence of Viktor's work on this field was the firm establishment of the fact that embryonic motility or behavior is mediated by the endogenous activity of neurons in the central nervous system with little, if any, contribution from sensory input. Although this notion of motor primacy is now considered commonplace, at the time it generated considerable controversy as indicated in his 1973 review of this topic, reprinted here. The idea that behavioral development in the embryo occurred without benefit of sensory input challenged one of the most cherished beliefs of comparative and developmental psychologists who held that motor patterns were gradually acquired in the embryo by a kind of trial-and-error process mediated by sensory experience. By using the simple, but powerful and elegant techniques of experimental embryology, Viktor and his colleagues put this matter to rest once and for all. In addition to their scientific value, these behavioral studies of the chick also provided an inspiration to the philosophical side of Viktor's personality. In the F.O. Schmitt lecture, also reprinted here, he states: "What impressed me most in all phases of our investigations is the *primacy of activity* over reactivity or response. This, to me, has become symbolic of animal life, and perhaps of life in general. The elemental force that embryos and fetuses can express freely in their spontaneous motility, sheltered as they are in the egg or uterus, has perhaps remained, throughout evolution, the biological mainspring of creative activity in animals and man, and autonomy of action is also the mainspring of freedom" (1976, p. 32).

Many of the articles and essays published in Sections III, V, and VI provide a first-hand account of the events that established experimental embryology and neuroembryology as bona fide scientific disciplines in this century. Having known most of the central players in these events (and having been an active participant in some of them), Viktor is able to provide a unique and valuable perspective on the history and personalities of these two closely related fields. What is clear from this analysis, is that by building on the accomplishments of pioneers in the last century, such as K.E. von Baer, W. His, W. Roux, H. Driesch, and S. Ramon y Cajal, a mere handful of experimental embryologists (e.g., R. Harrison, S. Detwiler, G. Coghill, P. Weiss, H. Spemann, and Hamburger himself) were able, with only the simplest of tools and techniques, but with great analytical insight, to forge the foundations for modern developmental biology and neurobiology. Regrettably, what is missing from the papers reprinted here on these matters (but which is easily available from your local bookseller), is the magnificent new book by Viktor, entitled, *The Heritage of Experimental Embryology, Hans Spemann and the Organizer* (1988). Concentrating on a single theme, the discovery and subsequent development of the organizer concept, Viktor has managed to weave together a compelling history of biology, personal vignettes, and a critical review of almost 60 years of research on what was at the time considered the crowning achievement of experimental embryology. Conceived and written during the final years of his eighth decade, this magnum opus is testament to a mind that only improves with age.

In the three papers published in Section IV, one first learns of early attempts by Viktor and others (most notably his friend, W. Landauer) to use the techniques of experimental embryology to elucidate the role of specific genes in development, and then in the remaining two papers, thoughts on the reasons behind the neglect of evolution and genetics by embryologists earlier in this century. Although it was clear to the founders of the Modern Synthesis that embryological considerations were central to evolutionary theory, with the exception of R. Goldschmidt, I. Schmalhausen, and later, C.H. Waddington, few embryologists, geneticists, or evolutionary theorists were prepared to undertake the efforts required for a *truly* modern synthesis. Only recently, with the advent of molecular genetics, is this missing link in the Modern Synthesis being remedied. Hamburger's masterly treatment of this topic provides a compelling chapter to our understanding of the events that contributed to the lack of interest in genetics and evolution by experimental embryologists.

In the final two papers of this collection, one obtains a glimpse, but regrettably only a glimpse, of facets of Viktor's personality that are understandably difficult or impossible to discern in his other writings. In "An Embryologist Visits Japan" he describes aspects of Japanese science, culture, landscape, and character that would elude a less inquisitive and romantic mind. For instance, he begins the article this way: "It was my good fortune to be invited to Japan in Spring when the famous cherry and peach blossoms cover the countryside and invade even the serene temple gardens which symbolize the eternal life of nature and of the spirit and banish all other bright-colored flowers. Perhaps they remind you there of the evanescence of life." The second of the two essays in this section, Goethe's *Zur Farbenlehre* (Theory of Colors), not only provides Viktor with the opportunity to pay homage to one of Germany's greatest literary figures (and one of Viktor's favorites as well) but also allows him to discuss another of his interests, color vision. The focus here, however, is not on Goethe the writer but rather on the scientist and natural philosopher. As Viktor points out, despite his considerable scientific studies, Goethe set self-imposed limits on what he was willing to subject to scientific scrutiny. For Goethe, "there are questions—mysteries if you will—that should be left untouched (*by science*) in a spirit of humility." In a different context, Viktor has expressed similar concerns. As he put it: "Do we really believe that our efforts to reduce the biological phenomena to physical and chemical processes will answer all questions? We had better realize that the scientific approach altogether opens only a small window to the universe. We cannot expect our intellect to fathom all depths." (1969, p. 1125). It is fitting that having begun this collection with an exposition of the strengths of the experimental, reductionist approach to nature, it ends with a recognition of the limits of this approach.

As the 20th century draws to a close and the scientific method continues to dominate our views of many great issues of the time, a little more, not less, humility seems the order of the day. In this, as in most other matters that he has touched, Viktor Hamburger may once again serve as a pioneer and pathfinder.

R.W. Oppenheim
The Bowman Gray School of Medicine
Wake Forest University
Winston-Salem, North Carolina
November 15, 1989

References*

1. Hamburger V, Hamilton H: A series of normal stages in the development of the chick embryo. *J. Morphol.* 1951; 88: 42.
2. Hamburger V: *Manual of Experimental Embryology*, Univ. of Chicago, Chicago, 1942.
3. Hamburger V: *The Heritage of Experimental Embryology, Hans Spemann and the Organizer.* Oxford Univ. Press, New York, 1988.
4. Hamburger V: The journey of a neuroembryologist. *Ann. Rev. Neurosci.* 1989; 12: 1.
5. Narayanan CH, Fox MW, Hamburger V: Prenatal development of spontaneous and evoked activity in the rat. *Behaviour* 1971; 40: 100.
6. Nobel Committee (eds.) Nobel, the Man and His Prizes. Elsevier, New York, 1962.
7. Levi-Montalcini R: Interview. *Omni Mag.* 1988; 10: 70.
8. Oppenheim RW: The neuroembryological study of behavior: Progress, problems, perspectives. *Curr. Topics Dev. Biol.* 1982; 17: 257.
9. Purves D, Sanes JR: *The 1986 Nobel Prize in Physiology and Medicine. Trends in Neurosci.* 1987; 10: 231.

---

*This list does not include papers cited in the Introduction that are contained in the collection of papers reprinted here.

Bibliography of Viktor Hamburger

1925

[1] Über den Einfluss des Nervensystems auf die Entwicklung der Extremitäten von Rana fusca. *W. Roux' Archiv.*, **105**: 149–201.

1926

[2] Versuche über Komplementär-Farben bei Ellritzen (Phoxinus laevis). *Ztschr. Vergl. Phys.*, **4**: 286–304.

1927

[3] Entwicklungsphysiologische Beziehungen zwischen den Extremitäten der Amphibien und ihrer Innervation. *Naturwiss.*, **15**: 657–681.

1928

[4] Die Entwicklung experimentell erzeugter nervenloser und schwach innervierter Extremitäten von Anuren. *W. Roux' Archiv.*, **114**: 272–362.

1929

[5] Experimentelle Beiträge zur Entwicklungsphysiologie der Nervenbahnen in der Froschextremität. *W. Roux' Archiv.*, **119**: 47–99.

1934

[6] The effects of wing bud extirpation in chick embryos on the development of the central nervous system. *J. Exp. Zool.*, **68**: 449–494.

1935

[7] Malformations of hind limbs in species hybrids of Triton taeniatus × Triton cristatus. *J. Exp. Zool.*, **70**: 43–54.

1936

[8] The larval development of reciprocal species hybrids of Triton taeniatus (and Tr. palmatus) × Triton cristatus. *J. Exp. Zool.*, **73**: 319–373.

1938

[9] Morphogenetic and axial self-differentiation of transplanted limb primordia of two-day chick embryos. *J. Exp. Zool.*, **77**: 379–397.

1939

[10] The development and innervation of transplanted limb primordia of chick embryos. *J. Exp. Zool.*, **80**: 347–389.

[11] Motor and sensory hyperplasia following limb bud transplantations in chick embryos. *Physiol. Zool.*, **12**: 268–284.

1940

[12] (with M. Waugh) The primary development of the skeleton in nerveless and poorly innervated limb transplants of chick embryos. *Physiol. Zool.*, **13**: 367–380.

[13] (with D. Rudnick) On the identification of segregated phenotypes in progeny from Creeper fowl matings. *Genetics*, **25**: 215–224.

1941

[14] Transplantation of limb primordia of homozygous and heterozygous chondrodystrophic ("Creeper") chick embryos. *Physiol. Zool.*, **14**: 355–364.

[15] (with M.G. Brown and F.O. Schmitt) Density studies on amphibian embryos with special reference to the mechanism of organizer action. *J. Exp. Zool.*, **88**: 353–372.

1942

[16] The developmental mechanics of hereditary abnormalities in the chick. *Biol. Symposia*, **6**: 311–334.

[17] *A Manual of Experimental Embryology.* 213 pp. (Univ. Chicago Press).

1943

[18] (with K. Gayer) The developmental potencies of eye primordia of homozygous Creeper chick embryos tested by orthotopic transplantation. *J. Exp. Zool.*, **93**: 147–183.

1944

[19] Developmental physiology. *Ann. Rev. Physiol.*, **6**: 1–24.

[20] (with E.L. Keefe) The effects of peripheral factors on the proliferation and differentiation in the spinal cord of chick embryos. *J. Exp. Zool.*, **96**: 223–242.

1946

[21] Isolation of the brachial segments of the spinal cord of the chick embryo by means of Tantalum foil blocks. *J. Exp. Zool.*, **103**: 113–142.

1947

[22] (with Karl Habel) Teratogenetic and lethal effects of influenza-A and mumps viruses on early chick embryos. *Proc. Soc. Exp. Biol. Med.*, **66**: 608–617.

1948

[23]   The mitotic patterns in the spinal cord of the chick embryo and their relation to histogenetic processes. *J. Comp. Neur.*, **88**: 221–284.

1949

[24]   (with Rita Levi-Montalcini) Proliferation, differentiation and degeneration in the spinal ganglia of the chick embryo under normal and experimental conditions. *J. Exp. Zool.*, **111**: 457–502.

1950

[25]   (with Rita Levi-Montalcini) Some aspects of neuroembryology. In: *Genetic Neurology* (ed. P. Weiss) (Univ. Chicago Press) pp. 128–160.

1951

[26]   (with Rita Levi-Montalcini) Selective growth stimulating effects of mouse sarcoma on the sensory and sympathetic nervous system of the chick embryo. *J. Exp. Zool.*, **116**: 321–362.

[27]   (with Howard Hamilton) A series of normal stages in the development of the chick embryo. *J. Morph.*, **88**: 49–92.

1952

[28]   Development of the nervous system. *Ann. N.Y. Acad. Sci.*, **55**: 117–132.

1953

[29]   (with R. Levi-Montalcini) A diffusible agent of mouse sarcoma, producing hyperplasia of sympathetic ganglia and hyperneurotization of viscera in the chick embryo. *J. Exp. Zool.*, **123**: 233–288.

1954

[30]   (with Stanley Cohen, Rita Levi-Montalcini.) A nerve growth-stimulating factor isolated from sarcomas 37 and 180. *Proc. Nat. Acad. Sci.*, **40**: 1014–1018.

[31]   (with Rita Levi-Montalcini, Hertha Meyer) In-vitro experiments on the effects of mouse sarcomas 180 and 37 on the spinal and sympathetic ganglia of the chick embryo. *Cancer Res.*, **14**: 49–57.

1955

[32]   Trends in experimental neuroembryology. In: *Biochemistry of the Developing Nervous System. Proceed. First Internatl. Neurochem. Sympos.* (ed. H. Waelsch), pp. 52–73.

[33]   *Analysis of Development* (Co-editor with B. Willier and P. Weiss) (W.B. Saunders) 735 pp.

[34]   (With J. Holtfreter) Amphibians. In: *Analysis of Development* (ed. B. Willier, P. Weiss and V. Hamburger) pp. 230–296.

1956

[35] Developmental correlations in neurogenesis. In: *Cellular Mechanisms in Differentiation and Growth.* (14th Growth Symp., ed. D. Rudnick). (Princeton Univ. Press) pp. 191–212.

1957

[36] The life history of a nerve cell. *Amer. Sci.,* **45**: 263–277.
[37] The concept of "Development" in Biology. In: *The Concept of Development* (ed. D.B. Harris) (Univ. Minnesota Press). pp. 49–58.

1958

[38] Regression versus peripheral control of differentiation in motor hypoplasia. *Am. J. Anat.,* **102**: 365–410.

1960

[39] *Manual of Experimental Embryology.* Revised edition. (Univ. Chicago Press) 221 pp.

1961

[40] Experimental analysis of the dual origin of the trigeminal ganglion in the chick embryo. *J. Exp. Zool.,* **148**: 91–123.

1962

[41] Specificity in neurogenesis. *J. Cell. and Comp. Physiol.,* **60**: 81–92.
[42] An Embryologist visits Japan. *Amer. Zoologist,* **2**: 119–125.

1963

[43] Some aspects of the embryology of behavior. *Quart. Rev. Biol.,* **38**: 342–365.
[44] (with Martin Balaban) Observations and experiments on spontaneous rhythmical behavior in the chick embryo. *Devel. Biol.,* **7**: 533–545.

1964

[45] Ontogeny of behaviour and its structural basis. In: *Comparative Neurochemistry; Proceed. of 5th Internatl. Neurochemistry Sympos.* (ed. Richter) (Pergamon Press) pp. 21–34.

1965

[46] (with M. Balaban, R. Oppenheim, E. Wenger) Periodic motility of normal and spinal chick embryos between 8 and 17 days of incubation. *J. Exp. Zool.,* **159**: 1–13.

1966

[47] (with E. Wenger, R. Oppenheim) Motility in the chick embryo in the absence of sensory input. *J. Exp. Zool.,* **162**: 133–160.

1967

[48] (with R. Oppenheim) Prehatching motility and hatching behavior in the chick. *J. Exp. Zool.*, **166**: 171–204.

1968

[49] Emergence of Nervous Coordination. Origins of integrated behavior. 27th Sympos. of the Soc. for Devel. Biol., *Develop. Biol. Suppl.*, **2**: 251–271.

1969

[50] (with C.H. Narayanan) Effects of the deafferentation of the trigeminal area on the motility of the chick embryo. *J. Exp. Zool.*, **170**: 411–426.
[51] Hans Spemann and the Organizer Concept. *Experientia*, **25**: 1121–1125.

1970

[52] (with R.R. Provine, S.C. Sharma, T. Sandel) Electrical activity in the spinal cord of the chick embryo, in situ. *Proc. Nat. Acad. Sci.*, **65**: 508–515.
[53] (with S.C. Sharma, R.R. Provine, T.T. Sandel) Unit activity in the isolated spinal cord of thick embryo, in situ. *Proc. Nat. Acad. Sci.*, **66**: 40–47.
[54] Embryonic motility in vertebrates. In: *The Neurosciences Second Study Program* (F.O. Schmitt, editor-in-chief) (Rockefeller Univ. Press) pp. 141–151.

1971

[55] Development of embryonic motility. In: *The Biopsychology of Development* (E. Tobach, L.R. Aronson, E. Shaw, eds.) (Academic Press) pp. 45–66.
[56] (with C.H. Narayanan) Motility in chick embryos with substitution of lumbosacral by brachial and brachial by lumbosacral cord segments. *J. Exp. Zool.*, **178**: 415–432.
[57] (with C.H. Narayanan and M.W. Fox) Prenatal development of spontaneous and evoked activity in the rat (Rattus norwegicus albinus). *Behaviour*, **40**: 100–134.

1973

[58] Anatomical and physiological basis of embryonic motility in birds and mammals. In: *Studies on the Development of Behavior and the Nervous System 1* (ed. Gilbert Gottlieb) (Academic Press) pp. 51–76.

1974

[59] (with R. Skoff). Fine structure of dendritic and axonal growth cones in embryonic chick spinal cord. *J. Comp. Neur.*, **153**: 107–148.

1975

[60] Cell death in the development of the lateral motor column of the chick embryo. *J. Comp. Neur.*, **160**: 535–546.

[61]   Changing concepts in developmental neurobiology. *Perspectives in Biology and Medicine*, **18**: 162–178.

[62]   (with Anne Bekoff, Paul S.G. Stein) Coordinated motor output in the hindlimb of the 7-day chick embryo. *Proc. Nat. Acad. Sci., USA*, **72**: 1245–1248.

[63]   Fetal behavior. In: *The Mammalian Fetus* (ed. E.S.E. Hafez) (Charles C. Thomas, Publisher) pp. 68–81.

1976

[64]   (with M. Hollyday) Reduction of the normally occurring motor neuron loss by enlargement of the periphery. *J. Comp. Neurol.*, **170**: 311–320.

1977

[65]   (with M. Hollyday and J. Farris) Localization of motor neuron pools supplying identified muscles in normal and supernumerary legs of chick embryos. *Proc. Nat. Acad. Sci., USA*, **74**: 3582–3586.

[66]   The F.O. Schmitt Lecture in Neuroscience. The developmental history of the motor neuron. *N.R.P. Bulletin (Suppl.)*, **15**: 1–37.

[67]   (with M. Hollyday) An autoradiographic study of the formation of the lateral motor column in the chick embryo. *Brain Res.*, **132**: 197–208.

1979

[68]   (with J.K. Brunso-Bechtold) Retrograde transport of nerve growth factor in chicken embryo. *Proc. Natl. Acad. Sci. USA*, **76**: 1494–1496.

1980

[69]   Trophic interactions in neurogenesis: A personal historical account. *Ann. Review Neurosci.* **3**: 269–278.

[70]   S. Ramón y Cajal, R.G. Harrison, and the beginnings of neuroembryology. *Perspect. Biol. Med.*, **23**: 600–616.

[71]   Prespecification and plasticity in neurogenesis. *Pontificiae Academiae Scientiarum Scripta Vera*, **45**: 433–447, reprinted In: *Nerve Cells, Transmitters and Behaviour* (ed. R. Levi-Montalcini) (Elsevier-North Holland Biomedical Press, Amsterdam, Oxford, New York) pp. 433–447.

[72]   Embryology and the modern synthesis in evolutionary theory; Evolutionary theory in Germany, A Comment: both In: *The Evolutionary Synthesis. Perspectives on the Unification of Biology* (ed. E. Mayr and W.B. Provine) (Harvard Univ. Press, Cambridge, MA and London) pp. 97–112; pp. 303–308.

1981

[73]   (with J.K. Brunso-Bechtold, J. Yip) Neuronal death in the spinal ganglia of the chick embryo and its reduction by Nerve Growth Factor. *J. Neurosci.*, **1**: 60–71.

[74]   Historical landmarks in neurogenesis. *Trends in Neurosci.*, **4**: 151–155.

1982

[75]   (with Ronald W. Oppenheim) Naturally occurring neuronal death in vertebrates. *Neurosci. Comm.*, **1**: 39–55.

1984

[76]   Neurogenesis. In: *Medicine, Science, and Society*, edited by Kurt J. Isselbachet. (John Wiley and Sons) pp. 623–641.
[77]   (with Joseph W. Yip) Reduction of experimentally induced neuronal death in spinal ganglia of the chick embryo by Nerve Growth Factor. *J. Neurosci.*, **4**: 767–774.
[78]   Hilde Mangold, Co-Discoverer of the Organizer. *J. Hist. Biol.*, **17**: 1–11.

1985

[79]   Hans Spemann, Nobel Laureate, 1935. *Trends Neurosci.*, **8**: 385–387.

1988

[80]   *The Heritage of Experimental Embryology. Hans Spemann and the Organizer.* (Oxford University Press) 196 pages.
[81]   Ontogeny of Neuroembryology. *J. Neurosci.* **8**, 3535–3540.

1989

[82]   The Journey of a Neuroembryologist, *Ann. Rev. Neurosci.*, **12**: 1–12.
[83]   The Rise of Experimental Neuroembryology: A Personal Reassessment. The S. Kuffler Lecture, 1989. *Int. J. Develop. Neurosci.* **8**, 121–131. (Pergamon Press).

Miscellaneous Publications

1943

[A1]   Embryologia chemica Vera in Statu Nascendi. Review of *Biochemistry and Morphogenesis*, by Joseph Needham. *Quart. Rev. Biol.*, **18**, 263–268.

1945

[A2]   Biology in the Premedical Curriculum. *Science*, **102**: 511–513.

1960

[A3]   Individuality, Biological. *Encycl. Brit.*, 1 page.

1961

[A4]   Regeneration. *Encycl. Brit.*, 8 pages.

1963

[A5]   Embryology, Experimental. *Encycl. Brit.*, 7 pages.

1968

[A6]   Malpighi the Master. Review of *Marcello Malpighi and the Evolution of Embryology*, Vol. I-V, by H.W. Adelmann. *Quart. Rev. Biol.*, **43**, 175–178.

1970

[A7]   Review of *Organization and Development of the Embryo*, by Ross G. Harrison, edited by Sally Wilens. *Am. Sci.* **58**, 321–322.

[A8]   Von Baer, Man of Many Talents. Review of *Karl Ernst von Baer, 1792–1876 (Sein Leben und Sein Werk)* by Boris E. Raikov. *Quart. Rev. Biol.* **45**, 173–176.

1981

[A9]   Goethe's *Zur Farbenlehre (Theory of Colors)*. Friends of the Libraries of Washington University, 3 pages.

I. Developmental Neurobiology—Reviews

Entwicklungsphysiologische Beziehungen zwischen den Extremitäten der Amphibien und ihrer Innervation

(Developmental-physiological correlations between the limbs of amphibians and their innervation.)
V. Hamburger, Berlin-Dahlem; Kaiser-Wilhelm-Institut für Biologie.
Die Naturwissenschaften, *15*: 657–661; 677–681, 1927

Developmental physiological correlations

Animal development begins with the relatively simple formation of folds, evaginations and cavitations, condensation and dispersion of cell groups, and ends with the formation of complex organs. In order to guarantee the integration of the assembly, the processes going on in the different regions of the embryo have to be interlocked in space and time with great precision. Sometimes, the coordination is achieved in such a way that in each part all ongoing processes are programmed from the beginning of development with respect to form and velocity, and thus proceed side by side in a mosaic fashion. In other instances, morphogenesis is regulated in such a way that processes in particular areas are influenced by neighboring areas which are more advanced in their development. Such dependencies or "developmental correlations" are well known, particularly in amphibian development. They can operate by *determining* the fate of an area, that is, they cause the area to follow a course of development which is irreversible after the time of determination. The crystalline lens of some amphibians is determined [induced] by the optic vesicle. The determinative capacity goes beyond mere triggering, because such inducing regions can exert their inductive effects not only at the normal site but also after transplantation to another region of the embryo. In addition, there are correlations which are merely *triggering* devices. For instance, the thyroid hormone triggers metamorphosis in amphibians. It does not occur after extirpation of the thyroid gland. Regeneration in the limbs of urodeles is prevented, or ongoing regeneration is arrested, if certain parts of the innervation are eliminated. Hence, normally, nerve supply keeps the regeneration process going. We may add to the determining and triggering factors the directional forces which determine the direction of migration of mobile cells and the direction of outgrowth of cell processes such as nerve fibers. As early as 1894, C. Herbst had taken the theoretically important step of comparing these directional forces with tropisms and taxies, that is, the stimuli exerted by extrinsic factors such as light, gravity, etc., on the direction of migration of mobile forms. We include in "developmental correlations" between structures: the *determining* [inducing], *triggering* and *directional forces* through which parts of the embryo interact and which are thus involved causally in morphogenesis.

Translated by Viktor Hamburger.

We start with the general question: What is *the significance of the nervous system in development*? Since the nervous system plays such a significant role as the functional center of the organism in later stages, the thought suggests itself that it might be an important center for morphogenesis and perhaps even *the* cause for the integration of developmental processes. Such an idea had been expressed a century ago by the anatomist, Tiedemann (1816): "Does the nervous system, whose activity underlies all animal behavior, have a share in the production and formation of organisms? Does nervous activity perhaps determine and regulate the formation of the embryo which originates from the fertilized egg?" He answered in the affirmative on the basis of malformations in which the deficiency of organs was paralleled by a deficiency of their innervation. Though the idea of a morphogenetic center in the brain was attractive, neither was his deduction from his observations correct nor did later experiments confirm these suppositions. Spemann and H. Mangold (1924) [organizer experiment] have shown that, at least in the amphibian embryo, the axial organs and thus, in a certain respect, the individuality of the embryo, is determined by the upper blastoporal lip which later forms the archenteron roof. Therefore, the neural plate that is the primordium of the central nervous system, at this critical stage, is not the inducing structure but the one which is being induced. However, it is true that the neural primordium is not devoid of determinative capacity. Its anterior part, the prospective eye material, can induce a lens in ectoderm, and according to the results of Spemann and O. Mangold (Mangold and Spemann, 1927) neural material can induce a neural tube, when implanted beneath the gastrula ectoderm in the heart region. But these capacities do not differ from those of other embryonic structures. The same holds for the central nervous system in later stages. One can remove the brain and spinal cord in young tadpoles (Schaper, 1898; Wintrebert, 1903; Harrison, 1904). As long as these animals survive—for days or weeks—all their organs develop normally. It follows from all this that one cannot attribute to the central nervous system and its primordium, the neural plate, a significant role as a determinative embryonic structure.

But this in no way excludes the possibility that developmental correlations of sorts exist between the nervous system and the innervated target organs. For instance, the innervated limb, when considered strictly from the viewpoint of developmental physiology, poses the following three groups of problems: First, one inquires whether a limb can develop normally from the beginning if it is deprived of all its innervation, or part of it. Furthermore, one has to consider the related question of whether functional activity which is tied to innervation has some significance for its development. Second, one has to explore the question of whether the normal formation of the central nervous system depends on the presence of the organ which it innervates, that is, in our case, the limb. Finally, the peripheral nerves follow very characteristic patterns within the limb, and it is an important challenge to investigate the conditions for the origin of these nerve patterns.

The fundamental investigations of Braus and Harrison are directly connected with the last-mentioned problem. In order to examine experimentally the origin of nerve pathways, Braus (1904) devised the momentous experiment of the transplantation of limb buds, while Harrison, starting from the same basic question, designed the method of explantation and elaborated the experiment of extirpation of parts of the spinal cord (Harrison, 1904, 1910). Since then, numerous investigators have used preferentially the amphibian limb for the analysis of questions related to neuronal correlations. The following survey is confined to limbs, and I point out in passing that the developmental correlations between the nervous system and sense organs are of a different kind.

I. Effect of the Nervous System on Limb Development

A. Is there a determining [inductive] action?

Since it is improbable that the determination of the limb primordium which is completed in the neurula stage has any causal dependency on the medullary plate, any effects of the nervous system can be expected only in later stages. One could imagine determinative or triggering influences on partial processes. For instance, in an early stage a limb primordium could be determined as a limb in a general way, but not yet as a forelimb or hindlimb, and this decision could be made by the ingrowth of brachial or lumbosacral nerves, respectively. This is not the case. Braus found already in his first experiment that if limb primordia of axolotl or toad are transplanted to any region, they become innervated by adjacent regions. A transplanted forelimb in the hindlimb region is innervated by hindlimb nerves, and in the region of the *n. facialis* by this nerve. The possibility that the *species specificity* of morphogenesis might be disturbed by ingrowth of nerves from a foreign species can also be dismissed. Harrison (1924) found that in an experiment of interchange of limb buds between two species of salamanders the host has a strong influence on the size of the transplant, but Wieman (1926) showed that this effect does not emanate from the nerves. He exchanged the limb-innervating sections of the spinal cord between the same two species and thus obtained host limbs which were innervated by foreign nerves; this had no influence on the development of the limbs. I have shown the same, in a different way, for the two salamander species, *Triton* [*Triturus*] *taeniatus* and *Tr. cristatus*. Their forelimbs are distinctly different, particularly in larval stages. Those of *taeniatus* are stubby and the toes are short; those of *cristatus* are slender and the toes are very long and delicate. If one transplants the early limb anlage of a *cristatus* to the flank of *taeniatus*, then the transplant retains all its species characteristics, although, as sections show, it is innervated by host nerves.

B. Is there a triggering action?

Limb malformations following limb-, eye-, and midbrain-extirpations. It is conceivable that the nervous system is indispensable as a triggering factor, or a factor which sustains on-going development, comparable to its effect on limb regeneration in urodeles, as shown by Wolff and others. Duerken is of this opinion. In his book *Experimental Zoology* he states: "The normal development of limbs depends on the normal formation of the nerve centers" (1919, p. 114). This opinion is not based on nerve extirpation experiments but on unexpected results obtained in connection with other problems. In one experimental series (1911) he removed one or both barely visible leg buds of tadpoles of the frog, *Rana fusea*, using a hot needle or a small knife; he found different degrees

5

of malformations in the unoperated legs, ranging from deficiencies of toes to a complete stunting of the whole limb. The detailed examination of the nervous system led him to conclude that the limb bud extirpation had resulted in an atrophy of the ipsilateral nerve centers and that the atrophy had spread to the other side and then caused the limb malformations. In other words, the limb abnormalities were considered to be neurogenic in nature. In another series (1913), he extirpated one eye, and in 50% of the cases the same limb malformations were obtained. Here again, he could demonstrate deficiencies in the brain and spinal cord which he made responsible for the limb defects—I could confirm the latter results (1925). A repetition of the unilateral eye extirpation on the same material gave among 400 metamorphosed, one-eyed little frogs 49 animals (12%) with slightly malformed legs. Midbrain extirpations on larvae of the same early stages gave also a positive result. This experiment had been suggested by Dr. Spemann on the basis of a statement of Duerken that in all his experiments the midbrain was distinctly affected and could therefore be considered as the center of the developmental correlations. The positive outcome of my experiment seemed to support the contention of Duerken.

The eye extirpations had been repeated by several other investigators as well; but it turned out that, with increasing frequency, the effect failed to occur; all efforts to find an explanation for this failure have been unsuccessful. Negative results were reported by Andressen (Petersen, 1924) for 38 individuals and by Luther (1915/16) for 283 individuals, following unilateral eye extirpation. The latter investigator believed that perhaps the hot needle had resulted in a toxic effect and he had used glass needles for the operation. Half of my own above-mentioned experiments had been done by electro-cautering and half with the glass needle. Defective animals occurred in both series; hence the type of operation is of no significance. Duerken (1917) had tried to explain the discrepancy between his results and those of Luther in terms of local races. He had used material from Goettingen, whereas Luther's material came from Rostock. Half of my material came from Freiburg and half from Goettingen, and abnormalities occurred in both series. In the past year, another 700 one-eyed frogs were carried to metamorphosis; they came from Koenigsberg, Tuttlingen, Rostock and Berlin-Dahlem; this time none of them showed any defects. Hence local races [genetic differences] have nothing to do with the results.

Age differences also play no role. I operated on 5 different stages, from early tail bud to larvae with covered gills, that is, stages younger and older than those used by Duerken. More than 400 metamorphosed animals were again completely normal. The defects cannot be attributed to nutrition. All animals, including those with abnormal-ities, had been fed amply with algae and boiled meat; on the other hand, undersized frogs that were undernourished showed normal legs.

This survey shows that limb abnormalities following eye extirpations fail to occur in the majority of cases and that a number of important factors, analyzed so far (method of operation, local race, age, nutrition) cannot be responsible for the variability in the results. The experiments described so far do not permit a definite conclusion concerning either a direct (correlative) or an indirect involvement of the nervous system; they do not even prove beyond doubt, that the nervous system has anything to do with the malformations.

C. Experimentally produced nerveless limbs

One can hope to obtain a clear picture of the role of the nervous system in limb development in a simpler fashion by preventing the ingrowth of nerves into the limb, that is, by producing nerveless limbs. Wintrebert (1903) and Lebedinsky (1924) have observed normal development of nerveless limbs at least for some period of development and Harrison (1904) has reported one case in which the ingrowth of nerves into the limb had been prevented mechanically and the limb had developed normally up to metamorphosis. I chose for my own experiments (unpublished)[1] the legs of anurans. I set myself the goal of removing that part of the spinal cord from which the leg nerves emerge. The operation has to be done before outgrowth of the nerves. The most suitable stage was that of the just closing neural folds. At that stage, one can remove very exactly one half of the spinal cord sector without injuring the other half (Figure 1). I did also bilateral extirpations, but eventually the unilateral operation was preferred because the mortality was lower and one has the invaluable advantage that the other side—always the left one—serves as a control for size and state of development. The operations were done on the frog, *Rana fusca,* and the toad, *Bombinator pachypus,* using glass needles. In many cases of the unilaterally operated animals, a complete regeneration precluded any conclusions. Two-thirds of the operated animals were normal at metamorphosis, including the nervous system. In the remaining animals, the limbs on the operated side were retarded in their growth, from the early stage of the toe plate on, and they were more or less shortened at metamorphosis. In addition, they were completely paralyzed in one or more joints, and different degrees of muscle atrophy were observed. Twenty-one animals showed these symptoms to the highest degree (Figures 2, 3), but even in these, as in all others, the extremity was formed normally in all its parts and did not

---

[1] See Hamburger, 1928.

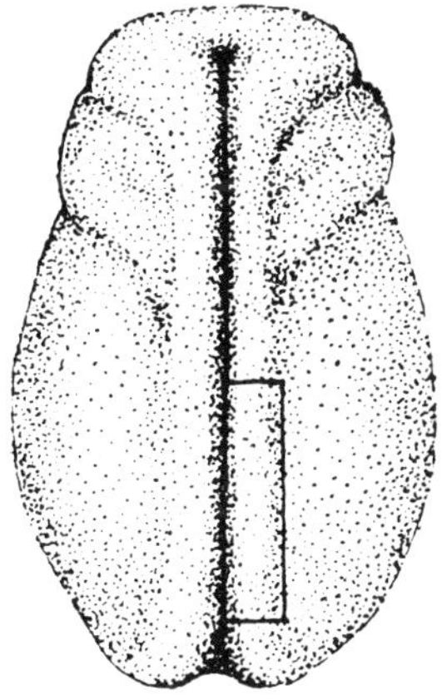

Figure 1. Embryo of *Rana fusca* at the stage of operation (neurula) The rectangle indicates the extirpated lumbar spinal cord in unilateral extirpation.

Figure 2. Toad (*Bombinator pathypus*) shortly after metamorphosis. Total extirpation of lumbar spinal cord. Both legs are highly atrophic, paralyzed, and shortened, but the segmentation in thigh, shank, and toes is normal.

Figure 3. Frog (*Rana fusca*) shortly after metamorphosis. The right half of the lumbar spinal cord had been extirpated at the neurula stage. The right leg is highly atrophic, shortened and paralyzed, but thigh, shank, and toes are normal. The left leg serves as a control.

display the slightest abnormalities. Sections of the spinal cord showed that in all mildly affected cases the leg nerves were missing in part, but that in the animals showing the most severe symptoms, the legs had been invaded only by very thin nerve branches, or they were completely free of nerves. The spinal cord and leg nerves were reconstructed from the sections in a semi-diagrammatic fashion. The branching points of the nerves, their lengths and diameters are reproduced exactly, but they are projected on a plane and their distances from each other are arbitrary. Figure 4 presents the nervous system of an animal with a highly atrophied right leg. The left normal leg is supplied by spinal nerves 8 to 11. On the right side, nerves 8–11 are absent, as was intended. Nevertheless, the leg was not entirely nerveless, since the 12th ganglion which disappears normally at metamorphosis together with the tail, remained intact; it is apparently hyperplastic and sends one of its 2 branches into the leg. This very thin bundle follows the typical course of the sciatic nerve and goes to the knee joint between femur and sciatic artery, where it gets lost. No branches to muscles were found. If they were overlooked, they could have been only extremely thin compared to those on the other side; they did not prevent complete muscle atrophy and paralysis. It is of crucial importance that *this leg has been very poorly innervated from the beginning of its development, and nevertheless it developed into a normally formed leg.*

In several instances, all nerves on the operated side were missing; nevertheless, even in these cases the leg was not entirely free of nerves. Quite unexpectedly, nerves from the contralateral plexus grew across the midline and entered the right leg. In Figure 5 a thin branch grew into the right thigh and shank; it sent several thin branches into muscles, but it could not prevent atrophy and complete paralysis. This extremely reduced innervation belongs to the completely normally formed right leg of Figure 3.

As a result of the amazing regulative capacity of the spinal cord I have not been able to obtain completely uninnervated legs after unilateral operation. Therefore, it seems that only bilateral extirpation can lead to success: but here high mortality is in the way. Animals with gaps in the spinal cord are naturally barely viable. Among 7 specimens with the highest degree of paralysis and muscle atrophy (Figure 2), one case was most probably and the other definitely nerveless. The latter animal had to be fixed as a weak larva. However, the thigh, shank and 5 toes were formed and all these parts were normal. The spinal cord in the operated region is a very thin strand from which no myelinated nerves entered the limb.

Although, due to the great difficulties encountered in the production of nerveless limbs, the present material is not extensive, it permits the conclusion that legs can undergo typical morphogenesis in spite of highly deficient innervation or in its total absence. Even the histological differentiation is normal; muscles in the most severely atrophic cases show distinct cross striation and only sporadic centers of degeneration. From the beginning, the limb primordium shows self-differentiation with respect to the nervous system. This result is in agreement with the work of Harrison (1918), according to which the limb anlage is a harmonious-equipotential, self-differentiating system, that is, a highly autonomous system which contains within itself the most important conditions for its morphogenesis. At the same time it follows that *functional activity* of the

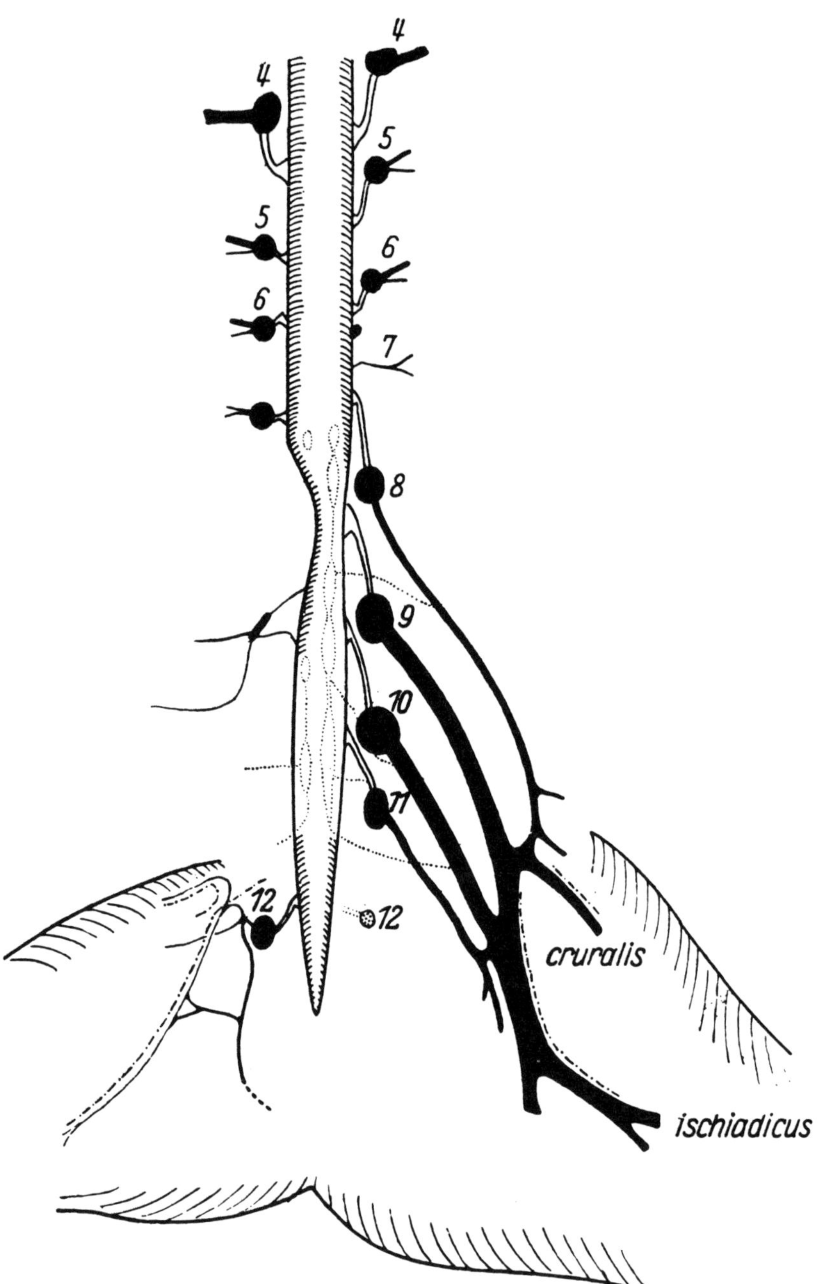

Figure 4. Semi-schematic reconstruction of the spinal cord and leg nerves of a metamorphosed frog (as in Figure 3), following extirpation of the right lumbar spinal cord. The innervation of the left leg is normal. The atrophied right leg is innervated only by nerve 12. Dotted lines = blood vessels.

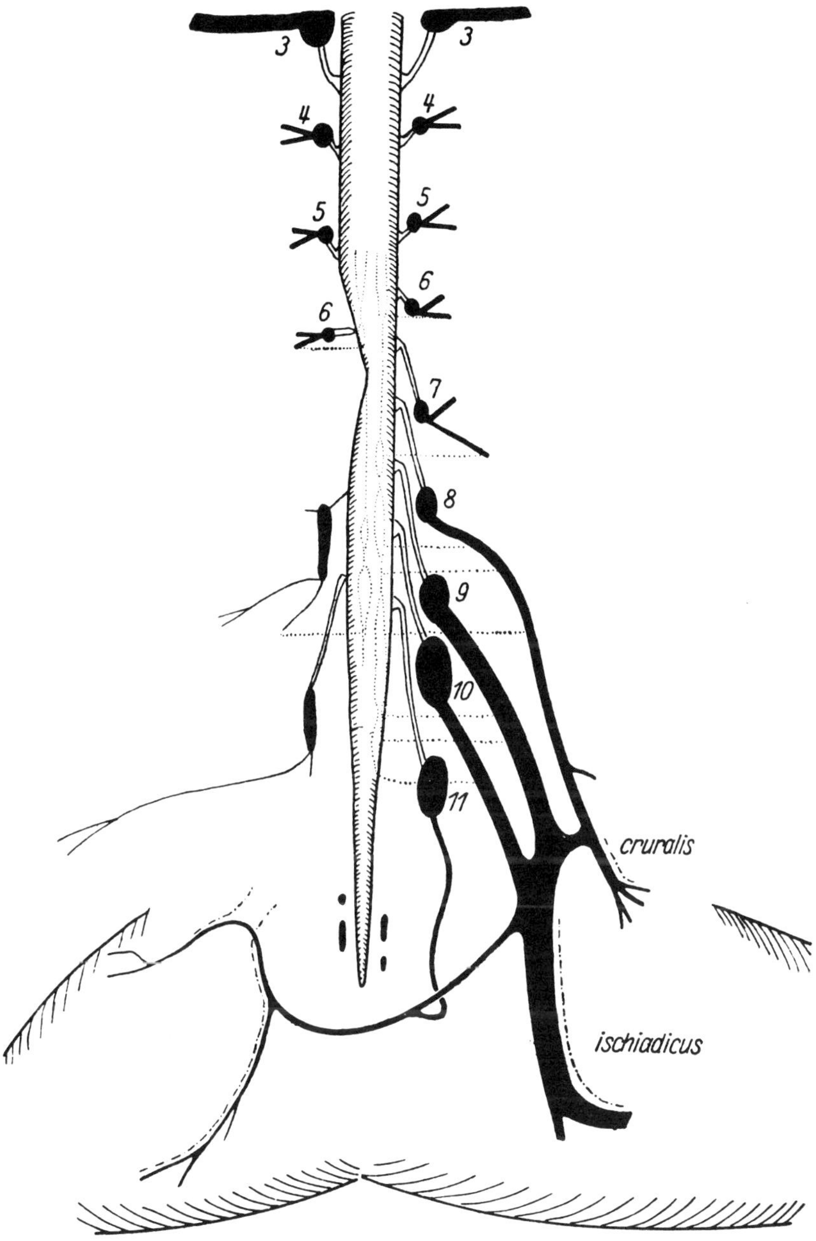

Figure 5. Innervation of the right leg of animal, Figure 3. No nerves from the right side enter the leg, but a thin branch from the left plexus crosses the midline and branches in the right leg. Dotted lines = blood vessels.

limb is not a necessary condition for its development; the completely or almost completely nerveless legs have never in their life performed a movement. All contentions which ascribe to functional activity an essential developmental-physiological role are erroneous, at least for the present instance.

It is difficult to reconcile the malformations described by Duerken and myself with these findings. One can make only the not very satisfactory statement that a limb is capable of normal development without any correlative relation to its innervation and that the particular conditions under which direct or indirect interference with the central nervous system impedes their development are unknown. There is also the strange contrast to limb regeneration which does not proceed in the absence of innervation.

The abnormalities which are actually produced by the absence of innervation, that is, shortening of the leg, paralysis and muscle atrophy, are not of a morphogenetic nature; they can be explained in terms of inadequate trophic conditions and as a result of lack of function and muscle tone. They are well known in clinical pathology, as for instance, in poliomyelitis and progressive muscle atrophy, resulting from degeneration of the anterior horn cells.

II. Effect of the Limb on the Nervous System

The second major question has been posed above as follows: Is the normal development of the central nervous system dependent on the presence of the peripheral organs? Again, the limb is particularly suitable for the analysis, since the sections of the spinal cord which supply the limbs usually have a larger diameter than those supplying the trunk or tail. The problem can be investigated by limb bud extirpation, or conversely, by "overloading" of the trunk with an extra limb. Among others, Shorey (1909) and Duerken (1911) found in extirpation experiments reductions in the spinal cord, and Duerken has traced the defects to the brain. More recently, Detwiler (1920) has combined both experiments in one and investigated the deficiencies in the spinal cord quantitatively. In an *Amblystoma* [salamander] embryo in the tail bud stage, he transplanted the anlage of the right forelimb 4–5 segments caudal to a region of the spinal cord which normally does not supply a limb. This segment was therefore "overloaded" with extremity, while the section which supplies the limb normally was devoid of its target. The left side served as the control. Precise counting of nuclei and weighing of [paper models of] the ganglia and of the spinal cord halves showed that the motor parts of the nervous system had been unaffected by both reduction and overloading, whereas in the absence of limbs the brachial sensory ganglia were reduced in size by 50% and the overloaded thoracic ganglia were enlarged by 50%. Again, the correlations are merely of a quantitative nature; typical morphogenetic characteristics such as the differentiation and patterning of sensory and motor cell groups are in no way affected.

III. Origin of Nerve Patterns

Spinal cord and leg are connected by spinal nerves 8–11. After they merge in the plexus, the leg nerve branches to form the *n. cruralis* and *n. ischiadicus*. Subsequently, both form a remarkably typical branching system in the leg which is embedded in the muscle and skeletal tissues in a typical fashion. This raises the developmental-physiological question: By which agents is the formation of these nerve pathways achieved?

This question was addressed experimentally by Braus and Harrison in 1904. At that time, it was connected very closely with the question of whether the nerves were outgrowths of the central neurons (His-Kupffer), or whether they are formed within the organs by plasma bridges (Hensen) or cell chains (Balfour) and subsequently become connected with the spinal cord. Harrison (1904, 1911) succeeded in demonstrating experimentally that the view of His is the correct one. To begin with, he showed that larvae whose spinal cord segments had been removed in the tail bud stage later on have nerveless muscles. When he removed only the dorsal half of the spinal cord, the sensory nerves were missing; when he removed the ventral half, the motor nerves were missing. Finally, he was able to observe the outgrowth of nerve processes with their characteristic growth cones in individual nerve cells which had been removed from the neural tube and cultivated *in vitro* under a coverglass. Hence, the nerve fibers are processes of nerve cells. Thus, one explanation for the origin of nerve pathways was ruled out: unlike muscle and cartilage, they are not formed *in situ*. There remain only two possibilities: Either the course of nerve paths is genetically fixed in the neurons; the time of outgrowth and every branching would be programmed exactly; and this space-time sequence would have to be coordinated exactly with that of the other structures of the limb. For instance, at the moment when a gap is formed between two muscle primordia, further nerve growth into this gap would occur. Of course, such a genetic competence could be attributed only to the specific leg-innervating neurons, because the adjacent neurons would be specialized for another task. Alternatively, the course and branching pattern would in some way be determined by the limb. The decision in favor of the latter view was provided by Braus (1904). He was the first to transplant a forelimb bud of a toad larva to a different position and he observed in the normally developing forelimb a typical forelimb pattern formed by nerves which were connected with the adjacent part of the central nervous system. For instance, after transplantation to the head, a *n. trigeminus* or *n. facialis* would form a typical limb pattern; after transplantation to the tail, a tail nerve would do it. Hence, the typical pathway formation is not limited to normal limb nerves but it can be performed by any nerve. The causes for the stereotypy must therefore be located within the limb itself.

By which means does the limb anlage create its typical innervation? The complex of problems involved can be dealt with best by dividing the formation of nerve pathways— from the moment of outgrowth of fibers to the terminal branching at the periphery— into several phases which can be analyzed separately.

1. The nerve process emerges from the spinal cord.
2. The processes join to form mixed nerves and they traverse the space between spinal cord and limb.

3.   The nerves mingle and form a plexus.

4.   The fibers sort out, regroup as limb nerves and form the typical limb pattern.

5.   Nerves form terminal connections with muscles and skin.

1. The tissue culture experiments of Harrison have shown that the first outgrowth of the nerve fibers is an autonomous property of the neurons which can be activated without extraneous help. However, observations of Ariens Kappers (1921), and Bok (1915) have made it probable that *in vivo*, stimuli are involved which derive from bypassing fibers within the central nervous system. Time of outgrowth and initial direction seem to be co-determined by intracentral stimuli. Nothing is known about extracentral stimuli (Tello, 1923; Herrick, 1925).

2. No investigations have been done on the union of sensory and motor fibers. At the time when the fibers grow from the spinal cord to the myotomes these structures are closely adjacent to each other. The distance between them is probably not much longer than the length of outgrowth which Harrison had observed in his tissue cultures. Hence, the capacity for active, undirected outgrowth could be a sufficient explanation of this performance. The experiences with limb transplantations also seem to imply that fibers simply continue to grow. As a rule, transplanted limbs are innervated by adjacent regions. However, Detwiler (1922), using a particular experimental design, gave evidence that the outgrowth of limb nerves is not undirected; rather the young limb bud provides directional stimuli for the nerves. To demonstrate this, the ordinary method of implanting a fifth limb was not applicable, because in this instance all limb nerves are requisitioned by their own adjacent limb buds. Therefore, he shifted the forelimb bud of an *Amblystoma* tail bud stage in the same embryo from its normal position below somite 3–5 either backward by the length of 1 to 4 somites or forward by 1 to 3 somites. He showed that, contrary to expectation, the innervation was not supplied by the immediately adjacent segments; rather, the original limb innervating sectors of the spinal cord were distinctly preferred as the source of the transplant innervation. For instance, in a backward shift by 4 somites, the innervation was not derived from the adjacent segments 7–9, but from segments 5–7 or 5–8. Detwiler implicated a directional stimulus from the limb bud, to which the actual limb nerves respond more readily than any other nerves. The possibility that the nerves were towed backward mechanically was ruled out by forward transplantation. Here again, the actual limb nerves participated preferentially. Hence, it seems certain that directional factors play a role. It is more difficult to explain the preference [for brachial nerves].

Detwiler assumed a limited affinity between limb buds and limb nerves, perhaps of a chemical nature, which, however, is not rigorously specific, since in ordinary transplantations any nerve will grow into the transplant. Another explanation first mentioned by Hoadley (1925) seems simpler: The preference of limb nerves would result from the fact that the anterior nerves grow out first. They follow the completely unspecific stimulus and saturate the limb before the adjacent more posterior nerves begin to grow out. However, the result of forward transplantation is difficult to explain in this way.

In my own experiments, the directional effect of the limb bud is borne out convincingly. In all cases in which the unilateral spinal cord extirpation had resulted in the

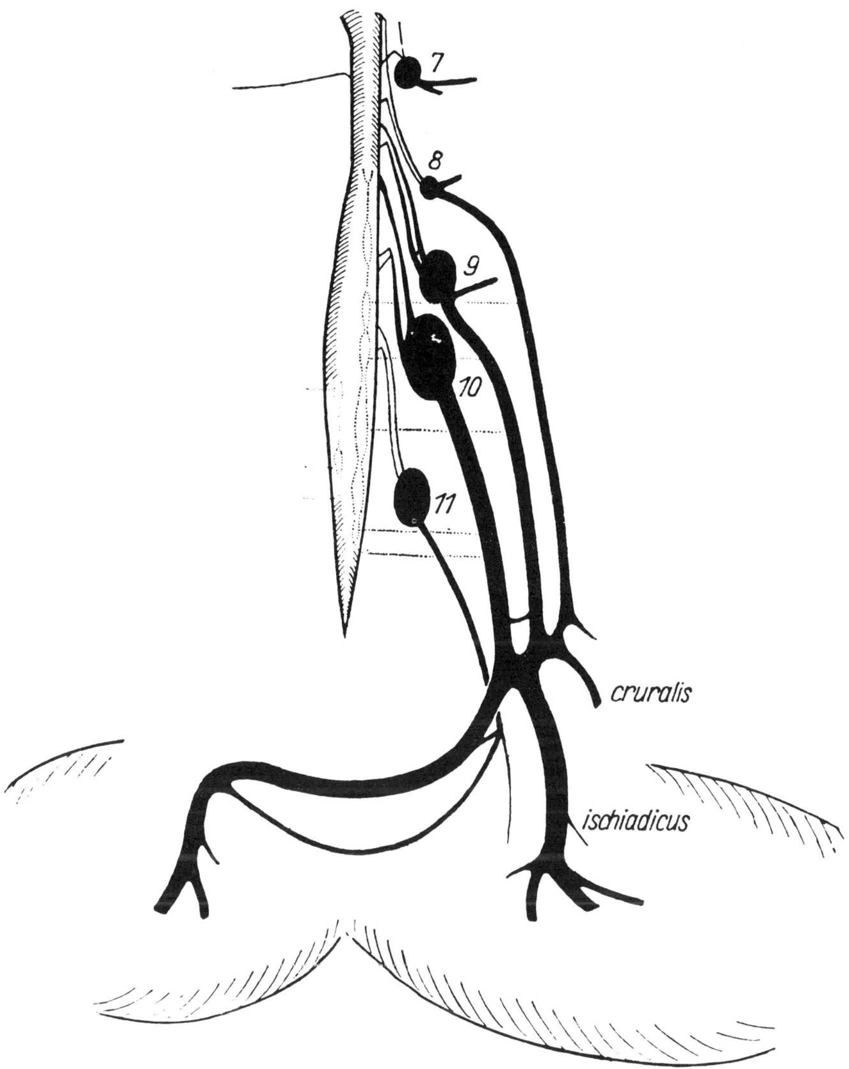

Figure 6. Spinal cord and leg innervation of a frog shortly after metamorphosis. Extirpation of right half of the lumbar spinal cord at the neurula stage. No innervation from the right side, but a strong nerve branch from the left side prevented complete atrophy.

complete absence of the limb nerves on the right side, a fiber tract of varying diameter grows from the left side across the median into the right limb; in one case, this happened even though some right nerves were present. The strongest transverse nerve is shown in Figure 6. It originates in the left plexus, is joined by the 11th nerve and branches into a crural and a sciatic nerve. The left ganglia 10 and 11 are apparently hyperplastic. The limb was well innervated and not distinctly atrophic. An analogous situation was found in a reciprocal experiment of Duerken (1911). In three cases of extirpation of the left leg bud, the left limb nerve which was deprived of its own target had grown across the midline and joined the right sciatic nerve.

Thus, the effectiveness of directing forces in the growth of nerves from spinal cord to limb was demonstrated by three different experiments. One cannot make a statement concerning their nature. We know several forces which exert a directional effect on outgrowing nerve fibers. For instance, Harrison (1910) observed in his tissue culture experiments that the fibers cannot grow out in a liquid medium but they require a support of coagulated fibers or spider webs along which they grow. Apparently, they respond to *stereotropic* stimuli. Forssman (1901) was able to direct outgrowing fibers by degenerating nerve substance, and he called this reaction *"neurotropism"*. Hoadley (1925) following a method of Danchakoff, implanted a piece of midbrain of a 48-hour chick embryo onto the chorio-allantoic membrane of an older embryo and placed myotomes in its vicinity. In general, only short fibers emerged from the midbrain, but thick and long strands grew into the muscle tissue. It seems that we are dealing with a *chemotropic* action. Finally, Ingvar (1920) was able to show that if, in tissue culture, neurons are exposed to a very weak electric current, the outgrowing fibers orient themselves in the electric field. However, so far all these forces have been observed only in tissue culture, and new experiments are required to answer the question of which of these operate in the embryo. In my case one might expect chemotropic or *galvanotropic* effects of the limb bud.

3. The question of the causes of plexus formation has not been dealt with experimentally.

4. From the plexus emerge the two main nerves, the sciatic and the crural nerve; they branch in a specific fashion. The branching can be performed by purely sensory fibers (facialis). Normal nerve patterns can be formed not only by foreign nerves but also by quantitatively very reduced nerves. In one case of my material of the spinal cord extirpations, the 8th nerve, which normally forms the crural nerve, was missing. A typical crural nerve was formed by the 9th and 10th nerves. The greatest achievement in my material was that of a single motor root (of the 8th nerve) which formed all typical branches. Hence, the specific nerve pattern formation in the limb is to a high degree independent of the origin and quantity of the available nerves.

Little can be said concerning the causes of nerve pattern formation. Harrison (1911), following W. His, thought that, during the formation of the limb, clefts and spaces are formed into which the nerves grow simply according to the principle of least resistance. However, it is possible that directional forces play a role also within the limb. To some extent, the blood vessels seem to assume a directive role. Normally, the sciatic nerve grows along the sciatic artery and the crural nerve grows along the arteria femoralis. Of course, one cannot derive a causal relation from this fact. But it is suspicious that if one finds only very thin nerves [in an experimental case] they are always attached to the sciatic artery, which they follow in some instances to the shank without giving off branches to muscles. The influence of blood vessels becomes even more probable in those cases in which the branching process of a nerve coming from the other side is clearly related to the branching pattern of blood vessels. In Figure 5, the contralateral nerve branches at the point where it meets the sciatic artery. The one branch continues distally as a typical sciatic nerve with typical further branching; the other branch follows the blood vessel in the opposite direction, crosses over to the arteria femoralis and

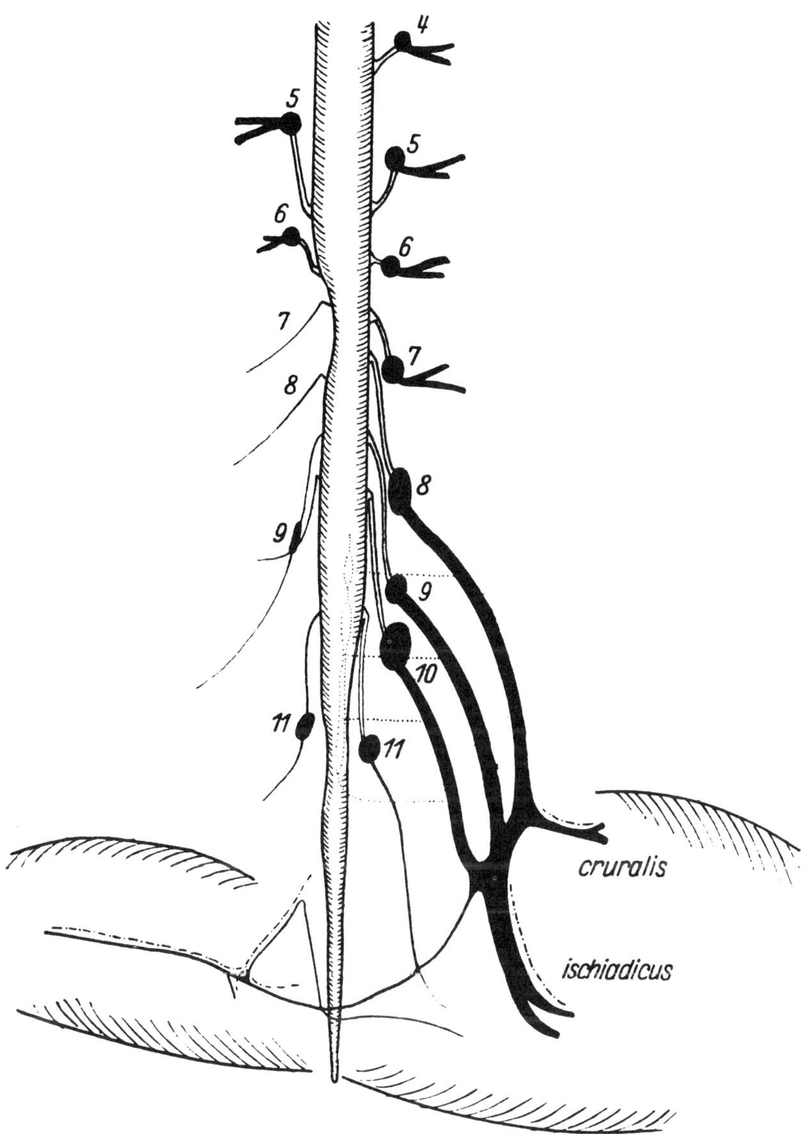

Figure 7. Nerve pattern of a metamorphosed frog after early extirpation of the right lumbar spinal cord. The right leg is innervated by a thin branch emerging from the left plexus. Dotted lines = blood vessels.

continues as a crural nerve. Interestingly enough, the nerve emerging from the 12th ganglion in Figure 4 behaves identically. In Figure 7, I found the same branching; however, the centripetal branch did not become a crural nerve; rather, it crossed back to the left side. Furthermore, another atypical branching coincides with the branching of a blood vessel. The main strand then runs along the sciatic artery to the knee, as a typical sciatic nerve, without further branching. The role of blood vessels can be only that of a general guidance; it cannot be responsible for the details of the nerve pattern, since that pattern does not replicate that of the blood vessels.

5. The simplest explanation for the end arborization of motor fibers in muscles has been given by Harrison (1911) though it has not yet been tested by experiments. The nerves grew out in the elongating and differentiating limb; they formed, so to speak, a reservoir, out of which the muscles drew their supply. However, they can do that only at a particular stage of their differentiation; when they have reached a certain state of maturity. Since the proximal muscles differentiate first, they become saturated first: the residual nerve bundle continues to grow distally and serves as a source for the innervation of distant muscles.

In summary, we obtain the following picture of the development of nerve pathways in the limb: The initial outgrowth is independent of extracentral factors but perhaps dependent on intracentral factors. Subsequently, the limb bud exerts an attracting effect on the outgrowing fibers. Once the nerves have arrived at the base of the limb, they grow in and with the limb, perhaps with the utilization of blood vessels, and they give off terminal branches, perhaps according to the principle of "saturation of mature muscle primordia." The way the sensory fibers find their way to the skin [and muscle spindles] is unknown.

The overall picture of developmental physiological correlations between amphibian limbs and their innervation is still very incomplete. However, the following points can be made with some certainty:

1. The nervous system exerts no determining or triggering effect of any kind on the development of the limb. The effects following the removal of innervation are trophic: shortening, paralysis, atrophy and eventually degeneration of muscles. Morphogenesis is undisturbed by the loss of innervation. The limb is self-differentiating with respect to the nervous system.

2. Likewise, no morphogenetic effect is exerted on the central nervous system by the developing limb. The effects of the lack of the target are of a purely quantitative nature and concern only the sensory parts. Hence, the nervous system is self-differentiating with respect to the limb.

3. However, the limbs have a strong developmental-physiological influence on the formation of the peripheral nerve pathways. The typical pattern originates as the result of directional forces which are exerted on the outgrowing nerves by limb buds and limb tissues.

References

Ariens-Kappers CU: On structural laws in the nervous system. The principles of neurobiotaxis. Brain. 1921; *44*: 125.

Bok ST: Stimulogenous fibrillation as the cause of the structure of the nervous system. Psych. en neurol. Bladen. 1915; *19*: 281.

Braus H: Einige Ergebnisse der Transplantation von Organanlagen bei *Bombinator*. Verh. Anat. Ges. 1904; *18*: 53.

Braus H: Experimentelle Beiträge zur Frage nach der Entwicklung peripherer. Nerven. Anat. Anz. 1905; *26*: 433.

Detwiler S: On the hyperplasia of nerve centers resulting from excessive peripheral overloading. Proc. Natl. Acad. Sci. (USA). 1920; *6*: 96.

Detwiler S: Experiments on the transplantation of limbs in Amblystoma. Further observations on peripheral nerve connections. J. exp. Zool. 1922; *35*: 115.

Duerken B: Ueber frühzeitige Extirpation von Extremitäten-Anlagen beim Frosch. Zeitschr. f. wiss Zool. 1911; *99*: 180.

Duerken B: Ueber einseitige Augenextirpation bei jungen Froschlarven. Zeitschr. f. wiss. Zool. 1913; *105*: 192.

Duerken B: Ueber Entwicklungskorrelationen und Lokalrassen bei *Ranz fusca*. Biol. Zentrbl. 1917; *37*: 127.

Duerken B: Einführung in die Experimental-Zoologie. Berlin, Julius Springer, 1919.

Forssman J: Zur Kenntnis des Neurotropismus. Beitr z path Anat. u. allg. Pathol. 1900; *27*: 407.

Hamburger V: Ueber den Einfluss des Nervensystems auf die Entwicklung der Extremitäten von *Rana fusca*. Roux'. Arch. f. Entw. mech. 1925; *105*: 149.

Hamburger V: Die Entwicklung experimentell erzeugter nervenloser und schwach innervierter Extremitäten von Anuren. Roux'. Arch. f. Entw. mech. 1928; *114*: 272.

Harrison RG: An experimental study of the relation of the nervous system to the developing musculature in the embryo of the frog. Am. J. Anat. 1904; *3*: 197.

Harrison RG: The outgrowth of the nerve fiber as a mode of protoplasmic movement. J. exp. Zool. 1910; *9*: 787.

Harrison RG: The stereotropism of embryonic cells. Science. 1911; *34*: 279.

Harrison RG: Experiments on the development of the forelimb of *Amblystoma*, a self-differentiating harmonious-equipotential system. J. exp. Zool. 1918; *25*: 413.

Harrison RG: Some unexpected results of the heteroplastic transplantation of limbs. Proc. Natl. Acad. Sci. (USA). 1924; *10*: 69.

Herbst C: Ueber die Bedeutung der Reizphysiologie für die kausale Auffassung von Vorgängen in der tierischen Ontogenese. Biol. Zentrbl. 1894; *14*: 657.

Herrick CJ: Morphogenetic factors in the differentiation of the nervous system. Physiol. Rev. 1925; *5*: 112.

Hoadley L: The differentiation of isolated chick primordia in chorioallantoic grafts. III. On the specificity of nerve processes arising from the mesencephalon in grafts. J. exp. Zool. 1925; *42*: 163.

Ingvar S: Reactions of cells to the galvanic current in tissue cultures. Proc. Am. Soc. Exp. Biol. Med. 1920; *17*: 198.

Lebedinsky NG: Eine neue Methode zum Erzielen nervenloser Extremitäten. Arch. f. mikr. Anat. u Entw. mech. 1924; *102*: 101.

Luther A: Ueber angebliche echte Entwicklungskorrelationen zwischen Auge und Extremität bei den Anuren. Oversigt. Finsks. Vetensk. Soc. Förhandl. 1915–1916; *58*: 18.

Mangold O, Spemann H: Ueber Induktion von Medullarplatte durch Medullarplatte im jüngeren Keim. Roux'. Arch. f. Entw. mech. 1927; *111*: 341.

Petersen H: Entwicklungsmechanik des Auges. Ergebn. d Anat. u Entw. gesch. 1924; *25*: 623.

Schaper A: Die frühesten Differenzierungsvorgänge im Zentralnervensystem. Roux'. Arch. f. Entw. mech. 1895; *5*: 82.

Shorey ML: The effects of the destruction of the peripheral areas on the differentiation of the neuroblasts. J. exp. Zool. 1909; *7*: 25.

Spemann H, Mangold H: Ueber Induktion von Embryonalanlagen durch Implantation artfremder Organisatoren. Roux'. Arch. f. Entw. mech. 1924; *100*: 599.

Tello F: Gegenwärtige Anschauungen über den Neurotropismus. Verhandl. u Aufs. über. Entw. mech. 1923; *33*: 1.

Tiedemann F: Anatomie und Bildungsgeschichte des Gehirns im Foetus des Menschen. Nürnberg Steinische Buchhandlung, 1816.
Wieman HL: The effect of heteroplastic grafts of the spinal cord on the development of the limb in *Amblystoma*. J. exp. Zool. 1926; *45*: 355.
Wintrebert P: Influence du système nerveux sur l'embryogénèse des membres. Comptes Rend. Acad. Sci. 1903; *137*:

Ann. NY Acad. Sci. 55: 117–132.

DEVELOPMENT OF THE NERVOUS SYSTEM

By V. Hamburger
Washington University, St. Louis, Missouri

The nervous system of the chick embryo has become increasingly important for neuro-embryological studies. It is more highly organized than that of the Amphibia which has been the classical object for the pioneer investigations in this field. At the same time, it is equally accessible to experimental analysis, which gives it a definite advantage over that of the mammals. The rapid development of the chick embryo and the excellent results which one obtains with silver impregnation techniques are additional assets. Since many centers are well circumscribed, and not excessively large or complex, simple quantitative methods can be applied, such as cell counts and area measurements.

In a previous symposium, D. Rudnick (1948) has given a lucid account of the origin of the medullary plate. Therefore, we shall pass over this first chapter of neurogenesis and take the neural tube of the two-day embryo (15–25 somites) as the starting point of our discussion. The cells of the neural epithelium are closely packed and interdigitated, and they do not show any visible differentiation. The main parts of the brain, however, are already blocked out and set off from the spinal cord.

The Organization of the Neural Tube

The spinal cord of this stage has been the object of a number of experiments which have revealed a considerable degree of organization in this structure. A rather rigidly determined pattern of regional differences was found in heterotopic transplantation experiments in which the cervical, brachial and thoracic sectors of the cord were substituted for each other (B. Wenger, 1951; Shieh, 1951), following the classical procedure which Detwiler (1923, 1936) inaugurated in Amphibia. The cell types and cellular arrangements characteristic of each level developed in the normal fashion, and not according to the site of implantation. For instance, the thoracic or cervical segments, which were implanted in the brachial level, innervated the wings and their nerves formed plexuses approaching the normal brachial nerves, but the impact of the peripheral overloading failed to induce a lateral motor column in the cervical and thoracic cord segments. On the other hand, the brachial cord retained the region-specific lateral motor column when transplanted to other levels. This example illustrates the high degree of inherent regional organization. A notable exception which was found in the motor system of the cervical cord will be discussed below. The transplantation of isolated parts of the cord, taken from different levels and implanted in the lumbosacral region between the somites and the hind limb bud of the host, gave essentially the same results (Bueker, 1943, 1945). The typical configuration of the nerve centers in the transverse plane is likewise fixed at these early stages (E. Wenger, 1950). In contrast to the situation in Amphibia, no regeneration or regulation occurs in the early neural tube. The extirpation of lateral and dorsal halves of the brachial level in embryos of 15–25 somites resulted in a wide range of deficiences in the mantle which could be correlated with corresponding deficiencies in the ependymal layer. It was found that a mosaic of at least six separate units exists in the early neural epithelium, along the dorso-ventral axis. Each sector gives rise to only those specific types of neurons which originate from it in normal development. No interactions that could be called "embryonic inductions" were observed between the different parts of the neural epithelium.

These early established patterns represent a general framework of organization. Many details within this framework are not yet rigidly determined, and very complex interactions between the different elements of the central nervous system and between nervous and non-nervous elements operate during the following period of progressive differentiation to establish the final configuration. In particular, the quantitative aspects of neurogenesis are subject to control by intrinsic and extrinsic agents.

The Spinal Cord at Eight Days

The nervous system of the eight- to ten-day embryo shows a remarkable increase in visible complexity which approaches the organization of the adult system (FIGURES 1, 2). The proliferative phase is nearly terminated, which means that the spinal cord of the eight-day embryo contains practically all potential neurons which will be present in the adult cord. The different cell groups have migrated to their destinations in the mantle and have formed the main centers. The size differences of the different types of neurons are apparent. The peripheral nerves have grown out and established provisional connections with non-nervous structures, and the intracentral fiber tracts either have formed their specific synaptic connections or are in the process of doing so. Obviously, the experimental analysis of the mechanisms of neurogenesis must concentrate on this period between the second and ninth to tenth days of incubation.

The analysis is facilitated by the fact that the three main aspects of progressive development, namely, proliferation, differentiation of neuroblasts, and cell growth, are quite well separated in space and time, and a fourth component, morphogenesis, does not enter into the picture, at least not in the spinal cord, which remains a simple tube. On the other hand, the situation is complicated by the fact that the unit of nerve tissue organization is not the nerve cell but the nerve center, although differentiation, proliferation and growth are functions of individual cells. Hence, the investigator is constantly confronted with events which happen on two levels, the cellular and a supercellular level. Limitation of space permits the discussion of only a few general correlations and concepts which have emerged from descriptive and experimental studies.

Mitotic Activity

It is logical to begin with the first component, proliferation. It is well known that all mitoses are located in the lining of the central canal. In an extensive study of proliferation between the third and eighth days of incubation (Hamburger, 1948), mitoses were counted separately in the dorsal and ventral halves of the spinal cord.   It was found that the distribution

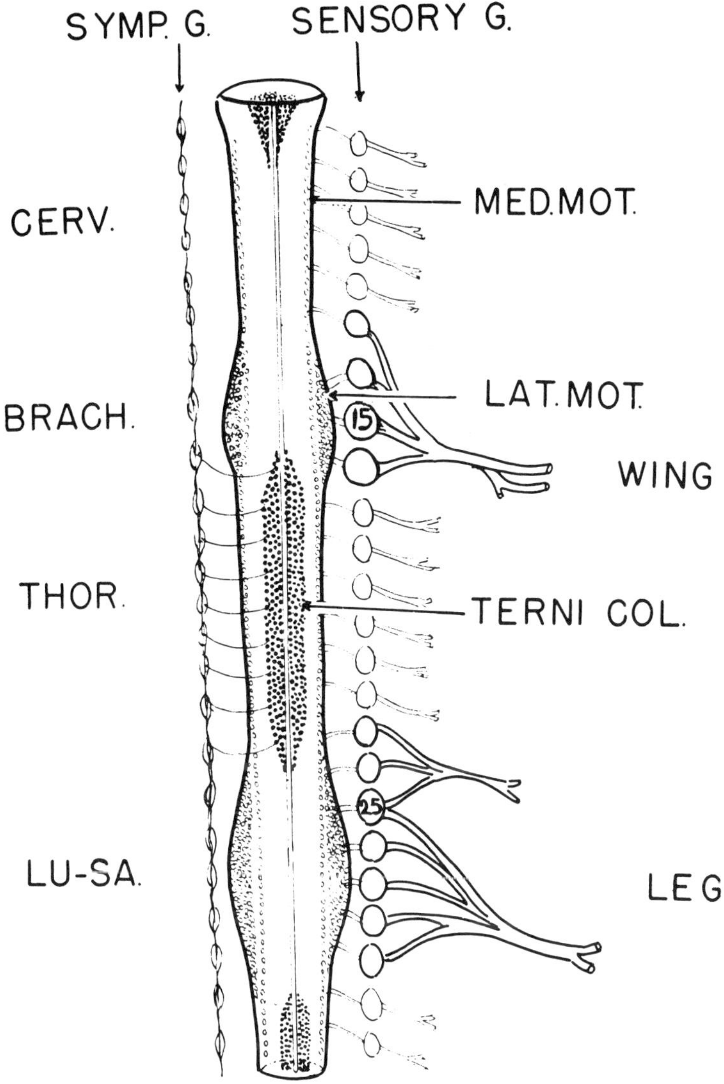

FIGURE 1. Diagram of spinal cord of eight-day embryo.   Brach., brachial level; Cerv., cervical level; Lat. Mot., lateral motor column; Lu-Sa., lumbo-sacral level; Med. Mot., median motor column; Terni Col., preganglionic column of Terni; Thor., thoracic level.

curves in the alar and basal plate, respectively, represent very different and, in some respects, contrasting, features (FIGURE 3).   The peak of mitotic activity in the basal plate is at three days, and that of the alar plate is at six to seven days.   Between three and six days, the mitotic activity rises in the alar plate and declines in the basal plate; between six and nine

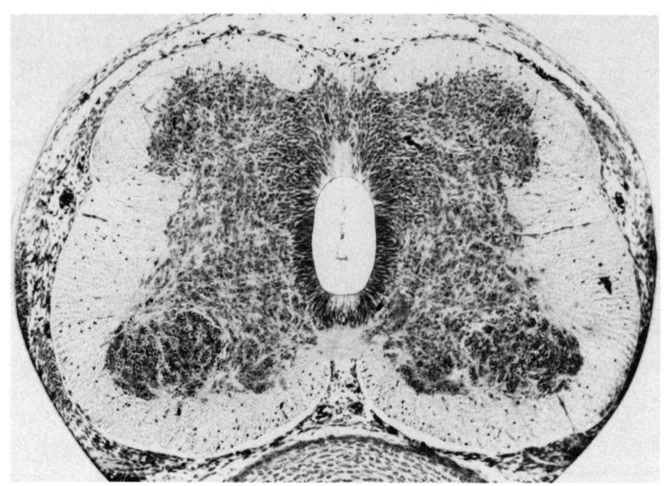

FIGURE 2. Cross section through brachial level of spinal cord of eight-day embryo.

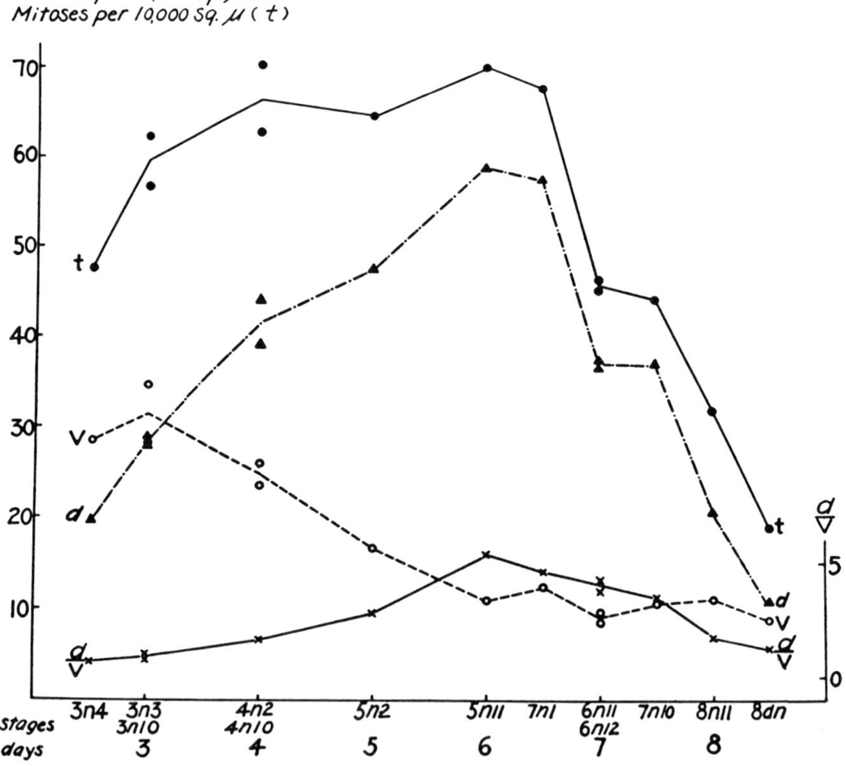

FIGURE 3. Time pattern of mitotic activity in spinal cord of chick embryo (averages for segments 10–20). *Abscissa*, stages of spinal cord differentiation and estimated chronological age; *ordinate* (left), average number of mitoses per unit area of lining of central canal; *ordinate* (right), ratio of mitoses in dorsal half (d) to mitoses in ventral half (v). (From Hamburger, 1948, J. Comp. Neurol., Vol. 88, Figure 5.)

days, proliferation in the alar plate declines sharply, but in the basal plate it retains a uniform, though low, average. The absolute figures are consistently higher in the alar plate than in the basal plate. For instance, the spinal cord of a six-day embryo contains approximately 20,000 mitotic figures, of which only 20 per cent are located in the basal plate. Generally speaking, the mitotic activity in the spinal cord is a patterned process, and proliferations in the basal and alar plates are independent of each other, both quantitatively and with respect to time patterns. This implies that agents which stimulate or control the proliferative activity of the neural epithelial cells must be distributed in a rather complex fashion along the dorso-ventral axis of the neural tube (for further discussion see Hamburger, 1948).

The time pattern of mitotic activity can be correlated rather closely with the subsequent histogenetic events. For instance, the motor columns make their appearance three to four days in advance of the dorsal sensory columns and this time sequence is reflected in the mitotic pattern. The peaks of mitotic activity in the dorsal and ventral halves are approximately $3\frac{1}{2}$ days apart.

Origin of Regional Differences in the Motor System

In another respect, no such correlation was found where it might have been anticipated. Since the motor columns are much larger in the limb levels than in other levels, one might have expected peaks of mitotic activity in these regions. The proliferative activity, however, was found to be rather uniform along the entire cephalo-caudal axis of the spinal cord, in all stages. This seemingly paradoxical stituation found its explanation when the origin of regional differences in the motor system (FIGURE 4) was studied in detail (Levi-Montalcini, 1950). It was found that the motor system in its earliest stages (3–$4\frac{1}{2}$ days) is represented by a ventro-lateral column which is of uniform size throughout the cord (FIGURE 4a). The regional differences are brought about in the following way: in the cervical level, a considerable number of neuroblasts undergo cytolysis and disappear; and in the thoracic level, a large group of neuroblasts migrate from the ventro-lateral column in a medio-dorsal direction (FIGURE 6, P). These cells settle near the central canal where they form the preganglionic column of Terni. In the limb levels, the column is not depleted, either by migration or by cytolysis (FIGURE 4, B, C). In other words, it is not the differential mitotic activity, but local depletions by cellular breakdown and migration, which are instrumental in the final regional specifications. These findings caution us against an all too common mistake: to attribute all quantitative differences, indiscriminately, to differences in proliferation. These differences bring cytolysis and migration into focus as instruments of pattern formation. The neurologist will be interested in another point, namely, that the visceral and somatic motor systems have a common morphological origin, although it is not contended that the early motor column is structurally homogeneous. It is conceivable that the somatic and visceral elements are already determined at that early stage.

The disappearance of part of the cervical motor system is rather puzzling, and one wonders why a similar breakdown does not occur in other parts of the column. A study of neuroblasts before their disappearance showed that they were connected by rami communicantes with a temporary sympathetic chain, and the idea suggested itself that we may be dealing with an

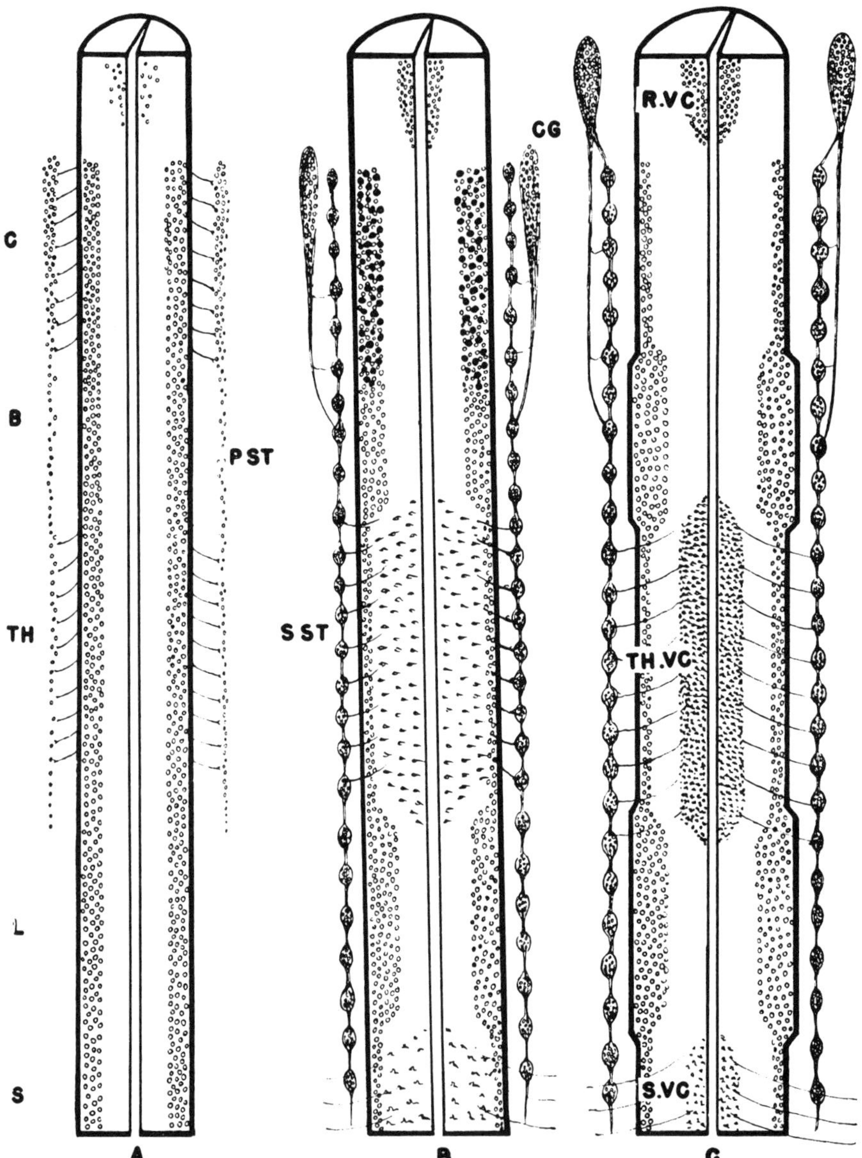

FIGURE 4. Diagrammatic frontal sections of the spinal cord of chick embryos, showing the emergence of regional differences in the motor system from a morphologically uniform system. A, 4 days; B, 5 days; C. 8 days. B, brachial level; C, cervical level; CG, cervical ganglion; L, lumbar level; PST, primary sympathetic trunk; R.VC, rhombencephalic visceral center; S, sacral level; SST, secondary sympathetic trunk; S.VC, sacral visceral center; TH, thoracic level; TH.VC, thoracic visceral center (nucleus of Terni). (From Hamburger and Levi-Montalcini, 1950, Genetic Neurology (ed. P. Weiss).)

abortive and short-lived cervical preganglionic system.   It was then thought that if the cervical and thoracic levels are akin in their potentiality to form preganglionic systems, it might be possible to activate this inherent capacity of the cervical cord by transplanting it in the place of the thoracic cord. In this position, optimal conditions would exist for the actual differentiation of a preganglionic system, mainly because an opportunity would be given for synaptic connections with the normal paravertebral sympathetic chain. This experiment (FIGURE 5) was performed by Shieh (1951).   A migration

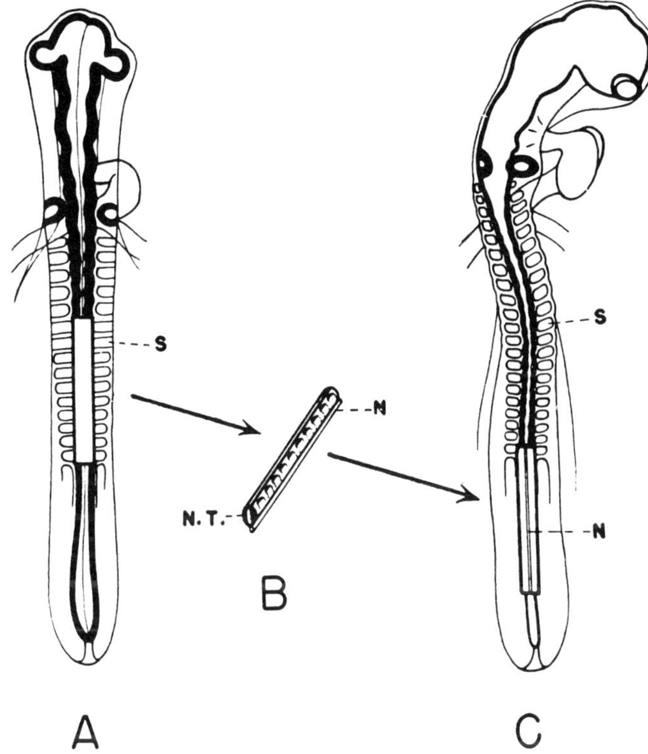

FIGURE 5. Transplantation of cervical spinal cord to the thoracic level. A, donor with cervical spinal segments including the notochord removed; B, transplant; C, host with implantation site; N, notochord; NT, neural tube; S, somites. (From Shieh, 1951, J. Exp. Zool., Vol. 117, Figure 1.)

of motor cells was actually observed in the posterior segments of the transplanted cervical cord (FIGURES 6, 7).   As a result, preganglionic cell groups formed which resembled closely the nucleus of Terni, both in position and in mode of origin.   In a few instances, neurites of these cells were traced to rami communicantes.   This experiment confirms the assumption of an abortive preganglionic system in the cervical cord.   It reveals, at the same time, an exception (so far the only one) to the statement made above, that regional differences in the cord are rather rigidly fixed in early stages.   Since we do not know which particular local conditions in the thoracic level call forth the neoformation of preganglionic cells, we prefer not to label it as an "embryonic induction."

Cytolysis

Differential cytolysis plays a role in the creation of quantitative differences of other parts of the nervous system. For instance, it was found that the size differences of spinal ganglia are due, in part, to the operation of this mechanism (Hamburger and Levi-Montalcini, 1949). During the

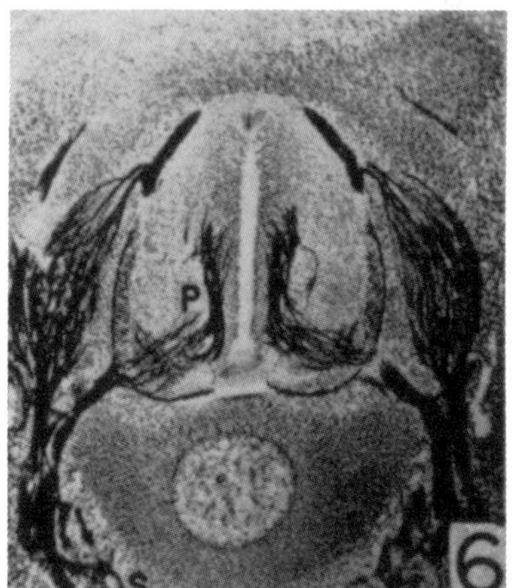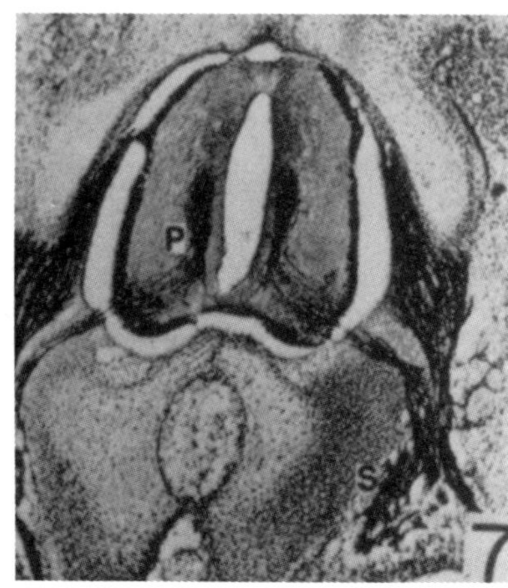

FIGURE 6. Cross section through the median part of the normal thoracic spinal cord of 6½-day embryo. P, preganglionic nucleus of Terni; S, sympathetic ganglion. (From Shieh, 1951, J. Exp. Zool. V117, Figure 12.)

FIGURE 7. Cross section through the middle part of a cervical segment transplanted to the thoracic level. Embryo fixed at 6½ days. P, preganglionic column of Terni; S, sympathetic ganglion. (From Shieh, 1951, J. Exp. Zool. Vol. 117, Figure 13.)

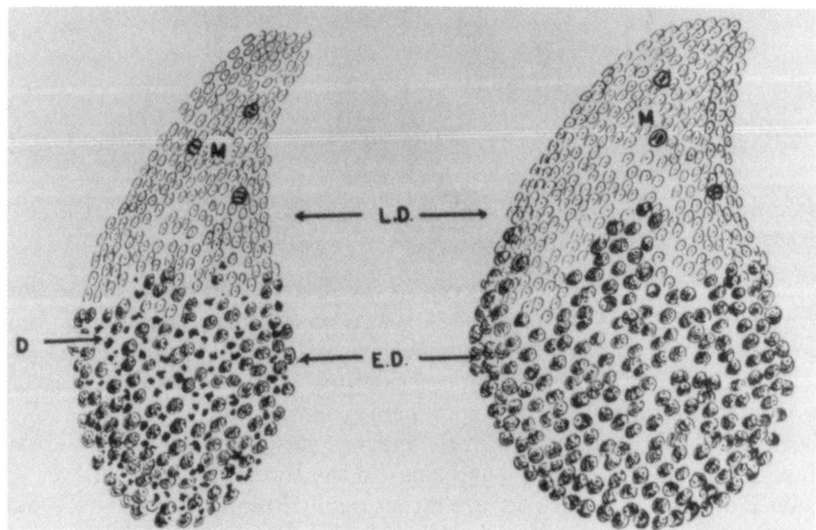

FIGURE 8. Cervical ganglion (left) and brachial ganglion (right) of six-day embryo. D, degenerating neuroblasts. ED, early differentiating, ventro-lateral neuroblasts; LD, late differentiating, medio-dorsal neuroblasts; M, mitoses.

fifth and sixth days, a large-scale degeneration of early differentiated neurons occurs in the cervical and thoracic ganglia, but none occurs in the limb-innervating ganglia (FIGURE 8).   Differences were also found in the mitotic activity, and the final regional differences in the size of spinal ganglia are the result of a combination of these two factors.   It is of considerable interest to note that the regressive changes produced in limb ganglia as a reaction to limb extirpation are identical with changes which occur in cervical and thoracic ganglia as part of the normal pattern.   The overproduction of neuroblasts, with a subsequent partial breakdown, seems to be a rather wide-spread feature in normal neurogenesis.   Cell death is known to be a common phenomenon in embryonic development and an integral part of many morphogenetic and histogenetic processes (see Glücksmann, 1951); but its involvement in the creation of numerical differences between cell groups seems to be rather unusual.   In this connection, the point should be stressed that, in all instances of cytolysis observed so far in the nervous system, the degeneration affects neuroblasts which have already sent out their neurites.   Therefore, the future analysis of the factors which are responsible for the breakdown of nerve cells will have to take into consideration not only the immediate environment of the doomed cells but also the situation at the axon terminals.

Migration

The significance of migrations of undifferentiated neural epithelial cells and of neuroblasts in neurogenesis has been discussed elsewhere (Hamburger and Levi-Montalcini, 1950).   Individual cell migrations and cell group migrations are of greatest importance for the establishment of the topographic patterns of nerve centers, and we cannot hope to come to a deeper understanding of the origin of the stratification of the brain and of the spinal cord until we approach the difficult analysis of the factors which are responsible for the activation and the direction of these migrations.

Trophic Relations

The unique property of the nervous tissue to spin out nerve fibers presents to the embryologist a number of special problems.   Foremost among them are those related to the directional outgrowth of nerve fibers and to the establishment of the highly selective synaptic connections.   A review of this wide field of investigations is not within the scope of the present discussion (see reviews by Piatt, 1948; Sperry, 1951; Weiss, 1941, 1951). We shall focus our attention to some other embryological aspects of nerve fiber outgrowth and synapse formation which may be referred to, collectively, as "trophic correlations", although they represent physiologically heterogeneous phenomena.   We shall discuss first the relations between primary nerve centers and non-nervous end organs and then the relations of nerve centers to each other.

Direct Effects of the "Remote Milieu" on Nerve Centers

As soon as the neurite emerges from the neuroblast, a new chapter is opened in the life history of the nerve cell.   It reaches out into a new sphere of influence and, at the same time, it comes under the control of the conditions to which the growing tip of the neurite is exposed. It acquires, so to speak, a second milieu, and some nerve centers become as sensitive to changes in their "remote milieu" as they are to changes in the immediate milieu which surrounds the pericaryon. In many instances, the successful completion of cell differentiation depends on the maintainance of normal trophic relations with the periphery, and the structural integrity of the neuron is threatened if this connection is interrupted.

This trophic dependence on the periphery is not a specifically embryonic property of the primary nerve centers. It persists throughout the entire life span of the neuron. However, it has very significant repercussions on developmental processes in the nervous system, whereas its role in the adult system is overshadowed by the functional aspects of impulse transmission. Furthermore, reactions of the embryonic nerve cells to peripheral changes are more rapid and more radical than those of the adult nerve cell and, finally, experimental embryology offers a variety of experimental approaches which are not readily available in the adult. For instance, in the embryo, one can disturb these relations without cutting the nerve, by the simple expediency of changing the periphery before the nerve grows out. In the adult, one deals largely with regressive changes following nerve transection, and with the restoration of the normal condition; whereas, in the embryo, one can enhance the growth processes beyond the normal range by enlarging the peripheral field of innervation. The chick embryo has proved to be an excellent material for such investigations.

If one surveys the effects of the decrease or increase of the peripheral area on the primary nerve centers, it is found that different centers behave differently. Some are much more sensitive than others. Following limb bud extirpation, the size of the somatic motor centers is reduced already in early phases of its differentiation (Hamburger and Keefe, 1944), and a hypoplasia, amounting to 90 per cent, can be obtained within eight days after operation (Bueker, 1943). On the other hand, the sympathetic chain ganglia differentiate normally and remain intact at least up to eight days of incubation following the same operation, and regressive changes do not occur until later (Simmler, 1949). The trochlear nucleus is also formed in the normal fashion and normal size in the absence of the primordium of the superior oblique muscle, and the effects of the operation are limited to the slow disappearance of differentiated cells, at later stages (Dunnebacke, 1952). In the spinal ganglia, cell groups lying side by side respond differently to peripheral changes.

The responses of the spinal ganglia were analyzed in detail, and it was found that the hypo- and hyperplasia, following the decrease or increase, respectively, of the peripheral area are the result of a combination of different factors (Hamburger and Levi-Montalcini, 1949). We have distinguished between the responses of differentiated neurons, which are connected directly with the periphery by their neurites, and the responses of undifferentiated cells, which are affected indirectly. We shall discuss the former reactions first.

In the spinal ganglia, one can distinguish two groups of neuroblasts by their topographic position and the time of their differentiation: a ventrolateral group which differentiates very early and rapidly, and another group which begins to differentiate later and develops more slowly and which occupies the central and mediodorsal regions of the ganglion (FIGURE 8). During the greater part of the incubation period, the former cells are conspicuously larger than the latter, but this size difference disappears later. In the absence of the limb, a considerable number of early differentiating neuroblasts differentiate and form neurites, but shortly after they have done so, that is at 5 and 6 days of incubation, they undergo a sudden and rapid degeneration, and most of them have disappeared completely by 8 days. The late-differentiating cells also undergo their initial differentiation in a normal fashion. During the later part of incubation, they undergo a slow atrophy but never a complete cytolysis. In the reverse experiment of the enlargement of the periphery by implantation of an additional limb, one might expect a cellular hypertrophy. However, cell size differences were not obvious and no cell measurements were made. Later on, a much more powerful promoter of the development of spinal ganglia was discovered in mouse tumors (Bueker, 1948), and, in this instance, we have observed not only a striking cellular hypertrophy but also an acceleration of the differentiation process (Levi-Montalcini and Hamburger, 1951, see FIGURE 9).

Embryologists resort occasionally to the "time pattern" of developmental processes in order to explain differential responses of primordia. The differences which one finds in the time of development of different nerve centers would make such an interpretation particularly attractive in the present instance. Our experiences with tumors, however, indicate that very subtle biochemical affinities exist between different types of nerve fibers and different peripheral structures (see the contribution of Levi-Montalcini in this monograph), and such differences seem to be of much greater importance for the interpretation of the different reactions of nerve centers than time patterns.

The trophic relation between the embryonic nerve cell and its "remote milieu" has its counterpart in the adult nerve cell. We recall the regressive changes in neurons following nerve transection and the atrophic and degenerative changes which nerve centers undergo if they are permanently separated from their end organs. The parallelism is even more striking if the regenerative process is compared with the process of embryonic differentiation. Recent studies of P. Weiss and his collaborators and of J. Z. Young and his collaborators have shown that the end result of nerve regeneration is very different, depending on whether or not the regenerating nerve is allowed to re-establish normal terminations in the end organ (P. Weiss, 1951; J. Z. Young, 1948, 1951). One finds, for instance, that in a regenerating motor nerve which is prevented from entering the muscle, the caliber size of fibers is smaller than in a nerve which has re-entered the muscle. "Maturation by increase in diameter and reduction in number of fibers only proceeds when contact with terminal sense organs and muscles is allowed. . . . Although new material is presumably produced by the cell body, the amount produced is profoundly affected by the periphery" (J. Z. Young, 1948, pp. 68, 69).   Young then continues: "Some message or influence must be transmitted up the nerve fibers from muscle or sense organ to produce this change."   We think that it is not necessary to assume the transport of some agent from the end organ to the pericaryon.   Since the neurite is part of the nerve cell, a direct metabolic exchange between its terminal part and its surroundings would certainly change the physiological condition of the entire cell.   Another alternative would be based on the

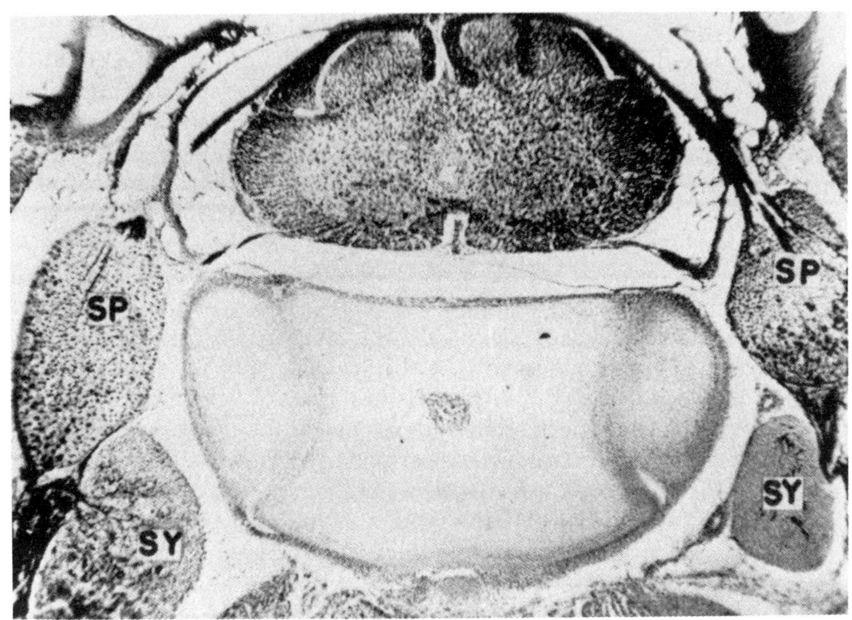

FIGURE 9. Embryo of 11½ days, with intra-embryonic tumor which is invaded by nerve fibers from right (apparent left) spinal ganglion (sp) and sympathetic ganglion (sy).   Note hyperplasin of ganglia and hypoplasia of lateral motor column.   (From Levi-Montalcini and Hamburger, 1951, J. Exp. Zool., Vol. 116, Figure 13.)

findings of Weiss and Hiscoe (1948) that a continuous production of axoplasm by the pericaryon and its continuous flow along the neurites is part of the normal physiological activity of the neuron.   An inhibition of the terminal growth process, by mechanical or other means, would then suffice to upset the metabolism of the cell.

The studies of Edds (1950) and Hoffman (1950) have revealed another interesting aspect of trophic nerve-muscle relations.   These authors found that if a muscle is partly denervated, the residual motor nerve fibers begin to develop subterminal collateral sprouts which are guided to adjacent denervated motor endplates.   In this case, the disturbance of a trophic equilibrium at the periphery reactivates the growth process of a nerve.

To summarize: The nerve cell establishes a subtle metabolic equilibrium with its "remote milieu" through its neurite.   The relation is of considerable consequence for the physiology of nerve cell growth, differentiation and regeneration, and its maintenance is a major concern of the neuron throughout its life span.

Indirect Effects of the Periphery

In the embryo, the peripheral milieu has additional, indirect effects. It influences, in some instances, the proliferative activity and the early differentiation of undifferentiated cells and controls in this way the quantitative development of entire nerve centers. The effects on mitotic activity were demonstrated directly by mitotic counts for spinal ganglia (limb extirpation, limb and tumor implantation) and inferred from cell counts, in the case of sympathetic ganglia (tumor implantation). On the other hand, no such effect was found in the basal plate of the brachial spinal cord which gives rise to the somatic motor system (Hamburger and Keefe, 1944), nor in the trochlear system (Dunnebacke, 1952). Negative results, however, are not always unequivocal. It is conceivable that one would obtain positive results under other experimental conditions.

The cells which are formed in excess of the normal complement, in cases of peripheral overloading, appear later on as supernumerary differentiated neurons, as was shown by cell counts in spinal and sympathetic ganglia, using silver impregnated material. From this we conclude that the indirect effects of the periphery extend also to the process of initial differentiation of undifferentiated cells into neuroblasts.

We have referred to these effects which are transneuronal, and, in a sense, supercellular, as "indirect" effects, because their targets are cells which have no direct connections of their own with the periphery. There is evidence to show that those neurons which differentiate first and which do establish peripheral connections are the mediators of this "remote control." The latter would be affected first and their cell bodies, in turn, would influence adjacent undifferentiated cells. This effect would spread in the fashion of an "assimilative induction" (Barron, 1943, 1946; Hamburger and Keefe, 1944).

Intracentral Interactions

Finally, we shall discuss the mutual relations between nerve centers during neurogenesis. We have much less information on this matter than on the peripheral relations; and we shall discuss only a few points which seem to be well established.

It has been postulated that outgrowing fiber bundles create an electric field around them, and thus stimulate the differentiation of neurites and dendrites in cell groups which they pass along their route. The direction of outgrowth of these processes would be perpendicular to the fiber bundles ("Stimulogenous fibrillation," Bok, 1915). This contention was ruled out by three different types of experiments, all of which show that long-range longitudinal fiber tracts have no such effects. Bueker (1943) isolated pieces of the neural tube of $2\frac{1}{2}$-day embryos by transplanting them to the flank. Levi-Montalcini (1945) eliminated descending tracts in the cord by extirpation of a number of cervical segments; and Hamburger (1946) blocked the entrance of both ascending and descending fiber tracts to the brachial cord by insertion of pieces of tantalum foil. In no instance were proliferation, initial differentiation or growth of the respective parts of the cord affected. Short-range local fiber tracts are equally ineffective (E. Wenger, 1950).

Do outgrowing intracentral tracts have any control over the development or maintenance of those centers towards which they grow, and in which they establish their terminal synaptic connections? There is no doubt that such effects are wide-spread. For instance, the extirpation of the limb bud affects not only the primary sensory and motor centers. The secondary sensory centers in the dorsal horn and the intermediate gray matter also become hypoplastic (Hamburger, 1934; Bueker, 1947). Similar transneural effects occur in the mammalian embryo (Barron, 1944). We refer also to numerous experiments in Amphibia which show the repercussion of the extirpation of sense organs on secondary sensory centers (reviews in Detwiler, 1936; Piatt, 1948). For the chick embryo, we have a detailed analysis of the effects of otocyst extirpation on the cochlear and vestibular centers (Levi-Montalcini, 1949). This operation removes also the eighth ganglion and deprives the medullary centers of their afferent root fibers. The vestibular centers remained unaffected up to the end of incubation, with the exception of the nucleus tangentialis. The cochlear centers underwent a normal initial differentiation up to eleven days, but, in later periods, they showed hypoplastic effects in varying degrees, which resulted, in part, from an arrest of differentiation and, in part, from a regression and breakdown of neurons. An interesting quantitative relation suggested itself. Those centers were most severely affected which receive no other synaptic connections than afferent root fibers, whereas centers which receive synapses from other sources in addition to root fibers are affected less severely, or not at all.

It seems that, at least in this group of secondary sensory centers, the important first phase of development which establishes the centers is not controlled by incoming fiber tracts, while the neurons require the presence of synaptic connections for their maintenance. In one exceptional case, namely, that of the nucleus tangentialis, the initial differentiation and migration of its component cells were blocked by the absence of the afferent root fibers and this center failed to form altogether.

These investigations which require supplementary studies of other systems indicate that trophic relations exist between nerve centers and that they are as varied and as complex as those between nerve centers and non-nervous end organs. Again, we refer to parallel conditions in the adult nervous system.

In conclusion, we wish to emphasize that all correlations which we have discussed reveal properties of the nervous system which are distinct and separable from its functional property of nerve impulse transmission. In fact, most of the early embryonic interactions occur before the involved nerve centers enter their functional phase. Neurogenesis does not make use of impulse transmission and its electrical and other correlates as an instrument of differentiation.

Bibliography

BARRON, D. H.   1943.   The early development of the motor cells and columns in the spinal cord of the sheep.   J. Comp. Neur. **78:** 1.

BARRON, D. H.   1944.   The early development of the sensory and internuncial cells in the spinal cord of the sheep.   J. Comp. Neur. **81:** 193.

BARRON, D. H.   1946.   Observations on the early differentiation of the motor neuroblasts in the spinal cord of the chick.   J. Comp. Neur. **85:** 149.

BOK, S. T.   1915.   Die Entwicklung der Hirnnerven und ihrer zentralen Bahnen. Die stimulogene Fibrillation.   Folia Neurobiologica.   Bd. IX.

BUEKER, E. D.   1943.   Intracentral and peripheral factors in the differentiation of motor neurons in transplanted lumbo-sacral spinal cords of chick embryos.   J. Exp. Zool. **93:** 99.

BUEKER, E. D.   1945.   The influence of a growing limb on the differentiation of somatic motor neurons in transplanted avian spinal cord segments.   J. Comp. Neur. **82:** 335.

BUEKER, E. D.   1947.   Limb ablation experiments on the embryonic chick and its effect as observed on the mature nervous system.   Anat. Rec. **97:** 157.

BUEKER, E. D.   1948.   Implantation of tumors in the hind limb field of the embryonic chick and the developmental response of the lumbo-sacral nervous system.   Anat. Rec. **102:** 369.

DETWILER, S. R.   1923.   Experiments on the transplantation of the spinal cord in Amblystoma, and their bearing upon the stimuli involved in the differentiation of nerve cells.   J. Exp. Zool. **37:** 339.

DETWILER, S. R.   1936.   Neuroembryology: 213   Macmillan, New York.

DUNNEBACKE, T. H.   1952.   The effects of the extirpation of the superior oblique muscle on the trochlear nucleus in the chick embryo.   J. Comp. Neur.   In press.

EDDS, MAC V., JR.   1950.   Collateral regeneration of residual motor axons in partially denervated muscles.   J. Exp. Zool. **113:** 517.

GLÜCKSMANN, A.   1951.   Cell deaths in normal vertebrate ontogeny.   Biol. Rev. **26:** 59.

HAMBURGER, V.   1934.   The effects of wing bud extirpation on the development of the central nervous system in chick embryos.   J. Exp. Zool. **68:** 449.

HAMBURGER, V.   1946.   Isolation of the brachial segments of the spinal cord of the chick embryo by means of tantalum foil blocks.   J. Exp. Zool. **103:** 113.

HAMBURGER, V.   1948.   The mitotic patterns in the spinal cord of the chick embryo and their relation to histogenetic processes.   J. Comp. Neur. **88:** 221.

HAMBURGER, V. & E. L. KEEFE.   1944.   The effects of peripheral factors on the proliferation and differentiation in the spinal cord of chick embryos.   J. Exp. Zool. **96:** 223.

HAMBURGER, V. & R. LEVI-MONTALCINI.   1949.   Proliferation, differentiation and degeneration in the spinal ganglia of the chick embryo under normal and experimental conditions.   J. Exp. Zool. **111:** 457.

HAMBURGER, V. & R. LEVI-MONTALCINI.   1950.   Some aspects of neuroembryology. Genetic Neurology (ed. P. Weiss).   Pp. 128–160.   U. of Chicago Press.   Chicago.

HOFFMAN, H.   1950.   Local re-innervation in partially denervated muscle: A histophysiological study.   Austral. Journ. Exp. Biol. Med. Sci. **28:** 383.

LEVI-MONTALCINI, R.   1945.   Correlations dans le développement des differentes parties du système nerveux. II. Correlations entre le développement de l'encephale et celui de la moëlle épinière dans l'embryon de poulet.   Arch. Biol. **56:** 71.

LEVI-MONTALCINI, R.   1949.   The development of the acoustico-vestibular centers in the chick embryo in the absence of the afferent root fibers and of descending fiber tracts.   J. Comp. Neur. **91:** 209.

LEVI-MONTALCINI, R.   1950.   The origin and development of the visceral system in the spinal cord of the chick embryo.   J. Morph. **86:** 253.

LEVI-MONTALCINI, R. & V. HAMBURGER.   1951.   Selective growth stimulating effects of mouse sarcoma on the sensory and sympathetic nervous system of the chick embryo.   J. Exp. Zool. **116:** 321.

PIATT, J.   1948.   Form and causality in neurogenesis.   Biol. Rev. **23:** 1.

RUDNICK, D.   1948.   Prospective areas and differentiation potencies in the chick blastoderm.   Ann. N. Y. Acad. Sci. **49:** 761.

SHIEH, P.   1951.   The neoformation of cells of preganglionic type in the cervical spinal cord of the chick embryo following its transplantation to the thoracic level.   J. Exp. Zool. **117**: 359.

SIMMLER, G. M.   1949.   The effects of wing bud extirpation on the brachial sympathetic ganglia of the chick embryo.   J. Exp. Zool. **110**: 247.

SPERRY, R. W.   1951.   Handbook of Experimental Psychology (ed. Stevens).   Pp. 235–280.   John Wiley & Sons, Inc.   New York.

WENGER, B.   1951.   Determination of structural patterns in the spinal cord of the chick embryo studied by transplantation between brachial and adjacent levels.   J. Exp. Zool. **116**: 123.

WENGER, E.   1950.   An experimental analysis of relations between parts of the brachial spinal cord of the embryonic chick.   J. Exp. Zool. **114**: 51.

WEISS, P.   1941.   Nerve patterns: the mechanics of nerve growth.   Third Growth Symposium **5**: 163.

WEISS, P.   1951.   An introduction to genetic neurology.   Genetic Neurology (ed. P. Weiss).   Pp. 1–39.   U. of Chicago Press.   Chicago.

WEISS, P. & H. HISCOE.   1948.   Experiments on the mechanism of nerve growth.   J. Exp. Zool. **107**: 315.

YOUNG, J. Z.   1948.   Growth and differentiation of nerve fibres.   Symposia Soc. Exp. Biol., No. II, Growth.   Pp. 57–74.

YOUNG, J. Z.   1951.   The determination of the specific characteristics of nerve fibers.   Genetic Neurology (ed. P. Weiss).   Pp. 92–104.   U. of Chicago Press.

Trends in Experimental Neuroembryology *

VIKTOR HAMBURGER

Washington University, St. Louis, Missouri

I. Introduction

In order to understand past and present trends in experimental neuro-embryology, one should realize that this field did not originate as an offshoot of neurology: it has its roots in classical experimental embryology, from which it derived its problems, its concepts and its methodology. It is true that the founder of modern neurology, Ramon y Cajal, encompassed in his broad approach both neurogenesis and nerve regeneration. Yet, it was the general trend of inquiry into the *causal factors* of embryonic development with the aid of the analytical experiment which was inaugurated by W. Roux and gathered momentum in the first decades of this century that brought experimental neurogenesis into existence. Harrison and his associates brought this approach to bear on the special problems that are inherent in the nervous system, and their pioneer work laid the foundation to the investigations which will be in the center of the present discussion.

* Some of the investigations of the author and his collaborators were supported by a grant from the Rockefeller Foundation and by Grants No. B463 and L-1801 from the National Institutes of Health, Public Health Service.

52

37

II. Induction of the Nervous System

At the same time, H. Spemann's experimental analysis of the earliest embryonic stages confronted him with a basic neuroembryological problem, namely, the first origin or "determination" of the nervous system as a distinct organ entity. Spemann was mainly concerned with the factors which initiate divergent trends of differentiation. One of his major contributions is the general concept that frequently the localization and the initiation of specific cellular differentiations are due to contact interactions with adjacent material, or *"embryonic inductions"* (1). These processes occur at very early stages of development, in our instance in the phase of gastrulation during which the germ layers, the precursors of organ primordia, are created. Spemann and H. Mangold discovered that the nervous system owes its origin to such a contact interaction. It shares with the epidermis a common origin from the ectodermal germ layer. That part of the ectoderm which during the process of gastrulation is brought into contact with the underlying axial mesoderm becomes neural tissue, and the part of the ectoderm which does not receive an inductive stimulus during this critical period becomes epidermis. The process of neural induction which is part of the complex activities of the so-called organizer has been recognized as the prototype of induction in general, and as such it has been the object of intensive analysis since its discovery in 1925 (1–4). The pertinent experiments were made on amphibian embryos, using techniques of microsurgery especially designed for this purpose. The crucial experiment was the transplantation of a small piece of axial mesoderm of the early gastrula to a position which brought it in contact with an area of ectoderm that would normally form epidermis. Under the impact of the underlying transplant, this area formed a neural plate which later developed into a typical nervous system. The significance of this finding from the neurological viewpoint is obvious: it is this process which initiates in a circumscribed area the specific chemical and structural changes that lead to nerve protein synthesis, to nerve tissue differentiation and to the organization of brain parts and spinal cord, setting this area apart from all others.

There is no doubt that neural induction involves a chemical interaction between the inductor and the reacting system. The inductive capacity of the organizer is not lost if it is killed by heat, cold, or alcohol treatment. Many adult tissues, both living and devitalized, derived from a variety of vertebrates and invertebrates, were found to have neuralizing effects on gastrula ectoderm. Extensive efforts have been made to isolate neuralizing agents and to identify them chemically. Tissue extracts and pure chemicals were tested by including them in a carrier, such as agar or egg albumen, and bringing them in contact with gastrula ectoderm. Steroid compounds, fatty acids, nucleic acids, and proteins were found to be active in varying degrees. We cannot evaluate these experimental results here (2–4); however a methodological predicament should be pointed out which makes it very difficult to prove that any one of these agents is identical with the natural inductor. The gastrula ectoderm is the only test tissue that is available for a bio-assay of neuralizing agents; however, gastrula ectoderm is among those tissues which when killed can induce a neural plate in living ectoderm. Hence, it contains the neural inductor in a masked form, and it is impossible to decide whether a substance which is to be tested is itself a neuralizing agent or whether it acts indirectly by unmasking or releasing this agent in the ectoderm through a slightly toxic effect.

The latter interpretation applies, for instance, to an experimental fact that seemed at first paradoxical and out of line with previous concepts of neural induction. It was found that the gastrula ectoderm of some amphibian species can form typical neural tissue without being brought in contact with a solid inductor, simply by exposing it to a slightly acid or alkaline or hypertonic medium. It is assumed now that such media have a slightly injurious effect, leading to an increase in the permeability of the surface membrane or to other cell surface changes, which in turn call forth the liberation of the intracellular neural precursors (neurogens). In other words, the materials and chemical machinery for neural differentiation seem to be present in all ectodermal tissue, and it has been suggested that the common denominator for all neuralization might be a nonspecific stimulus applied to the cell surface (2).

This trigger hypothesis may suffice to explain neuralization under exceptional experimental conditions. However, the normal induction of the nervous system involves more complex and more specific interactions. This is clearly indicated by the following findings: First, the normal, living organizer does not merely induce nonspecific neural tissue but it imparts to the overlying ectoderm a *regionally specific organization*. The rostral part of the axial mesoderm induces forebrain and other structures characteristic of the anterior head region. The adjacent mesoderm induces hindbrain and otocysts, and a more caudal region induces somites, spinal cord and other trunk-tail structures (archencephalic, deuterencephalic, spinocaudal inductors). Second, most adult tissues that have neuralizing capacity fall into two groups: those which induce primarily archencephalic structures (as for instance, liver of guinea pig), and those which induce primarily mesodermal trunk-tail structures (as for instance, guinea pig kidney). It is of particular interest that spinocaudal inductors when exposed to short heat treatment lose their capacity to induce trunk and tail and become converted into archencephalic inductors. The question has not been settled whether we are dealing with two different inducing agents, or with two forms of the same agent.

Fractionation studies of such tissues are actively in progress in several laboratories: they afford one of the few promising approaches to the chemical aspects of neural induction.

What are the first detectable changes in the reacting tissue? So far, the first indication that neural induction has occurred was found in changes of shape and kinetic properties of the affected cells. The thickening of the medullary plate, which is the first visible sign of neuralization, is brought about by a columnarization of the dorsal epithelium and a compensatory flattening of the ventral epithelium, without involvement of volume changes (5). These transformations of cell form are the expression of differences in the kinetic properties of individual neural and epidermal cells, which they acquire in the process of induction. When cells of the medullary plate are isolated in a culture medium, they assume an elongate, cylindrical shape and tendencies for ameboid movements and adhesiveness; if the cell body remains attached to a solid surface, pseudopodial projections are spun out and form branching processes. Epidermal cells, on the other hand, become spherical and flatten out when adhering to a solid surface (6). It remains to be determined to what extent these newly acquired properties are due to changes in the composition and molecular arrangements of the cell surface or of internal cytoplasmic components.

On the supracellular level, neural induction involves the blocking out of the major subdivisions of the nervous system, such as forebrain, hindbrain, spinal cord, and also of the neural crest, which is the precursor of spinal and sympathetic ganglia and of Schwann cells. Experiments done mainly on amphibian embryos show that these areas represent so-called *morphogenetic fields*. They are not yet sharply delimited against each other and not patterned in detail. This is clearly demonstrated in defect experiments, which reveal remarkable regulative capacities within each field. For instance, the extirpation of a large part of the forebrain-eye field results in its complete restoration, whereby eyes are formed from material that would not normally have formed them. The detailed structuration becomes gradually stabilized, but we know very little of the forces by which the definitive primordia such as eyes, forebrain, the alar and basal plate of the spinal cord, become segregated from each other (see ref. 7).

III. Dependencies of the Nervous System on Peripheral Structures

1. Introduction

At the core of experimental neuroembryology, in a more restricted sense, is the analysis of the factors which control the growth and differentiation of the central nervous system, the sense organs and ganglia, once these primordia are established. In addition, the unique property of nerve cells to spin out fibers poses a host of new problems concerning their outgrowth and the establishment of nerve patterns and of specific terminal connections. As in all embryogenesis, one is confronted at each step of the analysis with these basic questions: To what extent is a given process determined by factors intrinsic in the primordium and to what extent by extrinsic factors? Which specific extrinsic factors are at work, and how do they operate? It is clear from this formulation of the problem that the answers which we can provide with our standard methods of extirpation, transplantation, and isolation *in vitro* are given in terms pertaining to supracellular and cellular levels. Since the answers are in many instances reasonably specific, the ground is thus prepared for an attack of these problems by biochemical, histochemical and cytochemical methods. A comprehensive survey of neuroembryology is not within the scope of this discussion; I have selected a few topics which I think carry us rather close to such an approach.

In the early, formative period, in which the foundation and the framework of neuroembryology were created by Harrison, Detwiler, and others, amphibians were used almost exclusively as experimental material. An excellent summary of this earlier work may be found in Detwiler (8). In the following we shall be concerned primarily with the chick embryo, whose major advantage is the higher organization of its nervous system.

2. "Neurotrophic" Effects of the Periphery on Embryonic Nerve Centers

a. Experimental Reduction or Increase of the Area of Innervation. The interplay of intrinsic and extrinsic factors in neurogenesis has been studied extensively in the spinal cord and spinal ganglia of amphibian (8) and chick embryos. The basic structural patterns in the chick spinal cord and spinal ganglia are pre-established long before they become visibly manifest. Yet, the full realization of the local growth potentials depends on chemical interactions with nonnervous peripheral structures, and the outgrowing nerve fibers act as intermediaries in this process. Any change in the peripheral field of innervation of a nerve has profound repercussions on the development of the primary nerve center from which it emerges. The first evidence for such effects was obtained in a series of experiments in which the limb primordia of salamander embryos were excised and re-implanted at some distance posterior or anterior to their normal position. In this way, the normal limb nerves were deprived of their peripheral field, and trunk or head nerves entering the transplants were overloaded. Hypoplasia or hyperplasia of the respective spinal ganglia ensued, but the motor system appeared to be unaffected (9). A detailed study of the effects of wing or leg extirpation in chick embryos, prior to nerve outgrowth (in 2- to 3-day embryos), revealed that in this form the responses are varied and complex and they include the motor system. There are several types of reactions to peripheral changes, and each center has its own characteristic mode of response. The reactions of the spinal ganglia may illustrate the situation (10). They contain two groups of cells: large, early differentiating neuroblasts, probably exteroceptive in nature, and smaller cells which differentiate later. Following limb extirpation, the former grow out in a normal fashion; about two days after the operation they reach the amputation surface where their further outgrowth is blocked. Together with motor fibers they form a neuroma at that point. Shortly thereafter, that is about 3 days after the operation, a massive cell-degeneration of the large, early differentiating neuroblasts takes place, and at seven days of incubation this group has all but disappeared. The adjacent group of cells which differentiates later is less seriously affected; their differentiation is impaired, and the pericarya are atrophic, but they do not break down completely.

It is reasonable to assume that in both instances the nerve fibers mediate the effects. When the terminal growth cones encounter unfavorable conditions for continued growth and branching, the axons are likely to undergo slight pathological changes which would spread centripetally and eventually reach the pericaryon. It is of considerable interest that reactions were discovered also in cells which had no direct fiber connections with the periphery. Mitotic counts showed that the proliferative activity is reduced, whereby the total cell number is further decreased; and there are indications that the early phases of the differentiation process are impaired. In these instances, the peripheral influence is apparently indirect. It has been suggested that the deficiencies of the primarily affected neurons create abnormal physiological conditions in the entire ganglion and that dividing and differentiating cells are particularly sensitive to such changes (11, 12).

These diverse types of responses combine to produce the over-all result of severe *hypoplasia*. The reverse experiments, namely the increase of the peripheral area by implantation of a supernumerary limb in the distribution area of limb or trunk nerves, results in *hyperplasia,* which is due to an increase in cell numbers by enhanced proliferative activity and the subsequent neural differentiation of the supernumerary cells.

It should be emphasized that all these effects are brought about without injury to nerve fibers. Evidently the peripheral milieu controls the differentiation, proliferation, and maintenance of the nerve centers by providing adequate conditions for the peripheral growth of their axons.

The same types of reactions that occur in spinal ganglia are encountered in other centers; the response pattern is different in each center, however, and no generalizations are possible. For instance, the motor center in the lumbosacral cord reacts to leg bud extirpation by a rapid degeneration of neurons (unpublished), but very little degeneration occurs in the brachial motor center following wing extirpation. The early extirpation of the primordium of the superior oblique muscle leaves the proliferation and early differentiation of the trochlear center completely undisturbed; but in later stages, there occurs a gradual loss of differentiated neurons which amounts to 80% at 15 days (13).

Is it possible to define more concisely the interaction between the outgrowing nerve fiber and its milieu? Two hypotheses have been suggested: It is conceivable that the continued spinning-out of axoplasm at the terminal part of the axon is a necessary prerequisite for the physiological integrity of the axon and its pericaryon and that an interference with terminal branching during the growth period upsets the metabolic equilibrium of the entire neuron resulting in impaired growth and differentiation and, in extreme cases, in its complete breakdown. This hypothesis takes into account the finding that there is a constant centrifugal flow of axoplasm along the axon (14).

Alternatively, one can assume that an actual metabolic exchange takes place at the growing sector of the axon and that the milieu provides a substance or substances which are required for nerve growth and differentiation. They could be designated as "neurotrophic" agents.

b. The Tumor Agent. Accepting the second alternative as a working hypothesis, we tried to design an experiment in which we would substitute for the normal peripheral distribution area of nerves, such as the limb, another tissue that would meet the trophic requirements of the outgrowing fibers yet be more homogeneous and better suited for a chemical analysis than the heterogeneous tissues of the limb. Several tissues, such as liver and brain, have failed, but mouse sarcomas 180 and 37 were found to be well suited to our purpose (15, 16). Small pieces of the tumor were implanted near the base of the leg bud of 3-day embryos. The tumors grow rapidly and expand in the coelomic cavity. They are invaded by large masses of fiber bundles whose origin can be traced to adjacent spinal and sympathetic ganglia, whereas motor fibers do not enter the tumor. The ganglia become hyperplastic to a degree never observed before; at 13 to 15 days of incubation their volume increase amounts to as much as 600%. Both an increase in cell number and hypertrophy of neuroblasts contribute to this enlargement; in addition, the differentiation of the nerve cells was accelerated. Other tumors, such as adenocarcinoma and neuroblastoma, had no such effects. We have concluded that the sarcomas produce a neurotrophic agent (or agents) which stimulates selectively the growth and differentiation of sympathetic and sensory neurons.

Some observations suggested that the agent might be diffusible. To test this point, the tumors were implanted onto the extraembryonic membranes, where they become vascularized and grow well (17). The hyperplastic responses of sympathetic and spinal ganglia were at least as strong as in the case of intraembryonic tumors. Since the tumors were beyond the reach of nerve fibers, the agent must have reached the embryo by direct diffusion or, more probably, via the circulatory system. From the hyperplastic ganglia emerge an overabundance of nerve fibers which invade and literally flood the adjacent viscera, such as meso- and metanephros, gonads, thyroid and parathyroid gland. These organs are normally not yet neurotized in corresponding developmental stages. Furthermore, nerve fibers frequently penetrate the endothelia of veins and form neuromas of all sizes; some of them are so large that they occlude the lumen of the blood vessel. We are confronted here with a new situation. Whereas some of the tumor effects, such as the hyperplasia of the ganglia, could still be considered as the result of an extreme peripheral overloading, comparable to the effects of limb transplantation, the excessive hyperneurotization of the viscera and the infiltration of blood vessels indicate that the tumor agent calls forth pathological changes which have no parallel in the normal relations between nerve centers and their periphery. Furthermore, the mode of action of the tumor agent is fundamentally different from that of normal peripheral organs: the effects of the latter are mediated by nerve fibers (see above, p. 57), whereas the allantoic grafts have shown that the tumor does not require nerve connections for its action on the ganglia.

Our next aim was an attempt to characterize the tumor agent chemically. In order to avoid tedious injection experiments, the tissue culture method was applied (18, 19). When small pieces of living tumor are placed at a short distance from an isolated spinal or sympathetic ganglion, in an ordinary hanging-drop culture, a dense "halo" of radially oriented nerve fibers grows out within 24 hours; control cultures show little outgrowth in this period. This effect was then used as a bio-assay for the effectiveness of tumor extracts (20). They were tested by adding one drop of the substance to the ganglion at the moment of the coagulation of the plasma clot. The effects were recorded after 24 hours, in terms of a four-grade scale, based on the length and density of fibers.

Differential centrifugation of homogenates in a 0.25 M sucrose solution at pH 7.4 eliminated nuclei and mitochondria, which are inactive. All activity was retained in the microsome fraction. The active fraction was not dialyzable and was heat-sensitive at 75°C. If this fraction was treated with streptomycin sulfate (0.02 M; pH 7.2), the precipitate contained the full ribonucleic acid (RNA) concentration, as is shown by absorption spectra, and it retained full activity. The supernatant was inactive. After removal of the streptomycin by dialysis, the precipitate was treated with chloroform and centrifuged at moderate speed (8500 $\times$ g). The new precipitate is inactive;

it contains some but not all proteins. The lipids are dissolved in the chloroform. The clear, slightly opalescent supernatant still contains approximately 90% of the RNA of the original microsome fraction as shown by absorption spectra. It has retained at least 50% of the activity of the original homogenate and possibly more. It represents 4% of the dry weight of the original homogenate. An analysis of this material gave: proteins, 66%; RNA, 27%; DNA, 0.2%. The agent then is, or is associated with, a ribonucleoprotein.

There is no doubt that in the tissue culture experiments the nerve cells are the direct target of the tumor agent. If we are dealing with the same agent *in vivo* and *in vitro*—and there are strong arguments in support of this view—then there is good reason to believe that it operates *in vivo* in the same fashion. On the other hand, this assumption does not invalidate in the least the substantial evidence which can be marshalled in favor of the view that, in other instances, the nerve fibers mediate peripheral effects on their centers.

This dual control mechanism of nerve centers by peripheral agents is well illustrated by experiments on the mesencephalic V nucleus of amphibians. This nucleus consists of a relatively small number of neuroblasts which are conspicuously larger than their neighbors. It is an intramedullary sensory nucleus concerned with proprioceptive sensations from the jaw musculature. A study of the growth pattern of its cells showed that they undergo their most conspicuous growth shortly before or during metamorphosis, which suggested an influence of the thyroid hormone on their maturation. A direct test was applied by implanting small pellets of crystalline thyroxin anterior to the midbrain of larvae (21). The nearby mesencephalic V cells attained their postmetamorphic size very precociously, which proves that the pericarya are susceptible to direct stimulation by a humoral agent. Thyroxin, like the tumor agent, can operate by local diffusion, as in this experiment, or by remote control via the circulation, as in normal metamorphosis.

On the other hand, the development of the mesencephalic V nucleus is also subject to control by its peripheral area, via its nerve fibers. The unilateral extirpation of the mandibular primordium in young salamander larvae results in a reduction of the number of mesencephalic V cells amounting to 40% (22). A peripheral overloading can be accomplished by the implantation of the mandibular primordium of a large species to the appropriate site in the head of a small species. This results in a small but consistent increase of the cell number, amounting to 16%.

The main considerations of this chapter may be summarized thus: basic structural patterns are laid down very early, in connection with the induction of the central nervous system by its mesodermal *Unterlagerung*. However, the subsequent proliferation, differentiation, growth rate, and maintenance of nerve centers are controlled in their quantitative aspects by factors residing in the nonnervous milieu, including humoral agents. Peripheral factors operate directly, aiming at the pericarya as their target, or indirectly by mediation of the nerve fibers which connect the center with the periphery. The latter mechanism is thought of in terms of a metabolic exchange between the terminal part of the nerve fibers and the structures which they traverse.

3. The Role of Peripheral Factors in Nerve Regeneration

a. Influence on Maturation of Nerve Fibers. Neurotrophic and similar relations between nerve cells and the structures which they innervate persist throughout the life span of a neuron. They become manifest when these connections are permanently severed or, more impressively, when growth is revived in the adult fiber, in the process of nerve regeneration.

If a nerve fiber is transected and prevented from the re-establishment of synaptic connections, as for instance in limb amputation, then both axon and pericaryon atrophy. The atrophic condition may persist for long periods, but in many instances the regressive changes lead eventually to a breakdown of the neuron. Apart from clinical observations, there are a few data on animal experiments available. Thoracic ganglia of the white rat which were permanently disconnected by capping of the nerve stump suffered the loss of approximately 50% of all cells in the course of a year, and the surviving cells were severely atrophic. Chromatolysis became a chronic condition in this instance (23). Clearly, the structural integrity of the adult primary neuron depends on its connection at the periphery.

When a severed nerve is allowed to regenerate, then peripheral factors exert control over certain quantitative aspects of this process, just as they do in embryogenesis. To illustrate this point, we shall discuss briefly the role of peripheral connections in determining the restoration of fiber size, which has been studied carefully in its quantitative aspect (24–26). It is well known that the fiber spectrum (i.e., the ratio of fibers of different diameters), which is characteristic of each nerve, is profoundly disturbed in the early phases of regeneration. If one examines a cross section of a regenerating nerve at some point distal to the cut surface, some time after transection, he finds that it is composed of an excessively large number of fine fibers, all of a very small caliber. They originate by branching of the nerve fibers at the cut end of the proximal stump. Under normal circumstances, the original fiber spectrum is gradually restored by the growth and myelin formation of some fibers and the elimination of others (maturation). The successful maturation depends on a number of factors, among them the establishment of synaptic terminal connections. The following experiment gives quantitative data on this point (27, 28). A pure motor nerve in the leg of the rabbit (median gastrocnemius nerve) was chosen. This nerve has a bimodal distribution of fiber size with peaks at 7μ and 16μ. If regeneration is initiated by nerve crushing some distance from the muscle, the original fiber spectrum is restored almost completely after a period of 100 days. Crushing provides for optimal conditions for regeneration, since only the axons are interrupted, whereas the continuity of their neurilemmal sheaths is maintained and the regenerating fibers are guided by their sheaths to their former motor end plates. If, in addition to proximal crushing, the nerve is completely severed near its entrance into the muscle, the maturation process is considerably delayed because the fibers have to cross the gap between proximal and distal stump without guidance by their sheaths. After 100 days, the fiber spectrum is still shifted toward smaller calibers, the bimodal distribution is not yet restored, and the number of fibers is still excessive. If, in a third experiment, the nerve is transected and in addition deflected so as to prevent the re-establishment of synaptic connections, then its maturation is checked completely; after 100 days the distal part of the nerve still exhibits the typical postoperation pattern, that is, an excessively large number of very small fibers.

The reverse condition can be created by overloading of a nerve; in this instance, the caliber size of the nerves increases. This result was obtained in the soleus nerve of the white rat by denervating some of the synergists of the soleus muscle (29). In this instance, the structural continuity of the muscle was not interrupted. Hypertrophy ensued in the soleus muscle and nerve, and careful measurements showed an increase in the diameter of all nerve fibers, from the smallest to the largest. After 100 days, the mean diameter of all fibers was increased by 16.6%.

In speculating on the mechanism involved in these relationships, one has to take into account the possible role of functional activity. It is known that increased functional activity enhances the synthesis of neuroplasm in the pericaryon (30), and inactivity impairs it. The above-mentioned experiments could be interpreted in this way, but the alternative hypothesis that the muscle tissue has a direct trophic effect on the nerves is equally tenable. Perhaps both factors play a role. A third possible explanation, which was considered already in embryonic neurogenesis, must be kept in mind: As long as a regenerating fiber is growing actively, its metabolic energy is spent in this activity. When growth in length ceases after the establishment of synaptic connections, this energy might be shifted to growth in diameter, until an equilibrium is reached (25).

b. Collateral Nerve Regeneration. The extraordinary sensitivity of the adult nerve fiber to changes in its chemical milieu is most impressively demonstrated by the phenomenon of collateral nerve regeneration (see review, 31). It had been observed in clinical and experimental studies that muscles rendered paretic by partial denervation often attain a higher degree of structural and functional recovery than could be accounted for by hypertrophy of the intact muscle fibers that had retained their innervation. Several investigators had suggested that the residual nerve fibers might have formed collaterals which had reneurotized paretic muscle fibers, and some indirect evidence supported this inference. Such collaterals were actually observed by Edds (32) and Hoffman (33) on gold-impregnated, teased preparations of paretic muscles of the rat. Since muscle innervation is plurisegmental, it is feasible to denervate a leg muscle partially by destroying one or two of the ventral roots which feed into the sciatic plexus. As early as 4 to 10 days after the operation, fine collateral sprouts were found emerging from many residual motor fibers, either at a node of Ranvier or in an internodal segment. These collaterals are always intramuscular and subterminal; they manage to penetrate into adjacent empty neurilemmal sheaths left behind by degenerated fibers, and many of them establish functional synaptic connections with denervated motor end plates. Careful counts of terminal nerve fibers and of the end plates associated with them gave unequivocal evidence that each residual nerve fiber had reneurotized several vacant motor end plates. In the particular muscle that was investigated (anterior serratus) the normal terminal innervation ratio was 1:1.1, that is, each terminal fiber innervated one or, rarely, two motor end plates. Following a severe denervation amounting to 64%, the terminal innervation ratio rose to an average of 3 and in some parts of the muscle to 4.7. In an exceptional case, an old terminal fiber innervated as much as 30 times its normal complement of end plates (32). There is a direct correlation between the degree of denervation and the degree of branching, resulting in a fairly complete saturation of the empty motor end plates. This adjustment of supply and demand raises some difficult problems which will not be discussed here (see ref. 31).

It is remarkable, indeed, that an intact, uninjured terminal fiber should be stimulated to collateral branching when a nearby muscular junction is broken. The stimulus can hardly be other than chemical, and the search for a specific substance and its source of origin has begun (33–35). Intramuscular injections of saline homogenates of rat sciatic nerve and muscle were negative. Positive results were obtained with ether extracts of spinal cord and white matter of the brain of ox, muscle, and egg yolk. The fraction containing unsaturated (glyceride) fatty acids was found to retain a high activity, and it was suspected that the myelin of degenerating nerves in partially denervated muscles, rather than the denervated muscle itself, is the natural source of the agent.

How does the agent operate? Since the newly formed collateral fibers represent a very small fraction of the total volume of the axoplasm of an axon, it is hardly necessary to postulate a special growth-promoting effect (31). Hoffman and Springell (35) have made the suggestion that "the penetration of the fatty acid or its glyceride into the axon, locally disturbing the plasmalemmal structure and liquefying the axogel, results in an outflow of axosol which then reversibly gels at the point of breakdown, thus forming a new pseudopodial sprout."

The discovery of the phenomenon of collateral regeneration has opened up another field in which the chemical aspects of the relationship between the nerve fiber and its milieu can be explored.

I have given some attention to problems of nerve regeneration in the adult in order to emphasize that the primary neurons retain throughout their life span not only their potentialities for growth and differentiation but also their responsiveness to their peripheral milieu. Although the problems which the adult neuron faces in this respect are different from those of the neuroblast, a synoptic view of embryonic and regeneration phenomena will foster the understanding of both.

IV. Intracentral Correlations

1. Introductory Remarks

The study of intercellular relationships in neurogenesis has concentrated on the primary neurons for the obvious reason that their peripheral distribution patterns and their connections with nonnervous structures are readily accessible to experimental analysis. Yet the majority of all neurons are connected with other neurons. Their intracentral interrelations confront the experimental embryologist with formidable problems whose solution has eluded him so far. Mechanisms of utmost precision must be at work which guarantee the appropriate interconnections of nuclei. Complex track systems must be visualized which guide the fiber tracts each to its specific destination, often by way of decussations and other devious routes. Subtle selective affinities and disaffinities must induce a fiber bundle to by-pass one center and to choose another for synaptic union. The difficulties are somewhat simplified by the consideration that these interconnections are built up gradually, step by step, and that the complexities of the mature patterns emerge from much less complex initial conditions. Nevertheless, information on embryonic fiber tracts is almost completely lacking, with one creditable exception: the single pair of giant Mauthner fibers in amphibians, which offer a unique opportunity for the study of the determination of intracentral pathways, have been the object of a number of experimental investigations (review in ref. 36). However, limitation of space makes it necessary to set aside this matter in favor of other topics which are more closely related to the questions of "neurotrophic" relationships that have been dealt with so far. (See also review by Sperry in this volume.)

2. Neurobiotaxis

The topography of the nerve centers in brain and spinal cord is determined very early, long before fibers begin to differentiate. In the chick embryo, the first tract systems that form, the tectospinal tract and the fasciculus longitudinalis medialis, emerge from their centers in the di- and mesencephalon toward the end of the third day, and other tracts follow in succession according to a strict time schedule. The idea suggested itself that outgrowing fiber bundles might play a role in stimulating in some way the differentiation of neuroblasts which they pass along their course, or of centers in which they are destined to terminate.

Kappers (see ref. 37) and Bok (38) have strongly advocated the former idea and elaborated it in their theories of "neurobiotaxis" and "stimulogenous fibrillation." It was thought that outgrowing fiber tracts, by virtue of their functional activity, create an electrical field around them. Neuroblasts adjacent to their pathway would be polarized in such a way that their dendrites would be formed on the side facing the tract and be directed galvanotropically toward it, whereas the axon would form on the opposite side and grow in the opposite direction. Moreover, it was envisaged that functionally active fiber tracts would attract nerve cells and direct the migration of entire neuroblast groups toward them; this mechanism was supposed to explain both phylogenetic and ontogenetic shifts in the position of nuclei. Unfortunately, these ingeniously simple concepts did not stand the experimental test. Sectors of the brain or spinal cord of the chick embryo were found to be capable of normal differentiation when completely isolated, that is, in the absence of descending and ascending fiber tracts (39–42). Even the normal sequence of events is not compatible with the theory: in crucial instances, the dendrite and axon differentiation of a center was found to precede the arrival of adjacent fiber tracts (43). And, furthermore, evidence is lacking for the contention that outgrowing nerve fibers are sensitive to galvanotropic stimulation (44).

3. Transneuronal Effects in the Optic and Acousticovestibular Systems

The idea that outgrowing fiber tracts might control the differentiation processes of the centers with which they are destined to form synaptic connections has been verified by a number of experiments. Most of the work has been done on the sensory centers in the brain, since the primordia of the large sense organs—eye, ear, nose—are most readily accessible to extirpation and transplantation. A few selected instances of such trans-neuronal correlations will be presented.

Early work, dating back to the beginning of the century, has shown that the extirpation of an optic vesicle in early amphibian embryos results in a conspicuous hypoplasia of the contralateral optic center in the mid-brain. That we are dealing with true growth control and not merely with pathological after-effects of the operation was demonstrated by producing the reverse effect, hyperplasia. Harrison (45) did this by an ingenious experimental design: He substituted the optic vesicle of a large species of salamander, *Amblystoma tigrinum*, for that of a small species, *A. punctatum*, prior to the outgrowth of the optic nerve. In the cases in which the larger optic nerve established normal connections with the midbrain of the host, the latter became hyperplastic on the contralateral side.

Since the decussation of the optic nerve is complete in amphibians, one can obtain precise qualitative and quantitative data on these transneuronal effects by comparing the two sides in unilaterally operated cases. A careful analysis of the optic tectum hypoplasia in the frog has revealed that all components of the differentiation process are affected: the mitotic activity is reduced to as little as 45% (46), and the migration of neuroblasts from the ependymal layer to the more superficial layers is impeded (47). Both factors combine to produce a conspicuous numerical deficiency in cells which amounts in the superficial layers to over 50% at the end of metamorphosis, and in the deeper layers up to 38%. It was calculated that the reduction in mitotic rate can fully account for the cellular deficiency; hence, it is not necessary to assume that, in addition, a breakdown of neurons occurs (46). The neuroblasts which do reach the superficial layers remain atrophic, and their dendritic processes fail to undergo complete differentiation; they remain short and undergo less profuse branching than normally (48). Altogether it can be stated that proliferation, cell differentiation, cell growth, and migration are under the control of the optic fiber bundles which penetrate into the superficial layers of the tectum. It is remarkable that the response pattern is very similar to the one that was found in primary centers, for instance, in the spinal ganglia of the chick embryo, following limb bud extirpation, although the mechanisms which operate in the two instances cannot be the same. The deficient differentiation, atrophy, or degeneration of the primary neuron was ascribed to a direct milieu effect on the tip of its axon, which then transmits it to the pericaryon. In the present instance, it is not the axon but the pericaryon of the optic tectum neuroblast itself which is directly affected by the failure of optic nerves to synapse with it. The significant implication that the normal development of a neuroblast depends on its synaptic connections will be discussed presently. But this mechanism accounts for only part of the observed phenomena. It remains to be explained how the inflow of optic fibers at the superficial layer of the optic tectum controls the proliferation in the deepest layer, the ependyma, and the migration of cells from there to more superficial layers. We have encountered the same difficulty when we had to explain the effects of peripheral factors on proliferation and initial differentiation in spinal ganglia, that is, the responses of cells which are not directly connected with the periphery. At present, we have no satisfactory explanation for this correlation.

All neurons, with the exception of the sensory neurons in spinal ganglia, are the recipients of synaptic terminations. There is now enough experimental evidence available to state as a general rule that the establishment of synaptic connections on the surface of a neuroblast or on its dendrites is a necessary condition for the completion of its growth and differentiation and for the maintenance of its structural integrity. For a further illustration of this principle we refer to experiments on the cochlear and vestibular systems of the chick embryo (49). The otocyst, which includes the primordium of the eighth ganglion, was extirpated in 2-day embryos; the root fibers, whose outgrowth begins normally 1½ to 2 days later, were thus eliminated. The repercussions on the cochlear and vestibular centers were studied in closely timed series, and cell counts and area measurements were made on the two major cochlear centers, the nucleus angularis and the nucleus magnocellularis. It was found that their development proceeds normally up to 11 days, at which time they are numerically complete. Hence, in this instance, the mitotic activity is not affected by the absence of the root fibers. From this stage on, cell growth and cell differentiation are impaired on the operated side, resulting in atrophy of neurons. Furthermore, cell counts on the 11th and the 21st day showed a considerable loss of cells during this period. This implies that cells which had reached a considerable degree of maturity must have broken down and disappeared.

This response pattern is distinctly different from that of the optic tectum. In the latter there prevails an impairment of the initial phases of development: proliferation, migration, and initial differentiation; in the cochlear centers the regressive changes of advanced neurons dominate the picture. It will be remembered that similar differences in the type of response were found in the relations between nerve centers and periphery.

However, the reactions of the different cochlear and vestibular centers to the elimination of their root fibers are not uniform; on the contrary, there are striking differences in *degree*. The total cell loss in the nucleus angularis, between 11 and 21 days, amounts to 80%; that of the nucleus magnocellularis amounts to only 32%; and a third cochlear center, the nucleus laminaris, as well as the majority of the vestibular centers, shows no cell loss at all. How can this differential susceptibility be accounted for? The following hypothesis was suggested: It is assumed that a quantitative relationship exists between the degree of hypoplasia and the number of synaptic connections which a nerve center receives. If it is the recipient of fibers from only one source, namely the root fibers, as in the case of the nucleus angularis, then such a center will be very severely affected. If it receives, in addition, synapses from other, intracentral, sources, then these synapses will "protect" it against damage from the loss of root fibers. The degree of "protection" would be correlated with the number of accessory synapses which a given cell receives (49).

Transneuronal effects of this type are not limited to the sensory brain centers. For instance, conspicuous secondary reactions were observed in the dorsal and intermediate grey of the spinal cord following limb extirpation in the chick embryo (50, 51) and in the sheep fetus (52). It is assumed that the substantia gelatinosa and the intermediate grey are under the control of the inflowing sensory fibers. It has not been established to what extent the hypoplasia is due to a block of cellular differentiation or to a secondary breakdown of differentiated cells.

Similar transneuronal effects were found in the adult nervous system. For instance, decortication in the opossum is followed by cellular degeneration in the thalamic centers in which the corticothalamic tracts terminate (53).

To summarize: Neurons exert a metabolic control over the differentiation of those other neurons with which they are destined to synapse. This seems to be a general rule. Studies on the optic system show that, in addition, the influx of a fiber bundle into its prospective field of termination may control the proliferation and migration of undifferentiated cells in that center. However, this correlation does not seem to occur universally; it is not in effect in the acousticovestibular system (except for the nucleus tangentialis) (see ref. 39), nor, apparently, in the spinal cord. The elucidation of the physiological mechanisms involved remains for the future. Of particular interest is the question whether or not these metabolic or "neurotrophic" activities of neurons are linked with the process of impulse transmission. However, it will be very difficult to dissociate these two experimentally.

V. Differentiation and Neurotization

In the preceding chapter, the concept was developed that the nerve cell is not only the recipient of growth stimuli but that it controls, in turn, the differentiation and the metabolic equilibrium of other neurons. Can this concept be generalized? Is neurotization a significant factor in the differentiation of other nonnervous structures? Explantation experiments on early embryonic primordia of amphibians show that the initial differentiation of many organs and tissues proceeds normally in the absence of nerves (54). Observations over longer periods were made of the development of nerveless limbs in amphibians (55) and birds (56). The former were obtained by removing the limb level of the neural tube in early stages, and the latter by implanting early limb primordia into the coelomic cavity at some distance from the central nervous system. In both instances, the general pattern of limb development is not disturbed, but some developmental processes are impaired. The over-all size is reduced both in nerveless amphibian and chick limbs, and ankylosis is found frequently in the latter. Chondrogenesis and osteogenesis proceed normally, but in the chick, the musculature breaks down when it has barely reached the stage of cross striation (57).

On the other hand, nerve supply is indispensable for limb regeneration in adult and larval amphibians (58). No regeneration blastema is formed in the absence of nerves. If, in larval salamanders, the amputation stump is kept denervated for some time, a spectacular regression of the stump ensues and this degeneration process continues until the entire stump has disappeared, down to the base of the limb (59, 60). Obviously, neurotization plays an important role in the maintenance of a physiological equilibrium at the site of regeneration. It is known that the formation of a regeneration blastema is preceded by a short transitory phase of regression of the distal part of the stump. It seems that the nerve fibers produce an agent which checks this regression, but their influence extends probably to the progressive phase of blastema formation. (For detailed discussion see ref. 58.) The old controversy whether the stimulation is due to sensory, motor or sympathetic fibers has been resolved recently. The requirement is not of a qualitative but of a quantitative nature. Any combination of fibers is effective, provided their number exceeds a certain minimum threshold requirement (58).

It is well known that the maintenance of some adult tissues, especially musculature, depends on their nerve supply. Although this phenomenon is closely related to the topic of the present discussion, space does not permit discussion in detail (61, 62).

VI. Concluding Remarks

We had to omit from our discussion several important aspects of neuroembryology. For instance, we have not touched upon the extensive studies devoted to problems of the specificity of nerve pathways and terminal connections. Dr. Sperry will deal with some facets of this intriguing subject.

We have tried to concentrate on the data from which the concept has emerged that the nervous tissue, aside from its ordinary metabolic requirements, depends for its differentiation and maintenance on specific inductive and "neurotrophic" agents, supplied by its immediate and its remote milieu, and that, in turn, nervous tissue liberates specific agents which play a corresponding role in the differentiation of other nerve cells and in the maintenance and regeneration of some nonnervous structures.

That nerve fibers liberate neurohumoral substances in the process of impulse transmission is generally recognized. Furthermore, we have become familiar with the neurosecretory activity of special cell groups in the central nervous system of invertebrates and vertebrates. The ground is therefore prepared for an integration of the neuroembryological data into a wider physiological framework. However, this challenging task transcends the analytical resources of the experimental embryologist.

REFERENCES

1. H. Spemann, "Embryonic Induction and Development." Yale Univ. Press, New Haven, 1938.
2. J. Holtfreter, *Symposia Soc. Exptl. Biol.* **2,** 18 (1948).
3. J. Brachet, "Chemical Embryology." Interscience, New York, 1950.
4. J. Holtfreter and V. Hamburger, "Analysis of Development" (B. Willier *et al.,* eds.), p. 230. Saunders, Philadelphia, 1955.
5. R. Gillette, *J. Exptl. Zool.* **96,** 201 (1944).
6. J. Holtfreter, *J. Morphol.* **80,** 57 (1947).
7. H. B. Adelmann, *Quart. Rev. Biol.* **11,** 161 (1936).
8. S. R. Detweiler, "Neuroembryology." Macmillan, New York, 1936.
9. S. R. Detweiler, *Proc. Natl. Acad. Sci. U. S.* **6,** 96 (1920).
10. V. Hamburger and R. Levi-Montalcini, *J. Exptl. Zool.* **111,** 457 (1949).
11. D. H. Barron, *J. Comp. Neurol.* **78,** 1 (1943).
12. V. Hamburger and E. L. Keefe, *J. Exptl. Zool.* **96,** 223 (1944).
13. T. Dunnebacke, *J. Comp. Neurol.* **98,** 155 (1953).
14. P. Weiss and H. B. Hiscoe, *J. Exptl. Zool.* **107,** 315 (1948).
15. E. D. Bueker, *Anat. Record* **102,** 369 (1948).
16. R. Levi-Montalcini and V. Hamburger, *J. Exptl. Zool.* **116,** 321 (1951).
17. R. Levi-Montalcini and V. Hamburger, *J. Exptl. Zool.* **123,** 233 (1953).
18. R. Levi-Montalcini, *14th Intern. Zool. Congr.,* Copenhagen (1953).
19. R. Levi-Montalcini, H. Meyer and V. Hamburger, *Cancer Research* **14,** 49 (1954).
20. S. Cohen, R. Levi-Montalcini and V. Hamburger, *Proc. Am. Assoc. Cancer Research* **1,** (2), 9 (1954).
21. J. J. Kollross, V. Pepernik, B. Hill and J. C. Kaltenbach, *Anat. Record* **108,** (abstr.) 565 (1950).
22. J. Piatt, *J. Exptl. Zool.* **102,** 109 (1946).
23. M. W. Cavanagh, *J. Comp. Neurol.* **94,** 181 (1951).
24. P. Weiss and A. C. Taylor, *J. Exptl. Zool.* **95,** 233 (1944).
25. P. Weiss, M. V. Edds, Jr. and M. W. Cavanaugh, *Anat. Record* **92,** 215 (1945).
26. F. K. Sanders and J. Z. Young, *J. Exptl. Biol.* **22,** 203 (1946).
27. J. T. Aitken, M. Sharman and J. Z. Young, *J. Anat.* **81,** 1 (1947).
28. J. Z. Young, *Symposia Soc. Exptl. Biol.* **2,** 57 (1948).
29. Mac V. Edds, Jr., *J. Comp. Neurol.* **93,** 258 (1950).
30. H. Hydén, *Symposia Soc. Exptl. Biol.* **1,** 152 (1947).
31. Edds, M. V., Jr., *Quart. Rev. Biol.* **28,** 260 (1953).
32. Edds, M. V., Jr., *J. Exptl. Zool.* **113,** 517 (1950).
33. H. Hoffman, *Australian J. Exptl. Biol. Med. Sci.* **28,** 383 (1950).
34. H. Hoffman, *Australian J. Exptl. Biol. Med. Sci.* **30,** 541 (1952).
35. H. Hoffman and P. H. Springell, *Australian J. Exptl. Biol. Med. Sci.* **29,** 417 (1951).

36. A. Stefanelli, *Quart. Rev. Biol.* **26**, 17 (1951).
37. C. U. Ariens Kappers, G. C. Huber and E. C. Crosby, "The Comparative Anatomy of the Nervous System of Vertebrates, Including Man." Macmillan, New York, 1936.
38. S. T. Bok, *Folia neuro-biol.* **9**, 475 (1915).
39. R. Levi-Montalcini, *Arch. biol. (Liège)* **56**, 71 (1945).
40. V. Hamburger, *J. Exptl. Zool.* **103**, 113 (1946).
41. E. D. Bueker, *J. Exptl. Zool.* **93**, 99 (1943).
42. R. Rhines and W. F. Windle, *Anat. Record* **90**, 267 (1944).
43. W. F. Windle and M. F. Austin, *J. Comp. Neurol.* **63**, 431 (1936).
44. P. Weiss, *J. Exptl. Zool.* **68**, 393 (1934).
45. R. G. Harrison, *Wilhelm Roux' Arch. Entwicklungsmech. Organ.* **120**, 1 (1929).
46. J. J. Kollros, *J. Exptl. Zool.* **123**, 153 (1953).
47. O. Larsell, *J. Comp. Neurol.* **48**, 331 (1929).
48. O. Larsell, *J. Exptl. Zool.* **58**, 1 (1931).
49. R. Levi-Montalcini, *J. Comp. Neurol.* **91**, 209 (1949).
50. V. Hamburger, *J. Exptl. Zool.* **68**, 449 (1934).
51. E. D. Bueker, *Anat. Record* **97**, 157 (1947).
52. D. H. Barron, *J. Exptl. Zool.* **100**, 431 (1945).
53. D. Bodian, *J. Comp. Neurol.* **77**, 525 (1942).
54. J. Holtfreter, *Wilhelm Roux' Arch. Entwicklungsmech. Organ.* **138**, 522 (1938).
55. V. Hamburger, *Wilhelm Roux' Arch. Entwicklungsmech. Organ.* **114**, 272 (1928).
56. V. Hamburger and M. Waugh, *Physiol. Zool.* **13**, 367 (1940).
57. H. L. Eastlick, *J. Exptl. Zool.* **93**, 27 (1943).
58. M. Singer, *Quart. Rev. Biol.* **27**, 169 (1952).
59. O. Schotté and E. Butler, *J. Exptl. Zool.* **87**, 279 (1941).
60. E. Butler and O. Schotté, *J. Exptl. Zool.* **112**, 361 (1949).
61. S. S. Tower, *Physiol. Revs.* **19**, 1 (1939).
62. S. S. Tower, D. Bodian and H. A. Howe, *J. Neurophysiol.* **5**, 388 (1941).

DISCUSSION

S. S. KETY: Is it permissible to conclude from the very pretty cultures that you showed, and the way in which the nerve fibers grew out radially in all directions—not only in the direction of the sarcoma itself—that the sarcoma produces a trophic substance which fosters the development of nerve fibers but does not control the direction in which they grow?

V. HAMBURGER: This is exactly the conclusion which we have drawn. The tumor agent diffuses into the ganglion and stimulates fiber outgrowth more or less simultaneously in all cells.

S. S. KETY: Why do you say it is so obvious, in the case of transplanted limbs, that the mediator there is the nerve fiber itself and not a neurohumoral agent, which may be a specific humoral agent transported through the circulation?

V. HAMBURGER: The strongest argument is the following: Only those centers are affected from which we can trace fibers into the transplant, whereas other centers, which are as close as these but not connected, do not respond.

Reprinted from JOURNAL OF CELLULAR AND COMPARATIVE PHYSIOLOGY
Supplement 1 to Vol. 60, no. 2, October 1962

Specificity in Neurogenesis

VIKTOR HAMBURGER

Department of Zoology, Washington University, St. Louis, Missouri

A consideration of neurogenesis touches on several basic problems of cell differentiation and cell interactions. The great diversity of neuron types supplies us with an abundant material for the study of the origin of cell strain specificity. Cell interactions occur during neurogenesis in a great variety of situations, and in all of them, problems of specificity and selective affinity assume a major importance.

The main source of nerve cells is the embryonic neural tube (fig. 1a) which originates by upfolding of the medullary plate. In nerve cell formation, individual cells disengage from the epithelial lateral walls of the neural tube. They migrate to their final positions and aggregate in supercellular formations, in strata, columns or brain nuclei which are the architectural units of the central nervous system. For instance, the cells which migrate outward from the ventro-lateral part of the neural tube, in figure 1a, gather to form the lateral motor column in figure 1b. A second source of nerve cells is represented by the so-called "neural crest." This is a cluster of cells outside of the neural tube (fig. 1a).

These cells move ventrally and aggregate to form the sensory and sympathetic ganglia. Both cell migrations and aggregations are characterized by features of selective affinity. Cells move along specific pathways. The cues guiding a motor neuroblast to its destination are of necessity different from those guiding an adjacent internuncial neuroblast. Selective affinities are inherent in the aggregation of cells with others of their own kind, as in the formation of the motor column. Disaffinities are observed in segregation processes that occur, for instance, in the motor column of the chick embryo when the preganglionic Terni cells break away from the somatic motor cells (Levi-Montalcini, '50). An entirely different set of problems of selective affinity is inherent in the relation between the axon and its environment. This filamentous outgrowth of the neuroblast is dependent on the matrix on which it grows, not only for mechanical support but for directional guidance toward its destination. The actual termination, that is, the attachment of the axon tip to another nerve cell in synaptic junction, or to

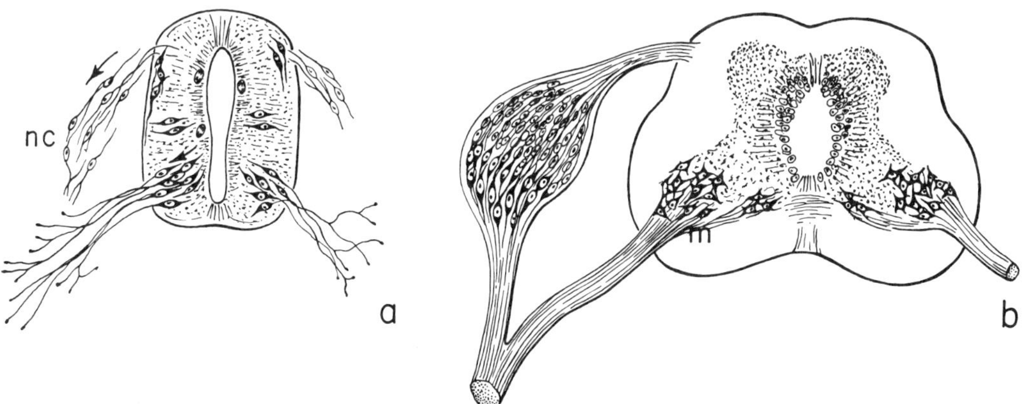

Fig. 1 Embryonic Neural Tube and Spinal Cord. (a) Neural tube and neural crest (nc) of 2-day chick embryo. Arrows indicate migration of cells. (b) Spinal cord and spinal ganglion of 8-day embryo. m, Lateral motor column.

81

a non-nervous structure, is predicated on a new and different set of specific cues. We have experimental evidence to show that the mechanisms guiding the nerve fiber are different from those which operate when selective terminal connections are established. Two other instances of selective affinity in axon development may be mentioned: (a) "Selective fasciculation" (Weiss, '41), an axon-to-axon relation that refers to the formation of bundles of nerve fibers of the same kind and (b) the unique partnership of axons with Schwann cells resulting in the production of the myelin sheath.

This paper will be limited to two points: the origin of cell-strain specificity, and some aspects of the formation of nerve patterns.[1]

NEUROBLAST-STRAIN SPECIFICITY

In the earliest period of neurogenesis, we can distinguish clearly two phases: (a) an epithelial phase, in which all prospective neuroblasts and spongioblasts are incorporated in the medullary plate or its derivative, the neural tube and (b) a phase in which individual cells are disengaged from the epithelium and operate as individuals while they proliferate and migrate to their destination. We know very little of the mechanisms which create neuroblast specification, but we can approximately fix the stage at which basic changes occur: It is most likely the epithelial phase prior to the "individualistic" phase. The medullary plate is characterized by its regulative capacity and the interchangeability of parts. The neural induction mechanism which creates the medullary plate initiates progressive differentiation by blocking out regional fields; it narrows the differentiation tendencies but does not carry the channeling process to the point of cell specification. On the other hand, the cells emerging from the neural epithelium give definite signs of specificity. As was mentioned, each cell, when it migrates to its location, follows its characteristic course and aggregates with cells of its own kind. Evidence for cell specification in the neural tube comes from regeneration experiments. After unilateral extirpation of parts of the brain or spinal cord, gross morphological structure

can be restored in amphibian embryos (Detwiler, '47) and, to a much lesser extent, in chick embryos (Watterson and Fowler, '53), but the regenerated halves are deficient in a number of cell types such as spinal motor cells and the giant Mauthner cells (Holtzer, '51; Stefanelli, '51; Wenger, '50). In one instance, that of the Mauthner cells, the time of determination could be traced back to the early medullary plate stage (Stefanelli, '51). These findings suggest that precursors of specific neuroblast strains have become irreversibly fixed in the neural epithelial stage, and their loss can no longer be repaired. It is possible that the cells are already specified when they enter into mitosis. Of course, it is realized that the specification process is not completed when the cell emerges from the epithelium. The well-known experiments of Weiss and Sperry on "myotypic function" of muscles, and on local sensory signs (reviewed in Sperry, '51 and Weiss, '55) have shown that specification is eventually carried to the point where nerves to individual muscles and to circumscribed skin areas or optic tectum regions become distinct from each other, and that this individualization is imprinted on the nerves by their end organs, after terminal connections have been established.

Experiments with cranial ganglia

The analysis of the determination processes that go on in the neural epithelium meets with great difficulties because a multitude of neuron types is produced by each sector of the tube, and we have no adequate device, at the moment, to tag individual cells while they are part of the epithelium. Tagging experiments with tritiated thymidine have been done by Sauer and Walker ('59) and Uzman ('60). In the search for a simpler neurogenic system, we have turned to the epidermal placodes of the head. They represent a third source of neuroblasts, supplementing the neural tube and the neural crest. They contribute to the somatic and visceral sensory ganglia of the head, the so-called cranial ganglia. Some of these ganglia, as, for instance, the trigeminal ganglion of the chick embryo,

[1] Our own experimental work was supported by research grants nos. B-1937 and B-3143 of the Institute for Neurological Diseases and Blindness of the U. S. Public Health Service.

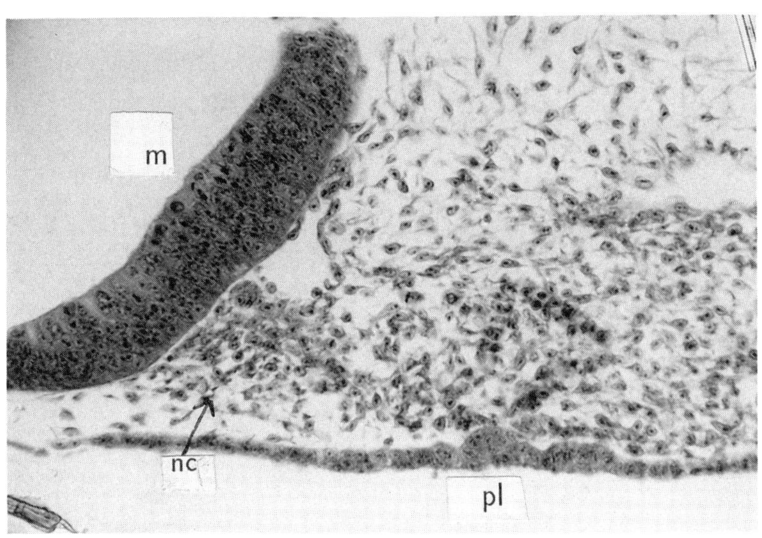

Fig. 2   Regions of 2-day embryo (25 somites) from which prospective neuroblasts emigrate: Medulla (m), trigeminal neural crest (nc), and trigeminal epidermal placode (pl). Note that the two cell populations are about to join.

which is located between the eye and the ear near the anterior part of the medulla, have a dual origin — from epithelial placodes and from the adjacent sector of the neural crest. In early stages, cells from these two sources migrate toward each other and join in the formation of the ganglion (fig. 2). Normally, the identity of the two components is lost when they have met. It is possible to extirpate one component, when they are still separate, and leave the other intact (Hamburger, '61). Similar experiments had been done in amphibians by Stone ('22), and in chick embryos by Yntema ('44) and Levi-Montalcini and Amprino ('47). It was known that neural crest ganglia can be obtained after placode extirpation, and placodal ganglia after extirpation of the anterior medulla, which includes the trigeminal neural crest.

There was a special reason for choosing the trigeminal ganglion of the chick embryo. Levi-Montalcini and Amprino ('47) had observed that this ganglion is composed of two cell populations, a core of small neuroblasts, and a marginal zone of much larger cells which become heavily impregnated with silver from early stages on (Hamburger, '61 and fig. 3a). One can also detect differences by histochemical methods. For instance, the two populations differ in the content of acetylcholinesterase. Here, then, we are dealing with only two cell types that are clearly distinct from early stages on. Our first aim was to relate these two populations to the two embryonic primordia.

The outcome of the experiment of extirpation of the preotic medulla was unequivocal (fig. 3b). In all instances, one or two pairs of placodal ganglia were formed. They are composed of a uniform population of large cells which are identical cytologically with the large marginal cells of the normal ganglia. They were found at a considerable distance from the epidermis and had sometimes fused in the midline (fig. 5 in Hamburger, '61). Hence they had migrated toward their normal location in the typical fashion. However, the frequent occurrence of two separate clusters and of cells strung along a nerve, indicates that some condition which normally leads to the aggregation of the large cells in one compact group is missing. Perhaps the neural crest component serves normally as a rallying point.

The reciprocal experiment of placode extirpation (fig. 3c) resulted in the formation of neural crest ganglia that were composed mostly of small-type cells. Clusters of large

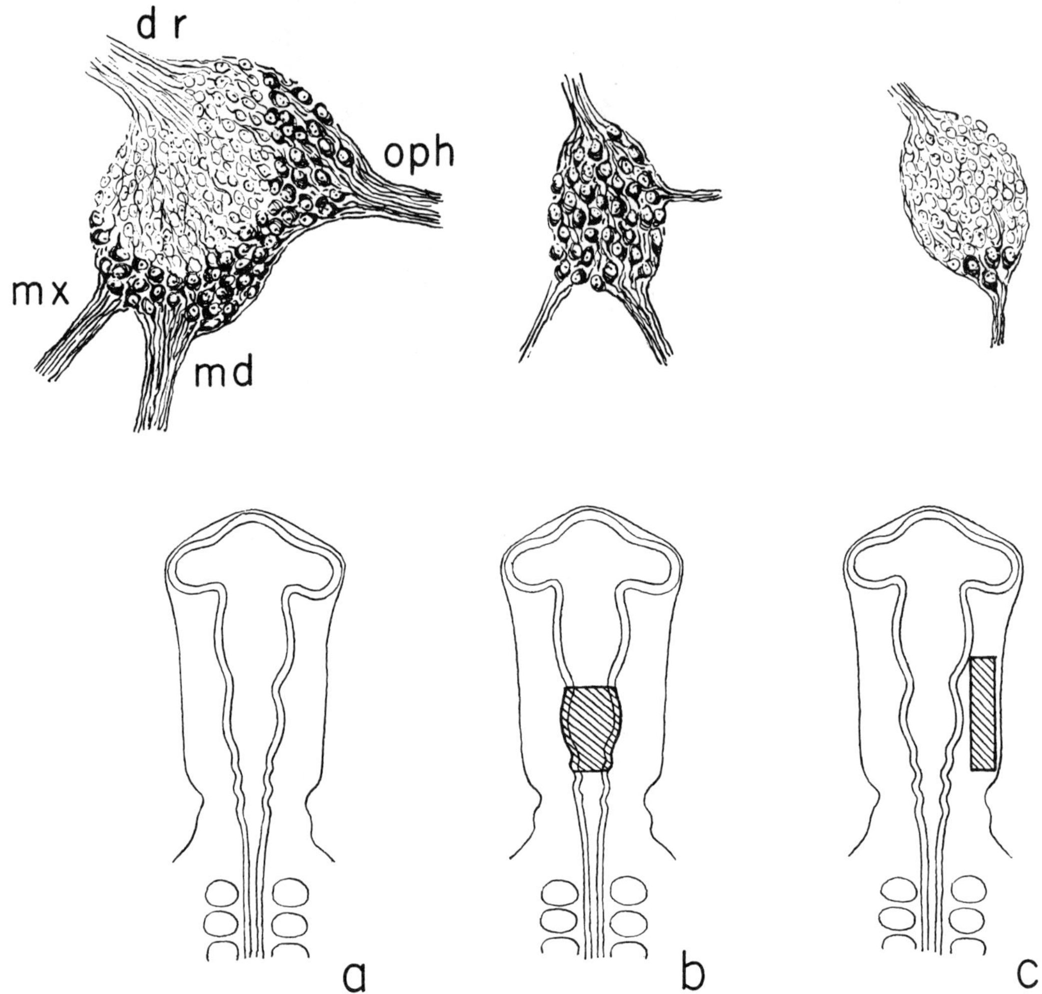

Figure 3

Normal and experimental embryos (below) and resulting trigeminal ganglion (above).

a   Normal embryo of 16 somites (less than 2 days), and normal trigeminal ganglion at later stage (above). dr, Dorsal root; md, mandibular ramus; mx, maxillary ramus; oph, ophthalmic ramus.

b   Extirpation of rostral part of medulla (below) including neural crest, and placodal ganglion resulting from this operation (above).

c   Extirpation of trigeminal placode (epidermis), and neural crest ganglion resulting from this operation (above).

cells were found frequently attached to the margin of these ganglia, in varying amounts and positions. These cells probably are not products of the neural crest but are regenerated placodal cells. We have indirect evidence to support this contention, but the crucial experiment — namely, heterotopic transplantation of the medulla — will have to decide whether or not the neural crest is capable of forming large cells in addition to small-type.

At any rate, we are sure that the placodes can produce only large cells and that all small cells are derived from the neural crest. In other words, the trigeminal placode is a primordium that gives rise to a pure population of nerve cells of one specific type, in contrast to the pluripotential neural tube.

Differences between cranial ganglia

These findings assume a deeper significance when viewed in a broader perspective, that is, in relation to other cranial ganglia. Dr. Levi-Montalcini (unpublished data) has observed that in the chick embryo all placodal ganglion cells, namely those of ganglion VII (geniculate), VIII (vestibular), IX (petrosal) and X (nodosum) are composed of large-type cells, and all cranial neural crest ganglia are pure populations of small cells. The situation in these ganglia is unequivocal, since, in contrast to the trigeminal, the neural crest and placodal components do not fuse but remain separate ganglia. It is clear that the placodal epidermis produces a type of cell that has a few basic characteristics in common: large size, early affinity for silver, early production of demonstrable quantities of acetylcholinesterase. A basic physiological difference between placodal and neural crest derivatives was found in their reaction to the nerve growth factor (see Levi-Montalcini, '58). Only the neural crest derivatives can be stimulated to fiber outgrowth *in vitro;* the large cells are refractory. This difference was demonstrated for the trigeminal ganglion in an experiment in which the two populations were separated surgically and exposed to the nerve growth factor *in vitro.* The ganglion nodosum, which is composed of a pure population of large cells, is also unaffected (Levi-Montalcini, unpublished data). The large cells of the spinal ganglia (see fig. 1b and Hamburger and Levi-Montalcini, '49) share all these characteristics with the placodal cells, except that they are derived from the neural crest, together with the small cells. Their lateral position in the ganglion suggests that they might originate at the more lateral parts of the neural fold, where the fold continues into the epidermis, and that they are, therefore, akin to epidermal derivatives.

The fact that the placodal neuroblasts have significant characteristics in common cannot be interpreted to mean that all placodal derivatives come from a common stem cell. The different ganglia serve entirely different functions; and cytological differences; as, for instance, cell size and silver affinity can be recognized in very early stages. In fact, the behavior of the cells at the very moment they leave the epidermis is already different in various placodal areas, and one can identify the placodes at the moment cells emerge. For instance, trigeminal cells migrate out singly or in small clusters, in moderate numbers at a time. They acquire their affinity for silver at a somewhat later stage. In contrast, the cells of the petrosum and nodosum placodes swarm out in large numbers, and are consolidated later in the two ganglia. These cells acquire their affinity for silver very early. There are other indications of a very early specification of each placode.

The problem of the origin of strain specificity

At this point we may ask a few questions about the origin of strain specificity. How does an epithelial system break up into a group of strain-specific cells? Do the cells acquire their characteristics gradually, beginning with a common stem cell that undergoes a sequence of dichotomous steps (as has been postulated for the derivatives of the neural epithelium); or does the specification process lead directly to a high degree of specificity, without the interposition of intermediate steps? The data presented above for the neural epithelium are in favor of the second alternative. However, as pointed out before, it is difficult to carry the analysis of this system further, because each small sector of the neural epithelium gives rise to a number of different types of neuroblasts. The situation is more favorable with respect to the placodes. The observations presented above indicate that each placode gives rise to only one neuroblast type, and that the different placodes possess specific information concerning the characteristics of their products at the time they begin to disperse their constituent cells. It is perhaps feasible to test this idea in the chick embryo by exchange of prospective placodal epidermis. Similar experiments might give insight into the factors that are responsible for the determination of the different placodes. It is tempting to speculate that specific inductor systems may be instrumental in the determination of the cranial ganglia, analogous to the determination of other placodal derivatives, such as nasal epithelium, lens and otocyst.

SPECIFICITY IN THE FORMATION OF NERVE PATHWAYS

The trigeminal ganglion gives rise to three rami (fig. 4). The ophthalmic ramus which runs in the orbit innervates the upper beak. The maxillary ramus goes to the upper jaw, and the mandibular ramus to the lower jaw. A characteristic feature is the emergence of the ophthalmic branch from the anterior end, whereas the other two emerge from the posterior end. In the stages under consideration the maxillary nerve is not yet consolidated in one ramus but is represented by five separate roots

which can be identified individually by their typical peripheral distribution. The mandibular ramus is the only mixed nerve; the other two are purely exteroceptive sensory to the skin and oral cavity.

Extirpation experiments

All placodal ganglia that were formed in the experiment of medulla extirpation (fig. 3b) invariably gave rise to a complete set of all three rami (Hamburger, '61). Their branching pattern was entirely typical, except for depletion or absence of some terminal branches where the ganglia were very small. The precision with which the normal peripheral pattern is replicated, in strict bilateral symmetry, is particularly impressive when the ganglia were dislocated and in asymmetrical positions, as a result of the severe derangement caused by

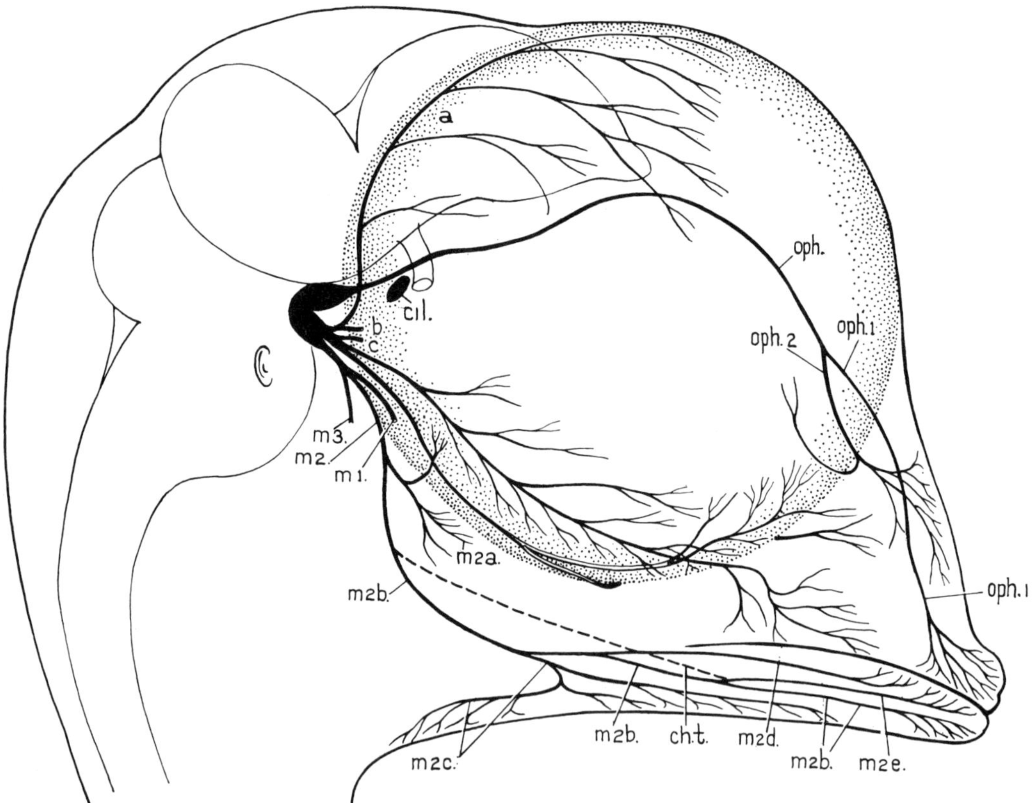

Fig. 4 Diagram of peripheral distribution of trigeminal nerves, based on dissections and serial sections of 10- to 14-day embryos. oph, ophthalmic nerve, located in orbit behind the eye, innervating the dorsal anterior head and upper beak; a–e, major branches of maxillary nerve to upper jaw; ml–m3, major branches of mandibular nerve to lower jaw.

the extirpation of the medulla. The ophthalmic ramus always emerged from the anterior-most cells of the anterior ganglion, and the other nerves originated in their typical sequence at the posterior ganglia. Even very small clusters managed to produce a typical pattern.

The faithful replication of the normal pattern is not so surprising, because the more distal parts of the beak and jaws, that is, the area of peripheral distribution of these nerves, had been left intact. It is generally acknowledged that nerves follow preneural pathways which are prepared by the structures to be innervated, though it is not known how these pathways are formed. The basic branching patterns are laid down remarkably early; they were observed in early stages of the limb bud (Taylor, '43) and jaws, where at best, slight differences in the density of the mesenchyme are detectable. Blood vessels and other structures that are contacted by nerves in later stages, play no role in the initial patterning.

One other observation deserves a comment: the strictly selective affinity of placodal axons and the refractoriness of other fibers for the trigeminal pathways. In the experiment of placode extirpation (fig. 3c) the axons of the small neural crest cells did not form ophthalmic or maxillary rami, although the ganglion is located in the normal position, with undisturbed access to the periphery (Hamburger, '61). The presence of the mandibular ramus in this series is explained by the fact that it is largely composed of motor and other fibers emerging from the intact medulla.

The strong positive affinity of one axon type, and the refractoriness of the other, is strikingly illustrated in a number of special cases from the placode extirpation series. In one instance, all three rami were actually observed in a neural crest ganglion (see fig. 8 in Hamburger, '61) in apparent contradiction to our previous statement. The pattern shows another unorthodox feature. The ophthalmic branch emerged from the posterior end of the ganglion, together with the other two. This is contrary to the normal pattern and also to all findings on experimental placodal ganglia, where this branch invariably emerges from the anterior-most end of the anterior

pair of ganglia. Closer inspection showed a cluster of large-type cells at the posterior margin of the otherwise uniform population of small cells. It was this cell group which gave rise to the ophthalmic and maxillary nerves. As was pointed out before, these cells are considered to be regenerated placodal cells that have managed to join the neural crest ganglion. In other cases, we found a variety of different arrangements and amounts of such large-type cells attached to the neural crest ganglion, and there was invariably a close correlation between the position of these cells and the emergence of the skin nerves.

A unique situation was found in another group of the placodal extirpation series — two separate ganglia had formed side by side, one essentially composed of neural crest cells, and the other of a small cluster of placodal cells (fig. 9, Hamburger, '61). As it happens occasionally, a combination of uncontrollable circumstances had created a result that could not have been duplicated by the most skillful experimental design. Again placodal cells had regenerated and started to migrate, but had not quite reached the neural crest ganglion. There was a neat division of labor as far as peripheral innervation is concerned. The placodal ganglion formed the ophthalmic and maxillary branches, and fibers from both ganglia joined the motor root in the formation of the mandibular nerve.

Whereas the purely exteroceptive sensory quality of the large-type placodal cells is very clearly established, we were not able to determine the peripheral distribution and functional significance of the axons from the small neural crest cells. The identity of these axons is lost when they join the motor and proprioceptive fibers of the mandibular nerve. They must play a significant role, since they make up half of the total population of the normal ganglion. Perhaps they supply proprioceptive innervation to distal jaw muscles.

Selective affinities of axons

Our material represents a striking illustration of the principle of selective affinities of axons for their pathways, an ordering principle that guarantees normal nerve distribution both intracentrally and peripherally. Obviously, it would be of major interest to understand the mechanism underlying the axon-matrix relations. The phenomenon is closely related to cell affinities and disaffinities, a subject discussed by other contributors to this volume. Both systems pose similar problems that have to be formulated in terms of ultrastructural or molecular matching. The pioneer work of Weiss in this problem of the axon-matrix relation has been reviewed by him ('41, '55) and I shall not elaborate on it here. The basic principle he established is that of *"contact guidance."* It is based on the earlier observations of Harrison and others that nerve fibers require a solid substrate on which to grow out. If patterns of orientation are imposed on the ultrastructural formed elements of the matrix (by experimental design in tissue culture, or by orienting forces in the embryo), then they can serve as guides for directional axon outgrowth. For instance, parallel alignment of micellae can create a direct bridge between the central nervous system and an organ primordium. However, it is generally acknowledged that additional specifications have to be postulated to account for selective axon affinities. A case in point is the situation in which the peripheral trigeminal pathways are equally accessible to all fibers but actually followed only by one type (Hamburger, '61). These discriminating specifications are considered to be biochemical differentials in the substrate. This means that preneural pathways have to be regarded as sites of biochemical interactions between axon and substrate and not merely as mechanical track systems. One must assume that the axon plays an active role in this process. In this connection, it is of interest that several investigators have called attention to the role of cell exudates in cell interactions (Weiss, '45; Grobstein, '54; Moscona, '60; and others). More and more significance is being attributed to cell-produced extracellular materials, in induction, in selective adhesions of cells, and in directional cell migration. The properties of these cell products are under active investigation. Moscona (this volume) has reported evidence for type-specificity of exudates of retina cells. One could imagine that different axon types produce specific exu-

dates, with properties matching those of the matrix on which they grow.

Random and directional outgrowth

A final consideration will be devoted to the problem of random outgrowth versus general directional outgrowth of nerve fibers in the initial phase of nerve pattern formation. We distinguish usually four phases in the over-all process: (1) undirected outgrowth of pioneer fibers in a random fashion, (2) localization of preneural pathways by some pathfinder fibers which then follow these pathways, while "unsuccessful" fibers are withdrawn, (3) reinforcement of the fibers that have "arrived," by the process of "selective fasciculation," and (4) establishment of terminal connections. A close inspection of the initial phase of fiber formation we have studied gives little indication of random outgrowth, but shows from the beginning a rather definite orientation of individual fibers. The first ophthalmic and maxillo-mandibular fibers that grow out show this behavior as do the initial stages in the formation of intracentral fiber tracts. Undoubtedly, ultrastructural organization of the substrate plays an important role, but it is likely that the biochemical specifications of structural pathways, which have been postulated above, already enter the picture in these early stages. I think of them as general directional or deflecting forces, in distinction to narrowly channeling devices. The existence of such agencies has been demonstrated for intracentral fibers. I shall mention two examples. Holtzer ('52) has studied the regeneration of longitudinal fiber tracts following the unilateral extirpation of parts of the spinal cord in *Ambystoma*. In some instances, dorsal root fibers from dislocated spinal ganglia entered the regenerating cord laterally or ventrally rather than dorsally. Invariably, they were deflected dorsad, and they reached eventually their normal location in the dorsal funiculus, where they bifurcated in a typical fashion. The actual pathways showed considerable individual variations, indicating some general directional agency to which the fibers responded individually. Conversely, when ventral funicular fibers entered the cord accidentally in a dorsal position, they were deflected toward and into the ventral motor funiculus where they belong.

The most interesting and most thoroughly investigated case of a general directional guidance, in distinction from specific channeling along track systems, is that of the Mauthner fibers. These are giant fibers that originate in a pair of very large cells in the medulla of aquatic vertebrates. Normally, they decussate near their origin and then follow a straight course caudad, in ventral position, near a major fiber tract, the fasciculus longitudinal medialis. Different experimental designs were used to study factors determining their pathways: (1) supernumerary Mauthner cells and fibers were created by medulla transplantation (Oppenheimer, '41; Piatt, '43), (2) the medulla was rotated (Stefanelli, '51), (3) foreign tissues were inserted along their pathways (Stefanelli, '51), or (4) the Mauthner fibers were forced to traverse atypical, regenerated spinal cord regions (Holtzer, '52). The major course of the displaced Mauthner fibers, with few exceptions, was straight caudad, even when they had to cross foreign territory (fig. 5). We are not concerned with this problem of polarized growth. What interests us in the present context is the fact that the fibers, no matter where they start, eventually reach a position in the ventral cord, very near, or somewhere in the neighborhood of, their normal, typical location (fig. 6). Their course varies greatly in individual cases and no consistent relation was found to any one structure, such as the fasciculus longitudinalis medialis. Obviously, directional forces with a wide spread rather than specific narrow pathways guide them. Since diffusion gradients and other "actions at a distance" have been ruled out (see Weiss, '55) we have to postulate in all these cases a biochemical conditioning or infiltration of the substrate over a considerable range, perhaps in a gradient fashion.

We have no experimental evidence for the operation of a similar agency in the initial orientation of peripheral fibers, but it is likely that this principle of biochemical specification of ultrastructural pathways that leads to orientation in a general direction is more important in initial

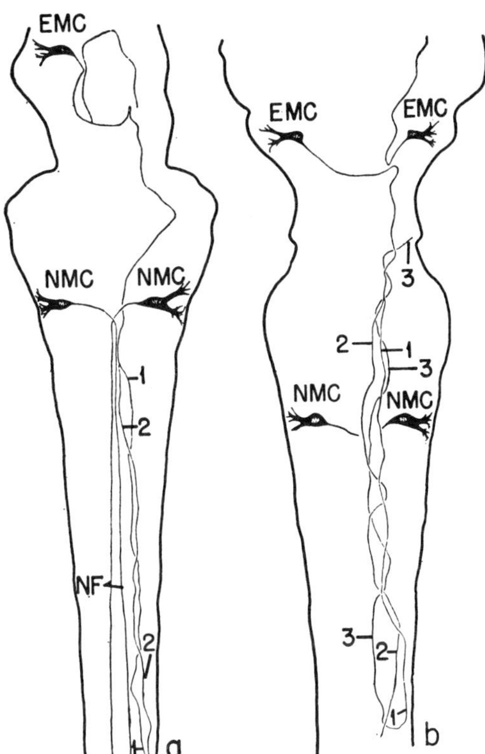

Fig. 5 Reconstruction of brain and spinal cord of *Ambystoma*. An extra medulla had been transplanted in front of the normal medulla. EMC, ectopic Mauthner cells in transplant; NMC, normal Mauthner cells of host. (From Piatt, '43.)

phases of axon outgrowth than random search of pioneer fibers.

I have refrained from an attempt to relate the selective mechanisms in neurogenesis to functional aspects. It is obvious that the normal functioning of the nervous system is predicated on the precision with which these selective and directional steps, which operate during its development, are synchronized and integrated. The structure of the nervous system is built in forward reference to its function, according to a blueprint, part of which is impressed very early on each individual, prospective nerve cell. It is worthwhile to visualize the amount of information which a nerve cell has to acquire at the beginning of its individual existence. This information includes not only specifications for its growth rate and structural differentiation, but also instructions for "recognition" of cells of its own type, the competences to re-

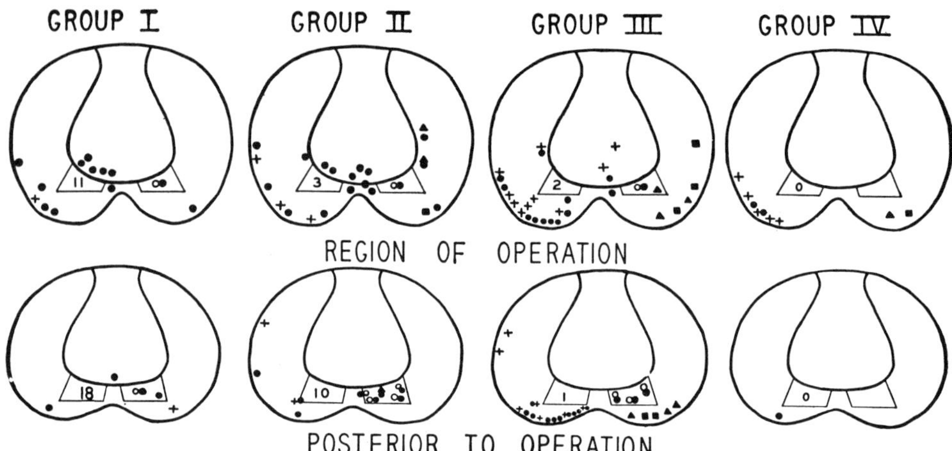

Fig. 6 Position of regenerated Mauthner axons in experiments of unilateral extirpation of brachial spinal cord. The apparent left side of the spinal cord is regenerated. The trapezoid area circumscribes the normal position of Mauthner axon. Circles indicate normal position of left Mauthner axon, dots indicate regenerated axons. Groups I–IV refer to age groups at the time of operation. Comparison of upper and lower row shows that fibers approach their normal position while growing caudad. (From Holtzer, '43.)

spond to the sequence of specific cues that the environment provides to guide the cell's own migration, and the directional outgrowth of its axon and dendrites, and finally the capacity to make the right choice of its functionally appropriate terminal connection.

OPEN DISCUSSION

CHAIRMAN WEISS[1] : Thank you very much Dr. Hamburger. First, I think this idea of the great diversity of cell types is something which should be firmly engrained in anyone who likes to make either experiments or statements about the problem of differentiation in higher forms.

Second, on the point of guidance of nerve fibers, we do not know just where what we call the nerve tip really is — studies by Fernandez-Moran and by ourselves have proved the inadequacy of even electron microscopic judgment.

SALOME WAELSCH[2] : Dr. Hamburger, will you elaborate on the biochemical specification of structural pathways?

If you change the pathway of the nerve, then how can you expect the same biochemical specification that you get in the normal pathway? If you assume interaction of this microexudate, as you call it, with the structural substrate, then, how do you get that same interaction if the nerve travels along a different, experimentally produced pathway?

HAMBURGER: There are degrees of specificity. For instance, trunk nerves can very well enter into the limb, and follow the preneural limb pathways; from that we have to conclude that there are certain matching properties which nerve fibers of different origin and function have in common. However, the pattern formed by trunk nerves in the salamander limb is never an entirely typical limb pattern. Cranial nerves can also form a limb pattern but they cannot form functional terminal connections. Neither is a limb innervated by trunk nerves functional. This limitation is due to regional specificity in the central action system which Weiss has explored. We have to postulate degrees and differentials of specificity in three situations: nerve pathways, terminal connections and central action systems.

SETO[3] : Is it possible to explain the directed growth of fibers in terms of a system of differential cell gradients in which the direction of fiber growth is influenced less by specific cell exudates and more by con-

[1] P. A. Weiss, The Rockefeller Institute.
[2] Salome G. Waelsch, Albert Einstein College of Medicine.
[3] Frank Seto, Berea College.

tacts with the surface of cells of increasing specificity?

HAMBURGER: Long-range diffusion gradients are completely ruled out; Weiss is the main witness for this statement.

CHAIRMAN WEISS: Diffusion in the general term of diffusion in bulk, as in solutions, is not really a very relevant or significant type of transport mechanism in cells. We now have definite evidence that chemical compounds are guided along linear or planar interfaces so that, instead of being dissipated in bulk, they remain concentrated on the way from source to destination. Such chemical transport can be expected by a sort of molecular "bucket brigade" along interfaces, even in a diphasic, solid-state system, like electron transport in semiconductors.

RUNNSTRÖM[4] : It might be useful to go to simpler systems when searching for analogies to the behavior of outgrowing nerve fibers. Tryggve Gustafson and collaborators have made time-lapse studies of the morphogenetic movements in sea urchins. The immigration and attachment of the primary (skeleton-forming) mesenchyme cells were for example analyzed (Gustafson and Wolpert, Exptl. Cell Research 24: 64, '61) in this way. The mesenchyme cells carry out pseudopodial movements, and random contacts are formed with ectoderm cells but the final pattern of arrangement of the cells is determined by those regions of the ectoderm with which the most stable pseudopodial contacts are established. One has thus the impression that the "affinity" of the primary mesenchyme cells is greater to certain than to other regions of the ectoderm. The contacts of the outgrowing nerve fibers may also primarily be random but gradually fine specificities may bring about a selection of the most stable contacts.

OPPENHEIMER[5] : In some of the cases where the ganglia were formed principally by the small cells, there were some large cells associated with them. What is the origin of those large cells?

HAMBURGER: I have never found a single case in which the neural crest formed an absolutely pure population of small cells. I have very convincing evidence to show that the large cells in these ganglia are not of neural crest origin but regenerated placodal cells which had managed to migrate to the ganglion, and that the neural crest is capable of producing only small cells.

There is one crucial experiment to test this assumption which is planned: that is, to transplant the medulla to another position and allow it to form a neural crest and see whether one or two kinds of cells will be produced. I will report on that when the experiment is done.

CHAIRMAN WEISS: Is it possible that the crucial dichotomy actually takes place early in development at the border of the epithelial and neural plates? There is a common mass of cells which is destined, just as in Runnström's case, to migrate out of the epithelial cell plate. These cells form the neural crest. Now, if we assume that these cells in the area of the spinal ganglia stay together, while in the area of cranial ganglia, apparently, one part moves down early from the crest and the other part stays and shifts with the ectoderm and moves out only later, then we have two rather distinct and experimentally accessible groups. Actually, could what you find in the exceptional cases be merely an expression of some sort of variance or indeterminacy in the breaking up of the clusters (leading to various amounts of contamination from neural crest in the placode, or vice versa) rather than a regeneration?

HAMBURGER: In the trunk and tail ganglia actually the cytological picture is similar to the trigeminal; a group of large cells are ventro-lateral and a group of small cells medio-dorsal. I conceive of this situation in the trunk and tail ganglia, as Weiss has indicated. The large cells derive from more lateral parts of the region where neural crest merges with epidermis in the neural plate stage.

The fact that the large cells are always at the outerpart of the ganglion, perhaps, permits us to trace them back to a lateral part of the crest, and the inner ones to an inner part. However, the trigeminal placode is a considerable distance lateral to the neural crest.

---

[4] John Runnström, University of Stockholm, The Wenner-Gren Institute.
[5] Jane M. Oppenheimer, Bryn Mawr College.

AUERBACH[6]: Perhaps the use of mutants might be helpful in this direction. A number of mouse mutants are known, in which the neural crest is either absent or reduced.

HAMBURGER: They are very suitable for this kind of an investigation.

LITERATURE CITED

Detwiler, S. R. 1947 Restitution of the brachial region of the cord following unilateral excision in the embryo. J. Exptl. Zool., *104:* 53–68.

Grobstein, C. 1954 Tissue interaction in the morphogenesis of mouse embryonic rudiments *in vitro.* In, Aspects of Synthesis and Order in Growth, ed. D. Rudnick. Princeton University Press, Princeton, N. J. pp. 233–256.

Hamburger, V. 1961 Experimental analysis of the dual origin of the trigeminal ganglion in the chick embryo. J. Exptl. Zool., *148:* 91–124.

Hamburger, V., and R. Levi-Montalcini 1949 Proliferation, differentiation and degeneration in the spinal ganglia of the chick embryo under normal and experimental conditions. J. Exptl. Zool., *111:* 457–502.

Holtzer, H. 1951 Reconstitution of the urodele spinal cord following unilateral ablation. I. Chronology of neuron regulation. J. Exptl. Zool. *117:* 523–558.

———— 1952 Reconstitution of the urodele spinal cord following unilateral ablation. II. Regeneration of the longtiudinal tracts and ectopic synaptic unions of the Mauthner's fibers. J. Exptl. Zool., *119:* 263–302.

Levi-Montalcini, R. 1950 The origin and development of the visceral system in the spinal cord of the chick embryo. J. Morphol., *86:* 253–284.

———— 1958 Chemical stimulation of nerve growth. In, Chemical Basis of Development, ed. McElroy and Glass. Johns Hopkins Press, Baltimore, Maryland, pp. 645–664.

Levi-Montalcini, R., and R. Amprino 1947 Recherches expérimentales sur l'origine du ganglion ciliaire dans l'embryon de Poulet. Arch. Biol., *58:* 265–287.

Moscona, A. A. 1960 Patterns and mechanisms of tissue reconstruction from dissociated cells. In, Developing Cell Systems and Their Control, ed. D. Rudnick. Ronald Press Co., pp. 45–70.

Oppenheimer, J. M. 1941 The anatomical relationships of abnormally located Mauthner's cells in *Fundulus* embryos. J. Comp. Neurol., *74:* 131–167.

Piatt, J. 1943 The course and decussation of ectopic Mauthner's fibers in *Amblystoma punctatum.* J. Comp. Neurol., *79:* 165–183.

Sauer, M. E., and B. E. Walker 1959 Radioautographic study of interkinetic nuclear migration in the neural tube. Proc. Soc. Exptl. Biol. Med., *101:* 557–560.

Sperry, R. W. 1951 Mechanisms of neural maturation. In, Handbook of Experimental Psychology, ed. Stanley Smith Stevens. John Wiley & Sons, Inc., pp. 236–280.

Stefanelli, A. 1951 The Mauthnerian apparatus in the Ichthyopsida; its nature and function and correlated problems of neurohistogenesis. Quart. Rev. Biol., *26:* 17–34.

Stone, L. S. 1922 Experiments on the development of the cranial ganglia and the lateral-line sense organs in *Amblystoma punctatum.* J. Exptl. Zool., *35:* 421–496.

Taylor, A. C. 1943 Development of the innervation pattern in the limb bud of the frog. Anat. Rec., *87:* 379–413.

Uzman, L. L. 1960 The histogenesis of the mouse cerebellum as studied by its tritiated thymidine uptake. J. Comp. Neurol., *114:* 137–148.

Watterson, R. L., and I. Fowler 1953 Regulative development in lateral halves of chick neural tubes. Anat. Rec., *117:* 773–804.

Weiss, P. 1941 Nerve Patterns: The mechanics of nerve growth. Growth, Third Growth Symposium, *5:* 163–203.

———— 1945 Experiments on cell and axon orientation *in vitro:* the role of colloidal exudates in tissue organization. J. Exptl. Zool., *100:* 353–386.

———— 1955 Nervous System (Neurogenesis). In, Analysis of Development, ed. Willier, Weiss and Hamburger. W. B. Saunders Co., pp. 346–401.

Wenger, E. L. 1950 An experimental analysis of relations between parts of the brachial spinal cord of the embryonic chick. J. Exptl. Zool., *114:* 51–86.

Yntema, C. L. 1944 Experiments on the origin of the sensory ganglia of the facial nerve in the chick. J. Comp. Neurol., *81:* 147–167.

---

[6] Robert Auerbach, The University of Wisconsin.

Neurosciences Research Program Bulletin Vol. 15 (Suppl.) 1977

THE F.O. SCHMITT LECTURE IN NEUROSCIENCE 1976

THE DEVELOPMENTAL HISTORY
OF THE MOTOR NEURON

Viktor Hamburger
Washington University
St. Louis, Missouri

The F.O. Schmitt Lecture and Prize in Neuroscience was established by the Associates of the Neurosciences Research Program in 1973 to mark the seventieth birthday of Francis O. Schmitt, founder of the organization. The purpose of the award is to recognize, encourage, and advance the achievement of excellence in neuroscience.

John Z. Young was the first recipient of the award in 1973. His lecture, "Sources of Discovery in Neuroscience," was delivered as part of a two-day symposium honoring Professor Schmitt; it was later published as a chapter in the book resulting from this symposium. *The Neurosciences: Paths of Discovery* (The MIT Press, 1975). The Second F.O. Schmitt Lecture and Prize was awarded jointly to Solomon H. Snyder and Leslie L. Iversen. Their lectures were entitled respectively "The Opiate Receptor" and "How Do Antipsychotic Drugs Work?" The third recipient of the award was Vernon B. Mountcastle in 1975 and his Prize lecture was entitled "The World Around Us: Neural Command Functions for Selective Attention."

Viktor Hamburger was presented the F.O. Schmitt Prize for 1976 in recognition of his fundamental contributions to developmental neurobiology. The award was made with the following citation:

"Most widely recognized for his definitive studies of the development of the chick spinal cord and of the spinal and trigeminal ganglia, he has also made basic contributions to developmental genetics, to experimental teratology, to the initial discovery of the nerve growth factor, and, in recent years, to the ontogeny of behavior. His career has spanned the entire era of modern experimental embryology, beginning with his graduate student days in Spemann's laboratory when the "organizer" was discovered. Throughout the intervening half-century, his pioneering achievements have increased our knowledge and understanding of the development of the vertebrate nervous system and mark him as a distinguished leader in developmental neurobiology."

Preface

I hope you will not consider it presumptuous if I deal with my topic in a semi-autobiographical way, rather than as a literature survey.* This makes it possible to cope with a large and heterogeneous subject and nevertheless preserve some continuity.

My personal predilection for the motor system dates from one of the most significant events in my life: my transplantation from Germany to the United States, in 1932, to the Zoological Laboratory of Dr. F. R. Lillie at the University of Chicago. This event, in turn, determined an equally important change in my scientific activity: the shift from the amphibian embryo, the celebrated hero in the laboratory of Dr. H. Spemann, where I took my Ph.D. and stayed as a Privatdozent, to the chick embryo, which reigned supreme in Chicago and has remained my loyal mentor for more than half a century; it has revealed to me many secrets except one: how to remain eternally young.

Frank Schmitt will remember that when I arrived at the Zoology Department of Washington University in St. Louis in 1935, I had not yet quite shed my past and my link to the amphibian embryo. Frank was then an upcoming young Associate Professor and, just as today, open to new adventures. It was not difficult to arouse in him an interest in the fascination of developmental processes and in the concepts of experimental embryology, the organizers, fields, gradients, and inductors, which must have struck him as semi-mythical notions. Our lively discussions and seminar sessions over several years led to what is probably unknown to most neurobiologists: a transient but serious engagement of Frank in embryology, and in actual research cooperation. But, as many others have experienced who have had encounters with Frank, it was *I* who was the main beneficiary. While he probably had never seen an embryo before, my unawareness of his biophysical and biochemical outlook was even more profound, and he opened up vistas that were entirely new to me. He was then far ahead of his time and a molecular biologist before this term was coined. And his biophysical and ultrastructural approach put its stamp on our joint investigation. It dealt with changes in cell shape during the upfolding of the neural plate in amphibian embryos subsequent to its induction by

*I apologize to my colleagues in neuroembryology, who use amphibian embryos, for the "benign neglect" of their work. I refer to the expert reviews of Kollros (1968), A. F. W. Hughes (1968) and Prestige (1970). My own research has been supported by grants from the NINCDS of the N.I.H. and the Muscular Dystrophy Association of America through the Neuromuscular Research Center of Washington University.

the organizer. Frank's special gift has always been the immediate grasp of the fundamental issues in biological phenomena and his faculty of synthesis, where others rarely get beyond analysis. With his typical bold imagination he tried to explain the upfolding, which involved a change from flat epithelial to columnar or wedge-shaped cells, as a result of changes of adhesive properties of protein molecules in membranes; and he presented his theory of surface interaction, properly dubbed the "*zipper theory*," at the Third Growth Symposium in 1941 (Figure 1).

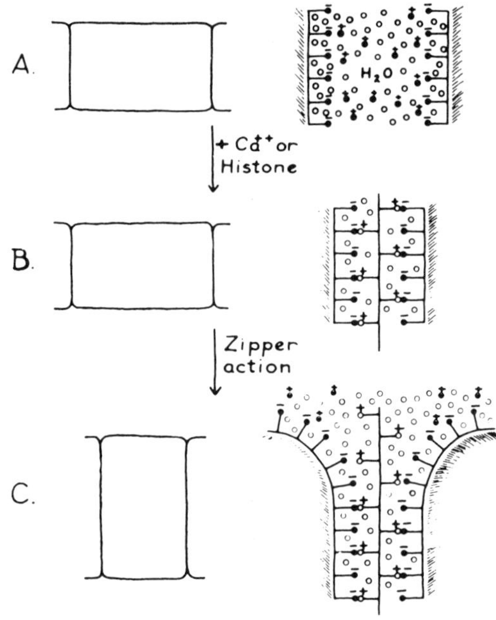

Figure 1. The "zipper" theory, as applied to upfolding of the neural folds. See text. [Schmitt, 1941].

In case the zipper theory should not hold water—and it did not—he held in reserve, as an alternative hypothesis, the notion of contractile protein microfibrils in the surfaces of adjoining cells. It took embryologists over two decades to catch up with Frank's second model, which seems to come closer to the truth.

In reminiscence of those exciting days, I am particularly grateful for being given this opportunity to honor Frank Schmitt, the transient embryologist and life-long leader of neurobiologists.

v

Introduction

The center of the stage in developmental neurobiology, at present, is preempted by the problem of the *origin of specific synaptic connections,* and rightly so, because the precision of the wiring is the *sine qua non* of the proper function of the nervous system. But to the neuroembryologist, synapse formation is a terminal event. It is the final step in a sequence which Ramón y Cajal in his delightful *Recollections of My Life* (1937, p. 73) once outlined rather poetically, yet precisely, in these words:

> "What mysterious forces precede the appearance of the processes, (axons), promote their growth and ramification, stimulate the corresponding migration of the cells and fibres in predetermined directions as if in obedience to a skillfully arranged architectural plan, and finally establish those protoplasmic kisses, the intercellular articulations which seem to constitute the final ecstasy of an epic love story?"

In more prosaic words, we try to analyze a sequence of processes that begins with proliferation in the neural tube, continues with migration of neuroblasts to the mantle zone where they form *cytoarchitectonic units* such as motor columns, is followed by the outgrowth of axons along highly stereotyped pathways, and ends with synaptogenesis (Figure 2). In addition, at least the smaller and

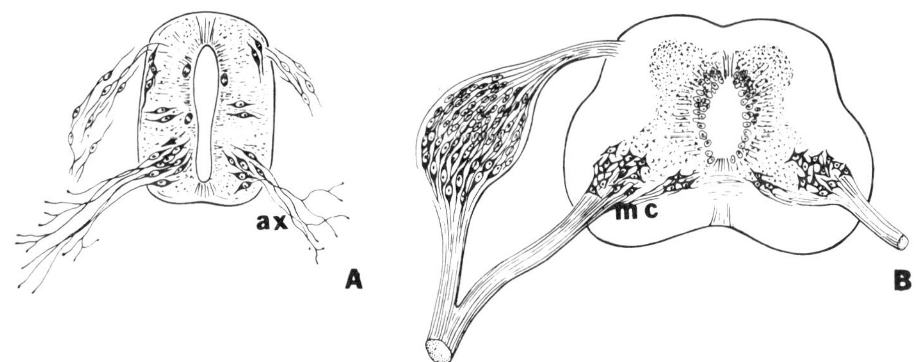

Figure 2. Schematic cross sections of spinal cord of brachial level. A. Three-day embryo; B. Eight-day embryo, ax, motor axons; mc, lateral motor column.

medium-sized units have rather precisely fixed population sizes, and the mechanisms that regulate population numbers will be included in our

1

discussion. All this is a big order, and we neuroembryologists would be well off if synaptogenesis were our only concern.

The *motor system of the chick embryo* has proved to be favorable for the analysis of some of the major problems. The embryo is easily accessible through a window in the shell for microsurgical operations that are not yet feasible in mammalian embryos (Figure 3);

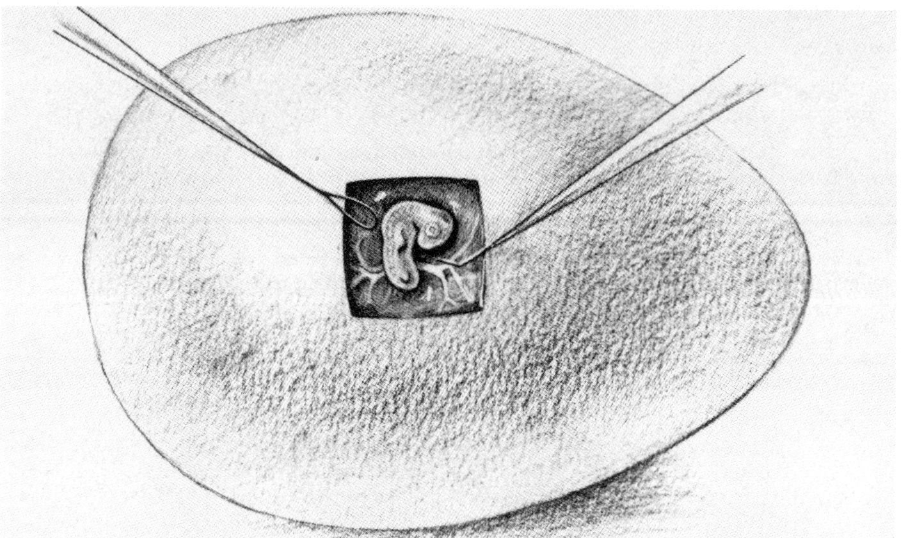

Figure 3. Three-day embryo through window in the shell; glass needle for microsurgery and hair loop (to hold embryo in place).

the higher organization of its nervous system, compared to amphibians, is an advantage; the lateral motor column (lmc) with which we will be much concerned is of manageable size—the lumbar lmc contains about 17,000 cells (Hamburger, 1975).

Origin of Strain Specificity

Ultimately the ancestral lines of all neurons and glia in the central nervous system can be traced back to the neural epithelium of the neural tube whose ventral part gives rise to the spinal motor system. Since in vertebrates neurons operate in assemblies of like cells, or *strains,* our first problem is actually the origin of *cell strain specificity.* At present the origin of motor neurons, or, for that matter, of other neuron strains, is unknown; the best we can do is to set up models as working hypotheses. I shall present briefly two models.

Strains could be *clones* originating from individual neuro-epithelial cells. Exploration of the clonal approach would require the invention of techniques for permanently marking individual neuro-epithelial cells in vivo or the study of the progeny of isolated neuroepithelial cells in vitro. Both approaches present formidable difficulties and remain challenges for the future.

Let me sketch briefly another conceptual approach developed by Harrison (1918), Weiss (1926), and others. I refer to the well-known *"morphogenetic fields."* According to this notion, progressive differentiation and specialization begin with the lay-out of regional areas that become distinct from adjacent areas (e.g., limb, forebrain-eye fields), while their *cellular components* still remain uncommitted. The well-known regulative properties of embryonic fields preclude an early specification of individual cells. Within the borders of a field, subunits are sorted out, such as muscle and skeletal precursors in the limb field and forebrain and optic vesicles in the forebrain-eye field. In this model, the fate of cells or cell strains is a function of their relative position in the subunit.

If we apply the field concept to the early neural tube, we envisage progressive segregation of smaller and smaller *subunits,* or *regionalization* along both the *rostrocaudal* and the *dorsoventral* axis. The terminal step would be the sorting out of groups or clusters of like cells, or specific strains. In this model, the strain would not derive as a clone from a single progenitor cell but represent a supercellular unit from start.

Rostrocaudal Regionalization

In the motor column rostrocaudal regionalization can be demonstrated at remarkably early stages, preceding any visible structural differentiation by several days. In the chick embryo, rostrocaudal regionalization of the spinal cord refers to cervical, brachial, thoracic, lumbar, and sacral segments (Figure 4).

In the amphibian embryo, rostrocaudal regionalization of brain parts and the spinal cord can be traced back to regional inductive differentials in the organizer, that is, the mesoderm that underlies and

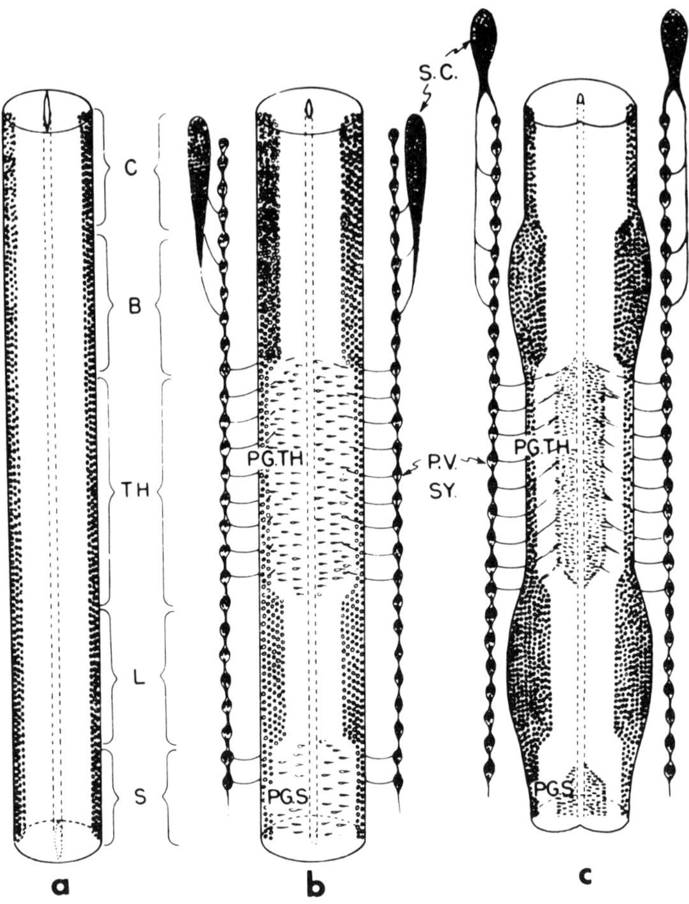

Figure 4. Diagrammatic illustration of the emergence of regional differences from a morphologically uniform system in the spinal motor column of the chick embryo. a, 3-day embryo: The
motor column is of uniform width from the cervical to the sacral level. b, 5-day embryo: The
majority of the differentiating neuroblasts in the cervical segment of the motor column undergo
degeneration. They are represented as solid black circles. In the thoracic and sacral segments the
migration of the preganglionic columns is under way. c, 8-day embryo: The degenerated nerve
cells in the cervical segment of the motor column have disappeared. The remaining nerve cells in
this segment form the slender medial motor columns. Note the size increase of the brachial and
lumbar motor columns innervating the limbs. In the thoracic and sacral segments, the two
preganglionic columns have reached their definitive position adjacent to the central canal. The
two slender columns in a peripheral position represent the medial motor columns. C, cervical
level; B, brachial level; L, lumbar level; PGS, preganglionic sacral center; PG.TH, preganglionic
thoracic center; P.V.SY, paravertebral sympathetic ganglia; S, sacral level; S.C, superior cervical
ganglion; TH, thoracic level.

induces the neural plate. In the spinal cord of the 2-day chick embryo,
that is shortly after the closure of the neural tube, the 5 major sections

of the motor system are already irreversibly fixed in their fate. If at this stage (stages 12–14, Hamburger and Hamilton, 1951) brachial segments are substituted for cervical or thoracic segments, or vice versa, then the transplants express the cytoarchitectural characteristics of their site of origin uninfluenced by their new environment (Wenger, 1951). Even 2 days later, in 4-day embryos, the motor column still has uniform thickness along its entire length without indication of regional differences.

The *visible* patterning in subsequent stages and its underlying mechanisms have been partly revealed by the investigations of Levi-Montalcini (1950) (Figure 4). At the cervical level, a massive degeneration reduces the column to a narrow strand of motoneurons subserving the neck musculature. In the brachial and lumbar segments, the conspicuous *lateral motor columns* (lmc) composed of large cells are formed. Of particular interest is the fate of the thoracic motor segments that turn out to be a mixed population of somatic and visceral neurons. Levi-Montalcini has described how part of this population migrates in a mediodorsal direction from 4 ½ to 7 days and forms the preganglionic sympathetic column, the so-called "column of Terni." Even before the migration of Terni cells begins, their axons have reached the sympathetic chain ganglia, and during their migration, the axons elongate, giving the impression that the Terni neurons drag their axons behind them (Figure 5). These findings were among the earliest

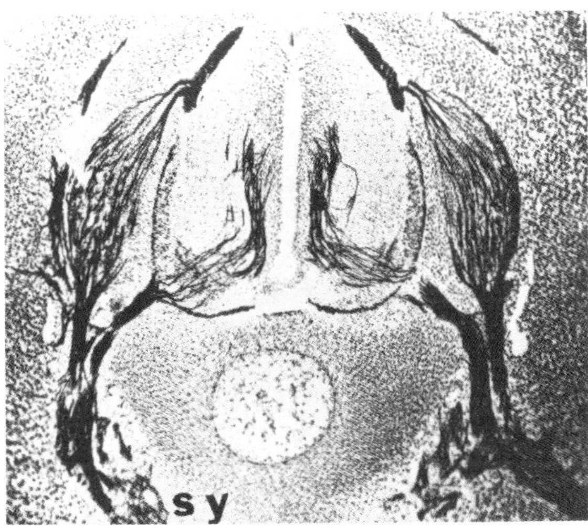

Figure 5. Preganglionic (Terni) neuroblast in migration (6-day embryo). Silver impregnation. Note connection of axons with sympathetic ganglia (sy).

demonstrations of the significant role played by *natural neuronal death* and by secondary migrations of differentiating neuroblasts in the molding of cytoarchitectonic details in neurogenesis.

As was shown first by Detwiler (1936) in amphibians, the very early regionalization includes the *central circuitry* for locomotor pattern generators. The capacity to develop the circuitry for coordinated limb movements exists only in the brachial and lumbar segments, but not in thoracic segments, and is built in even before axons are formed and long before any kind of functional activity is feasible. Straznicky (1963) and Narayanan and Hamburger (1971) have interchanged brachial and lumbar segments in 2 ½-day chick embryos (Figure 6) and have found that after hatching there is a close coupling of wing and leg movements in these embryos with 2 brachial and 2 lumbar regions, respectively; and we found also that the brachially innervated legs never showed stepping movements, only synchronous flexion and extension as in wing flapping. Thoracic segments can produce, at best, slight twitches in legs transplanted to the flank.

Dorsoventral Regionalization

If one takes an overview of brachial spinal cord development in cross sections from 2 ½ to 8 days of incubation (Figure 7), one striking feature becomes obvious: the progression of maturation from the ventral part, the basal plate, to the dorsal part or alar plate. For instance, at 6 days (7nl), the ventral motor column is already numerically complete, the proliferative capacity of the neural epithelium of the ventral region is nearly exhausted, and the epithelium is already transformed into a specialized ependymal layer, while in the dorsal half proliferation and maturation are still in full swing. Again, this chronological sequence is preprogrammed long before it becomes visibly manifest. It can be traced back to the earlier mitotic phase (Hamburger, 1948; Corliss and Robertson, 1963). The peak of mitotic activity in the basal plate, which produces the motoneurons and adjacent interneurons, precedes by 3 to 3 ½ days the peak in the alar plate (Figure 8). This implies that proliferation is strictly programmed in forward reference to the cytoarchitectural features. In other words, the motor part of the spinal cord proliferates and differentiates 3 to 4 days earlier than the dorsal part, which receives sensory input. The chronological primacy of the motor over the sensory system will become a major theme later in this discussion.

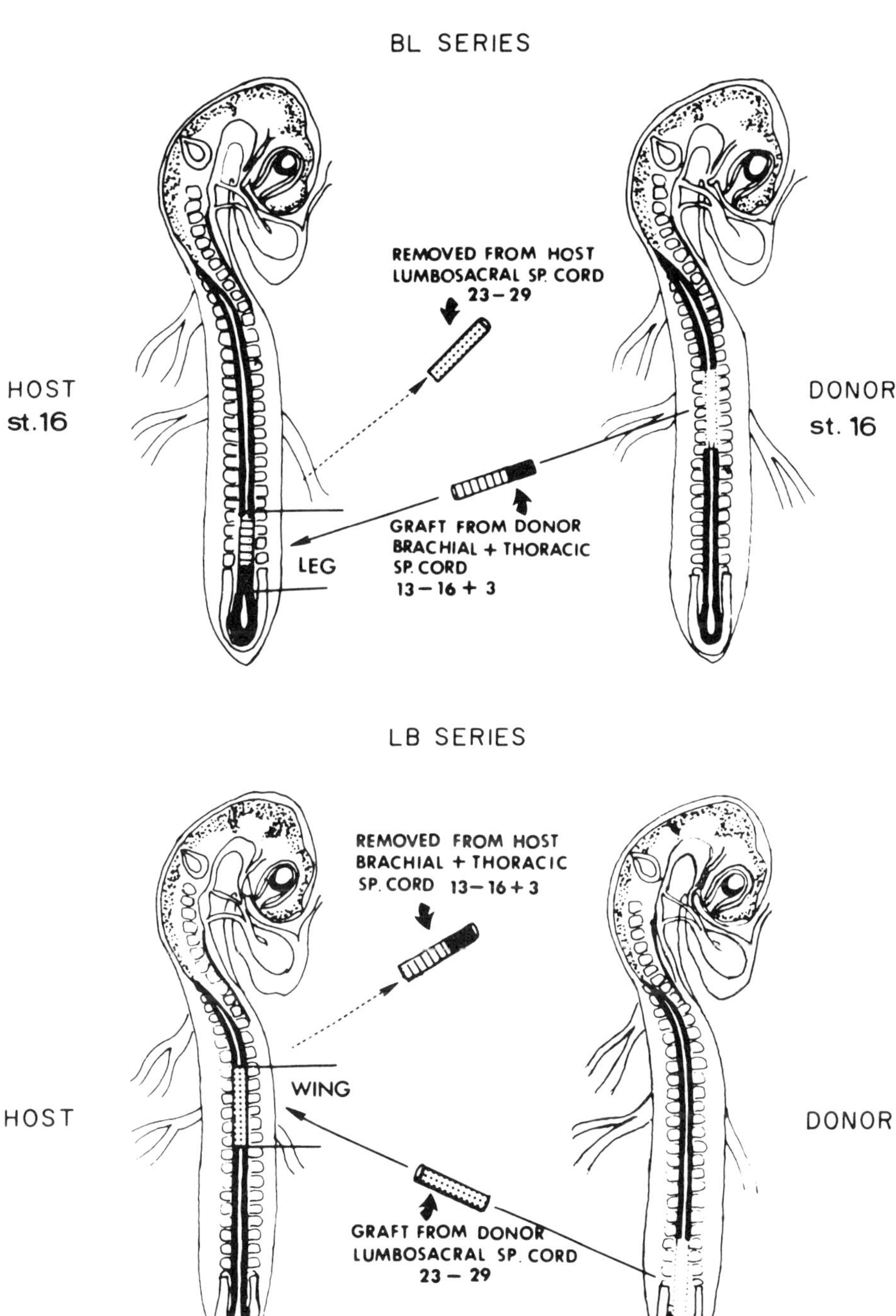

Figure 6. *Above:* transplantation of brachial to position of lumbar spinal cord; *below:* transplantation of lumbar to position of brachial spinal cord, in 2½-day embryos. [Narayanan and Hamburger, 1971]

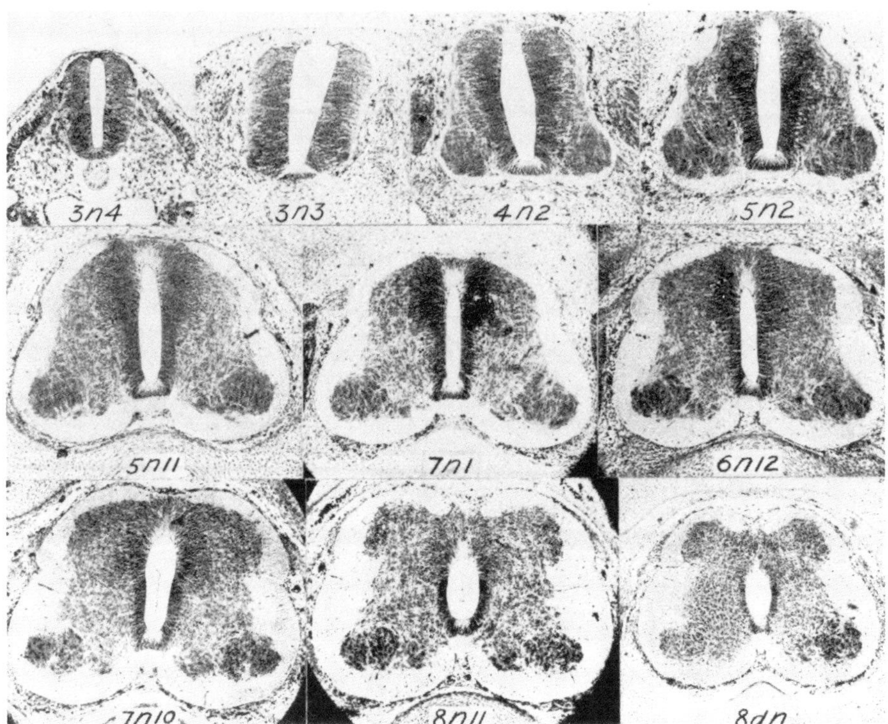

Figure 7. Cross sections through spinal cords at brachial levels. 3n4 = 2½ days (stage 16); 4n2 = 4½ days (stage 24); 7n1 = 6½ days (stage 29); 8n11 = 8 days (stage 34). Note progression of differentiation in ventro-dorsal direction. In 7n1, the basal plate is almost completely differentiated and an ependymal layer has formed, whereas in the alar plate proliferation and early migration are in progress. [Hamburger, 1948]

Determination and Stabilization of Population Size in the Lateral Motor Column

As mentioned before, the lumbar lateral column consists of a fairly constant number of about 17,000 motoneurons when its formation is completed at about 5 ½ to 6 days. This implies a rather precise programming of mitotic cycles; but the mechanism that terminates the cycling and thus the population size is not yet known. However, we can achieve a more modest goal. The autoradiographic technique permits us to fix the date of the terminal DNA duplication, which is designated as the *birthdate*. Like the human birthday, it is a significant landmark in the life cycle of the motor neuron. After the terminal mitosis, the cell is released from its epithelial constraints, it

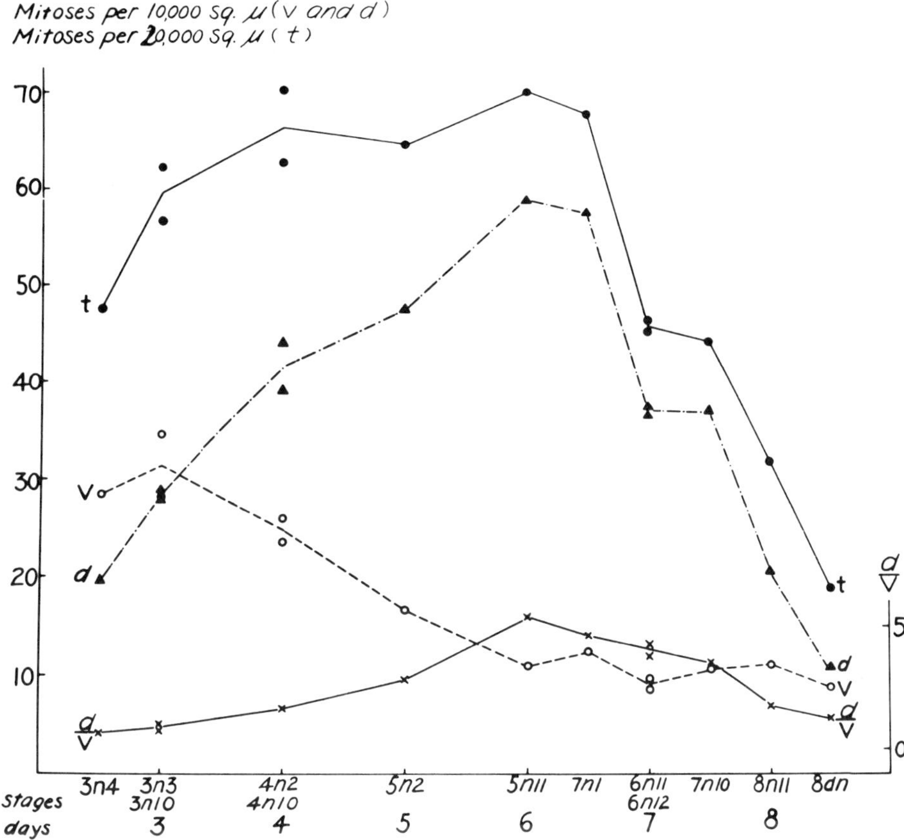

Figure 8. Time pattern of mitotic activity in chick spinal cord. Mitoses near the lining of the central canal were counted and "mitotic density," as defined on top of the left ordinate, was calculated. *Abscissa*, stages and age in days. *Left ordinate*, mitotic density; *right ordinate*, ratio of mitotic density in alar and basal plate. d = mitotic density in dorsal half of neural tube (alar plate); v = mitotic density in ventral half (basal plate), t = total mitotic density (d + v). Note peak of v around 3 days and peak of d around 6 to 7 days. [Hamburger, 1948]

becomes an independent, freely moving cell, now properly called a *neuroblast*, and begins its migration to the mantle and cyto-differentiation; but its phenotypic characteristics have been pro-grammed much earlier.

Dr. Hollyday and I (1977) have determined the birthdates of the lateral motor columns; the first lmc neuroblasts are born in 2-day embryos; the brachial neurons a few hours earlier than the lumbar neurons (stages 15 and 17, respectively). Rostrocaudal chronological gradients were found also within each column. A similar gradient has been observed in the amphibian lmc (Prestige, 1973). It is remarkable that about 95% of the total number are born within 2 days and the remaining 5% during the following 2 days. We do not know whether the slowdown is caused by a lengthening of the cell cycle or a near-exhaustion of the proliferating population. As in most other systems, the lm columns are formed in an inward-outward progression. The oldest neuroblasts settle down at the future median border of the column, and the younger cells migrate across them and build up the columns in the mediolateral sequence.

Shortly after the lmc is assembled and numerically complete (at 5 ½ to 6 days), a very conspicuous *degeneration process* sets in which depletes the lumbar lmc by over 40%. This happens rather precipitously within 3 days (between 6 ½ and 9 ½ days of incubation (Hamburger, 1975) (Figure 9). Obviously, the permanent population size of

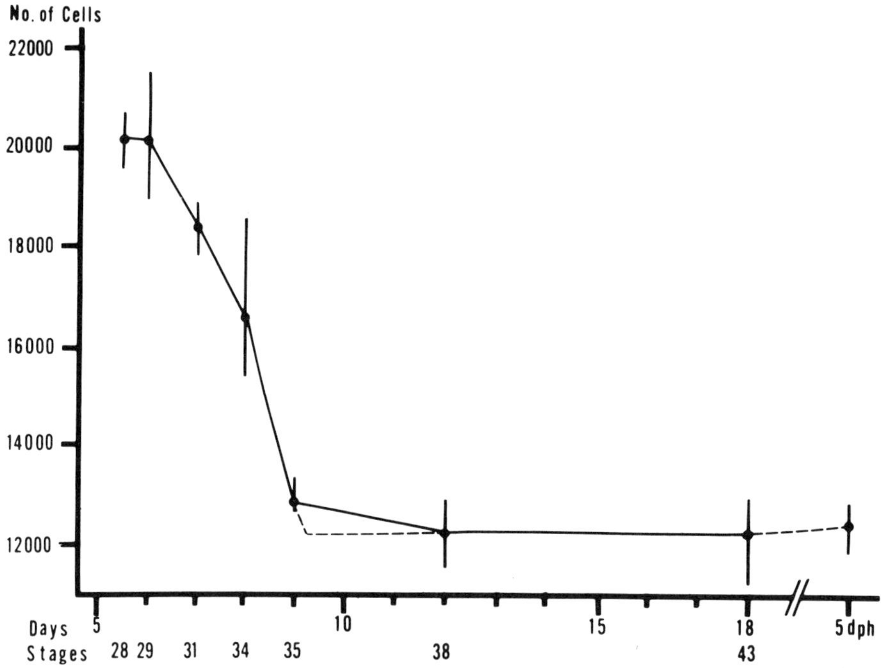

Figure 9. Normally occurring cell loss in the lumbar motor column of the chick embryo. Abercrombie correction for double counting reduces cell numbers to those given in the text. [Hamburger, 1975]

approximately 10,300 motoneurons is attained in two steps. The phenomenon of naturally occurring neuronal death, for which we have used the term *neurothanasia,* is widespread if not universal in neuronal units. In the frog, *Xenopus laevis,* 3 out of 4 motoneurons are said to die (Hughes, 1968). The significance of the overproduction and the related question of the causation of the secondary degeneration are now widely discussed (see Cowan, 1973). The consensus among developmental neurobiologists seems to be that we are dealing with a *competition process* among axons at the site of their projection fields, at the time when they establish contacts with their target structures. Failure to do so would result in retrograde degeneration. The premise that originally all axons, including the unsuccessful ones, grow out and reach their target areas has been verified for several systems (Landmesser and Pilar, 1974b; Clarke and Cowan, 1975), including the lmc of the chick embryo (Chu-Wang and Oppenheim, 1975). But the question: "Competition for what?" is still not settled. Most neuro-embryologists take it for granted that the competition is for synaptic sites. They argue that a highly specific matching chemoaffinity between particular growth cones and their target cells is a prerequisite for successful synapse formation and that considerable trial and error and mismatching are to be expected; hence a surplus of axons at the target site would be advantageous.

While this model seems very plausible, an alternative model, competition for a trophic *"maintenance factor,"* should not be lost sight of. Such a factor would presumably be produced by the target area and its effects be comparable to the nerve growth factor (NGF) (see Hamburger, 1958; Hughes, 1968; Prestige, 1970). Limitation in the availability of a "trophic" substance would also create a scenario for competition. If wishful thinking were permitted, we would speculate that the synaptic site might be the site of production or release of the hypothetical trophic agent; thus we could avoid a sharp distinction between the two notions of competition for synaptic sites or for a maintenance agent.

Trophic Dependence of Motor Neurons on their Target Area

My emphasis on the possibility that we are dealing with competition for a trophic agent is based on the clear experimental evidence for the very remarkable dependence of apparently all young neuroblasts on their target structures. This phenomenon has been a classic topic in neuroembryology since its beginnings. The discovery of the dependence of motor neurons on peripheral factors dates back to an experiment performed in 1909 by Miss Shorey in the laboratory of Dr. F. Lillie in Chicago. She extirpated limb primordia in amphibian and chick embryos and observed subsequently a conspicuous hypoplasia of the lateral motor columns; but she did not pursue the topic any further. When I repeated the extirpation experiment on chick embryos 25 years later in the same laboratory, with improved techniques, I confirmed her findings and noticed that the amount of motor neuron loss was proportional to the amount of muscle tissue removed. On the other hand the concurrent hypoplasia in spinal ganglia varied independently of the motor hypoplasia. I concluded that the motor and sensory nerve centers are dependent on their own respective target areas and that the peripheral influence is mediated by their respective axons and not by diffusion, as had been suggested by Miss Shorey (Hamburger, 1934). The recent confirmation of retrograde axonal transport (Kristensson and Olsson, 1974; LaVail and LaVail, 1974) has given this notion a substantive basis. A later reinvestigation (Hamburger, 1958), which involved radical extirpation of the wing or leg bud in 2- to 2 ½-day embryos (Figure 10), added information on

Figure 10. Ten-day chick embryo, whose right leg primordium was extirpated at 2½ days (operation stage of embryo, see Figure 12). [Hamburger, 1958]

two important points. First, no effects were detected until after the lmc was assembled and numerically complete. That is, the initial processes of proliferation, migration, motor column formation, and axon outgrowth to the periphery are not under peripheral control. Second, the hypoplasia results from a cataclysmic degeneration of neurons which wipes out practically the entire lmc within 3 days, between incubation days 5 ½ and 8 ½ (Figure 11). This implies that the

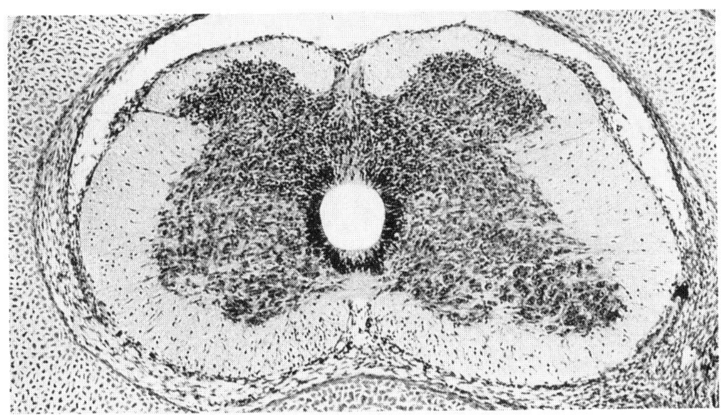

Figure 11. Cross section of lumbar spinal cord of 9-day embryo whose right leg primordium was extirpated at 2½ days. Note: complete absence of lmc on operated (apparent left) side. [Hamburger, 1958]

embryonic motoneuron can sustain itself metabolically for only a short period. It then becomes critically dependent on some unidentified conditions residing in the limb tissue. Several other systems that have been investigated in this respect, such as the trochlear nucleus (Dunnebacke, 1953; Cowan and Wenger, 1967), spinal ganglion neurons (Hamburger and Levi-Montalcini, 1949), the ciliary ganglion (Landmesser and Pilar, 1974a), all in the chick embryo, and the motor columns and other centers in amphibians (see Hughes, 1968; Prestige, 1970) undergo similar hypoplasia when their targets are removed. It should be pointed out that in all these experiments, the target structures were extirpated before axon outgrowth had begun. In this respect, the situation is different from the repercussions of nerve transection in older stages or adult animals.

There is a striking parallel between the circumstances accompanying total removal of the target area and naturally occurring cell death. In the lmc of the chick embryo, both events occur at corresponding stages, and in both instances the neuronal degeneration period coincides approximately with the time at which axonal growth cones distribute themselves among the incipient muscle primordia. The same holds for other systems, such as the trochlear nucleus (Cowan and Wenger, 1967) and the ciliary ganglion (Landmesser and Pilar, 1974a). One is strongly tempted to assume that the same mechanism operates in both instances. In the case of the limb extirpation, where the complete absence of the target area precludes any competition, the outgrowing axons form a neuroma-like strand and eventually degenerate. One could consider the complete breakdown of the lmc in this case as naturally occurring depletion carried to its extreme.

But the extirpation experiment, by its nature, has a low analytical resolution power. In the present case, the competition hypothesis would compete with a number of alternative explanations for the regression of the lmc (see Hamburger, 1958). In this respect, the complementary experiment, the enlargement of axonal projection fields, is analytically more rewarding because it challenges the nerve centers to give a positive response. This experimental design also has a long history. Detwiler (1920) was the first to transplant forelimb primordia of the salamander, *Ambystoma,* to the flank and to demonstrate a *hyperplasia* in thoracic spinal ganglia that innervated the supernumerary forelimb. For reasons that need not be discussed here, Detwiler found no hyperplastic response in the motor system of the salamander; but May (1933), using the experimental design of implanting a supernumerary leg adjacent to the normal leg in the frog, *Discoglossus pictus,* obtained a clear enlargement of the lmc compared to the normal side. In the chick embryo, the same experiment of transplanting a supernumerary wing or leg bud, respectively, closely adjacent to the normal one (Hamburger, 1939a,b) (Figure 12), resulted in very conspicuous and consistent hyperplasia of overloaded ganglia but only in an inconspicuous and not consistent surplus of moto-neurons on the operated side. This might have been due to a scanty ingrowth of motor fibers into the transplant; the experimenter has no control over the transplant innervation. This is a serious shortcoming of this method, but a better one is not available. Our prayers for a motor NGF have not yet been granted. At any rate, even if we had obtained a clear-cut hyperplasia, we would have found ourselves in an embarrassing situation: at that time it would have been difficult to explain. The most obvious explanation, namely an increase in proliferation in the neural epithelium, was already questionable at that time, and it has meanwhile been ruled out. An alternative hypothesis that motor neurons were

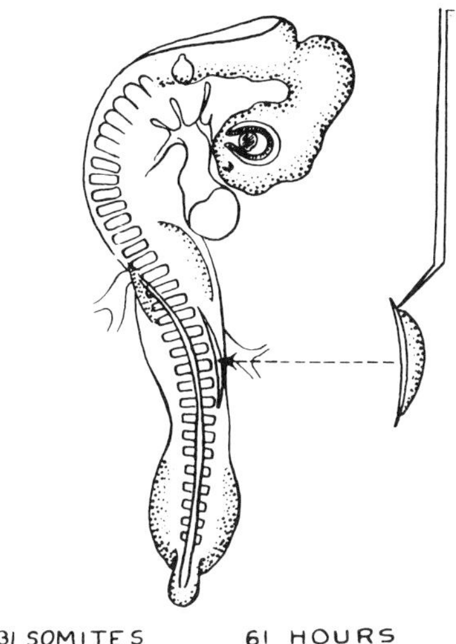

Figure 12. Limb transplantation in 2½-day chick embryo. Arrow points to slit in body wall, into which the transplant, cut out from another embryo, will be inserted. [Hamburger, 1939a]

recruited from a (hypothetical) pool of uncommitted cells was likewise shown to be untenable (Hamburger, 1958).

In the meantime, neurothanasia in the *normal* development of the lmc had been discovered, and this phenomenon suggested an alternative mechanism: that some of the motor neurons that normally would have died could be saved by enlarging the axonal target area. To test this hypothesis, Hollyday repeated the transplantation experiment. She obtained well-developed leg transplants which indicated by their vigorous spontaneous motility that they had received adequate motor innervation from the normal lumbar plexus (Figure 13B). Motor neuron counts were made on 6-day embryos with well-developed transplants (Figure 13A), before the onset of normal cell loss and on 12- or 18-day embryos, after termination of the normal depletion process. A significant and consistent surplus of motor neurons on the experimental side, ranging from 11% to 27.5%, was found in the older embryos. Although only rostral lmc nerves supplied the transplant (Figure 14), the excess of cells was not limited to the rostral segments of the lmc but was spread over the entire column (Figure 15). Apparently, axons of rostral motoneurons that had been deflected into the transplant had left their innervation sites in the normal leg unoccupied, and this gave a larger number of fibers originating in more caudal segments a chance for survival. In other words, redistribution of axons in the absence of reorganization of the lmc had taken place (Hollyday and Hamburger, 1976).

These findings have some important implications. First, the

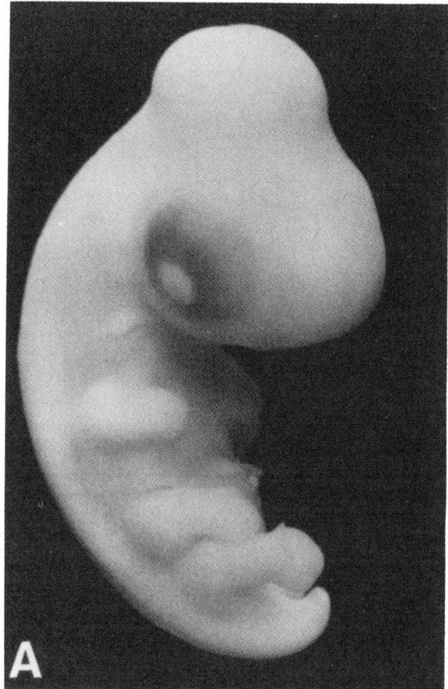

Figure 13. A. Six-day (stage 28) chick embryo with right supernumerary leg. Transplantation was done at 2½ days (stage 17½, see Figure 12). B. Twelve-day (stage 38) embryo with motile right supernumerary leg. (Normal left leg is not visible.) [Hollyday and Hamburger, 1976]

term "hyperplasia" is obviously inappropriate. If the term *"neurothanasia"* is adopted for normally occurring death, then the term *"hypothanasia"* would be suitable for this mechanism. Second, by validating a prediction from the competition hypothesis, the latter is given strong support independent of previous findings and inferences. Third, and perhaps most importantly, by resolving the apparent paradox of an excess of motor neurons after peripheral overloading in the absence of an increase in mitotic activity, we provide a unitary explanation for neurothanasia and the effects of both reduction and enlargement of the periphery: the hypoplasia is an accentuation of normally occurring cell death and the apparent hyperplasia is actually a reduction of normal lmc depletion.

Coming back to the question of determination of population size, we conclude that the original, preneurothanic population size of 17,000 is determined by intrinsic factors in the neural epithelium which program a group of precursor cells, set the clock, and fix the termination of mitotic cycling. The final population size is satisfactorily explained by one or another version of the competition hypothesis. In other words, conditions at the target area determine the final population size.

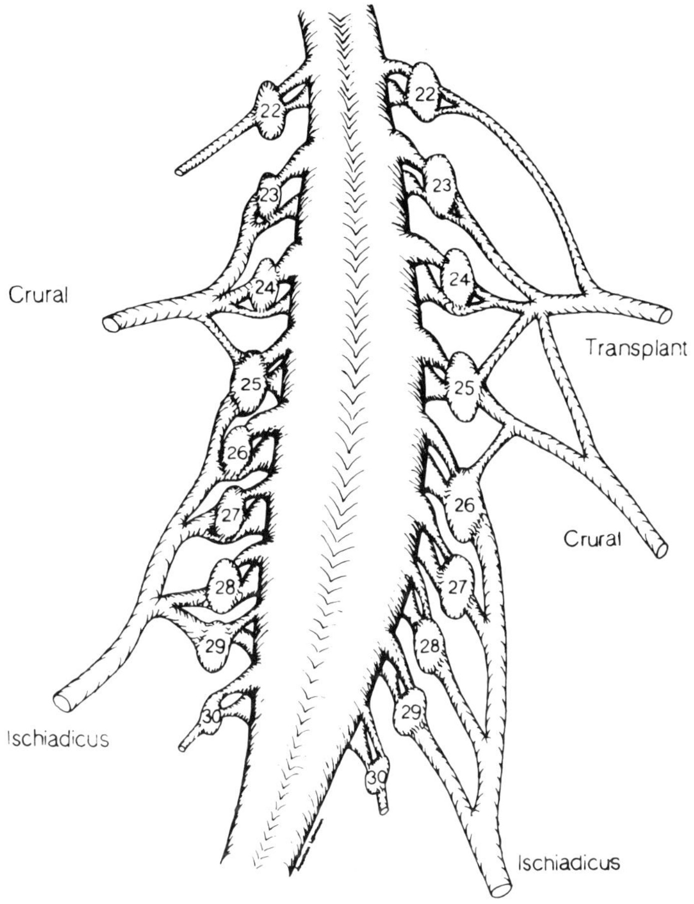

Figure 14. Reconstruction of lumbar plexuses in 12-day embryo with transplanted leg on the right side. Note transplant innervation by 1 thoracic and 3 lumbar segmental nerves which unite in one transplant nerve. [Hollyday and Hamburger, 1976]

Motor Axon Pathways and Synaptogenesis

We should discuss next the problem of how motor nerves from a particular motor pool find their way to their specific muscle, and the problem of selective synaptogenesis, or neuromuscular specificity. The former is one of the most neglected and the latter one of the most controversial topics in neuroembryology. Since time does not permit an adequate discussion of either, I shall limit myself to a few general remarks and the discussion of some recent new data.

Concerning the formation of stereotyped nerve patterns, for instance those in the limbs that have been studied most thoroughly, the

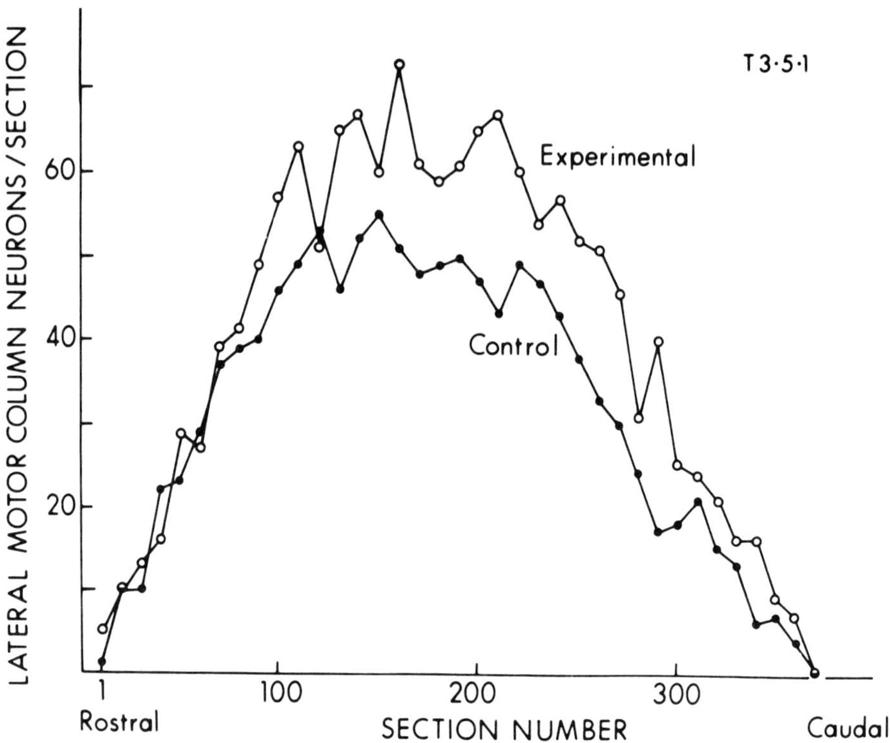

Figure 15. Graph of lateral motor cell counts in lumbar spinal cord of embryo of Figure 13B, on right (experimental) and left (control) lmc side. Note that the extra cells on the transplant side are distributed along the entire rostrocaudal extent of the line. [Hollyday and Hamburger, 1976]

earliest limb transplantation experiments of Braus (1905) and Harrison (1908) indicated that limb tissues provide *preneural pathways.* In the words of Harrison: "The experiments in transplanting limbs show . . . that we must seek in the limb itself for the factors which influence the distribution of the ingrowing nerve; for any nerve at all, in whose way a limb may be implanted, may enter the latter and become distributed in a manner normal for that limb" (1908, p. 409). Weiss (see 1941), taking up from there, was particularly instrumental in establishing the present-day view that nerve growth cones do not seek their targets, nor are they attracted by chemotropism or other agents acting at a distance, but are guided by cues supplied by the preneural pathways. His originally strongly mechanical concept of *"contact guidance"* was soon refined by him and others to include specific chemical cues; and today we are ready to extend Sperry's principle of *chemoaffinity,* which was originally conceived in the context of synapse formation, to path-finding. That is, we ascribe to growth cones and to the substrate on which they grow matching chemical affinities.

Opinions concerning pathfinding of motor axons range all the way from the assumption of very rigidly preformed and prespecified pathways to individual muscles, implying a high degree of discrimination or chemoaffinity between growth cones from particular motor pools and the preneural tracks that guide them to their particular muscle, to the other extreme view that axons are unspecified and receive their specification from the muscle ("myotypic specification," see reviews in Fambrough, 1976; Gaze, 1970; Landmesser, 1976; Purves, 1976). Some experiments, for instance those dealing with pathfinding of Mauthner cell axons, suggest an intermediate position. It is likely that we are dealing with a spectrum of degrees of pathfinding specificity in different systems (Hamburger, 1962). Without entering the debate, I merely wish to point out that most conclusions are based on *regeneration* experiments in lower vertebrates in which the growth cone is confronted not with *preneural* but with *formerly neural* pathways that have had previous contact with nerves.

Let me present briefly two recent well-authenticated experiments on the motor innervation of the leg of the chick embryo in which the nerves enter virgin territory, that is, experiments dealing with initial embryonic innervation. At first sight the two experiments seem to confront us with the same dilemma of prespecification versus plasticity; but in the end, this system may turn out to be very suitable for in-depth analysis of our problem. I refer first to the electro-physiological study of Landmesser and Morris (1975), who stimulated in succession the 8 leg-innervating motor roots and recorded from identified individual muscle nerves. They found that each muscle has a consistent pattern of innervation derived from a fixed combination of motor roots, and they constructed a topographic map of motor root-muscle relations for most leg muscles. It is of crucial importance that this specific pattern "is evident from the earliest time that muscles can be caused to contract by nerve stimulation" (p. 307), that is, as early as day 6 to 6½. The recordings indicated that aberrant or random outgrowth requiring corrections had not occurred. The inference is that motor pools are prespecified for their respective muscles from the beginning and that their axon growth cones recognize their appropriate preneural tracks and make correct choices at branching points and eventually make correct synaptic connections. The simplistic notion of a temporal outgrowth pattern, in the sense that the more rostral nerves saturate proximal muscles first and leave the more distal muscles to more caudal motoneurons, was ruled out effectively by the same authors (see also Landmesser, 1976).

The question then arose whether motor neuron prespecification is modifiable by experimental manipulation, in the sense that functional connections with foreign muscles can be achieved in primary embryonic innervation. The experimental design was again the transplantation of a supernumerary leg primordium rostral to the host leg (Figure 13B), and the rationale was to identify in fully functional transplants the origin of the nerves innervating particular, identified muscles. For the first test, the gastrocnemius (primarily an ankle extensor) was chosen, because its normal motor pool is located in caudal segments of the lmc according to the map of Landmesser and Morris (1975), whereas the transplants are always innervated by rostral plexus nerves. If prespecification is unmodifiable, then one would expect that the caudally located gastrocnemius motor pool would supply both the host and the transplant gastrocnemius. This would require the assumption that some caudally originating axons would find their way across the plexus and enter rostral nerves. However, this is not the case; the transplant gastrocnemius is actually innervated functionally by nerves from rostral segments. This was shown first by Morris (1975) electrophysiologically by stimulation of successive individual motor roots. The motor pool that gives rise to the innervation of the transplant gastrocnemius was identified by Hollyday, using the method of retrograde axonal transport of horseradish peroxidase (HRP). She injected the transplant and the host gastrocnemius with HRP in 12-day embryos, either simultaneously in the same embryo or in different embryos. The normal location of the gastrocnemius motor neuron pool was mapped in normal embryos and is shown on the left of Figure 16, confirming the electrophysiological map of Landmesser and Morris (1975). She found the motor pool for the transplant muscle located in a median topographic position in the lmc in rostral segments 23 to 25 separated by a considerable gap from the host gastrocnemius motor pool in the more caudal segments 27–29 with no overlap (Figure 16). Clearly, rostral motoneurons that normally supply a thigh muscle have changed their destination; if prespecification exists, it is modifiable under experimental conditions.

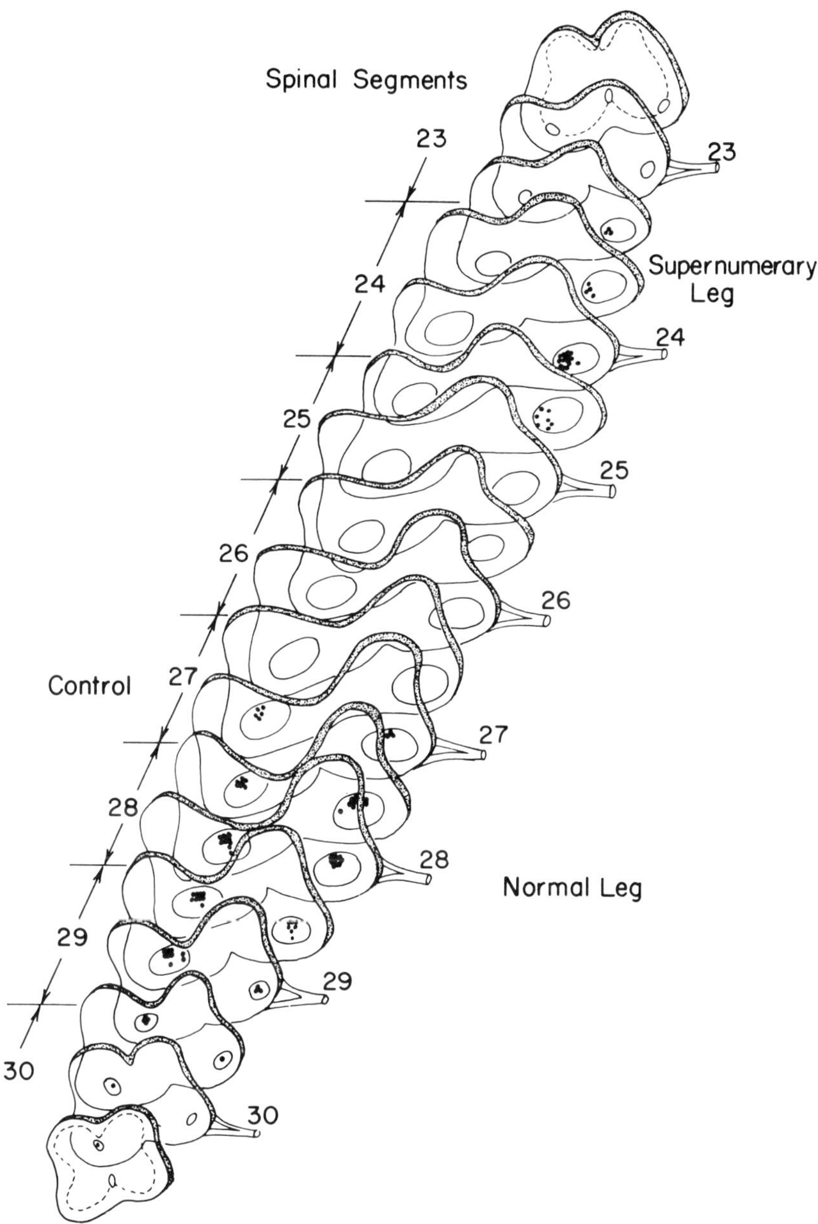

Figure 16. Localization of gastrocnemius motor neuron pools by injection of HRP into gastrocnemius muscle of normal and of transplanted (supernumerary) leg in 12-day embryos, indicated by black dots. Note localization of motor pools for *normal* left and right legs in segments 27–30 of lmc; and motor pool for gastrocnemius of *transplant* in segments 24–25. [Hollyday et al., 1977]

The most important aspect of the HRP injection experiments seems to me the consistency with which, in all 6 cases studied so far, the nerves to the transplant gastrocnemius originated in the *same* identifiable rostral motor pool, located medially in the lmc rather than in randomly scattered motoneurons. In other words, a particular motor pool, normally innervating a thigh muscle, has been assigned to a distal transplant muscle. This suggests that the reorganization of the innervation pattern, instigated by the transplant, follows certain *rules* (Hollyday et al., 1977). Of course, this first impression will have to be validated by a systematic extension of the analysis to other muscles; this work is now in progress.

To summarize: There is ample and long-standing evidence, mainly derived from regeneration experiments on adult amphibians, that limb muscles can be innervated functionally by foreign nerves, the only constraint being that the latter originate in a limb plexus (see, for instance, Weiss, 1955). By using the embryonic limb innervation as a model and by exploiting the presently used experimental design of limb transplantation or modifications of it in conjunction with the powerful tools of electrophysiology and mapping of motor pools by HRP retrograde transport, we hope to obtain deeper insight into the *general rules* governing the specification of normal nerve *pathways* and neuromuscular *synapses,* two aspects of neurogenesis that are more closely linked than is generally recognized in present-day neuro-embryology. Such rules may include the role of positional information, of gradients, and perhaps a preferential innervation of flexor muscles from foreign flexor (rather than extensor) motor pools, or a greater affinity of abductor motoneurons for abductor (rather than adductor) muscles, and the like. In other words, we hope that this embryonic system will make additional contributions to the elucidation of the general problem of prespecification versus plasticity in neuromuscular pathway and synapse formation.

Embryonic Motility

The shaping of the structure of the nervous system, which has preoccupied us so far, is in a way only the prelude to its *functional activation.* The origin of embryonic motility became the object of scientific investigation a century ago through the efforts of a then famous German physiologist and psychologist, William Preyer. In his classical book *Die Specielle Physiologie des Embryo* (1885), he reported among many other original observations the fundamental discovery that in the chick embryo motility begins on the fifth (actually the fourth) day of incubation, but that it is impervious to any kind of stimulation for several days. The first responses to tactile or other stimulation could not be elicited by him until the eleventh day (actually tactile stimulation becomes effective on the eighth day) (Figure 17). Preyer himself considered the existence of a *prereflexogenic period* as "one of the most important facts in the whole area

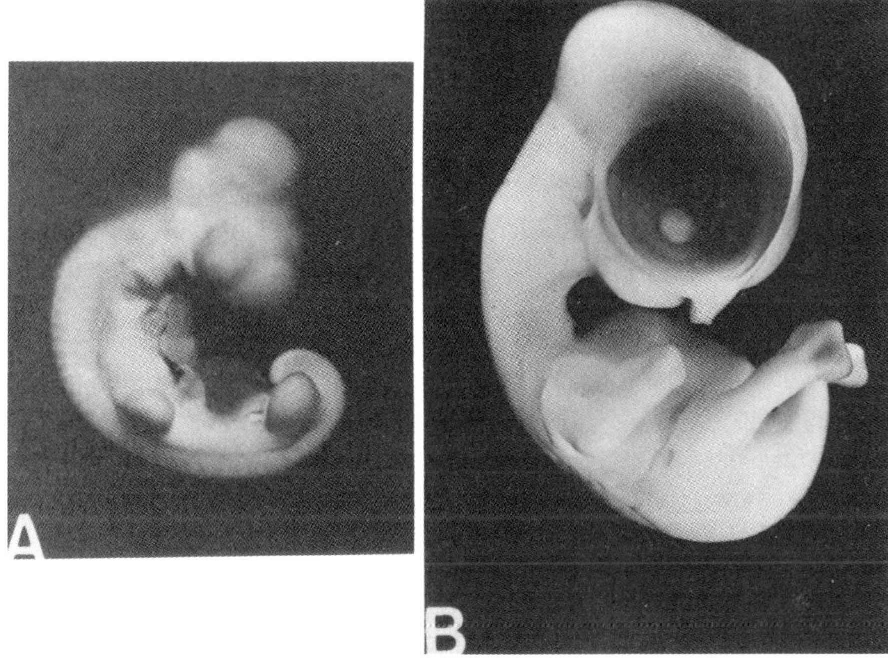

Figure 17. The prereflexogenic period in the chick embryo. A. Beginning of motility at 3½ days (stage 21); slight movements of head. B. First response to tactile stimulation of head, at 7½ days (stage 31). (See Hamburger and Balaban, 1963, and Hamburger and Narayanan, 1969)

of the physiology of the embryo" and he continues: "I have made a very large number of observations and experiments, before I convinced myself that the sensibility of the embryo starts later than the motility. At first, movements occur only from internal physical causes (so-called 'impulsive movements') in the absence of peripheral stimuli which remain without effects when they occur. Only much later can skin sensibility be demonstrated by reflex movements" (1885, pp. 470–71). Seventy-five years later I could show that in a prophetic way he had actually anticipated the key to the understanding of all embryonic behavior of higher vertebrate embryos and fetuses.

But first I want to call your attention to the striking parallel between the prereflexogenic behavior period and the structural maturation pattern in the spinal cord. I thus return to a theme that I had touched upon before: that proliferation and cytodifferentiation occur in the ventral motor part of the spinal cord 3 to 4 days earlier than in the dorsal or alar plate (Figures 7, 8). The time span between onset of motility and initiation of response to stimulation coincides approximately with the interval between the first neuromuscular synapse formation and the inflow of dorsal root fibers into the dorsal gray matter.

Our own investigations of embryonic motility started on the behavioral level and led to the insight that the *spontaneous nonreflexogenic nature* of motility is not limited to the initial period identified by Preyer but remains the outstanding, characteristic feature throughout the incubation period until around the seventeenth day of incubation when it is superseded by another type of motility that marks the preparation of the embryo for hatching.

Before I give experimental evidence for this claim, I shall describe briefly the peculiar manifestations of *overt, spontaneous motility* (Hamburger, 1970). It begins around 3 ½ days with a slight nodding of the head and in subsequent days extends to the whole body. The most prominent features of embryonic motility are abrupt, convulsive-type, jerky movements and the apparent lack of coordination. Movements of different body parts, such as head, trunk, wings, legs, tail, occur independently of each other in unpredictable random combinations. There is little change in this unorganized performance, except that the amount of motility increases to reach a peak around 13 to 14 days (Figure 18). Until then movements are performed periodically, activity phases alternating with inactivity phases; and eye and eyelid movements and beak clapping are added to the repertory as their muscles mature. From days 13 to 17, motility is almost continuous. This rather bizarre performance is perhaps to be expected in the absence of any guidance from sensory feedback.

We then used the tools of microsurgery to obtain experimental evidence that sensory input is not a prerequisite for embryonic motility. Of the several *deafferentation* experiments that we have performed, I shall describe only the leg deafferentation in some detail (Hamburger et al., 1966). The dorsal half of the lumbar spinal cord was extirpated in 2-day embryos. This included the neural crest, which is the precursor of the spinal ganglia. To eliminate any input from rostral

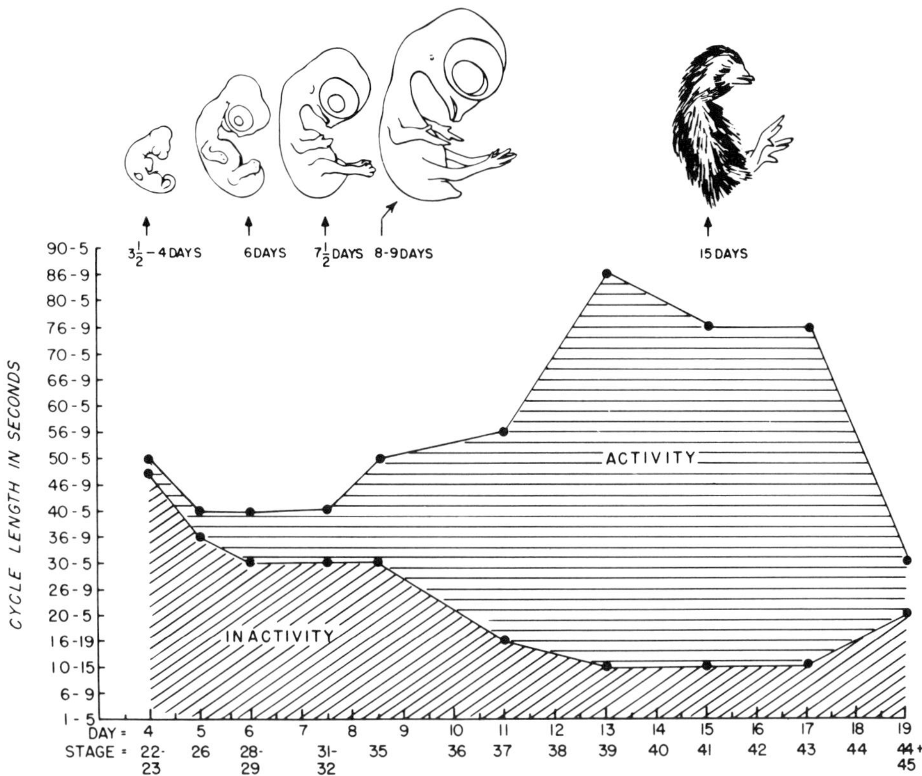

Figure 18. Changes in amount of total spontaneous activity (*top line*) during the incubation period. An inactivity phase is defined as a period of inactivity lasting 10 sec or less. A cycle is the period between the onset of two successive activity phases (in sec). [Hamburger et al., 1965]

parts of the nervous system, several segments of the thoracic spinal cord were also removed, thus effectively isolating the ventral lumbar cord (Figure 19). We then recorded spontaneous motility and compared it with that of controls in which only the dorsal gap had been made. This was necessary because the thoracic gap itself lowers the overall activity, at least when our method of quantification of motility is used (but see Oppenheim, 1975 for a different interpretation). Experimental and control embryos showed no difference in spontaneous motility, both qualitatively and quantitatively, up to 15 to 17 days (Figure 20), when the spinal cords of the experimental embryos began to deteriorate. Of course, the legs showed no response to tactile stimulation. Since both extero- and proprio-ceptive input had been eliminated with the spinal ganglia, we could rule out the claim that self-stimulation, as for instance by brushing of the legs against head or body, or proprioception, plays

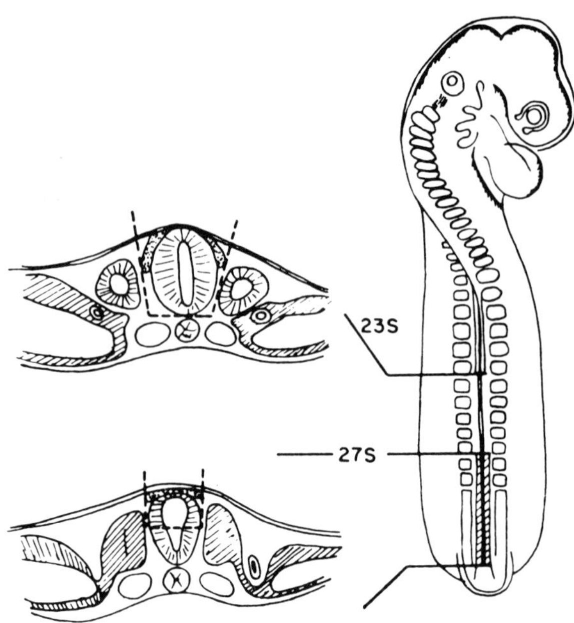

Figure 19. Early embryonic deafferentation of both legs in 2- to 2½-day chick embryos. Total removal of spinal cord between somites 23 and 27, and removal of dorsal half of spinal cord, including neural crest (primordium of spinal ganglia) caudal to somite 27. [Hamburger et al., 1966]

an important role in causing embryonic movements (Kuo, 1967; Gottlieb and Kuo, 1965).

The early bilateral extirpation of the trigeminal ganglion (Hamburger and Narayanan, 1969) deprives the entire head and beak of exteroceptive sensory input, and the extirpation of both otocysts (Decker, 1970) eliminates vestibular input. In both instances, spontaneous motility was unchanged until before hatching.

So far, it has not been possible to perform deafferentation experiments in mammalian fetuses. But a careful study of the motility of 16- to 20-day rat fetuses, which were exteriorized with the placenta intact and the mother immobilized by low spinal transection, revealed spontaneous periodic and completely unintegrated, unorganized behavior, very similar to that of the chick embryo (Narayanan et al., 1971). Therefore we believe that the spontaneous, nonreflexogenic, unorganized type of motility is characteristic of all embryos and fetuses of higher vertebrates.

I then postulated that spontaneous motility results from autonomously generated electrophysiological activity of neurons in the ventral spinal cord which drive motoneurons and that, since all parts

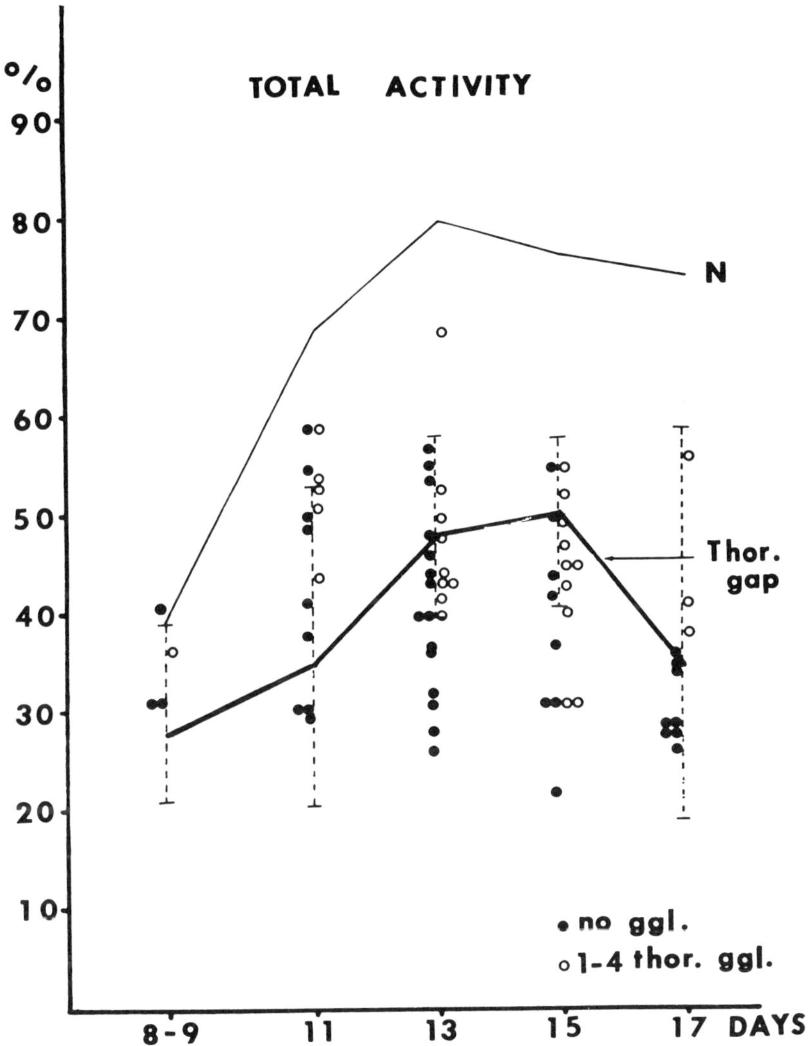

Figure 20. Spontaneous motility of completely deafferented legs (solid and open circles; embryos with open circles had remnants of thoracic, but no lumbar, ganglia). All deafferented legs were checked histologically for absence of ganglia and behaviorally for absence of responses to tactile stimulation. The heavy dark line (thor. gap) represents spontaneous motility in controls, in which only the total spinal cord extirpation between somites 23–27 was performed. Spontaneous motility in unoperated embryos is shown in top line, N. [Hamburger et al., 1966]

are simultaneously in motion during an activity phase, the discharges spread indiscriminately in the spinal cord. I should add that embryos with chronic cervical spinal gaps had shown that input from brain centers plays a minor role up to 17 days. (For brain influence see Decker and Hamburger, 1967; Oppenheim, 1975.)

To test this hypothesis, techniques for extracellular recording in vivo from the ventral cord were developed by Provine and Sharma, with the generous support of Dr. T. Sandel, Chairman of the Washington University Psychology Department (Provine et al., 1970). Provine, partly in collaboration with Ripley, was then able to show that polyneuronal bursts rather than single-unit discharges of neurons (probably interneurons) in the ventral cord are the electrophysiological correlates of overt motility. The chronological coincidence of burst patterns and motility patterns between 4 and 21 days of incubation is very remarkable indeed (Figure 21). In these tests, uncurarized embryos

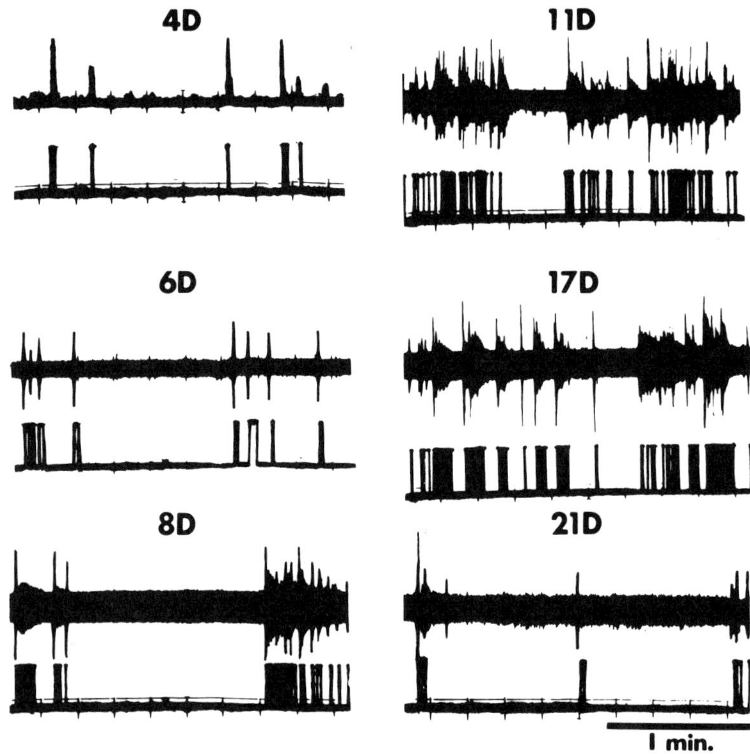

Figure 21. Comparison of polyneuronal burst discharges in spinal cord of freely moving embryos (*upper trace*) with visually observed spontaneous movements of same embryo (*lower trace*). Recordings were made from lumbar cord. D = days. [Ripley and Provine, 1972]

were used. To exclude movement artifacts, simultaneous recordings were made in the spinal cord and the sciatic nerve in curarized embryos. They gave the same results. The clinching evidence that the bursts are the primary events and that they cause the motility came from experiments in which 15-day embryos were curarized for 15 min and burst activity was recorded for three 15-min periods: just before, during, and just after curarization. Burst activity continued throughout the 45-min period. Simultaneous recordings from different regions of the spinal cord confirmed the prediction that the discharges would spread near-synchronously in the ventral cord (Figure 22). (For review see Provine, 1973.)

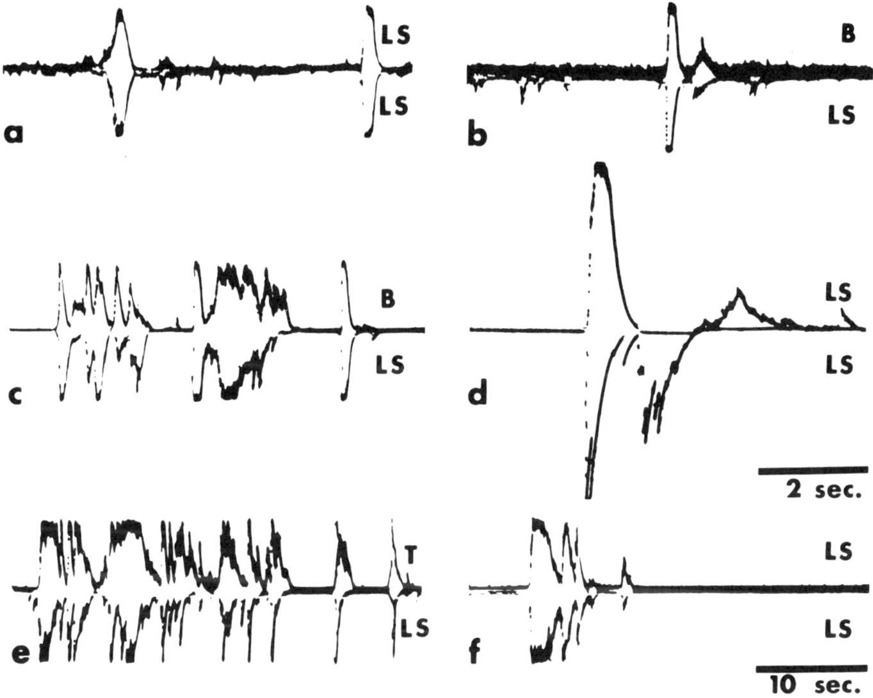

Figure 22. Records of integrated polyneuronal burst discharges simultaneously recorded from pairs of spinal cord sites. Activity from one region is inverted and placed base-to-base with the activity of the other region to facilitate comparison. B, brachial region; LS, lumbosacral region; T, thoracic region. [Provine, 1973]

The spontaneous motility of the embryo is a weird performance, not comparable to any postnatal behavior, except perhaps the fidgeting of the human infant. One asks how the highly *organized postnatal behavior acts* such as walking or pecking can be performed shortly after birth, although we have failed to observe any antecedents or a gradual build-up. The obvious answer is that the spinal cord and brain centers for integrated activity are structurally prepared and gradually perfected without becoming overtly manifest; on the contrary, their maturation is actually concealed by the overt, unintegrated, behavior type of the embryo, which dominates the picture.

There are two ways by which the existence of covert organization and its maturation can be demonstrated: (1) During inactivity phases local reflexes can be elicited from day 7 ½ on. They are evidence of the establishment of reflex arcs long before natural stimuli become an integral part of behavior. (2) One can explore the central organization by monitoring spontaneous motor output, either by recording from muscle nerves or by EMG recording. Dr. Anne Bekoff, with the generous cooperation of Dr. P. Stein in our Department, has made a beginning, using EMG recording for an analysis of the origin and refinement of intralimb muscle coordination. She recorded simultaneously from a pair of synergistic, or a pair of antagonistic, identified leg muscles in vivo, using suction electrodes. To our surprise, she found that, almost as early as muscles become identifiable as individuals and begin to show spontaneous contractions (that is, at day 7, Figure 17B), signs of coactivation of synergists can be detected while antagonists are activated at different times with a phase lag (Figure 23). The alternation of antagonists implies that there exist practically from the start built-in presynaptic pattern generators for each motor pool which are linked with each other in an orderly fashion, involving excitatory and inhibitory synapses (Bekoff et al., 1975). It is of interest that the first synapses including flat vesicles (supposed to be inhibitory) were found in embryos of the same age (Oppenheim et al., 1975).

Concluding Remarks

I think neuroembryologists can claim that in the last decades they have been able to sort out the essential problems in neurogenesis and have brought them sharply into focus. They have gained insight into some of the embryological interactions that operate on the cellular and supercellular levels. This should be of help to biophysicists, biochemists, and molecular biologists who are now taking over. We leave to them, as a challenge, the solution of such problems as the intrinsic regulation of proliferation, the selective activation of genes determining strain specificity, the identification of metabolic "maintenance factors," and the foremost problem: to give to the concept of "chemoaffinity," which is crucial for both nerve pattern formation and synaptogenesis, a concrete, substantive content. It is quite clear that the answer to this latter problem will be in terms of membrane surface chemistry, and several laboratories are pioneering in this direction.

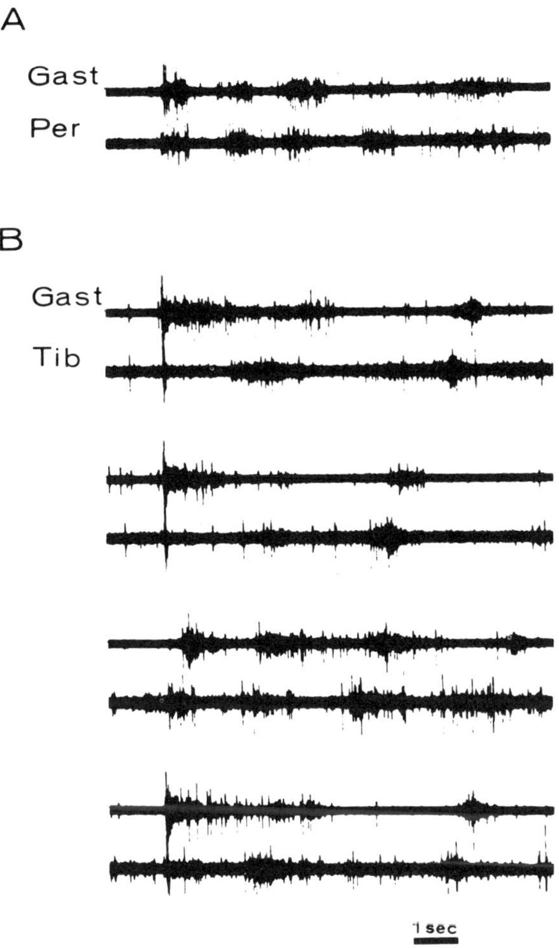

Figure 23. Electromyographic recordings of spontaneous activity from pairs of ankle muscles in a 7-day embryo (see Fig. 17B). A. Simultaneous recordings from ankle extensors, gastrocnemius (Gast) and peronexus (Per). B. Four sets of simultaneous recordings from ankle extensor, gastrocnemius (Gast) and ankle flexor, tibialis (Tib). The 4 sets were obtained during 4 successive activity periods. [Bekoff et al., 1975]

Concerning my *general theoretical position,* I think I am in agreement with most neuroembryologists that the cytoarchitecture and the central circuitry for organized motor output is built according to the basic rules of epigenetic development, in which manifold interactions between neuroblasts, between multicellular units, and between neural units and their target structures play a dominant role, and that the maturation proceeds in forward reference to integrated activity, but with little or no benefit from sensory feedback. Similar views, expressed long ago by Coghill, Weiss, Sperry, and others, based on their classical experiments, have clearly stood the test of time. Of course, this position in no way challenges the contributions of sensory information in the postnatal period.

Let me end with a personal note. What has impressed me most in all phases of our investigations is the *primacy of activity* over reactivity or response. This, to me, has become symbolic of animal life, and perhaps of life in general. The elemental force that embryos and fetuses can express freely in their spontaneous motility, sheltered as they are in the egg or uterus, has perhaps remained, throughout evolution, the biological mainspring of creative activity in animals and man, and autonomy of action is also the mainspring of freedom.

G. E. Coghill, the great neuroembryologist of the early part of the century, who had a strong philosophical bent, once said in a dialogue with his friend, C. Judson Herrick: "The intrinsic control of creative capacity in the practical affairs of life by acquisition of new skills and inventions and by imaginative art and philosophy gives mankind a larger measure of freedom; for the rational approach to the problem of freedom is through the conception of the individual as primarily a system of action" (Herrick, 1949, p. 222). And Goethe's Faust, searching for a meaningful translation of the first verse of the gospel according to St. John, not satisfied with the rendering: "In the beginning was the word," finds his solution: "Am Anfang war die Tat " (In the beginning was the act).

BIBLIOGRAPHY

Bekoff, A., Stein, P.S.G., and Hamburger, V. (1975): Coordinated motor output in the hindlimb of the 7-day chick embryo. *Proc. Nat. Acad. Sci.* 72:1245–1248.

Braus, H. (1905): Experimentelle Beiträge zur Frage nach der Entwickelung peripherer Nerven. *Anat. Anz.* 26:433–479.

Chu-Wang, I.W. and Oppenheim, R. (1975): The natural death of motoneurons in the chick embryo: a quantitative and qualitative analysis of ventral roots by electron microscopy. *Neurosci. Abstr.* 1:752.

Clarke, P.G.H. and Cowan, M. (1975): Ectopic neurons and aberrant connections during neural development. *Proc. Nat. Acad. Sci.* 72:4455–4458.

Corliss, C.E. and Robertson, G.C. (1963): The pattern of mitotic density in the early chick neural epithelium. *J. Exp. Zool.* 153:125–140.

Cowan, W.M. (1973): Neuronal death as a regulative mechanism in the control of cell number in the nervous system. *In: Development and Aging in the Nervous System.* Rockstein, M., ed. New York: Academic Press, pp. 19–34.

Cowan, W.M. and Wenger, E. (1967): Cell loss in the trochlear nucleus of the chick during normal development and after radical extirpation of the optic vesicle. *J. Exp. Zool.* 164:267–280.

Decker, J.D. (1970): The influence of early extirpation of the otocyst on development of behavior in the chick. *J. Exp. Zool.* 174:349–364.

Decker, J.D. and Hamburger, V. (1967): The influence of different brain regions on periodic motility in the chick embryo. *J. Exp. Zool.* 165:371–384.

Detwiler, S.R. (1920): On the hyperplasia of nerve centers resulting from excessive peripheral overloading. *Proc. Nat. Acad. Sci.* 6:96–101.

Detwiler, S.R. (1936): *Neuroembryology.* New York: MacMillan.

Dunnebacke, T.H. (1953): The effects of the extirpation of the superior oblique muscle on the trochlear nucleus in the chick embryo. *J. Comp. Neurol.* 98:155–177.

Fambrough, D.M. (1976): Specificity in nerve-muscle interactions. *In: Neuronal Recognition.* Barondes, S.H., ed. New York: Plenum Press, pp. 25–67.

Gaze, R.M. (1970): *The Formation of Nerve Connections.* London: Academic Press.

Gottlieb, G. and Kuo, Z.Y. (1965): Development of behavior in the duck embryo. *J. Comp. Physiol. Psychol.* 59:183–188.

Hamburger, V. (1934): The effects of wing bud extirpation on the development of the central nervous system in chick embryos. *J. Exp. Zool.* 68:449–494.

Hamburger, V. (1939a): The development and innervation of transplanted limb primordia of chick embryos. *J. Exp. Zool.* 80:347–388.

Hamburger, V. (1939b): Motor and sensory hyperplasia following limb-bud transplantations in chick embryos. *Physiol. Zool.* 12:268-84.

Hamburger, V. (1948): The mitotic patterns in the spinal cord of the chick embryo and their relation to histogenetic processes. *J. Comp. Neurol.* 88:221-284.

Hamburger, V. (1958): Regression versus peripheral control of differentiation in motor hypoplasia. *Am. J. Anat.* 102:365–410.

Hamburger, V. (1962): Specificity in neurogenesis. *J. Cell. Comp. Physiol.* 60(Suppl. 1):81-92.

Hamburger, V. (1970): Embryonic motility in vertebrates. *In: The Neurosciences: Second Study Program.* Schmitt, F.O., editor-in-chief. New York: Rockefeller University Press, pp. 141-15.

Hamburger, V. (1975): Cell death in the development of the lateral motor column of the chick embryo. *J. Comp. Neurol.* 160:535-546.

Hamburger, V. and Balaban, M. (1963): Observations and experiments on spontaneous rhythmical behavior in the chick embryo. *Dev. Biol.* 7:533-545.

Hamburger, V., Balaban, M., Oppenheim, R., and Wenger, E. (1965): Periodic motility of normal and spinal chick embryos between 8 and 17 days of incubation. *J. Exp. Zool.* 159:1-13.

Hamburger, V. and Hamilton, H.L. (1951): A series of normal stages in the development of the chick embryo. *J. Morphol.* 88:49-92.

Hamburger, V. and Levi-Montalcini, R. (1949): Proliferation, differentiation and degeneration in the spinal ganglia of the chick embryo under normal and experimental conditions. *J. Exp. Zool.* 111:457-501.

Hamburger, V. and Narayanan, C.H. (1969): Effects of the deafferentation of the trigeminal area on the motility of the chick embryo. *J. Exp. Zool.* 170:411–426.

Hamburger, V., Wenger, E., and Oppenheim, R. (1966): Motility in the chick embryo in the absence of sensory input. *J. Exp. Zool.* 162:133-160.

Harrison, R.G. (1908): Embryonic transplantation and development of the nervous system. *Anat. Rec.* 2:367–410.

Harrison, R.G. (1918): Experiments on the development of the forelimb of *Amblystoma*, a self-differentiating, equipotential system. *J. Exp. Zool.* 25:412–462.

Herrick, C.J. (1949): *George Ellett Coghill. Naturalist and Philosopher.* Chicago: The University of Chicago Press.

Hollyday, M. and Hamburger, V. (1976): Reduction of the naturally occurring motor neuron loss by enlargement of the periphery. *J. Comp. Neurol.* 170:311–320.

Hollyday, M. and Hamburger, V. (1977): An autoradiographic study of the formation of the lateral motor column in the chick embryo. *Brain Res.* (In press).

Hollyday, M., Hamburger, V., and Farris, J.M.G. (1977): Localization of motor neuron pools supplying normal and supernumerary legs of chick embryos. *Proc. Nat. Acad. Sci.* (In press).

Hughes, A.F.W. (1968): *Aspects of Neural Ontogeny*. London: Logos Press, Academic Press.

Kollros, J.L. (1968): Order and control of neurogenesis (as exemplified by the lateral motor column). *Dev. Biol. Suppl.* 2:274–305.

Kristensson, K. and Olsson, Y. (1974): Retrograde transport of horseradish peroxidase in transected axons. *Brain Res.* 79:101–110.

Kuo, Z-Y. (1967): *The Dynamics of Behavior Development*. New York: Random House.

Landmesser, L. (1976): The development of neural circuits in the limb moving segments of the spinal cord. *In: Neural Control of Locomotion.* Herman, R.M., Grillner, S., Stein, P.S.G., and Stuart, D.G., eds. New York: Plenum Publishing Corp., pp. 707–737.

Landmesser, L. and Morris, D.G. (1975): The development of functional innervation in the hindlimb of the chick embryo. *J. Physiol.* 249:301–326.

Landmesser, L. and Pilar, G. (1974a): Synapse formation during embryogenesis on ganglion cells lacking a periphery. *J. Physiol.* 241:715–736.

Landmesser, L. and Pilar G. (1974b): Synaptic transmission and cell death during normal ganglion development. *J. Physiol.* 241:737–749.

Landmesser, L. and Pilar, G. (1976): Fate of ganglionic synapses and ganglion cell axons during normal and induced cell death. *J. Cell. Biol.* 68:357–374.

LaVail, J.H. and LaVail, M.M. (1974): The retrograde intra-axonal transport of horseradish peroxidase in the chick visual system. *J. Comp. Neurol.* 157:303–358.

Levi-Montalcini, R. (1950): The origin and development of the visceral system in the spinal cord of the chick embryo. *J. Morphol.* 86:253–283.

May, R.M. (1933): Réactions neurogéniques de la moelle à la greffe en surnombre ou à l'ablation d'une ébauche de patte posterieure chez l'embryon de l'anoure, *Discoglossus pictus.Bull. Biol. Fr. Belg.* 67:1–23.

Morris, D. (1975): Development of motor innervation in supernumerary hindlimbs of chick embryos. *Neurosci. Abstr.* 1:753.

Narayanan, C.H., Fox, M.W. and Hamburger, V. (1971): Prenatal development of spontaneous and evoked activity in the rat *(Rattus norvegicus albinus). Behavior* 40:100–134.

Narayanan, C.H. and Hamburger, V. (1971): Motility in chick embryos with substitution of lumbosacral by brachial and brachial by lumbosacral spinal cord segments. *J. Exp. Zool.* 178:415–432.

Oppenheim, R. (1975): The role of supraspinal input in embryonic motility. A re-examination in the chick. *J. Comp. Neurol.* 160:37–50.

Oppenheim, R.W., Chu-Wang, I.W., and Foelix, R.F. (1975): Some aspects of synaptogenesis in the spinal cord of the chick embryo: a quantitative electron microscopy study. *J. Comp. Neurol.* 161:383–417.

Prestige, M.C. (1970): Differentiation, degeneration and the role of the periphery: quantitative considerations. In: *The Neurosciences: Second Study Program.* Schmitt, F.O., editor-in-chief. New York: Rockefeller University Press, pp. 73–82.

Prestige, M.C. (1973): Gradients in time of origin of tadpole motorneurons. *Brain Res.* 59:400–404.

Preyer, W. (1885): *Die Specielle Physiologie des Embryo.* Leipzig: Grieben's Verlag.

Provine, R.R. (1971): Embryonic spinal cord: synchrony and spatial distribution of polyneuronal burst discharges. *Brain Res.* 29:155–158.

Provine, R.R. (1973): Neurophysiological aspects of behavior development in the chick embryo. *In: Studies on the Development of Behavior and the Nervous System Vol. I.* Gottlieb, G., ed. New York: Academic Press, pp. 77–102.

Provine, R.R., Sharma, S.C., Sandel, T.T., and Hamburger, V. (1970): Electrical activity in the spinal cord of the chick embryo *in situ. Proc. Nat. Acad. Sci.* 65:508–515.

Purves, D. (1976): Long-term regulation in the vertebrate peripheral nervous system. *In: International Review of Physiology. Neurophysiology II. Vol. 10.* Porter, R., ed. Baltimore: Baltimore University Press, pp. 125–177.

Ramón y Cajal, S. (1937): *Recollections of My Life.* Cambridge, Mass.: M.I.T. Press.

Ripley, K.L. and Provine, R.R. (1972): Neural correlates of embryonic motility in the chick. *Brain Res.* 45:127–134.

Schmitt, F.O. (1941): Some protein patterns in cells. *In: Growth, Suppl. to Vol. 5 (Third Symposium on Development and Growth),* pp. 1–20.

Shorey, M.L. (1909): The effects of the destruction of the peripheral areas on the differentiation of the neuroblasts. *J. Exp. Zool.* 7:25–63.

Straznicky, K. (1963): Function of heterotopic spinal cord segments investigated in the chick. *Acta Biol. Hung.* 14:145–155.

Weiss, P. (1926): Morphodynamik. *Abhandl. theoret. Biol.* 23:1–46. Verlag Gebrüder Bornträger.

Weiss, P. (1941): Nerve patterns. The mechanics of nerve growth. *In: Growth, Suppl. to Vol. 5 (Third Symposium on Development and Growth)*, pp. 163–203.

Weiss, P. (1955): Nervous system (neurogenesis). *In: Analysis of Development.* Willier, B.H., Weiss, P.A. and Hamburger, V., eds. Philadelphia: W.B. Saunders Co., pp. 346–401.

Wenger, B.S. (1951): Determination of structural patterns in the spinal cord of the chick embryo studied by transplantation between brachial and adjacent levels. *J. Exp. Zool.* 116:123–149.

PONTIFICIAE ACADEMIAE SCIENTIARVM SCRIPTA VARIA

—45—

V. HAMBURGER

PRESPECIFICATION AND PLASTICITY
IN NEUROGENESIS

Abstract of the Proceedings of the Study Week on:
NERVE CELLS, TRANSMITTERS AND BEHAVIOUR

PONTIFICIA
ACADEMIA
SCIENTIARVM

EX AEDIBUS ACADEMICIS IN CIVITATE VATICANA

—

MCMLXXX

110

PRESPECIFICATION AND PLASTICITY
IN NEUROGENESIS

V. HAMBURGER

Washington University
St. Louis, Mo. (U.S.A.)

INTRODUCTION

The concepts "prespecification" and "plasticity" are often ill-defined: to make them useful, exact criteria as well as the stage of development to which they are applied have to be stated precisely. For instance, plasticity often refers to functional adaptation. But our discussion will deal only with structural modifiability. The term "prespecification" conveys the notion that a developmental process, or phenotypic characteristics of a particular neuron strain, are irreversibly programmed or determined from an early stage on. The neural tube which gives rise to the entire central nervous system represents an early stage. One can inquire whether the undifferentiated neural tube of a 2-day chick embryo has acquired already the *regional specification* for cervical, brachial, thoracic, lumbar sectors which can be characterized by several criteria in later stages. One widely used test for specification is the transplantation experiment borrowed from general experimental embryology. If prospective brachial and cervical or brachial and thoracic segments are exchanged in the 2-day neural tube, they differentiate according to their site of origin, rather than the site of implantation, as shown, for instance, by the presence or absence of a lateral motor column in the transplant (B. Wenger, 1951). Another criterion would be the circuitry for coordinated limb movements which develops only at limb levels. Transplantation of the brachial neural tube to the lumbar level in 2-day chick embryos results in the performance of synchronus leg movements akin to wing flapping. Alternating movements were never observed; hence segmental specification defined by this criterion is also fixed at that stage (Narayanan and Hamburger, 1971).

If similar experiments are performed on tail bud stages of *urodele* embryos, the outcome is different. Prospective thoracic segments transplanted to the forelimb level differentiate all characteristics of a limb-innervating level, including normal reflexes and support of normal ambulatory movements (Detwiler, 1923). Obviously, irreversible specification of these aspects occurs at a later stage. The data show that taxonomic status and stage of development are two important parameters which have to be identified precisely. Furthermore, the data suggest that in the chick embryo regional plasticity might still prevail in embryos younger than those in which the above-mentioned tests were performed.

A third fundamental parameter in considering the two concepts is level of organization. At a time when the focus is on the cellular and subcellular levels, this aspect deserves special consideration. One of the ground rules established by classical experimental embryology, at least for vertebrates, is the progression of differentiation from larger supercellular units to increasingly smaller subunits. The larger units are called "morphogenetic fields". They are defined operationally as being endowed with a high degree of plasticity or regulative capacity. Relatively large defects are restored; fields cut in half can produce whole organs. For instance, in the rostral part of the neural plate, a regulative forebrain-eye field can be identified; it is set off from an adjacent midbrain field. Transplantation of part of the field to the flank results in the formation of one or two eyes in the transplant and of two eyes in the regulating residual field (Adelmann, 1936). In subsequent stages, the field becomes segregated into two eye fields and a forebrain field, each no longer capable of substituting for the other, but still endowed with regulatory capacity within its boundaries. Following this, pigment epithelium, retina and retinal substructures become specified. Individual cells or small groups of like cells are the last units to acquire specification. The mechanism of this hierarchical sequence of segregation into smaller units, or "self-organization" is obscure. Exchange of information between differentiating areas and cells as well as gradients seem to be involved.

Implicit in this discussion is the insight that specification of definable supercellular aspects occurs before differences can be detected by other presently available methods than transplantation; and, furthermore, that the terms "prespecification" and "plasticity" are not mutually exclusive. The same substructure (field) can be considered as prespecified with respect to other fields and plastic within its own boundaries. This holds even for individual differentiating neurons (see below).

There are at least three areas in neurogenesis in which the analysis of particulars of prespecification and plasticity is in progress:

1) origin of strain specificity;

2) formation of the stereotyped central axonal pathways and peripheral nerve patterns;

3) selectivity of synaptic connections.

I shall omit the last topic which has been reviewed extensively in recent years (see books edited by Barondes, 1976; Gottlieb, 1976; also Jacobson, 1978). My discussion is limited to the origin of strain specificity and the determination of peripheral nerve patterns.

Origin of Strain Specificity

When and how do the hundreds of different neuronal strains acquire their specific phenotypic characteristics? To what extent are strain-specific characteristics irreversibly fixed and to what extent are they modifiable? How is the population size of neuronal units regulated? These are some of the questions that can be asked concerning strain specificity. We shall address briefly the first question.

With one exception to be mentioned below, different neuron strains are represented by populations of like cells. One can assume that each strain represents a polyclone (Crick and Lawrence, 1975) which is derived from a relatively small number of founder cells. The founder cells would be the smallest units formed in the segregation process discussed in the introduction.

Critical events in the establishment of strain specificity in the central nervous system seem to occur in the neural plate and the early neural tube. Ramón y Cajal (1890) who was the first to apply the silver impregnation method to embryos observed postmitotic differentiating neurons in the spinal cord of 2-day chick embryos: motor neuroblasts and commissural neuroblasts were clearly identifiable by their position and direction of outgrowth of their axons. The extensive and carefully analysed rotation experiments on the medulla level of the neural plate of the *Axolotl,* by C.-O. Jacobson (1964, 1976) have shown that the brain stem nuclei are fairly well stabilized at that stage. However, regulative capacity was observed within areas forming a particular motor nucleus, and the author envisages the neural plate with rising folds as a pattern of small, partly overlapping fields. In other words, we deal apparently with a critical stage of transition from a pattern of small subunits with restricted plasticity to a pattern of subunits representing specified strains. An early establishment of irreplaceable precursor cells for motor neurons in the spinal cord can be deduced from regeneration experiments in a slightly later stage, the neural tube of early tail bud stages of *Ambystoma* (Holtzer, 1951). Following extirpation of the lateral half of several neural tube segments, regulation occurs predominantly by overgrowth from the dorsal part of the intact side. Regulation is fairly complete histologically, with the exception of motor neurons which are missing. Obviously, plasticity still prevails for other neuron types, but the precursor of motor neurons have been sorted out already. It remains to be seen whether the model suggested by these experiments applies only to large, early differentiating neurons or whether it can be generalized.

There is a unique instance in which it has been possible to define precisely the time of specification of a neuron. I refer to the single pair of giant *Mauthner cells* (M cells) in the medulla of fishes and aquatic amphibian larvae. In a classical study, Stefanelli (1947, 1951) has analyzed the time of origin of these cells in frog embryos. In the late gastrula, the M cells are not yet specified. If the M-cell forming area is extirpated on one side and grown in tissue culture, no M cells differentiate in the explant, but the wound in the donor embryo is closed, and a normal M cell is differentiated. A few hours later, after the appearance of the neural plate, the same experiment results in the differentiation of one M cell in the explant and one in the donor. This, then, is the critical stage of incipient specification. The experiment shows that more than one cell, probably a small cluster, has the capacity to form M cells. In normal development, one cell only emerges as the M cell (perhaps one in central position in a small concentric gradient field). In the experiment, the cluster is divided between the explant and a residual group in the donor, and one cell in each becomes programmed for M cell differentiation. The same situation was found in the above-mentioned experiments of C.-O. Jacobson, using a different experimental design. If the edge of the transplant cuts across the M cell area, one M cell was found occasionally in normal position near the edge and another in the rotated piece. The experiments were done in neural plate stages of the *Axolotl*. Stefanelli found in experiments done at later stages that an M cell was formed only in the explant, indicating irreversible and irreplaceable specification by the time the neural folds close. An autoradiography study by Vargas-Lizardi and Lyser (1974) has shown that the specification follows shortly after the terminal mitosis in the late gastrula. Another set of large neurons, the transient intracentral sensory *Rohon-Beard* cells in amphibians likewise undergo their terminal mitoses in the gastrula stage (Spitzer and Spitzer, 1975).

At the other end of the spectrum are the neural derivatives of the *neural crest*. The latter originates as a wedge of cells at the point of closure of the neural tube which detach themselves from the epithelia and begin their migration laterally and ventrolaterally. They give rise to a remarkably diverse set of structures. In the head, they form mesenchyme, the trigeminal and root ganglia and their glia, and furthermore part of the cartilaginous visceral skeleton. In the trunk they produce melanophores, dorsal root ganglia, sympathetic para- and prevertebral and enteric ganglia, glia and adrenal medulla.

Experimental tests for specification were preceded by mapping of the migration routes and fates of different regions of the crest. Two methods are being used for mapping: replacement of a small sector by the same sector from a ^{3}H-thymidine labelled donor, or exchange of identical regions between chick and quail embryos, the quail cells being identified by a nuclear marker.

We encounter here again a progression of specification from larger to smaller subunits. In the chick embryo, in an early stage (1½ days), crest cells destined for neural derivatives can be interchanged with cells destined for mesenchyme or cartilage differentiation: they follow the migratory routes characteristic for their new topographic position and subsequently differentiate along the line of the cells for which they have been substituted. This was shown by Noden (1978) using the tritiated thymidine labelling technique. In a slightly later stage, plasticity in the different categories of neural derivatives was demonstrated by Le Douarin and others (1975) using the chick-quail exchange technique. Normally, the neural crest cells at the vagus level migrate into the intestinal tract and form the cholinergic Auerbach and Meissner ganglia and plexuses; crest cells at the level of trunk segments 18-24 form adrenergic paravertebral sympathetic ganglia and adrenal medulla. Transplantation of the prospective adrenal medulla segment of the neural tube and crest of a quail embryo to the vagus site of a 2-day chick embryo results in the migration of the transplanted crest cells into the intestine, where they form typical cholinergic Auerbach-Meissner plexus ganglia. The reciprocal experiment shows the same versatility of the transplants. The plasticity of the neural-crest derived neurons offers the opportunity to analyze the conditions which influence specification. Micro-environmental factors encountered during migration can have an influence (see Patterson, 1978), but on the other hand, the terminal site of ganglion cell differentiation can also decide the specification of the transmitter under conditions which exclude migration (Smith and others, 1977).

In recent years, extensive *in-vitro* studies in several laboratories have demonstrated that isolated embryonic or postnatal sympathetic neurons can shift from the production of adrenergic to the production of cholinergic transmitters under a variety of conditions. We shall not discuss these experiments in detail (see reviews by Patterson, 1978 and Varon and Bunge, 1978). Thus the plasticity of this particular category of differentiated neurons is demonstrated for at least one significant phenotypic trait: neurotransmitter synthesis.

Formation of Peripheral Nerve Patterns

Since the pioneer work of Harrison at the beginning of this century, the formation of the stereotyped nerve patterns in limbs has served as a convenient model for the analysis of nerve pattern formation in general. I shall consider primarily the motor component.

Harrison's tissue culture experiment, apart from its major objective to prove the axon outgrowth theory, showed also that the growth cone requires a solid substrate, that is, mechanical guidance. His limb transplantation experiment (in amphibian embryos) gave evidence that pathfinding is an interaction between growth cones and their microenvironment. The finding that foreign nerves can form a fairly typical nerve pattern in transplanted limbs led to the proposition that limb tissues form pathways which guide the nerves to their targets (Harrison, 1907). But it became obvious soon that mechanical guidance is only a permissive condition: *directional* outgrowth requires more specific cues. Sperry then directed our thinking along the line of *chemoaffinity* between the growth cone and its substrate. The chemoaffinity hypothesis was designed originally to explain selective synapse formation, but it was extended later to the formation of nerve patterns (Sperry, 1963). It postulates matching cytochemical affinities between growth cone and specific chemical cues provided by preneural pathways.

Other hypothesis have been proposed. The hypothesis of "myotypic specification" assumes an original random distribution of nerve fibers. The target muscles would then specify the uncommitted axons by imprinting on them their own biochemical tag. Another hypothesis also assumes random outgrowth, but specification would be imposed by functional validation: only fibers that have established functionally appropriate connections would survive. It should be mentioned that both hypotheses were based largely on nerve regeneration experiments. They can be ruled out for initial embryonic innervation, since initial outgrowth is not random, as will be shown presently. An argument against functional validation is the simple fact that the basic nerve pattern is formed before function begins.

We ask now the specific question: how do axon bundles from a particular motor pool find their way to their particular target muscle? Before the question can be approached experimentally, a map of motor centers for individual muscles has to be prepared. Landmesser and Morris (1975) have constructed a map based on successive stimulation of the lumbar segmental nerves and recording from individual muscle nerves, in chick embryos, at different stages. The results are unequivocal: the same combination of segmental nerves which innervates a given muscle at a fully functional stage does so at the earliest stage that gives responses, that is, at 6-7 days, when the first spontaneous movements were observed and the individual muscles have just become segregated from a common muscle mass. Stimulation by a "wrong" segmental nerve was not observed. Hence a strong case can be made against initial random distribution and for direct, selective outgrowth from the very beginning. Since the correct innervation pattern is established before onset of natural neuronal death, the latter phenomenon is not implicated in correcting fiber connections with wrong muscles. The elimination of large numbers of motor neurons after day 7 results from unsuccessful competition of fibers belonging to the same or a related motor pool. Miotypic specification is ruled out because the axon bundles find their way along selective routes before they synapse. The chemoaffinity hypothesis seems to give the best explanation of the data.

An experiment done by Dr. Anne Bekoff in the laboratories of Dr. P. Stein and myself strongly reinforces the case for very early specification of the limb motor system. It was mentioned that spontaneous leg motility begins at day 7. The movements are non-reflexogenic since the reflex circuits are not yet closed at that stage. The jerky, convulsive-like character of the movements suggested absence of muscle coordination. To test this point, Bekoff recorded EMGs with floating electrodes in 7-day embryos from a pair of synergists and also from a pair of antagonists, shortly after onset of leg motility. Contrary to expectation, she found that synergists are activated simultaneously and antagonists with a time lag, at that early stage (Bekoff and others, 1975). Coordination is still imperfect; it is refined in subsequent stages. The experiments dealt with intra-joint coordination; inter-joint coordination exists at least as early as day 9 (Bekoff, 1976). Thus, a different approach from that of Landmesser and Morris leads to the same conclusion: that the stereotyped mature nerve pattern and selective synapse formation are established by initial outgrowth along selective pathways and not by selection from randomly distributed fibers. The experiments go a step further: they demonstrate that the central circuitry for coordination is also prespecified. One has to postulate that at the onset of motility there exist already in the lumbar spinal cord particular interneuron centers which drive individual motor pools and, furthermore, that the centers for different motor pools are linked by excitatory and inhibitory synapses which account for the patterned motor output.

One can hardly expect stronger evidence for early prespecification in the establishment of nerve pathways and functional neuromuscular connections. At the same time, there exists an equally substantial set of data which seems to be incompatible with this notion. I refer to the above-mentioned limb transplantation experiments of Harrison which have shown that foreign nerves can grow along the typical pathways and form functional synapses with inappropriate muscles. How can we extricate ourselves from this dilemma?

Recent findings of Dr. M. Hollyday on chick embryos seem to open up a new road to a better understanding of pathway choice, by an analysis of the origin of nerves innervating identified muscles in limb transplants. A prerequisite was the preparation of a map of the motor pools in the lumbar motor column which would be more precise than the one based on motor root stimulation. The method using retrograde axoplasmic transport of horseradish peroxidase (HRP) was employed. HRP was injected into identified muscles and reaction product was localized in the lateral motor column (Hollyday, 1978, 1979). Landmesser (1978) independently obtained almost identical maps for the motor pools of most leg muscles. The maps established for advanced embryonic and post-hatching stages were identical with those found in 10- to 12-day embryos. Both investigators found that there is no correlation between the proximity of muscles in the leg and the topographic position of the motor pools innervating them; nor are the motor pools grouped according to the joints on which they act. However they did find a grouping of motor pools according to embryonic origin of the muscles (see also Ferguson, 1978). Romer (1927) had shown that the leg muscles are derived from two primordia, a dorsal and a ventral muscle mass, respectively. The motor pools for muscles derived from the ventral muscle mass have a medial position in the motor column, and those for muscles derived from the dorsal muscle mass have a lateral position.

The experimental design of Hollyday was as follows: Supernumerary leg and wing buds were transplanted immediately rostral to the right leg bud of the host, in 2½-day embryos. Normal leg innervation is by segmental nerves 23-30; transplant innervation was by segments 23-25, and right host leg innervation by segments 26-30. Homologous muscles in the host and transplant were injected with HRP at 12 days and the respective motor pools were identified. In the first test, the lateral gastrocnemius, an ankle extensor, was chosen because its normal motor pool is located in the caudal part of the column. Simultaneous injection of the host and transplant gastrocnemius was successful in several cases. The motor pool for the transplant gastrocnemius was found in a medial position in segments 23-25, separated clearly from the normal gastrocnemius pool which is in a medio-dorsal position in segments 27-29 (Fig. 1). The decisive point is the finding that in all 6 cases the same rostral motor pool supplied the innervation for the transplant gastrocnemius. The substitute motor pool normally innervates a thigh adductor; in the experimental situation it supplies probably both the foreign and its own target muscle. Two conclusions can be drawn: prespecification of motor pools and of pathways, and synapse specificity are not absolute but modifiable by experimental design. Furthermore, nerve ingrowth into a foreign muscle is not from randomly distributed motor neurons; rather, there is consistency in the atypical innervation pattern. We have used the term *selective mismatching* in this context (Hollyday and others, 1977).

In the meantime, the investigation has been extended to 4 other transplant leg muscles, and two transplant wing muscles and the picture was the same as in the first case: a particular atypical motor pool innervates selectively and reliably the same transplant muscle (Hollyday, personal communication). Obviously, we are dealing with a general rule: there exists a selective affinity of second order between transplant muscles and their atypical substitute motor pools. Axon bundles entering a foreign territory make preferential second choices and disregard other possible targets. One can look at the situation in terms of a hierarchy of neuronal specificities. Meyer and Sperry (1976) express a similar idea, in a discussion of retino-tectal specificity. They speak of "preferential graded affinities, not all-or-none specificity" (1976, p. 113).

The next step was to look for rules governing selective mismatching. Several possibilities suggest themselves: functional affinities could play a

GASTROCNEMIUS MOTOR NEURON POOLS

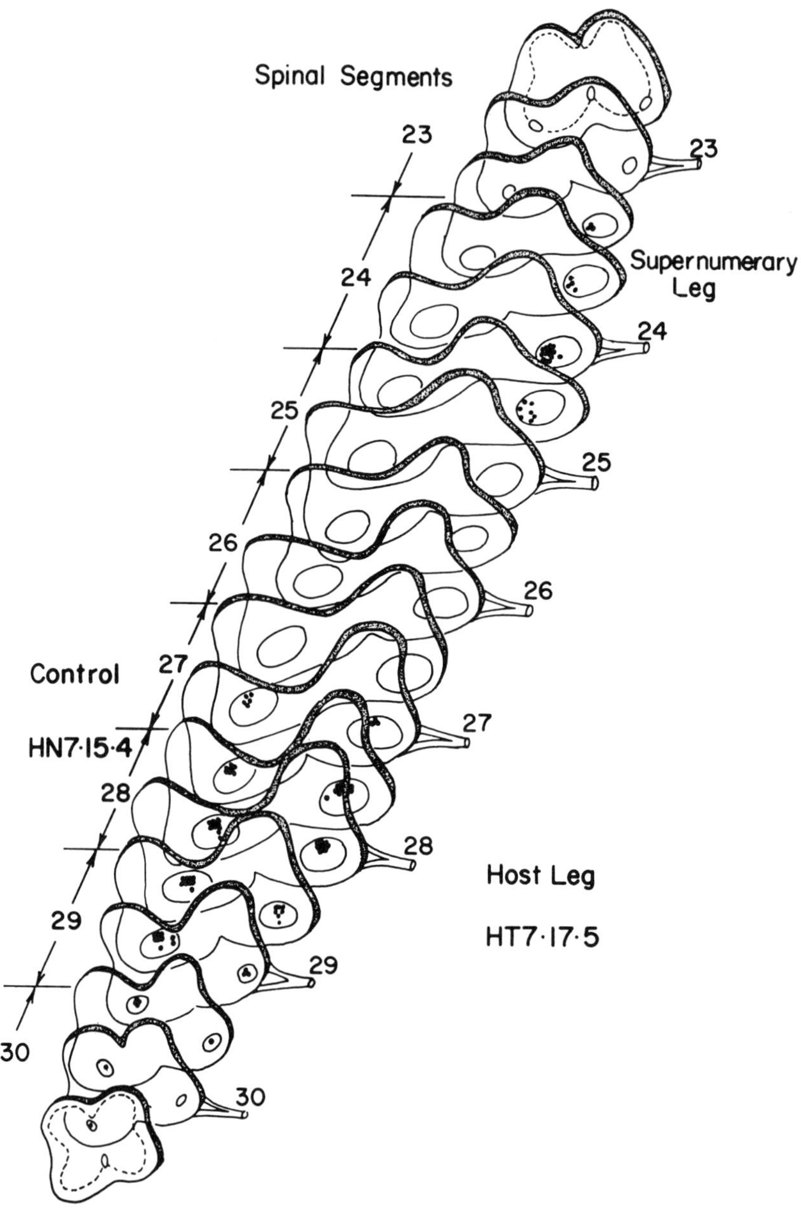

FIG. 1. HRP-labelled motor pools (dark dots) innervating the gastrocnemius of the host leg and of the transplanted (supernumerary) leg, respectively. Note similar medial position of the motor pools and gap between them (from HOLLYDAY, HAMBURGER, FARRIS, in "PNAS", 74, 3582, 1977).

role in the sense that motor pools tagged for a flexor preferentially seek out another flexor rather than an extensor or abductor. Or topographic position of muscles with reference to a joint could be involved. None of these possibilities are borne out by the data. However, one consistent relation was established: the mismatching is related to the above-mentioned embryonic origin of the target muscles. A derivative of the ventral muscle mass is always connected with an inappropriate motor pool which normally innervates another derivative of the ventral muscle mass; and the same holds for derivatives of the dorsal muscle mass. The validity of this rule is reinforced by the finding that it holds also for wing transplants placed in front of the host leg (Hollyday, 1978, 1979). An autoradiographic study had shown that medially located motor neurons are born earlier than laterally located motor neurons (Hollyday and Hamburger, 1977), hence selective mismatching may also be linked with the time of outgrowth of axons.

Since the axon bundle from a mismatched motor pool is not distributed randomly to all derivatives of the dorsal or ventral muscle mass, respectively, but ends up always in the same transplant muscle, this rule cannot cover the actual situation. An additional, more discriminating selection process must be in operation.

So far, the new results of Hollyday have referred to target specificity: they should now be related to pathway specificity. In her material, the pathways in the transplants were remarkably normal with respect to entrance points into the limbs and intralimb patterns. This is in agreement with the above-mentioned early findings of Harrison and many subsequent studies. However, occasionally nerves grew to their targets along atypical pathways.

To accommodate all data, the original chemoaffinity hypothesis requires an extension, in the same direction as suggested above, that is, in terms of a hierarchical order of specifications. The first choice which motor growth cones emerging from the plexures would be confronted with would be biochemical cues of a rather general nature, provided by the dorsal and ventral muscle mass, respectively. The next branching point would be determined by slight differentials in each of the two original cues, resulting from the segregation of the two muscle masses into four. Eventually each muscle would provide a rather specific cue. By the time a fiber bundle reaches its matching muscle primordium many non-matching fibers have already been split off; hence, the information content near the terminal branching would not have to be great. This consideration and the phenomenon of selective mismatching suggest that the biochemical

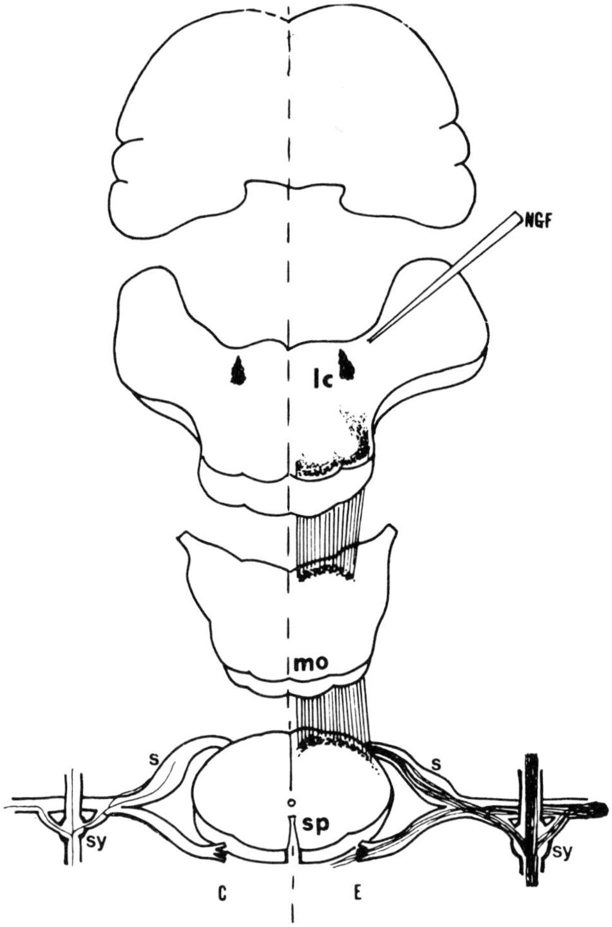

FIG. 2. Injection of NGF near *locus coeruleus* (lc). Note hyperplasia of sympathetic ganglion (sy) and fiber tract in the dorsal funiculus (from CHEN and others, in « Arch. Ital. Biol. », *116, 53,* 1978)

differentials which operate at branching points might be only quantitative, or partially shared molecular configurations. The pathfinding is facilitated 1) by the early invasion of fibers into the limb bud; the growth cones are already in the mesenchyme when it condenses into the 2 primary muscle masses; 2) the short distances between branching points; 3) the creation of mechanical tracks (preneural pathways) during the process of differentiation of limb tissues, as for instance along the interfaces of pre-muscle and pre-cartilage condensations.

Implicit in this model is a modified chemo- or neurotropic hypothesis in which "attraction at a distance" is replaced by guidance along biochemically tagged mechanical pathways. This notion finds strong support in recent experiments of Levi-Montalcini (1976; Chen and others, 1978). NGF was injected into the medulla of neonatal mice and rats near the *locus coeruleus.* NGF reached the paravertebral sympathetic ganglia, including the stellate and superior cervical ganglia, probably by diffusion along the dorsal and ventral roots. The sympathetic ganglia showed hyperplasia, and massive bundles of sympathetic fibers entered the spinal cord by way of dorsal roots. Within the cord, the fibers travelled to the injection site along the dorso-lateral funiculus, as demonstrated by histo-fluorescence (Fig. 2). Apparently, they followed an NGF gradient. However, they did not establish synapses and got lost near the injection site. Obviously, in this experiment, NGF provides directional or tropic cues to the sympathetic fibers, in addition to its role as a trophic agent. Levi-Montalcini rejects the notion of "action at a distance" and states: "the axonal tip of the fibers moves along gradients of diffusion of trophic and tropic factors released by end organs". C.-O. Jacobson has coined the term "bound gradient" in another context (1976). It can well be applied here.

The authors suggest that the trophic-tropic double role of target structures could serve as a general model of nerve pathfinding in embryos. The mode of outgrowth of embryonic motor fibers, as outlined above, can be harmonized readily with this model. We know that muscle tissue provides a trophic agent which is required for the maintenance of motor neuroblasts (Hamburger, 1958). And in the discussion above we have presented arguments in favor of the assumption that the differentiating muscle masses provide also chemical cues of different degrees of specificity which guide the growth cones to their destination. As a matter of parsimony it would be expedient, here again, to assign both trophic and tropic functions to the same agent. What is called for now is the discovery of the MGF.

Acknowledgements

Research done in the laboratory of the author has been supported by grants from the Nat. Inst. of Health (Public Health Service) U.S.A. and a grant from the Muscular Dystrophy Association to Washington University Medical School. I wish to thank Dr. M. Hollyday for making unpublished data available and Mrs. Irma Morose for typing the manuscript.

BIBLIOGRAPHY

ADELMANN H., *The problem of cyclopia*, in "Quart. Rev. Biol.", *11*, 161-182, *ibid.* 284-304 (1936).

BARONDES S. H. (ed.), *Neuronal Recognition*, Plenum Press, New York and London (1976).

BEKOFF A., *Ontogeny of leg motor output in the chick embryo: a neural analysis*, in "Brain Res.", *106*, 271-291 (1976).

BEKOFF A., STEIN P. S. G. and HAMBURGER V., *Coordinated motor output in the hindlimb of the 7-day chick embryo*, in "Proc. Natl. Acad. Sci. USA", *72*, 1245-1248 (1975).

CHEN M. G. M., CHEN J. S. and LEVI-MONTALCINI R., *Sympathetic nerve fiber ingrowth in the central nervous system of neonatal rodents upon intracerebral NGF injections*, in "Arch. Ital. Biol.", *116*, 53-84 (1978).

CRICK F. H. C. and LAWRENCE P. A., *Compartments and polyclones in insect development*, in "Science", *189*, 340-347 (1975).

DETWILER S. R., *Experiments on the transplantation of the spinal cord in Amblystoma, and their bearing upon the stimuli involved in the differentiation of nerve cells*, in "J. Exp. Zool.", *37*, 339-393 (1923).

FERGUSON B., *Effect of dorsal-ventral limb rotations on the development of motor connections*, in "Soc. for Neurosci.", Abstracts, *4*, 111 (1978).

GOTTLIEB G. (ed.), *Neural and Behavioral Specificity*. Studies on the Development of Behavior and the Nervous System, vol. 3, Academic Press, New York, San Francisco, London (1976).

HAMBURGER V., *Regression versus peripheral control of differentiation in motor hypoplasia*, in "Am. J. Anat.", *102*, 365-410 (1958).

HARRISON R. G., *Experiments in transplanting limbs and their bearing upon the problems of the development of nerves*, in "J. Exp. Zool.", *4*, 239-281 (1907).

HOLLYDAY M., *Target selectivity of motor pools in chick embryos*, in "Soc. for Neurosci.", Abstracts, *4*, 115 (1978).

HOLLYDAY M., *Motor neuron pools and their birth-dates for muscles of chick legs* (in press), (1979).

HOLLYDAY M., HAMBURGER V. and FARRIS J. M. G., *Localization of motor neuron pools supplying identified muscles in normal and supernumerary legs of chick embryo*, in "Proc. Natl. Acad. Sci. USA", *74*, 3582-3586 (1977).

HOLLYDAY M. and HAMBURGER V., *An autoradiographic study of the formation of the lateral motor column in the chick embryo*, in "Brain Res.", *132*, 197-208 (1977).

HOLTZER H., *Reconstitution of the model spinal cord following unilateral ablation*, in "J. Exp. Zool.", *117*, 523-558.

JACOBSON C. O., *Motor nuclei, cranial nerve roots, and fibre pattern in the medulla oblongata after reversal experiments on the neural plate of axolotl larvae. I. Bilateral operations*, in "Zool. Bidr. från Uppsala", *36*, 73-160 (1964).

JACOBSON C. O., *Same. II. Unilateral operations*, in "Zoon", *4*, 87-100 (1976).

JACOBSON M., *Developmental Neurobiology*, 2nd Edition, Plenum Press, New York and London (1978).

LANDMESSER L., *The development of motor projection patterns in the chick hindlimb*, in "J. Physiol.", *284*, 391-414 (1978).

LANDMESSER L. and MORRIS D. G., *The development of functional innervation in the hindlimb of the chick embryo*, in "J. Physiol.", *249*, 301-326 (1975).

LEDOUARIN N. M., RENARD D., TEILLET M. A. and LEDOUARIN G. H., *Cholinergic differentiation of presumptive adrenergic neuroblasts in interspecific chimaeras after heterotopic transplantation*, in "Proc. Nat. Acad. Sci. USA", *72*, 728-732 (1975).

LEVI-MONTALCINI R., *The nerve growth factor: its role in growth, differentiation and function of the sympathetic adrenergic neuron*, in "Progr. in Brain Res.", 45, 235-258 (1976).

MEYER R. L. and SPERRY R. W., *Retinotectal specificity: chemoaffinity theory*, in "Studies on the Development of Behavior and the Nervous System" (ed. G. Gottlieb), 3, 111-149 (1976).

NARAYANAN C. H. and HAMBURGER V., *Motility in chick embryo with substitution of lumbo-sacral by brachial and brachial by lumbosacral spinal cord segments*, in "J. Exp. Zool.", 178, 415-432 (1971).

NODEN D., *Interactions directing the migration and cytodifferentiation of avian neural crest cells*, in "Specificity of Embryological Interactions" (ed.) Dr. R. Garrod (Chapman and Hall, London), pp. 3-49 (1978).

PATTERSON P. H., *Environmental determination of autonomic neurotransmitter functions*, in "Ann. Rev. Neurosci.", 1, 1-17 (1978).

RAMÒN Y CAJAL S., *A quelle époque apparaissent les expansions des cellules nerveuses de la moelle épinière du poulet?*, in "Anat. Anzeiger", 5, 609-631 (1890).

ROMER A., *The development of the thigh musculature of the chick*, in "J. Morph.", 43, 347-385 (1927).

SMITH J., COCHARD P. and LEDOUARIN N. M., *Development of choline acetyltransferase and cholinesterase activities in enteric ganglia derived from presumptive adrenergic and cholinergic levels of the neural crest*, in "Cell Different.", 6, 199-216 (1977).

SPERRY R. W., *Chemoaffinity in the orderly growth of nerve fiber patterns and connections*, in "Proc. Natl. Acad. Sci. USA", 50, 703-710 (1963).

SPITZER J. L. and SPITZER N. C., *Time of origin of Rohon-Beard neurons in spinal cord of Xenopus laevis*, in "Am. Zool.", 15, 575 (abstr.) (1975).

STEFANELLI A., *I fenomeni della determinazione, della regolazione e del differenziamento del sistema nervoso*, in "Atti della Accad. Naz. dei Lincei", ser. 8, vol. 1, 27-114 (1947).

STEFANELLI A., *The Mauthnerian apparatus in the Ichthyopsida: its nature and function and correlated problems of neurohistogenesis*, in "Quart. Rev. Biol.", 26, 17-34 (1951).

VARGAS-LIZARDI and LYSER H. M., *Time of origin of Mauthners neuron in Xenopus laevis embryos*, in "Dev. Biol.", 38, 220-228 (1974).

VARON S. S. and BUNGE R. P., *Trophic mechanisms in the peripheral nervous system*, in "Ann. Rev. Neurosci.", 1, 327-361 (1978).

WENGER B. S., *Determination of structural patterns in the spinal cord of the chick embryo studied by transplantation between brachial and adjacent levels*, in "J. Exp. Zool.", 116, 123-146 (1951).

Neuroscience Commentaries
Vol. 1, No. 2, pp. 39–55
September 1982
Copyright © Society for Neuroscience

NATURALLY OCCURRING NEURONAL DEATH IN VERTEBRATES[1]

Viktor Hamburger* and Ronald W. Oppenheim†

** Department of Biology, Washington University, St. Louis, Missouri 63130*

and

† Division of Research, Department of Mental Health, Raleigh, North Carolina 27611, and Neurobiology Program, University of North Carolina, Chapel Hill, North Carolina 27514

1. Introduction and Historical Survey

Neuronal death during the development of the vertebrate nervous system is composed of a variety of categories, some of which are shared with other developing organ systems, while others are unique to the nervous system. Among the former is the disappearance of neural structures in the course of tail resorption in frog larvae at metamorphosis. At the same time the Mauthner neurons in the medulla, which are involved in the startle response during aquatic life, also regress or disappear. In early embryos of lower vertebrates one finds primitive sensory neurons in the dorsal part of the spinal cord, the Rohon-Beard cells. They undergo regression when the spinal ganglia take over their function (Hughes, 1957; Lamborghini, 1981). This is comparable to the disappearance of the mesonephros in mammalian development, which degenerates when the metanephros becomes functional. The term "phylogenetic death" has been applied to these phenomena (Glücksmann, 1951); but this designation with its implied inevitability should not distract one from the analysis of the proximate causes of these events. For instance, the thyroid hormone has been implicated in the disappearance of the Mauthner cells, but its actual role is still controversial (Kimmel and Model, 1978).

Another category, designated "morphogenetic cell death," occurs during tissue differentiation; it contributes to the morphogenesis of structures. Källen (1955) has described a zone of necrosis during the formation of the optic stalk and choroid fissure which he called the "suboptic death center" (see also Silver, 1978). This phenomenon reminds one of the so-called posterior necrotic zone which is found in early stages of limb formation in the chick embryo and in the formation of digits by the necrosis of interdigital tissue in the hands and feet of many terrestrial vertebrates (Saunders, 1966).

Finally, the continuous shedding and replacement of epithelial cells which occurs in the epidermis and parts of the lining of the intestine has its parallel in the nervous system in the turnover of the olfactory epithelium (Graziadei and Graziadei, 1978). Perhaps the small amount of sporadic death of individual cells which occurs at a very low rate in the developing nervous system as in all other embryonic tissues also deserves mentioning.

We shall not deal with these categories, because very few analytical data are available. Instead, we shall focus on a category of neuronal death which is unique for neural tissue and enjoys at present a great deal of attention. It is commonly referred to as "naturally occurring neuronal death" and is characterized by three distinct aspects: it consists of a substantial loss of embryonic neurons amounting to as much as 40–70% of a population; it is most clearly seen in distinctive units such as ganglia, motor columns or brain nuclei; and in each unit the degeneration process occurs during a relatively well-defined period. In other words, it displays a clear space-time pattern and is an important factor in the regulation of the population size of neuronal units.

It is of some interest to retrace the circuitous path that led to the discovery of this phenomenon. Neuroembryology developed as a branch of experimental embryology. One of the key issues in the latter was the role of embryonic tissue interactions, as in embryonic induction, in the determination and differentiation of structures. The origin of the nervous system is a prime example: the neural plate is induced by the subjacent mesodermal layer during a very early process of development, called gastrulation (Spemann, 1938). The subsequent development of the nervous system offered a wide field for the study of embryonic interactions, though they turned out to be different from the interactions involved in embryonic inductions. The early experimental embryologists also received repeatedly stimulating suggestions from the teratological literature. For instance, it had been reported that the congenital absence of an arm (abrachia)

---

[1] We would like to thank Peter Clarke, Lanny Haverkamp and Jerome L. Maderdrut for helpful discussion and comments.

was accompanied by a reduction of the brachial motor column and the dorsal root (sensory) ganglia. The causal relationship in this correlation is not immediately obvious, but the experimental embryologist has the tools to analyze it. It was precisely this issue which motivated the embryologist F. R. Lillie to suggest to his student, M. Shorey, the experiment of wing-bud extirpation in chick embryos. Her research (Shorey, 1909) demonstrated clearly that the *hypoplasia* of the embryonic nerve centers results from the reduction of their target area.

Detwiler (1920) confirmed the sensory hypoplasia following limb-bud extirpation in the salamander, *Ambystoma*, and discovered that the enlargement of the target area by transplantation of a limb bud to the flank of an early embryo resulted in the *hyperplasia* of the thoracic ganglia which innervated the transplant. Hamburger, in his limb extirpation and transplantation experiments on chick embryos (Hamburger, 1934, 1939) advanced the analysis of target-related hypo- and hyperplasia. But since regression and cell death had no place in the repertory of mechanisms studied by experimental embryologists, Detwiler and Hamburger interpreted their results in terms of an interference with ongoing differentiation. It was assumed that the target area somehow regulates proliferation and/or the recruitment of neurons from a (hypothetical) pool of undifferentiated precursor cells. The idea that hypoplasia might be due to cell degeneration did occur to Shorey, but she dismissed it: "It is impossible to demonstrate degenerating nerve cells or nerve fibers, and the defects must therefore be due to the failure of these to develop" (Shorey, 1909, p. 62).

Levi-Montalcini and Levi (1942, 1944), being less encumbered by an experimental embryological bias, were the first to interpret hypoplasia correctly in terms of neuron loss. They found that, following leg-bud extirpation in chick embryos, the number of differentiated neurons in a lumbar spinal ganglion declined between 6 and 19 days of incubation. But it took a third repetition of the Shorey experiment to substantiate this claim by the detection of actually degenerating neurons, this time in the brachial ganglia of 5- to 7-day embryos following wing-bud extirpation (Hamburger and Levi-Montalcini, 1949). Thus prepared for the unexpected, Levi-Montalcini discovered an equally severe degeneration process going on simultaneously in the cervical and thoracic ganglia of the same embryos, although these ganglia had not been affected by the operation. The fortunate coincidence of this manifestation of natural and experimentally induced neuron death, side by side, and during the same time span between 5 and $7\frac{1}{2}$ days, suggested immediately that the two phenomena were closely related. Moreover, the further observation that the degeneration period coincided approximately with the arrival of the axons from the ganglia at their target area gave the first clue to the explanation of natural neuronal death, namely "that in early stages cervical and thoracic VL cells send out more neurites than the periphery can support. [The term VL cells refers to the early-differentiating large neurons in a ventrolateral position in the ganglion.] The excess of

neurons would break down at the stage at which the VL cells are highly susceptible to environmental conditions" (Hamburger and Levi-Montalcini, 1949, p. 495). It was not until the discovery of the nerve growth factor (NGF) shortly thereafter, however, that the true nature of this relation was recognized. The "environmental conditions" were identified as a *trophic* agent supplied by the target area. Consequently, the neuronal death was attributed to unsuccessful competition for a trophic agent; however, it has only been in the last few years that the specific trophic role of NGF in the regulation of natural neuronal death has begun to be elucidated. Thus, our understanding of trophic and other mechanisms that may be involved in neuronal death, as exemplified in the discussion that follows, is still far from complete.

Despite the fact that natural neuronal death was first systematically described in the dorsal root ganglion, the subsequent discovery of NGF, and its role in the development of the sympathetic ganglia and dorsal root ganglia, quickly diverted attention from cell death to the hyperplasia in these structures. At about the same period, Levi-Montalcini (1950) also discovered a population of cells in the ventral horn of the cervical spinal cord of the chick in which massive cell death occurred; this phenomenon was tentatively thought to reflect an example of *phylogenetic* cell death. The belief that a massive, natural loss of motoneurons occurred only in the cervical spinal cord, and was apparently of the phylogenetic type, was partly responsible for the lack of a sustained interest in cell death in the chick embryo. Only following the work of Hughes (1968, review) and Prestige (1967) on motoneuron death in frog spinal cord did a gradual reawakening of interest in natural neuronal death occur (Cowan, 1973; Oppenheim, 1981a).

Since that time, it has been found that natural neuronal death occurs widely not only in spinal and sympathetic ganglia and motor neurons which have peripheral nonneural targets, but also in interneuronal units within the central nervous system. Further analyses have been primarily limited to neuronal centers and ganglia with a manageable population size, and we do not know whether units with very large populations, such as small interneurons in the cortex or spinal cord, also undergo a significant numerical depletion.

In contrast to the situation found in some invertebrates (see the accompanying article by Truman), it seems unlikely that the kind of neuronal death we are considering is genetically programmed. The basic assumption underlying the following discussion is that, at least in those populations in which cell death has been clearly demonstrated, neurons are capable of independent initial differentiation, but sooner or later become dependent on a trophic or maintenance factor which is supplied in limited amounts; hence, there is a competition resulting in the death of the losers. In other words, neuronal death is considered to be *probabilistic* or *stochastic* in nature and due to extrinsically imposed restrictions, rather than to intrinsic or genetic programming.

Nevertheless, we do not deny the possibility that, at

some stage in their development, differences become manifest among a population of neurons within a unit which may render some neurons more susceptible to degeneration than others. Such differences may even have a genetic basis. The fundamental difference between the genetically programmed and probabilistic models of cell death, however, is that whatever genetic differences may exist in the latter, they would find their expression in differential capabilities for competition.

In the discussion that follows, we shall present four examples in which the analysis of natural neuronal death has been fairly well advanced, and then we shall attempt to generalize and expand our understanding from these examples.

2. Analysis of Four Cases of Natural Neuronal Death

A. Spinal somatic motoneurons

As first reported by Hamburger (1975), and later replicated by others (Chu-Wang and Oppenheim, 1978a, b; Laing and Prestige, 1978), cell death among the lumbar motoneurons in the lateral motor column of the chick embryo extends from embryonic day 5 (E5) to E12 and involves a loss of 40–50% of the original neuronal population (Figure 1). This numerical depletion of lateral motor column neurons is accompanied by the appearance of a significant number of dying neurons, and the degeneration process itself has been characterized in detail at the ultrastructural level (Chu-Wang and Oppenheim, 1978a, b).

The onset of cell death in this population occurs subsequent to the cessation of proliferation (Hollyday and Hamburger, 1977) and overlaps little, if at all, with the migration of postmitotic neurons to the lateral motor column. Moreover, since the motoneurons initiate differentiation prior to the onset of cell death, all of the major early events of neurogenesis are independent of the target area. Since this is not the case for all neuronal populations, the chick spinal motoneurons are especially favorable for the analysis of neuronal death. Taken collectively, these various lines of evidence show unequivocally that there is a massive natural neuronal death of spinal motoneurons in the lumbar lateral motor column of the chick.

A similar loss has also been shown to occur in the brachial and thoracic motoneurons of the chick (Oppenheim and Majors-Willard, 1978; Maderdrut and Oppenheim, 1982) as well as among the electric organ motoneurons in fish (Fox and Richardson, 1981) and in frog (Prestige, 1967), mouse (Harris-Flanagan, 1969; Lance-Jones, 1982) and rat (Nurcombe et al., 1981) spinal motoneurons. It seems likely that massive motoneuron death is a common vertebrate characteristic.

Do the spinal motoneurons that will eventually die differ somehow in their ability to differentiate normally prior to the onset of frank degeneration? Unlike the situation with the mesenchymal cells in the posterior necrotic zone of the limb, neuronal death is not restricted to a spatially distinct subpopulation of motoneurons; rather, the cells that die appear to be randomly inter-

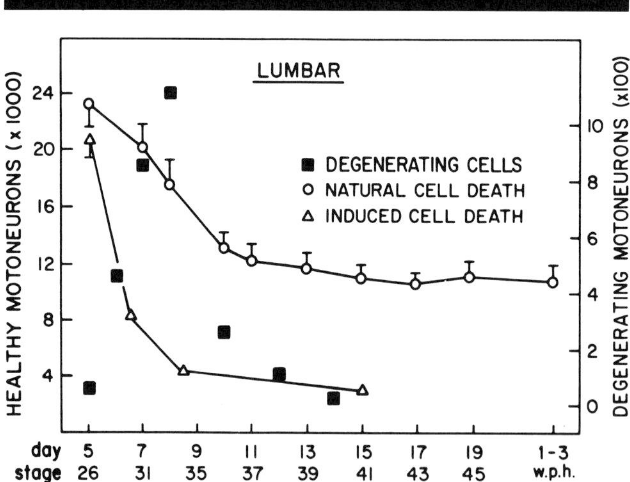

Figure 1. Cell number in the lateral motor column of the lumbar spinal cord in the chick. All healthy and degenerating motoneurons were counted according to previously described criteria (see Hamburger, 1975; Chu-Wang and Oppenheim, 1978a). Induced cell death reflects the effect of unilateral limb-bud removal on E2. Degenerating cell values were obtained from the same *unoperated* embryos as were used for counts of healthy cells. (Data from Hamburger, 1975; Oppenheim et al., 1978; Oppenheim, unpublished data.)

mingled with those that survive. Consequently, the differentiation of those cells which will undergo neuronal death cannot be easily studied independently of those which will survive; however, an analysis of cell size and other aspects of differentiation among a random sample of lateral motor column cells in normal embryos has failed to reveal a subpopulation which displays signs of impending degeneration. Furthermore, an analysis of differentiation of motoneurons prior to the onset of cell death (E5–6) in embryos in which, owing to unilateral limb-bud removal, virtually all of the peripherally deprived cells will eventually die, has also failed to reveal any obvious differences between the deprived motoneurons and those in normal embryos (Hamburger, 1958; Oppenheim et al., 1978).

In both the normal and the peripherally-deprived situations, all motoneurons have sent axons into the ventral root or beyond. Moreover, massive injections of horseradish peroxidase into the proximal part of the hindlimb of normal embryos on E5—i.e., prior to cell death—results in the labeling of approximately 95% of the motoneurons in the related segments of the lumbar spinal cord (Oppenheim and Chu-Wang, 1977; Chu-Wang and Oppenheim, 1978b). Similar results have been reported in other populations of neurons such as the chick isthmo-optic nucleus (Clarke and Cowan, 1976) and mouse (Lance-Jones, 1982) and rat (Nurcombe et al., 1981) spinal motoneurons. These data show that virtually all of the neurons in these populations send axons to their targets prior to neuronal death. Not only does this provide important evidence for the normal differentiation capacity of neurons which will eventually die, but it also implies that the critical factors involved in the competition process occur at the target site.

The results of the limb-bud removal experiments are also consistent with the important role ascribed to the target, since in this situation, as predicted, motoneuron death is increased from the normal 40–50% to 90% or more (Figure 1). An even more critical observation is that an experimentally induced *increase* in the size of the target by transplantation of a supernumerary hindlimb reduces the natural death of lumbar motoneurons from 40–50% to 25–35%. A greater reduction might have been expected, but because of physical constraints, a majority of lumbar motoneurons were unable to send their axons into the supernumerary leg (Hollyday and Hamburger, 1976). Before we can conclude that motoneuron death is exclusively regulated by competitive interactions at the target, however, the role of afferents must be considered. We know from other experiments that presynaptic neurons have a trophic influence on postsynaptic neurons (see below).

There is natural neuronal death of primary afferent neurons in the chick dorsal root ganglia, and this partially overlaps temporally with the period of motoneuron death (see below); limb removal also induces an increased cell loss in the dorsal root ganglia (Hamburger and Levi-Montalcini, 1949). Thus, it seems plausible that both natural as well as induced motoneuron death could

be controlled, in part, by primary afferents; descending intrinsic afferents from spinal and supraspinal regions could conceivably also be involved. Recent experiments, in which both intrinsic and extrinsic afferents were surgically removed prior to the onset of cell death in chick spinal motoneurons, show that natural neuronal death is not affected; deafferented embryos have normal numbers of motoneurons on E10–12, that is, after the cessation of natural cell death (Okado and Oppenheim, 1981). Clearly, then, the death of motoneurons is regulated by a competitive interaction at the target, and not by events acting centrally via afferents (also see Wenger, 1950). As we discuss below, however, afferents may be important in controlling the natural death of other neuronal units.

During the stages in which spinal motoneurons are dying in the chick, the limb muscles are in the process of segregating from the dorsal and ventral muscle masses. The incipient muscles at this time are composed of myoblasts, myotubes and other undifferentiated cells which are in clusters of 3–10 cells surrounded by a common basement membrane. Neuromuscular contacts are present, albeit few in number, especially during the early stages of cell death. Although a small amount of motoneuronal death occurs prior to the onset of neuromuscular function (as determined by the presence of limb movements), there is nonetheless a reasonably close correlation between the onset of limb motility and cell death.

A recent quantitative comparison of muscle cells in the anterior and posterior latissimus dorsi (ALD and PLD) muscles of the chick wing with the number of brachial motoneurons innervating these two muscles has shown the following: prior to the onset of cell death, when numbers of motoneurons are at their peak, the ALD and PLD muscles contain fewer than 10% of their mature complement of muscle cells (Chu-Wang and Oppenheim, 1982). Numbers of muscle cells reach adult values only at the end of the period of cell death (i.e., on E13–15). Thus, there is a rather striking quantitative mismatch between the number of motoneurons and the number of muscle cells during the cell death period. Since there is evidence that all of the motoneurons in the ALD and PLD motor pools have sent axons to these muscles at the time of neuronal death (Chu-Wang and Oppenheim, 1982), these data imply that the natural degeneration of motoneurons is closely associated with the availability of muscle cells at the target. At present this relationship is only correlative in nature: it remains to be seen whether this mismatch implies particular causal mechanisms of cell death.

The observation that the natural death of spinal motoneurons coincides with the onset of neuromuscular function in the limbs led to an examination of the role of synaptic and/or muscular activity in this process. Experiments in which neuromuscular activity was chronically blocked during the entire cell death period (i.e., from E5 to E10) by pharmacological agents has shown that motoneuronal death can be reduced or prevented entirely by this treatment (Pittman and Oppenheim, 1978, 1979;

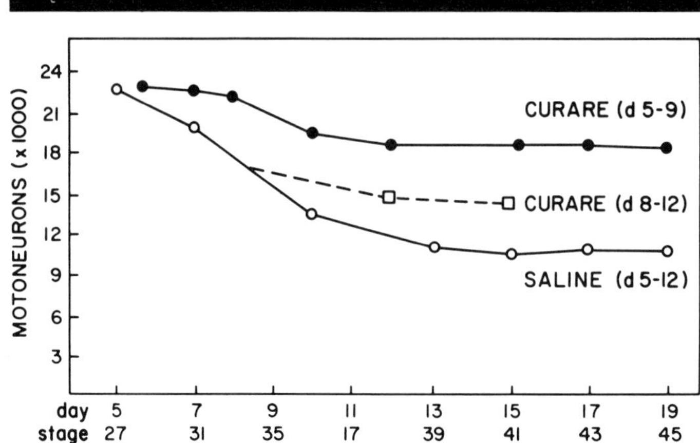

Figure 2. Cell number in the lateral motor column of the embryonic chick lumbar spinal cord. Two mg of curare in 0.2 ml saline were administered daily from either day 5 to day 9 or day 8 to day 12. Saline (0.2 ml) was used as a control. Cell number was evaluated as in Figure 1. (Data from Pittman and Oppenheim, 1979; Oppenheim, unpublished data). The E19 value for the curare d 5-9 group is retained for at least 4 days after hatching (Oppenheim, 1982).

Laing and Prestige, 1978). As long as neuromuscular activity is blocked, motoneurons fail to degenerate (Figure 2). Following the cessation of the blockade and the resumption of neuromuscular activity in the embryo, cell death ensues. Recently, however, it has been possible to study four chicks treated with curare from E5 to E9 which hatched and survived for 4 days. Despite the occurrence of considerable limb activity after hatching, the "excess" motoneurons were maintained (cf. Figure 2). Thus, after hatching, neuromuscular blockade no longer appears necessary for maintaining the "excess" motoneurons (Oppenheim, 1982). Beginning the blockade at any time prior to the cessation of natural neuronal death prevents the degeneration of cells that have not yet died. The reduction of cell death can be induced by either pre- or postsynaptic (both competitive and depolarizing) blockade of neuromuscular transmission (Pittman and Oppenheim, 1978, 1979; Oppenheim and Maderdrut, 1981).

The neuromuscular blockade does not affect cell number in either sensory (dorsal root) or sympathetic ganglia but rather appears to affect only cholinergic motoneurons in the spinal cord and brainstem (e.g., Creazzo and Sohal, 1979). Furthermore, *increased* muscular activity induced by chronic electrical stimulation of the chick hindlimb *in ovo* between E6 and E8 *increases* the number of dying motoneurons in the lateral motor column without any apparent effect on sensory cells (Oppenheim and Nunez, 1982). This result is consistent with our contention that the survival of embryonic motoneurons is related to muscular activity.

The neuromuscular blockade appears to induce an increased arborization of peripheral axons as well as sprouting of axonal terminals within the muscle (Chu-Wang and Oppenheim, 1982). Individual muscle fibers are multiply innervated, and each neuromuscular endplate is contacted by a greater-than-normal number of axon terminals. Muscle mass is greatly reduced following neuromuscular blockade. Virtually all of this reduction is due to the loss of muscle fibers; surviving muscle fibers appear to differentiate rather normally, although fiber size is more variable. Since more motoneurons are maintained despite the reduced muscle mass, we assume that reduced neuromuscular activity is associated with an increased supply of trophic agent.

The observation that a chronic neuromuscular blockade results in the prevention of natural motoneuronal death was an unexpected result; its underlying cellular mechanisms are not yet well understood. We have suggested that the level and/or distribution of acetylcholine receptors (AChR) in the muscle membrane is critically involved; muscular inactivity would alter muscle membrane properties (e.g., induce more AChR) so that a single muscle fiber is able to support a greater number of axonal terminals than it normally does, thereby increasing the likelihood of neuronal survival. Although it is conceivable that AChR's are *the* trophic agent for motoneurons, it seems more likely that the AChR merely defines the site at which the putative agent is taken up or released. Whatever the result of future analysis will be, we wish to underscore our present belief that the inhibition of neuromuscular function *per se* is not the primary cause of neuronal survival. Rather, normal synaptic and/or muscular activity, although of obvious importance, is merely a link in the chain of events regulating the availability of a target-derived trophic agent.

In summary, the present evidence on naturally occurring death of motoneurons in the chick is consistent with the probabilistic model of neuronal death and supports the competition hypothesis, according to which cell death is related to the limited availability of a putative trophic agent supplied by the targets of spinal motoneurons. Although there are data from other systems which appear inconsistent with the competition hypothesis (e.g., Lamb, 1980), recent reviews of this literature have concluded that, as a working hypothesis, a model emphasizing the role of competition is still valid for interpreting virtually all known cases of motoneuronal death (Oppenheim, 1981a; Purves, 1980).

B. Ciliary ganglion

We owe practically all our present knowledge about the development of the ciliary ganglion in the chick—especially aspects of cell death—to an elegant series of studies by Landmesser and Pilar (e.g., Landmesser and Pilar, 1978, review; Pilar et al., 1980). Consequently, the present synopsis is based almost entirely on their observations.

The ciliary ganglion in the chick is composed of two structurally and functionally distinct populations of cholinergic, parasympathetic neurons, the *choroid* neurons which innervate the smooth vascular musculature in the choroid coat of the eye, and the *ciliary* neurons which innervate the striated iris and ciliary muscles. All cells in this ganglion are derived from cranial neural crest from which cells migrate early to form a distinct ganglion by stage 9 (1.3 days of incubation). Cell proliferation continues up to stage 23–24 (3–4 days) and is complete by stage 25 (Landmesser and Pilar, 1978), at which time the ganglion contains approximately 6,000 neurons. This population remains numerically uniform until stage

34 (8 days). Depletion of neurons begins at that time and lasts until stage 40 (14 days), when only 3,000 neurons remain (Figure 3). Thus, there is a 50% loss of ganglion cells owing to natural neuronal death; loss of cells in both the ciliary and choroid populations occurs at the same time, and both populations are affected by about the same amount.

The ciliary ganglion appears to follow the same developmental rules as the spinal motoneurons: proliferation and migration are independent of the target; substantial initial differentiation of the neurons occurs prior to cell death and is not dependent upon influences from either the ciliary afferents (i.e., the accessory oculomotor nucleus, AON) or the ciliary targets in the eye; the numerical depletion is accompanied by neuronal and axonal degeneration; all neurons extend axons to the target prior to the onset of cell death at E8 (stage 34); the onset of cell death coincides closely with the establishment of functional peripheral connections; and removal of the ganglionic targets prior to cell death results in the loss of virtually all neurons (ciliary and choroid) in the ganglion (Figure 3). Additionally, it has been established that preganglionic axons from the AON form synapses which are structurally and functionally normal and that 100% of the ganglion cells transmit impulses prior to the onset of cell death.

We have pointed out above that chronic blockade of synaptic transmission from motoneurons to their targets greatly reduces naturally occurring motoneuron death. Although the same relationship may hold true for the chick ciliary ganglion, at present, no information is available on the effects of chronic neuromuscular blockade on cell survival in this population.

Observations on cell death in the ciliary ganglion are

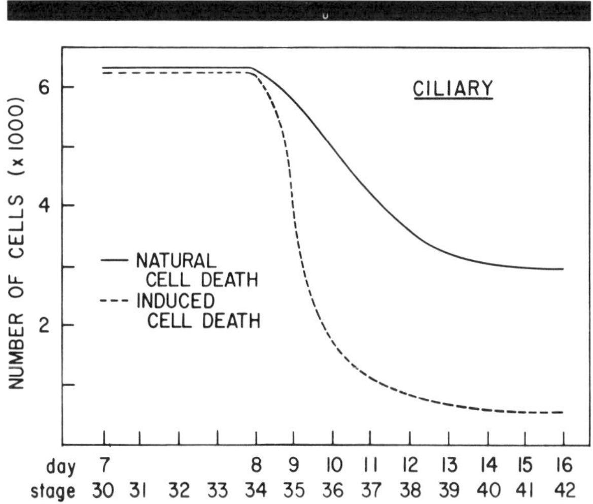

Figure 3. The number of neurons in the ciliary ganglion of control embryos (natural cell death) and in embryos following eye removal on E2-3 (induced cell death). (From Landmesser and Pilar, 1974, Figure 2.)

entirely consistent with the competition hypothesis. Particularly supportive is the evidence that innervation occurs prior to cell death and that target removal enhances, whereas target enlargement reduces (Narayanan and Narayanan, 1978), natural neuronal death. Furthermore, a recent experiment by Pilar, Landmesser and Burstein (1980) has provided substantial new support for the role of competition in ganglionic cell death. Cutting two of the three nerve branches that contain axons from only the *ciliary* neurons results in the axotomy-induced cell death of two-thirds of these neurons. By doing the surgery at stages 32–34 just prior to the normal cell death period, those investigators expected that competition among the remaining cells for the normal-sized target would be reduced. The results supported this expectation in that a substantial number of the remaining neurons that would have died were rescued (i.e., cell loss was reduced from 69% to 42%). Moreover, neurons supplying the intact branch expanded their peripheral innervation territory from 160° to 285° of the circular ciliary muscle and in several cases included the iris constrictor muscle, which they normally never innervate.

Finally, a novel outcome of this experiment was that the axons in the intact branch developed more rapidly than control axons, as assessed by axon diameter, conduction velocity, and degree of myelination. A similar effect on axonal maturation has been observed in spinal motoneurons following the reduction of cell death by neuromuscular blockade (Chu-Wang and Oppenheim, 1982). The implication in both of these cases is that competition normally retards the rate of neuronal (axonal) maturation. In both the lateral motor column and ciliary ganglion *cell size* was unaffected by the reduced competition.

C. Death of interneurons

The occurrence of natural death in interneuronal systems is well established (Oppenheim, 1981a). We shall discuss briefly two instances, both of which demonstrate that the same rules for competition at the target area hold if the targets are neurons rather than nonneural structures. This implies that neurons likewise produce trophic maintenance factors. Evidence for this notion has come from denervated muscles, sensory receptors and neurons which regress and may eventually die in some instances (Jacobsen, 1978, review), and from the observation that amphibian limb regeneration depends on nerve supply (Singer, 1959).

We consider first the comprehensive investigation of neuronal death in *retinal ganglion neurons* of the chick embryo by Rager (1978, 1980) and Rager and Rager (1978). Counts of axons in the optic nerve in thin sections have shown a maximum of 4 million axons at day 10 to 11 and a dramatic decline to 2.4 million by day 18, the most intense loss occurring between days 9 to 15. Since the optic axons form no collaterals, these data also represent neuronal death in the retinal ganglion cells, and degenerating perikarya have been observed. In ac-

cordance with the previous examples, the period of neuronal degeneration coincides rather precisely with the time of arrival and terminal arborization of the axons at the optic tectum, and it ends when all fibers have reached their targets. Furthermore, the arrival of axons at different regions of the tectum follows a distinct pattern, spreading from the center to the periphery (Crossland et al., 1974), and the wave of retinal degeneration follows the same pattern. On the basis of these and other data, such as increase of fiber diameter (which is correlated with spread of terminal arborization), the following hypothesis has been formulated: each axon terminal develops an arborization that occupies a certain extent of the tectum ("termination domain"). Furthermore, a certain minimum number of target contact sites is required for the survival of the neurons. Neuronal death results from an unsuccessful competition for adequate arborization space and terminal sites. The end result would be a quantitative matching of the two systems, retina and tectum. While further observations will be required, particularly with respect to the terminal structures, the available data support the competition hypothesis.

A second example is neuronal death in the *isthmo-optic nucleus* (ION) of the chick embryo. It is located at the medial edge of the optic tectum and projects to the contralateral retina, where it primarily synapses with amacrine cells. The studies of Cowan and Wenger (1968), Clarke and Cowan (1976) and Clarke et al. (1976) give the following picture: the nucleus is assembled at day 8 as a solid mass, and by day 11 it has attained its full numerical complement of approximately 22,000 cells. Between days 13–17, it suffers a loss of 60% and thus attains its definitive population of 10,000 neurons. During the same period, it is transformed into a convoluted laminar structure. Again, the clearly defined period of degeneration coincides in all probability with the formation of synapses by the isthmo-optic tract in the retina. This assumption is supported by several observations: injections of horseradish peroxidase into the eye at day 11, that is, before the onset of degeneration in the isthmo-optic nucleus, label virtually all contralateral ION neurons. Furthermore, removal of the optic vesicle on E2 results in the complete disappearance of the contralateral nucleus, and the degeneration process extends over the same period as naturally occurring death.

D. Dorsal root ganglia

A reinvestigation of neuronal death in the dorsal root ganglia (DRG) of the chick embryo (Hamburger et al., 1981) has given results that reflect interesting differences between these units and the three units discussed above. In the spinal ganglia of chick embryos one finds two populations of neurons, the early-differentiating, larger neurons in a ventrolateral (VL) position and the smaller neurons in a dorsomedial (DM) position, which differentiate later. In both brachial and thoracic ganglia, the two populations have clearly defined separate degeneration periods, each lasting for several days. The peaks of degeneration are at E5.5 (stage 27) for thoracic VL cells and somewhat later for brachial VL cells, and at about E8.5 (stages 34–35) for the DM populations. Considerable indirect evidence suggests that both populations of cells are dependent upon NGF for their survival and differentiation. Treatment with exogenous NGF during the major period of cell death greatly reduces the normal loss of both VL and DM cells (Hamburger et al., 1981).

The target areas of the two types of neuron are not known precisely, but it is clear that both populations contain proprioceptive and exteroceptive neurons (Honig, 1977). Silver-impregnated tissue shows that the VL cells which differentiate earliest are exteroceptive. An important difference between these neurons and the previously discussed spinal and ciliary motoneurons is that degeneration begins at a remarkably early stage of their differentiation at E4.5 (stage 25). Between E4.5 and E7, when the bulk of the VL cells die, they are immature bipolar neuroblasts that have not yet reached the more advanced pseudo-unipolar stage of differentiation (Pannese, 1976). At the same time, active neuronal proliferation goes on, and we find proliferating and degenerating cells side by side. According to Carr and Simpson (1978), proliferation in the VL cells continues to E6.5. Silver preparations show that bundles of sensory axons have reached the dermis at E4.5, and they enlarge in subsequent days. Moreover, VL cells which have sent axons to the skin of the hindlimb can be retrogradely labeled with HRP beginning at about E5 (stage 26–27) (Oppenheim and Heaton, 1975; Honig, 1977; Scott, 1981); DM cells can be first labeled at slightly later stages.

It is obvious that the time span between birth and death in those VL cells which die very early is quite short, and one may doubt that all fibers even manage to traverse the short distance between the ganglion and skin before their degeneration begins. This contrasts with the spinal motoneurons and ciliary ganglion cells, which connect with their targets and become functional before the inception of the degeneration process. While in these two systems—and in many others as well—*all* axons are in competition with each other, in the case of the *earliest* degenerating VL neurons of the dorsal root ganglia, the situation is such that at any given time probably only a limited number of axons compete with each other. Another difference is that the axons of spinal and ciliary neurons terminate at rather specific synaptic sites, whereas the distal processes of the exteroceptive neurons develop rather extensive arborizations in the dermis. Admittedly, however, there is much less information available concerning the details of peripheral innervation, the development of afferents to the central terminals of DRG cells, the onset of afferent (sensory) function as well as what defines a target for the DRG cells, compared to spinal and ciliary neurons.

These observations have led to the following considerations: if one adheres to the basic tenet that natural neuronal death results from a deficiency in an extraneously supplied trophic agent, then we encounter here a case of either very short supply of the trophic agent, or a pronounced dependency on this agent which manifests itself at very early stages of differentiation, or both. In other words, the capacity of the sensory neurons for self-sustenance is greatly inferior to that of, for instance, embryonic spinal or ciliary neurons. As a corollary, the emphasis is shifted from competition at the target area, and particularly from competition for synaptic contact sites among *all* axons of the population, to competition and survival under highly restrictive trophic conditions in their immediate surroundings. We have even suggested that some DRG neurons may die in their infancy, before or while they begin to sprout their axons. This suggestion is supported by the finding of Carr and Simpson (1982) that, following a single injection of [^{3}H]thymidine at E5.5, that is, at the peak of the VL degeneration period, and after sacrifice 2 hr later, up to 14% of labeled cells are undergoing degeneration. This implies that proliferating neuron precursors already can fall victim to trophic deprivation. We don't wish to imply that all cell death in the DRG is of this type. It is conceivable, for instance, that the death of DRG neurons destined to contact muscle spindles may obey rules more comparable to those controlling spinal motoneurons. In this regard, it is of interest that natural neuronal death in the *mouse* DRG occurs in two distinct waves: the first, which peaks at E13, occurs before the processes of DRG neurons have reached their targets, and the second, which peaks at E19–20, is coincident with target innervation (Scaravilli and Duchen, 1980). Furthermore, it seems likely that some small proportion of spinal motoneurons may also die very early and for reasons other than competition for synaptic sites at the target (e.g., Chu-Wang and Oppenheim, 1982; Oppenheim et al., 1978; Laing, personal communication).

In summary, the case of the DRG puts the issue of natural neuronal death in its proper perspective. It reminds us that the common denominator is the dependency of embryonic neurons on extraneous maintenance or growth factors, and that units like the somatic motoneurons and ciliary ganglion cells, with their conspicuous axon competition at the target, are perhaps special cases. In other words, competition for an agent which is in limited supply is the basic issue, and in a number of systems, the target area happens to be the primary arena where the struggle for survival is normally fought; other cases may be discovered, however, in which the competition for an agent supplied by *afferents* is preponderant in cell survival (e.g., Parks, 1979).

3. General Considerations

A. Range

A recent review of neuronal death (Oppenheim, 1981a) contains a list of the neural regions and cell types in which this phenomenon is known to occur. These include such a wide variety of cells, involving every major neuronal type (e.g., brain and spinal cord neurons; motor, sensory, and autonomic neurons; central and peripheral neurons; early- and late-proliferating neurons; neurons with both long and short axonal projections; and neurons with only one source of afferents as well as neurons with many sources), that there can be no question that this is a remarkably widespread and seemingly indiscriminate phenomenon. Notwithstanding this pervasiveness, however, in light of a few recent reports in which a numerical depletion in some neural units has not been observed (Armstrong and Clarke, 1979; Oppenheim, 1981a), we do not wish to imply that natural neuronal death is ubiquitous. For reasons that may turn out to be important, but which are presently obscure, some neural units may remain numerically constant following proliferation and migration.

B. Extent

The numerical magnitude of natural neuronal death exhibited by most systems for which quantitative data are available ranges from 15% or less in the avian cochlear nucleus magnocellularis (Rubel et al., 1976) to over 70% in frog spinal motoneurons (Prestige, 1967), 75% in the chick mesencephalic V nucleus (Rogers and Cowan, 1973) and 84% in the chick cochlear nucleus laminaris (Rubel et al., 1976). On the average, however, approximately one-half of the original population in a neural unit is lost. Virtually nothing is known of why some neural units exhibit normal death to a considerably greater or lesser extent than the average of about 50%.

C. Normal and experimentally induced death

The onset of neuronal death is generally coincident with target innervation—and in some cases with the onset of synaptogenesis and physiological function. In all cases that have been studied the natural neuronal death can be exaggerated by surgical removal of the target prior to cell death; and, the cell loss in both the natural and induced situations occurs over the same restricted time span, implying a common mechanism. Since it has been shown in several cases that neurons project axons to their targets prior to cell death, it is generally believed that in these cases it is some aspect of the axon interaction with the target that is of primary significance; the reduced cell death which occurs in several systems following target enlargement supports this contention.

We can deduce from all available evidence that the common denominator for normally occurring and experimentally induced death is the trophic dependence of embryonic neurons on the target. The former reflects a limited supply and the latter the complete absence of a trophic agent. As discussed in the following section, however, this is not meant to exclude the possibility that trophic agents supplied by afferents may also play a contributory role in the natural death of some populations.

D. Transneuronal degeneration

We have mentioned the double aspect of trophic relations in neurogenesis: while neurons become dependent on extraneous trophic agents, they supply trophic support to muscles, sense organs and other neurons on which they synapse. We shall deal briefly with the last-mentioned aspect, in which natural neuronal death in one population may have secondary effects on the population of their target cells. We refer to this situation as *anterograde transneuronal degeneration*. When the affected cells are presynaptic neurons we speak of *retrograde transneuronal degeneration* (Cowan, 1970). While little information is available concerning such secondary effects subsequent to *normal* neuronal death, both antero- and retrograde transneuronal degeneration are well documented in numerous *experimental* situations, and we can extrapolate from these to the possibility of such occurrences in normal neurogenesis. The multiple effects of the extirpation of the optic vesicle in chick embryos may serve as an illustration.

We have seen that as a result of this operation the great majority of ciliary cells undergoes degeneration. Subsequently, virtually all cells of the accessory oculomotor nucleus (AON) which synapse on ciliary neurons disappear (Cowan and Wenger, 1968). This is an example of retrograde transneuronal degeneration. By inference one can assume that subsequent to the naturally occurring neuronal death in the ciliary ganglion, which amounts to 50%, a corresponding number of accessory oculomotor cells will die. Conversely, the addition of a supernumerary optic primordium reduces the cell loss in both the ciliary ganglion and the AON (Narayanan and Narayanan, 1978).

Another consequence of eye extirpation in the chick is the nearly complete disappearance of the isthmo-optic nucleus, the nucleus of origin of centrifugal fibers to the retina. While there is no doubt that this effect is due primarily to the loss of its target area, the amacrine cells, another factor may be considered. This nucleus, which is located in the midbrain, receives afferent input from the tecto-isthmal tract, whose cells of origin are located in the deeper layers of the tectum. Since the tectum itself receives much of its input from the retina, eye extirpation also results in serious degeneration in several layers of the tectum (Kelly and Cowan, 1972). This is an example of anterograde transneuronal degeneration. If this includes the loss of the cells of origin of the tecto-isthmal tract, then the neurons of the isthmo-optic nucleus could be affected by secondary anterograde degeneration in the presynaptic fibers of the tecto-isthmal tract. In fact, partial lesions of the optic tectum result in localized defects in the isthmo-optic nucleus (Clarke et al., 1976). While details have not been worked out, it seems likely that eye extirpation puts the isthmo-optic nucleus in double jeopardy: a loss of retrograde trophic influence via the isthmo-optic tract and the loss of an anterograde influence via the tecto-isthmal tract. In other words, degeneration of its presynaptic input could contribute to the total loss of the isthmo-optic nucleus. Again by inference, the normal neuronal death in the ION, which, as mentioned, amounts to 60%, could result from a combination of competitive loss at the target area and loss of presynaptic input; since a normal cell loss occurs in both the retina (Rager and Rager, 1978; Hughes and McLoon, 1979) and optic tectum (Cantino and Sisto Daneo, 1972) of the chick, this would seem to be a likely possibility.

The reverse of the retrograde transneuronal degeneration, that is, retrograde transneuronal *rescue* of cells, has been achieved by treatment of chick embryos with NGF between E3 and E9. It was found that NGF produces a dose-dependent reduction of natural neuronal death in the thoraco-lumbar paravertebral sympathetic ganglia, thus increasing the target areas of the sympathetic preganglionic neurons in the column of Terni (CT). As a result, there was a dose-dependent reduction of cell death in the CT (Oppenheim et al., 1982).

E. Neuronal death and error correction

One of the intuitively most appealing suggestions for the adaptive value of neuronal death is that it serves to remove errors in the pattern of connections between neurons and their targets. Although some developmental errors of this sort undoubtedly occur (see Jacobson, 1978; Landmesser, 1980; Lamb, 1981, for examples), their magnitude is not sufficiently great to account for the massive cell loss that occurs in many neural units.

In two of the best-studied cases, the chick and frog hindlimb motoneurons, it has been shown that virtually all motoneurons, even those that will die, send axons to the limb and that the vast majority of these projections are correct with reference to the adult situation. That is, prior to cell death, each muscle is innervated by motoneuron pools that are located in the same position in the spinal cord as in the adult animal (Lamb, 1976; Landmesser, 1980; Hollyday, 1980). (For a somewhat different view on this issue see Pettigrew et al., 1979.) Furthermore, in the chick, experimentally induced inappropriate connections between motoneurons and muscles are maintained throughout the normal cell death period even when they are in competition with the appropriate connections in the same muscle (Lance-Jones and Landmesser, 1981). Finally, the prevention of virtually all motoneuron death in the chick spinal cord by neuromuscular blockade does not alter the position of motoneuron pools for specific fore- or hindlimb muscles (Oppenheim, 1981b). Consequently, in this case one must also conclude that the projection pattern of motoneuron pools was generally correct at the time that neuromuscular blockade was initiated on E5, that is, before the cell death period.

It has been suggested that neurons may die not because of a simple mismatch between the location of their soma and their target, but rather because of a mismatch between their target and presynaptic inputs (e.g., Lewis, 1980). According to this view, if the inputs to a motoneuron are not appropriate for producing functionally adaptive contractions of its target muscle, then it will be at a competitive disadvantage and is more likely to die. An argument against this possibility, however, is the observation that the nearly total removal of both intrinsic and extrinsic afferents to spinal motoneurons does not

alter natural neuronal death (Wenger, 1950; Hamburger et al., 1966; Okado and Oppenheim, 1981). It is also difficult to explain cell death in the ciliary ganglion in this way since the two populations of ciliary neurons, the ciliary and choroid, both receive highly specific and appropriate presynaptic inputs prior to cell death, yet 50% of the cells in each population subsequently die (Landmesser and Pilar, 1972).

Another way in which cell death might be involved as a means of error correction would be if neurons were somehow specified to form connections in specific regions *within* a grossly appropriate target. For instance, cells within a motoneuron pool might project topographically onto specific regions or fiber types of their target muscle (Burke et al., 1977; Swett et al., 1970), or neurons within the central nervous system might project to specific cell types or neuronal regions, such as dendrites versus soma, within their normal target. Although this is an interesting notion, whether significant errors of this type occur, and if so, whether they are eliminated by cell death remains to be substantiated.

A final case that needs to be discussed in this context is the isthmo-optic nucleus in the chick. The beautiful studies of Clarke and Cowan (1976) and Clarke et al. (1976) on the isthmo-optic nucleus are perhaps the most widely, and often inappropriately, cited evidence in support of the significant role of cell death as an error correction mechanism. (We might add that the authors, themselves, are more cautious in their conclusions on this issue.) We have already mentioned that 60% of the neurons within the normal confines of this nucleus die despite having projected axons to their normal target region, the contralateral retina. Prior to cell death, about 10–20% of all isthmo-optic neurons are located outside of, or ectopic to, the normal confines of the nucleus and another 1% or so send axons to the incorrect, ipsilateral eye. Originally it was reported that virtually all of the ectopic and ipsilateral cells degenerated. However, by the use of more sensitive retrograde tracer techniques it has recently been shown that a substantial number of these cells survive the normal period of cell death and are even retained in the adult chicken (Hayes and Webster, 1981; O'Leary and Cowan, 1981). Thus, of the almost 14,500 ION neurons that undergo natural death, little if any of this loss can be directly attributed to error correction, at least errors of the type being considered here.

In conclusion, in those few systems that have been examined in some detail, error correction appears to be either nonexistent or only minimally associated with natural neuronal death. Many more cases need to be examined, however, before any general conclusions can be reached on this issue.

F. Evolutionary significance of excess neuroblasts

This topic has been the major focus of an interesting article by Katz and Lasek (1978). The authors point out that the nervous system is organized in the form of sets of "matching populations"; that is, a proportional relation exists between a given neuron population and its target. If a single genetic change occurs only in one partner, which disturbs the match, then ontogenetic buffer mechanisms must operate to restore the match. Otherwise, the mutation is not adaptive.

The overproduction of neurons, which can be viewed as such a unilateral genetic change, has often been considered as a "safety factor" to guarantee the appropriate numerical innervation of the target. The authors go a step further and envisage in the overproduction a potential for evolutionary change. For instance, in the case of the neuromuscular system, the excess motoneurons can be considered as a reservoir that would permit a considerable enlargement of the musculature without the necessity of a concomitant genetic change in the nervous system.

From this point of view, Katz and Lasek suggest that embryological manipulations can be considered as experimental tests for evolutionary speculations. For instance, the transplantation of a supernumerary limb, resulting in the utilization of some of the excess motoneurons (Hollyday and Hamburger, 1976), would mimic, so to speak, an evolutionary enlargement at the target; partial limb removal, which results in an enhanced loss of motoneurons, would be the other side of the coin. Another example that can illuminate evolutionary considerations is an early experiment by Twitty (1932). One eye primordium of a small salamander species, *Ambystoma punctatum*, was replaced by an eye primordium of the larger species, *A. tigrinum*. In later larval stages, the population of cells in the optic tectum had increased to match the larger retinal cell population. However, in this instance, it is not known whether altered cell death is the mechanism by which the adjustment is achieved.

In a more general vein, neuronal death may only be the most extreme case of what is actually a continuum of regressive, neurogenetic events that have both ontogenetic and phylogenetic significance. Cellular overproduction, exuberant growth of axonal and dendritic processes and excessive synapse formation may reflect only the most obvious cases in which regression acts to mold the final product (for recent discussions of this view see Innocenti, 1982; Oppenheim, 1981c; Purves and Lichtman, 1980).

G. Embryonic neuronal death and neuropathology

It follows from the preceding section that if ontogenetic buffer mechanisms are defective and therefore incapable of neutralizing certain genetic changes in one partner of a matched population, then maladaptive, but viable, pathological conditions might result. It has, in fact, been suggested that certain kinds of neuromuscular disorders might involve a primary alteration in motoneuron numbers. Preliminary data from the mutant mouse Wobbler (*wr/wr*), a model for infantile spinal muscular atrophy in the human, indicated that there is a reduced number of spinal motoneurons (50%) on E18, which is near the end of the normal cell death period for mouse

spinal motoneurons (Hanson and Strominger, 1980). Similarly, the mutant mouse Sprawling (*Swl*), in which there is a significant numerical depletion of neurons in the DRG, has a marked increase in the number of pyknotic DRG nuclei during the normal period of cell death in this population, that is, between E11 and E20, implying that altered cell death is involved (Scaravilli and Duchen, 1980).

By contrast, the situation in the case of the avian model of forelimb muscular dystrophy is presently confusing in that one author reports *fewer* brachial motoneurons (Murphy, 1977), one reports *more* brachial motoneurons (Susheela et al., 1980), and our own studies have found no differences in the number of brachial motoneurons at any time from E6 to 3 weeks posthatching (Oppenheim, Rose, and Stokes, 1982). The reason for these discrepancies is not known.

The human disorder familial dysautonomia is a genetic disease which primarily affects individuals of Ashkenazic Jewish descent. This disease is manifest at birth and primarily involves deficits in sensory and sympathetic ganglia (Breakefield, 1981). It is pertinent that the number of neurons in the sympathetic ganglia and the intermediolateral column (i.e., the sympathetic preganglionic neurons) is reduced in patients with this disorder (Pearson and Pytel, 1978). Because of the parallel between these findings and those from mice treated with the antiserum to β-nerve growth factor (Levi-Montalcini, 1972), it was suggested that nerve growth factor might somehow be altered in these patients (Pearson et al., 1974). In fact, the protein from patients has only about 10% of the biological activity of the normal protein (Schwartz and Breakefield, 1980). Recent evidence implicating nerve growth factor as a trophic agent directly involved in the survival of embryonic sensory (Hamburger et al., 1981; Gorin and Johnson, 1980a, b) and sympathetic neurons (Oppenheim et al., 1982) and, indirectly, in the survival of sympathetic preganglionic neurons (Oppenheim et al., 1982) raises the possibility that altered cell death may be one of the effects of the nerve growth factor abnormality which characterizes this disease.

Although the evidence is still scanty, the possibility that altered natural neuronal death is a factor in various neuropathological disorders cannot be ignored. Genetic defects in the cell death process could alter cell numbers either directly or transneuronally (e.g., Oppenheim et al., 1982), or both, with the possibility of long-term repercussions on neuronal function. Environmental factors such as drugs given to pregnant females could also affect neuronal numbers by altering the normal cell death process. Conceivably, even benign neurobehavioral differences between "normal" individuals could, in some instances, be associated with differences in neuron numbers resulting from subtle genetic or environmentally induced alterations in neuronal death.

4. Concluding Remarks

The phenomenon of natural neuronal death is an integral part of a broader neuroembryological problem for which one can use the term *"systems-matching"* that has been coined by Gaze and Keating (1972) in a similar context. What is required is a quantitative matching of neuron populations with target areas: these can be synaptic sites on muscles or on other neurons, or arborizations. This need not be a one-to-one relation; however, it is assumed that a fixed quantitative relationship of some sort between the two systems is required for optimal functional activity.

How is systems-matching achieved? We have demonstrated in all our examples that there is an initial overproduction in the neuronal populations supplying the targets, followed by an elimination process. In other words, systems-matching is achieved in a two-step process. Although the two aspects are intricately interwoven in the attainment of the end result, nevertheless overproduction and elimination of neurons are regulated by two entirely different and independent developmental mechanisms. This point will be elaborated in the hope that it may lead to a better understanding of the curious indirect method by which systems-matching is accomplished.

We turn first to the question: how does the overproduction come about? In confronting this question, we should realize a crucial difficulty inherent in systems-matching: that as a rule the two systems which have to be matched develop topographically as separate entities, usually at some distance from each other; and they undergo critical initial steps of development, including proliferation, independently of each other. The important point is that, whatever the factors are which regulate proliferation of neuronal units, the target area is not involved. This has been demonstrated in several instances and it may be a general rule. A clear example is the way in which the lateral motor columns in the spinal cord of chick embryos are assembled. These neurons are derived from the ventral neuroepithelium (basal plate) in the brachial and lumbar segments of the neural tube. They are produced during a restricted period by a precisely controlled number of mitotic cycles. This we conclude from the fact that there is relatively little individual variation in the maximal (pre-elimination) population size (Hamburger, 1975). The proliferation process is independent of the target area, i.e., the limbs, since the full numerical complement of the lateral motor column can be formed in the absence of the limb buds (Hamburger, 1958; Oppenheim et al., 1978).

The same independence can be demonstrated for the proliferation process in the optic tectum of the chick embryo, which is unaffected by the early removal of the optic vesicle (Cowan et al., 1968), and it probably holds for other systems. The few claims that the target actually does influence the mitotic activity in neuronal units are probably not valid. In the case of a decrease of mitotic figures in chick spinal ganglia following wing extirpation, and increase following the addition of a transplanted wing (Hamburger and Levi-Montalcini, 1949), it is not possible to distinguish between proliferating neurons and glia precursors. Kollros (1953) has found that mitotic activity in the optic tectum of frog embryos is reduced

following extirpation of the optic vesicle. However, Currie and Cowan (1974), who repeated the experiment, presented good evidence that it is the production of glia that is affected by the absence of optic fibers (but see Kollros, 1982).

All available data indicate that the factors which regulate motoneuron proliferation reside in the neuroepithelium. In the case of ganglia, proliferation also seems to be regulated by intrinsic factors, although we have no direct proof for this contention. This implies that an understanding of the mechanisms that regulate initial cell numbers must be based on information concerning these intrinsic regulatory factors. In addition, such information must account for the very high rate of overproduction. It is not immediately obvious why there should be a redundancy of 50%, if it were primarily a matter of providing a safety factor to guarantee adequate nerve supply to the target.

Let us pursue this thought more explicitly, using again the formation of the lateral motor column as an example. Several factors determine the final maximum (pre-elimination) population size of the lateral motor column: the initial number of (hypothetical) founder or progenitor cells located in a segment of the ventral neural epithelium (basal plate), and the number of mitotic cycles which they undergo. The latter could be programmed within the founder cells, or else it is determined by a factor in their immediate environment which turns off the cycling. The production process is regulated very precisely, since, as we have seen, the maximum population number shows a remarkable constancy with little individual variation. The rigorous control of the production process must have evolved independently of, and without concern for, the actual later demand at the target. Let us assume for the sake of argument that at a given period there are 10,000 precursor cells, each with the capacity for another mitotic cycle, and that the requirement eventually will be for 13,000 motor neurons. If proliferation stops at 10,000, then the target would be undersupplied, and natural selection would eliminate such individuals with malfunctioning muscles. If, on the other hand, entering another cycle were an all-or-none proposition for all 10,000 cells, then the overproduction would be inevitable, and, indeed, the closest approach to fulfilling the actual demand. We are aware that this model is unrealistic, since the assumption of a synchronized simultaneous mitotic cycle of all 10,000 cells is improbable. The purpose of the model is to demonstrate that there is a way of understanding the high overproduction rate of 50% strictly in terms of developmental mechanisms intrinsic to the neural epithelium, without reference to the possible adaptive value of the redundancy. (The same explanation of the high rate of overproduction has been proposed independently by M. Katz and U. Grenander, personal communication.) Of course, once the excess numbers are produced, they are available for adaptive use, such as systems-matching and error elimination.

We have stated that systems-matching between neuron populations and their targets is achieved in two steps:

overproduction and elimination of unsuccessful neurons. We have shown that the first step is independent of the second, and we will show that the reverse is also true. The elimination process is predicated on two related premises. The first is that embryonic neurons can sustain themselves only for a limited period. Sooner or later, they become dependent on trophic factors which are provided by the targets. The second premise is that the trophic factors are available only in limited supply, and its corollary, that neurons are in competition for the agent. This notion is supported by strong indirect evidence, which may be summarized briefly again: the parallelism in the spatial-temporal pattern of normally occurring and experimentally induced death; the rescue of neurons which would have died, by experimental enlargement of the target; and the finding that normally occurring death in the dorsal root and sympathetic ganglia can be alleviated or completely prevented by supplementing the available trophic agent with exogenously introduced nerve growth factor.

The competition hypothesis has been challenged by Lamb (1979, 1980, 1981). In a recent experiment, he has reported that if in a frog embryo one limb bud is extirpated and its nerve supply is manipulated so as to grow into the contralateral intact limb, the total number of motor neurons in both lateral motor columns is significantly greater than the normally surviving number on one side. However, as Purves (1980) has pointed out, a simple linear one-to-one relation is not expected if neurons regulate the target property which they compete for (i.e., the production or release of the trophic factor). And, as mentioned previously, the prevention of motoneuron cell death by neuromuscular blockade in which a greatly reduced muscle mass is able to maintain many more neurons than normally, shows that experimental findings which at first sight seem to contradict the competition hypothesis can be easily accommodated in it (see discussion pp. 42–43.

Another challenge to the competition hypothesis, namely, that neuronal death serves to eliminate heterospecific, that is, grossly inappropriate, connections has also been repudiated. At best, cell death may only remove small-scale errors and these probably cannot account for the massive elimination which is the rule. This means that competition occurs within homospecific populations, such as the neurons of an individual motor pool, or the ciliary neurons of the ciliary ganglion.

Finally, one can ask: competition for what? One finds in the literature occasional references to "competition for synaptic sites *or* for a trophic agent." The apparent need for alternatives is probably motivated by the fact that dorsal root ganglia which are also subject to neuronal death have no peripheral synaptic contacts, although they do have terminal contacts with receptor cells, in addition to free nerve endings. We can readily avoid the notion of alternatives by assuming that wherever we are dealing with synaptic or terminal contacts, they are the sites at which trophic agents are internalized. In other words, competition is not *for synaptic sites* but for trophic

substances released and taken up *at synaptic or terminal sites*. Justification for this assumption comes from well-known experiments on sympathetic neurons in which it was shown that the nerve growth factor is actually internalized at adrenergic terminals and transported retrogradely (Hendry et al., 1974; review in Thoenen and Barde, 1980). Although at present nerve growth factor remains the only well-characterized trophic agent implicated in neuronal survival *in vivo*, efforts are under way in many laboratories to discover similar trophic agents for other types of neurons.[2]

In summary, we arrive at the *unifying principle* that competition for a trophic agent is the overriding factor in explaining natural neuronal death. Our experience with the nerve growth factor has taught us that an agent can reach an embryonic neuron by retrograde axonal transport or by diffusion. Of our examples, the motoneurons, the ciliary ganglion and the retinotectal system belong to the first category, and the early differentiating VL cells of dorsal root ganglia seem to take up the growth factor along both routes. Other cases may exist in which agents delivered by anterograde axonal transport via afferents are important in regulating cell survival.

References

Armstrong, R. C., and P. G. H. Clarke (1979). Neuronal death and development of the pontine nuclei and inferior olive in the chick. Neuroscience *4*:1635–1647.

Bennett, M. R., K. Lai, and V. Nurcombe (1980). Identification of embryonic motoneurons in vitro: their survival is dependent on skeletal muscle. Brain Res. *190*:537–542.

Breakefield, X. O. (1981). Altered nerve growth factor in familial dysautonomia: discovering the molecular basis of an inherited neurologic disease. Neurosci. Commun. *1*:28–32.

Burke, R. E., P. L. Strick, C. Kanda, and B. Walmsley (1977). Anatomy of medial gastrocnemius and soleus motor nuclei in cat spinal cord. J. Neurophysiol. *40*:667–680.

Cantino, D., and L. Sisto-Daneo (1972). Cell death in the developing optic tectum. Brain Res. *38*:13–25.

Carr, V. M., and S. B. Simpson (1978). Proliferative and degenerative events in the early development of chick dorsal root ganglia. I. Normal development. J. Comp. Neurol. *182*:727–740.

Carr, V. M., and S. B. Simpson (1982). Rapid appearance of labeled degenerating cells in the dorsal root ganglia after exposure of chick embryos to tritiated thymidine. Dev. Brain Res. *2*:157–162.

---

[2] Crude extracts or partially purified factors or agents have recently been isolated from various target tissues and have been shown to influence the *in vitro* growth and survival of *ciliary neurons* (Nishi and Berg, 1981; Hendry and Hill, 1980), other *parasympathetic neurons* (Coughlin et al., 1981), *spinal motoneurons* (Godfrey et al., 1980; Smith and Appel, 1981; Bennett et al., 1980; Pollack, 1980), *trigeminal sensory neurons* (Chan and Haschke, 1981), and a population of non-NGF-sensitive *sympathetic neurons* (Edgar et al., 1981). Although it remains to be seen whether these factors are involved in the regulation of neuronal survival *in vivo*, future prospects are favorable for the discovery and characterization of a variety of other NGF-like trophic agents.

Chan, K. Y., and R. H. Haschke (1981). Action of a trophic factor(s) from rabbit corneal epithelial culture on dissociated trigeminal neurons. J. Neurosci. *1*:1155–1162.

Chu-Wang, I.-W., and R. W. Oppenheim (1978a). Cell death of motoneurons in the chick embryo spinal cord. I. A light and electron microscopic study of naturally-occurring and induced cell loss during development. J. Comp. Neurol. *177*:33–58.

Chu-Wang, I.-W., and R. W. Oppenheim (1978b). Cell death of motoneurons in the chick embryo spinal cord. II. A quantitative and qualitative analysis of degeneration in the ventral root, including evidence for axon outgrowth and limb innervation prior to cell death. J. Comp. Neurol. *177*:59–86.

Chu-Wang, I.-W., and R. W. Oppenheim (1982). Cell death of motoneurons in the chick embryo spinal cord. VII. Development of brachial neurons, peripheral nerves, and wing muscles in normal embryos and following chronic neuromuscular blockade. J. Neurosci., submitted.

Clarke, P. G. H., and W. M. Cowan (1976). The development of the isthmo-optic tract in the chick, with special reference to the occurrence and correction of developmental errors in the location and connection of isthmo-optic neurons. J. Comp. Neurol. *167*:143–163.

Clarke, P. G. H., L. A. Rogers, and W. M. Cowan (1976). The time of origin and the pattern of survival of neurons in the isthmo-optic nucleus of the chick. J. Comp. Neurol. *167*:125–141.

Coughlin, M. D., E. M. Bloom, and I. B. Black (1981). Characterization of a neuronal growth factor from mouse heart-cell-conditioned medium. Dev. Biol. *82*:56–68.

Cowan, W. M. (1970). Anterograde and retrograde transneuronal degeneration in the central and peripheral nervous system. In: *Contemporary Research Methods in Neuroanatomy*, W. J. Nauta and S. O. E. Ebbesson, eds. Springer-Verlag, New York, pp. 217–251.

Cowan, W. M. (1973). Neuronal death as a regulative mechanism in the control of cell number in the nervous system. In: *Development and Aging in the Nervous System*, M. Rockstein, ed. Academic Press, New York, pp. 19–41.

Cowan, W. M., A. H. Martin, and E. L. Wenger (1968). Mitotic patterns in the optic tectum of the chick during normal development and after early removal of the optic vesicle. J. Exp. Zool. *169*:71–92.

Cowan, W. M., and E. L. Wenger (1968). Degeneration in the nucleus of origin of the preganglionic fibers to the chick ciliary ganglion following early removal of the optic vesicle. J. Exp. Zool. *168*:102–124.

Creazzo, T. L., and G. S. Sohal (1979). Effects of chronic injections of α-bungarotoxin on embryonic cell death. Exp. Neurol. *66*:135–145.

Crossland, W. J., W. M. Cowan, L. A. Rogers, and J. P. Kelly (1974). The specification of the retino-tectal projection in the chick. J. Comp. Neurol. *155*:127–164.

Currie, J., and W. M. Cowan (1974). Some observations on the early development of the optic tectum in the frog (*Rana pipiens*), with special reference to the effects of early eye removal on mitotic activity in the larval tectum. J. Comp. Neurol. *156*:123–142.

Detwiler, S. R. (1920). On the hyperplasia of nerve centers resulting from excessive peripheral loading. Proc. Natl. Acad. Sci. USA *6*:96–101.

Edgar, D., Y.-A. Barde, and H. Thoenen (1981). Subpopulations of cultured chick sympathetic neurons differ in their requirements for survival factors. Nature *289*:294–295.

Fox, G. Q., and G. P. Richardson (1981). Cell death and proliferation in the electric lobes of *Torpedo mamorata*. Soc. Neurosci. Abstr., Vol. 7, p. 293.

Gaze, R. M., and M. J. Keating (1972). The visual system and "neuronal specificity." Nature *237*:375–378.

Glücksmann, A. (1951). Cell deaths in normal vertebrate ontogeny. Biol. Rev. *26*:59–86.

Godfrey, E. W., B. K. Schrier, and P. G. Nelson (1980). Source and target cell specificities of a conditioned medium factor that increases choline acetyltransferase activity in cultured spinal cord. Dev. Biol. *77*:403–418.

Gorin, P. D., and E. M. Johnson (1980a). Effects of exposure to nerve growth factor antibodies on the developing nervous system of the rat: an experimental approach. Dev. Biol. *80*:313–323.

Gorin, P. D., and E. M. Johnson (1980b). Effects of long-term nerve growth factor deprivation on the nervous system of the adult rat: an experimental autoimmune approach. Brain Res. *198*:27–42.

Graziadei, G. A. M., and P. P. C. Graziadei (1978). Continuous nerve cell renewal in the olfactory system. In: *Handbook of Sensory Physiology*, Vol. 9, M. Jacobson, ed. Springer-Verlag, New York, pp. 55–85.

Hamburger, V. (1934). The effects of wing bud extirpation on the development of the central nervous system in chick embryos. J. Exp. Zool. *68*:449–494.

Hamburger, V. (1939). Motor and sensory hyperplasia following limb-bud transplantations in chick embryos. Physiol. Zool. *12*:268–284.

Hamburger, V. (1958). Regression versus peripheral control of differentiation in motor hypoplasia. Am. J. Anat. *102*:365–410.

Hamburger, V. (1975). Cell death in the development of the lateral motor column of the chick embryo. J. Comp. Neurol. *160*:535–546.

Hamburger, V., J. K. Brunso-Bechtold, and J. Yip (1981). Neuronal death in the spinal ganglia of the chick embryo and its reduction by nerve growth factor. J. Neurosci. *1*:60–71.

Hamburger, V., and R. Levi-Montalcini (1949). Proliferation, differentiation and degeneration in the spinal ganglia of the chick embryo under normal and experimental conditions. J. Exp. Zool. *111*:457–502.

Hamburger, V., E. L. Wenger, and R. W. Oppenheim (1966). Motility in the chick embryo in the absence of sensory input. J. Exp. Zool. *162*:133–160.

Hanson, P. A., and N. L. Strominger (1980). Intrauterine motor neuron death in normal mouse and in the Wobbler mutant. Soc. Neurosci. Abstr., Vol. 6, p. 669.

Harris-Flanagan, A. (1969). Differentiation and degeneration in the motor horn of foetal mouse. J. Morphol. *129*:281–305.

Hayes, B. P., and K. E. Webster (1981). Neurones situated outside the isthmo-optic nucleus and projecting to the eye in adult birds. Neurosci. Lett. *26*:107–112.

Hendry, I. A., and C. E. Hill (1980). Retrograde axonal transport of target tissue-derived macromolecules. Nature *287*:647–649.

Hendry, I. A., K. Stockel, H. Thoenen, and L. L. Iversen (1974). The retrograde axonal transport of nerve growth factor. Brain Res. *68*:103–131.

Hollyday, M. (1980). Motoneuron histogenesis and the development of limb innervation. In: *Current Topics in Developmental Biology, Neural Development*, Vol. 15, R. K. Hunt, ed. Academic Press, New York, pp. 181–215.

Hollyday, M., and V. Hamburger (1976). Reduction of the naturally occurring motor neuron loss by enlargement of the periphery. J. Comp. Neurol. *170*:311–320.

Hollyday, M., and V. Hamburger (1977). An autoradiographic study of the formation of the lateral motor column in the chick embryo. Brain Res. *132*:197–208.

Honig, M. G. (1977). Outgrowth of cutaneous and proprioceptive neurons from chick embryo dorsal root ganglia. *Soc. Neurosci. Abstr.*, Vol. 3, p. 108.

Hughes, A. F. (1957). The development of the primary sensory system in *Xenopus laevis* (Daudin). J. Anat. *91*:323–338.

Hughes, A. F. (1968). *Aspects of Neural Development.* Logos, London.

Hughes, W. F., and S. C. McLoon (1979). Ganglionic cell death during normal retinal development in the chick: comparisons with cell death induced by early target field destruction. Exp. Neurol. *66*:587–601.

Innocenti, G. (1982). Transitory structures as a substrate for developmental plasticity of the brain. In: *Recovery from Brain Damage*, M. W. Van Hof and J. Mohn, eds. Elsevier, Amsterdam, in press.

Jacobson, M. (1978). *Developmental Neurobiology.* Plenum Press, New York.

Kallën, B. (1955). Cell degeneration during normal ontogenesis of the rabbit brain. J. Anat. *89*:153–162.

Katz, M. J., and R. J. Lasek (1978). Evolution of the nervous system: role of ontogenetic buffer mechanisms in the evolution of matching populations. Proc. Natl. Acad. Sci. USA *75*:1349–1352.

Kelly, J. P., and W. M. Cowan (1972). Studies on the development of the chick optic tectum III. Effects of early eye removal. Brain Res. *42*:263–288.

Kimmel, C. B., and P. Model (1978). Developmental studies of the Mauthner cell. In: *Neurobiology of the Mauthner Cell*, D. S. Faber and H. Korn, eds. Raven Press, New York, pp. 183–220.

Kollros, J. J. (1953). The development of the optic lobes in the frog. I. The effects of unilateral enucleation in embryonic stages. J. Exp. Zool. *123*:153–187.

Kollros, J. J. (1982). Peripheral control of midbrain mitotic activity in the frog. J. Comp. Neurol. *205*:171–178.

Laing, N., and M. Prestige (1978). Prevention of spontaneous motoneurone death in chick embryos. J. Physiol. (London) *282*:33–34P.

Lamb, A. H. (1976). The projection patterns of the ventral horn in the hindlimb during development. Dev. Biol. *52*:82–99.

Lamb, A. H. (1979). Ventral horn cell counts in *Xenopus* with naturally occurring supernumerary hindlimbs. J. Embryol. Exp. Morphol. *49*:13–16.

Lamb, A. H. (1980). Motoneurone counts in *Xenopus* frogs reared with one bilaterally-innervated hindlimb. Nature *284*:347–350.

Lamb, A. H. (1981). Target dependency of developing motoneurons in *Xenopus laevis*. J. Comp. Neurol. *203*:157–171.

Lamborghini, J. E. (1981). Kinetics of Rohon-Beard neuron disappearance in *Xenopus laevis*. *Soc. Neurosci. Abstr.*, Vol. 7, p. 291.

Lance-Jones, C. (1982). Motoneuron cell death in the developing lumbar spinal cord of the mouse. Dev. Brain Res., in press.

Lance-Jones, C., and L. Landmesser (1981). Pathway selection by embryonic chick motoneurons in an experimentally altered environment. Proc. R. Soc. London B Biol. Sci. *214*:19–52.

Landmesser, L. (1980). The generation of neuromuscular specificity. Annu. Rev. Neurosci. *3*:279–302.

Landmesser, L., and G. Pilar (1972). The onset and development of transmission in the chick ciliary ganglion. J. Physiol. (London) *222*:691–713.

Landmesser, L., and G. Pilar (1974). Synapse formation during embryogenesis on ganglion cells lacking a periphery. J. Physiol. (London) *241*:715–736.

Landmesser, L., and G. Pilar (1978). Interactions between neurons and their targets during in vivo synaptogenesis. Fed. Proc. *37*:2016–2022.

Levi-Montalcini, R. (1950). The origin and development of the visceral system in the spinal cord of the chick embryo. J. Morphol. *86*:253–283.

Levi-Montalcini, R. (1972). The morphological effects of immunosympathectomy. In: *Immunosympathectomy*, G. Steiner and E. Schonbaum, eds. Elsevier, New York, pp. 55–78.

Levi-Montalcini, R., and G. Levi (1942). Les conséquences de la destruction d'un territoire d'innervation périphérique sur le développement des centres nerveux correspondants dans l'embryon de poulet. Arch. Biol. *53*:537–545.

Levi-Montalcini, R., and G. Levi (1944). Correlazioni nello sviluppo tra varie parti del sistema nervoso. I. Consequenze della demolizione dell'abbozzo di un arto sui centri nervosi nell'embrione di pollo. Comment. Pontif. Acad. Sci. *8*:527–568.

Lewis, J. (1980). Death and the neurone. Nature *284*:305–306.

Maderdrut, J. L., and R. W. Oppenheim (1982). Reduction of naturally-occurring cell death in the sacral and thoracolumbar preganglionic cell column of the chick embryo following blockade of ganglionic transmission. *Soc. Neurosci. Abstr.*, Vol. 8, in press.

Murphy, B. J. (1977). An analysis of myogenic and neurogenic influences on the development of muscular dystrophy in the chick embryo. *Dissertation*, Louisiana State University Medical Center.

Narayanan, C. H., and Y. Narayanan (1978). Neuronal adjustments in developing nuclear centers of the chick embryo following transplantation of an additional optic primordium. J. Embryol. Exp. Morphol. *44*:57–70.

Nishi, R., and D. Berg (1981). Two components from eye tissue that differentially stimulate the growth and development of ciliary ganglion neurons in cell culture. J. Neurosci. *1*:505–513.

Nurcombe, V., P. A. McGrath, and M. R. Bennett (1981). Postnatal death of motor neurons during the development of the brachial spinal cord of the rat. Neurosci. Lett. *27*:249–254.

Okado, N., and R. W. Oppenheim (1981). Developmental changes in the lateral motor column of the chick embryo following either spinal transection or neural crest removal. *Soc. Neurosci. Abstr.*, Vol. 7, p. 291.

O'Leary, D. D. M., and W. M. Cowan (1981). Further observations on the development of the nucleus of origin of centrifugal fibers in the avian retina. *Soc. Neurosci. Abstr.*, Vol. 7, p. 293.

Oppenheim, R. W. (1981a). Neuronal cell death and some related regressive phenomena during neurogenesis: a selective historical review and progress report. In: *Studies in Developmental Neurobiology: Essays in Honor of Viktor Hamburger*, W. M. Cowan, ed. Oxford University Press, New York, pp. 74–133.

Oppenheim, R. W. (1981b). Cell death of motoneurons in the chick embryo spinal cord. V. Evidence on the role of cell death and neuromuscular function in the formation of specific peripheral connections. J. Neurosci. *1*:141–151.

Oppenheim, R. W. (1981c). Ontogenetic adaptations and retrogressive processes in the development of the nervous system and behavior: a neuroembryological perspective. In: *Maturation and Development: Biological and Psychological Perspectives*, K. L. Connolly and H. F. R. Prechtl, eds. J. B. Lippincott, Philadelphia, pp. 73–109.

Oppenheim, R. W. (1982). Reduction of neuronal death by embryonic neuromuscular blockade persists after hatching. *Soc. Neurosci. Abstr.*, Vol. 8, in press.

Oppenheim, R. W., and I.-W. Chu-Wang (1977). Spontaneous cell death of spinal motoneurons following peripheral innervation in the chick embryo. Brain Res. *125*:154–160.

Oppenheim, R. W., I.-W. Chu-Wang, and J. L. Maderdrut (1978). Cell death of motoneurons in the chick embryo spinal cord. III. The differentiation of motoneurons prior to their induced degeneration following limb-bud removal. J. Comp. Neurol. *177*:87–112.

Oppenheim, R. W., and M. B. Heaton (1975). The retrograde transport of horseradish peroxidase from the developing limb of the chick embryo. Brain Res. *98*:291–302.

Oppenheim, R. W., and J. L. Maderdrut (1981). Pharmacological modulation of neuromuscular transmission and cell death in the lateral motor column of the chick embryo. *Soc. Neurosci. Abstr.*, Vol. 7, p. 291.

Oppenheim, R. W., J. L. Maderdrut, and D. J. Wells (1982). Reduction of naturally-occurring cell death in the thoracolumbar preganglionic cell column of the chick embryo by nerve growth factor and hemicholinium-3. Dev. Brain Res. *3*:134–139.

Oppenheim, R. W., and C. Majors-Willard (1978). Neuronal cell death in the brachial spinal cord of the chick is unrelated to the loss of polyneuronal innervation in wing muscle. Brain Res. *154*:148–152.

Oppenheim, R. W., and R. Nunez (1982). Electrical stimulation of hindlimb increases neuronal cell death in chick embryo. Nature *295*:57–59.

Oppenheim, R. W., L. L. Rose, and B. Stokes (1982). Cell death of motoneurons in the chick embryo spinal cord. VIII. The survival of brachial motoneurons in dystrophic chickens. Exp. Neurol., in press.

Pannese, E. (1976). An electron microscopic study of cell degeneration in chick embryo spinal ganglia. Neuropathol. Appl. Neurobiol. *2*:247–267.

Parks, T. N. (1979). Afferent influences on the development of the brain stem auditory nuclei of the chicken: otocyst ablation. J. Comp. Neurol. *183*:665–678.

Pearson, J., F. Axelrod, and J. Dancis (1974). Current concepts of dysautonomia: neuropathological defects. Neurology *21*:486–493.

Pearson, J., and B. A. Pytel (1978). Quantitative studies of sympathetic ganglia and spinal cord intermedio-lateral gray columns in familial dysautonomia. J. Neurol. Sci. *39*:47–59.

Pettigrew, A. G., R. Lindeman, and M. R. Bennett (1979). Development of the segmental innervation of the chick forelimb. J. Embryol. Exp. Morphol. *49*:115–137.

Pilar, G., L. Landmesser, and L. Burstein (1980). Competition for survival among developing ciliary ganglion cells. J. Neurophysiol. *43*:233–254.

Pittman, R., and R. W. Oppenheim (1978). Neuromuscular blockade increases motoneurone survival during normal cell death in the chick embryo. Nature *271*:364–366.

Pittman, R., and R. W. Oppenheim (1979). Cell death of motoneurons in the chick embryo spinal cord. IV. Evidence that a functional neuromuscular interaction is involved in the regulation of naturally occurring cell death and the stabilization of synapses. J. Comp. Neurol. *187*:425–446.

Pollack, E. D. (1980). Target-dependent survival of tadpole spinal cord neurites in tissue culture. Neurosci. Lett. *16*:269–274.

Prestige, M. (1967). The control of cell number in the lumbar ventral horn during the development of *Xenopus laevis* tadpoles. J. Embryol. Exp. Morphol. *18*:359–387.

Purves, D. (1980). Neuronal competition. Nature *287*:585–586.

Purves, D., and J. W. Lichtman (1980). Elimination of synapses in the developing nervous system. Science *210*:153–157.

Rager, G. (1978). Systems matching by degeneration. II. Interpretation of the generation and degeneration of retinal ganglion cells in the chicken by a mathematical model. Exp. Brain Res. *33*:79–90.

Rager, G. (1980). Development of the retino-tectal projection in the chicken. Adv. Anat. Embryol. Cell Biol. *63*:1–92.

Rager, G., and U. Rager (1978). Systems matching by degeneration. I. A quantitative electron microscopic study of the generation and degeneration of retinal ganglion cells in the chicken. Exp. Brain Res. *33*:65–78.

Rogers, L. A., and W. M. Cowan (1973). The development of the mesencephalic nucleus of the trigeminal nerve in the chick. J. Comp. Neurol. *147*:291–320.

Rubel, E. W., D. J. Smith, and L. C. Miller (1976). Organization and development of brain stem auditory nuclei of the chicken: ontogeny of N. magnocellularis and N. laminaris. J. Comp. Neurol. *166*:469–489.

Saunders, J. W. (1966). Death in embryonic systems. Science *154*:604–612.

Scaravilli, F., and L. W. Duchen (1980). Electron microscopic and quantitative studies of cell necrosis in developing sensory ganglia in normal and Sprawling mutant mice. J. Neurocytol. *9*:373–380.

Schwartz, J. P., and X. O. Breakefield (1980). Altered nerve growth factor in fibroblasts from patients with familial dysautonomia. Proc. Natl. Acad. Sci. USA *77*:1154–1158.

Scott, S. A. (1981). Development of the segmental pattern of skin sensory innervation in the chick hindlimb. *Soc. Neurosci. Abstr.*, Vol. 7, p. 464.

Shorey, M. L. (1909). The effect of the destruction of peripheral areas on the differentiation of the neuroblasts. J. Exp. Zool. *7*:25–64.

Silver, J. (1978). Cell death during development of the nervous system. In: *Handbook of Sensory Physiology*, Vol. 9, M. Jacobson, ed. Springer, Berlin, pp. 419–436.

Singer, M. (1959). The influence of nerves on regeneration. In: *Regeneration in Vertebrates*, C. S. Thornton, ed. Univ. of Chicago Press, Chicago, pp. 59–80.

Smith, R. G., and S. H. Appel (1981). Evidence for a skeletal muscle protein that enhances neuron survival, neurite extension and acetylcholine (ACh) synthesis. *Soc. Neurosci. Abstr.*, Vol. 7, p. 144.

Spemann, H. (1938). *Embryonic Development and Induction*. Yale University Press, New Haven.

Susheela, A. K., M. Seraydarian, and B. C. Abbott (1980). Increase of alpha motor neurons in chicken afflicted with muscular dystrophy. Exp. Neurol. *67*:453–458.

Swett, J., E. Eldred, and J. A. Buchwald (1970). Somatotopic cord-to-muscle relations in efferent innervation of cat gastrocnemius. Am. J. Physiol. *40*:762–766.

Thoenen, H., and Y.-A. Barde (1980). Physiology of nerve growth factor. Physiol. Rev. *60*:1284–1335.

Twitty, V. (1932). Influence of the eye on the growth of its associated structures, studied by means of heteroplastic transplantation. J. Exp. Zool. *61*:333–374.

Wenger, E. L. (1950). An experimental analysis of relations between parts of the brachial spinal cord of the embryonic chick. J. Exp. Zool. *114*:51–85.

NEUROGENESIS

VIKTOR HAMBURGER

Neurogenesis begins with the formation of a simple tube whose walls consist of a few thousand undifferentiated cells. The embryo accomplishes the remarkable feat of transforming the neural tube within a few weeks into the most intricate precision instrument ever created, the central nervous system. How does it do this? At the beginning, the mechanisms at its disposal are not different from those used for the formation of other organs. Morphogenetic processes, such as evagination, produce first the three brain vesicles— fore-, mid-, and hindbrain—then the optic vesicles in the telencephalon; they are transformed into optic cups by invagination. When the optic vesicle contacts the overlying epidermis, it induces a thickening, the lens placode, which invaginates and forms the crystalline lens. Further caudal, another placode (the otic placode) is induced that invaginates and is detached from the epidermis, like a lens. This vesicle, called the otocyst, undergoes complicated morphogenetic transformations leading to the formation of the semicircular canals and cochlea of the inner ear. The several billion neurons are produced by ordinary cell divisions. Those in the neural tube occur near the lumen of the central canal, the so-called ventricular

Hamburger V: Neurogenesis: pp 623–641. In: Isselbacher KJ: Medicine, Science, and Society. Churchill Livingstone, New York, 1984.

zone. Following their last mitotic division, they migrate radially to the outer part of the tube, where they settle down and arrange themselves in the lamina, brain nuclei, columns, and reticular formations that represent the cytoarchitecture of the central nervous system. At this time, or even earlier, the cytodifferentiation of neurons and glia begins. The differentiation of neurons is distinct from that of other cell types because of two features: (*a*) the production of a prodigious variety of neuron types differing from each other in many structural and biochemical characteristics, in contrast to the uniformity of bone or muscle cells; and (*b*) the capacity to spin out protoplasmic processes, the axons and dendrites. The axons follow stereotyped pathways and form the intracentral tract systems and peripheral nerve patterns. Eventually they establish synapses or sensory terminations at their targets.

All this is a monumental task paralleled only by our own efforts to understand the inner workings of the embryo. I have chosen not to deal with these elementary processes in their sequential order (for this, see Refs. 1 and 2), but to follow a historic approach. This is tempting because I had the privilege of witnessing and participating in this adventure almost from the beginning of neuroembryology. It is true, I was very young when Ross Harrison introduced the methods of experimental embryology to neurogenesis and established experimental neuroembryology as a special field. When I entered it, in the middle 1920s, he had moved on to other pursuits; but to me and others, he has remained a source of inspiration to this day. His most active student, S.R. Detwiler, continued the tradition: his laboratory was one of very few that pursued neuroembryologic problems, using primarily salamander (*Ambystoma*) embryos. Incidentally, Detwiler was briefly at Harvard, from 1924 to 1926, before he moved to the Anatomy Department of Columbia University in New York. During this short period, he produced no less than 16 publications! He was the author of the first book on neuroembryology (3).

I was guided to neuroembryology by my teacher, H. Spemann, who suggested as a PhD thesis the reinvestigation of a then controversial issue. A German embryologist, B. Duerken had observed limb malformations following the extirpation of one eye in young frog tadpoles (4). He claimed that the operation had caused the impairment of other parts of the nervous system including the lumbar spinal cord, which provides the leg innervation, and that the inadequate nerve supply in turn was the cause of the leg abnormalities. My own experiments were inconclusive, but they had the merit of earning me the PhD. I then decided on a crucial experiment: I extirpated the lumbar segments of the spinal cord in very early frog embryos long before nerves had entered the limb primordium. The nerveless limbs I obtained were normal in every respect, morphologically and structurally, although the musculature became atrophic. I concluded that limb development is independent of the nerve supply (5,6). The result was quite definitive, and there seemed nothing else to be done along this line; I turned to other pursuits and almost gave up neuroembryology.

TROPHIC RELATIONS

My return to the fold was linked to my transplantation to this country. Related to this was my shift to the use of chick embryos instead of amphibian ones. This was not a deliberate choice but one instigated by Spemann and the Rockefeller Foundation, which gave me a one-year fellowship (from 1932 to 1933) to work in the laboratory of Spemann's friend, Dr. Frank Lillie, at the University of Chicago. This was one of the few laboratories in which chick embryos were studied, but the only experimental technique then available was transplantations onto the chorio-allantoic membrane. By using Spemann's glass needle technique, I learned within a few months to perform limb bud extirpations and transplantations (Fig. 1). Wing bud extirpations confirmed earlier findings by Detwiler. He had shifted forelimb anlagen of salamander embryos caudally to the flank, where they were innervated by thoracic nerves. He observed that the brachial ganglia, which were deprived of their targets, were smaller than normal, that is, they were hypoplastic, and that the thoracic ganglia, which had an enlarged target area, were hyperplastic; yet, the motor centers seemed to be unaffected (3,7). My experiments showed not only hypoplasia of the brachial dorsal root ganglia (DRG), but they revealed that the lateral motor columns also were severely affected and had disappeared within six days after the operation (Fig. 2) (8). The trophic nature of this reaction was not clearly recognized at that time, however. The question of how the size reduction came about was resolved 15 years later in my first collaboration with Dr. Levi-Montalcini, when we repeated the limb extirpations and transplantations. The decisive clue came when numerous degenerating neurons were found in brachial ganglia subsequent to wing bud extirpation. A massive cell loss occurred during a limited period, approximately three to five days after operation. At that time, axons reach their target areas. We concluded that after a period of self-differentiation and axon outgrowth, the young neurons become critically dependent on a trophic maintenance factor produced by the target tissues (9). I later found that the motor neurons likewise can survive only for a few days unless they are sustained by the target (10). The search for the maintenance factor

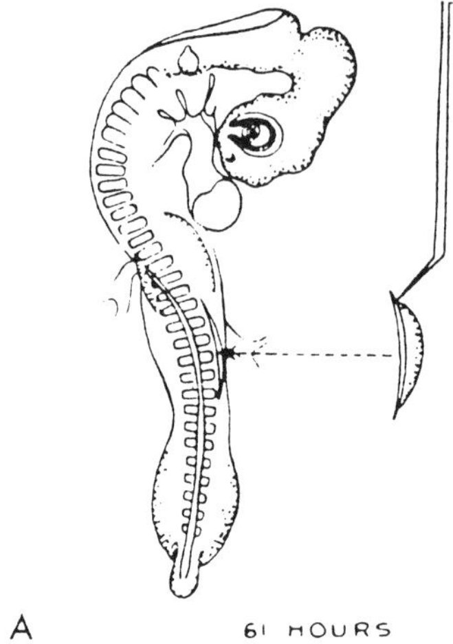

Figure 1. *Limb bud transplantation in chick embryo. (A) A limb bud is removed from a donor embryo (two and one-half days old) and implanted in a slit in the flank of a host embryo, using a glass needle. (B) Twelve-day embryo with supernumerary right leg. (The left host leg was removed.)*

led within a few years to what is now known as the nerve growth factor (NGF). Because the story of its discovery has been told repeatedly (11,12), I will limit myself to a few comments. It might not have happened had we not resorted to a bold experiment performed by my former student, E. Bueker. He implanted small pieces of a mouse sarcoma into the coelomic cavity of early chick embryos. The rapidly growing tumor was invaded by sensory fibers, although shunned by motor fibers, and the DRG involved were considerably hyperplastic (13). The repetition of the experiment by Dr. Levi-

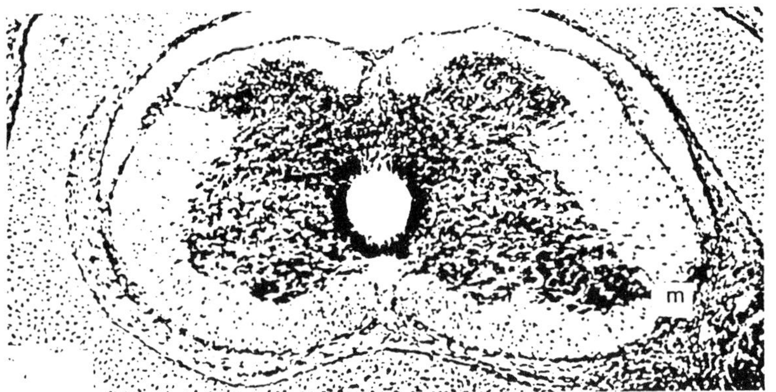

Figure 2.
Cross section through wing level of spinal cord of a nine-day embryo in which the wing bud had been extirpated at two and a half days. Note absence of lateral motor column (m) on the operated side. (From Ref. 10.)

Montalcini showed the same effect, although magnified, and revealed in addition an invasion of the tumor by sympathetic fibers and spectacular enlargement of the adjacent sympathetic chain ganglia (14). The evidence that the tumor actually produced a growth-promoting agent came from two crucial experiments. Small tumor fragments grown on the chorio-allantoic membrane had the same effects on spinal and sympathetic neurons; hence, we were dealing with a diffusible agent carried in the blood stream (15). But the most dramatic evidence came when Dr. Levi-Montalcini devised a simple and direct test using tissue culture: ganglia isolated from eight-day-old embryos were confronted with sarcoma fragments. Within 24 hours, a dense halo of fibers had been produced, whereas control ganglia showed little or no outgrowth (16). This bioassay enabled Dr. Stanley Cohen a few years later to test extracts, first from tumors then from mouse salivary glands, for neurite outgrowth-promoting activity, and eventually to identify the active agent as a protein (17). This assay is still in use.

Nerve growth factor has turned out to be a very versatile molecule that affects the neurons in a dozen or more diverse ways. But its basic attributes are these: it is specific for DRG and sympathetic ganglia, it stimulates metabolism in these cells, and it has a maintenance function in that it assures the survival of embryonic neurons that otherwise would die. In vitro, even dissociated embryonic ganglion cells can survive if standard medium is supplemented with NFG. And in vivo, we obtained striking evidence that the substantial degeneration of spinal ganglion neurons caused by limb extirpation can be prevented by NGF. In an experiment performed by Dr. Joseph

Yip, wing extirpation was combined with daily injections of NGF. This procedure rescued 50% of the population of early differentiating neurons and 100% of late differentiating neurons (18). This and other evidence strengthens my belief that NGF is perhaps identical with the maintenance agent for DRG that normally is produced by limb tissue.

The search for other growth factors is now underway. In these explorations, one must distinguish between agents with the capacity to either stimulate actual cell hypertrophy, maintain neurons, or promote neurite outgrowth. Some progress has been made with an agent released from cultured embryonic heart cells (heart-conditioned medium, HCM) (19,20). When HCM is deposited on the substrate in the culture dish, it elicits rapid neurite outgrowth by promoting the adhesion of growth cones to the substrate (21). Furthermore, HCM supports the survival of ciliary ganglia. Other efforts to identify target-produced growth factors are in progress (22).

PROLIFERATION, MIGRATION, AND DIFFERENTIATION

I should note here that I have omitted neuronal death as an integral part of neurogenesis because it is dealt with by Dr. Cowan in this symposium. But I will briefly consider the initial steps of neurogenesis—proliferation, migration, and initial differentiation—and show that they are tightly interlocked and proceed according to a precise spatiotemporal program. One day in the late 1940s, I was looking through the microscope of a student in an embryology class who was trying to understand the intricacies of cross sections of a 10 mm pig embryo (which was then a standard feature in such courses). I made the chance observation that in the spinal cord, all mitotic figures were conspicuously crowded in the dorsal part of the ventricular zone and practically absent in the ventral part, which looked altogether more mature. Following up on this, I found that indeed in the chick embryo, there was a clear ventral-to-dorsal time sequence in mitotic activity. In fact, the ventral, motor region was advanced over the dorsal region throughout early development. When the ventral motor columns were completely formed and the axons had innervated the muscles, the dorsal regions were still immature and preoccupied with proliferation and migration (23). Much later, in the early 1960s when I undertook an entirely unrelated study of overt motility in the chick embryo, a link to the earlier findings unexpectedly became evident. The most intriguing aspect of embryonic motility is that it is spontaneous, that is, generated in the spinal cord and entirely independent of sensory input. While the embryo becomes increasingly active, no amount of tactile stimulation will elicit a response until the reflex circuit is closed at seven and a half days of incubation. This means that for four and a half days, embryonic motility is nonreflexogenic. To prove this point experimentally, we have done radical deafferentation in early stages; this does not alter the motility pattern, even in much older embryos (24). Obviously, the rule of ventral-to-dorsal maturation already reflected in the mitotic pattern can be traced to the subsequent migration and differentiation patterns and to autonomous motility; all events are intricately linked. One day, it might be possible to detect these ventrodorsal differentials at still earlier stages.

The in-depth analysis of proliferation and migration patterns had to await the development of more sophisticated methods than were available in the 1940s. The breakthrough came with the introduction of ^{3}H-thymidine autoradiography by Sidman and Miale (25) and independently by Sauer and Walker (26). The use of tritiated thymidine permitted precise determination of the terminal mitotic cycle, or the "birthdate" of a neuron. An impressive illustration of the linkage of proliferation and migration was provided by Rakic in his autoradiographic study of the formation of lamina in the visual cortex of the rhesus monkey (27). Proliferation occurs, as usual, near the lumen of the ventricle. The earliest-born population migrates the shortest distance and settles down to form the deepest layer, VI. The upper layers are born and formed in sequential order. This implies that younger neurons migrate radially across earlier-formed horizontal strata. This layering is referred to as the inside-out sequence of neuron disposition. Other examples of this sequence of layering do exist, but other systems are built in an outside-in sequence. An example is the retina, where the layer of ganglion cells that are farthest away from the ventricular zone, are born first, and the others follow, again in strict sequential order (28). The most complicated situation was observed by La Vail and Cowan in the optic tectum of the chick embryo (29). Three waves of migration follow each other. In the first, the innermost layers of the tectum, near the ventricular zone, are formed in an outside-in sequence. Next, the outermost layers are formed, following an inside-out sequence. The middle zone is formed last, again in an inside-out pattern. Angevine observed a similar situation in the hippocampus of the mouse (30).

All these data provide some inkling of how the cytoarchitecture is created. We have dealt only with relatively simple systems; in others, neurons that have reached one position can undergo a secondary migration (31). The most elaborate and longest routes are taken by the cells of the so-called neural crest, which gives rise to the cranial and the spinal ganglia, to sympathetic and parasympathetic ganglia (32,33).

It is one thing to recognize the well-integrated sequence of events; it is another to gain insight in the factors and mechanisms that determine and regulate them. We have to find answers to questions like these: When and how do the progenitor cells of each neuron and glia strain acquire their specifications? How is mitotic cycling regulated? How is it terminated? How do migrating cells know that they have arrived at their destination? Interactions with the environment are undoubtedly an important part of the story. They can be analyzed more conveniently in later phases of neurogenesis, when experimental manipulation is easier.

ORIGIN OF PERIPHERAL NERVE PATTERNS

I have selected for special consideration one particular problem: the question of how the stereotyped peripheral nerve patterns are formed. How do nerves find their way to their targets? Again, progress in the last decade was due to a technical advance: the tracing of nerve origin and nerve connections by the use of horseradish peroxidase (HRP). When injected into a target, this enzyme is transported retrogradely in axons and can be detected in the neuron by a reaction product in the form of dark granules. One also can inject HRP into nerve centers; it is then transported anterogradely in the nerve.

In the early part of this century, Harrison introduced the limb transplantation experiment in amphibians to analyze limb innervation (34). He transplanted limb buds to the flank and made the important observation that foreign nerves can form a typical nerve pattern. He concluded that the tissues of the differentiating limb make a critically important contribution to the success of nerves in finding their targets. He stated: "The mode of segregation and growth of the individual structures of the limb determines the intrinsic distribution of its nerves," (34). Some later investigators have over-interpreted these findings and sought to explain limb nerve patterns solely in terms of a nonspecific mechanical guidance mechanism referred to as "contact guidance" by Weiss (35). But it has become evident that additional, more specific guidance mechanisms come into play.

Before I present evidence for this claim, I briefly will describe the normal innervation process in the leg of the chick embryo. The leg is innervated by eight segmental nerves that converge to form two plexuses, the crural and the sciatic plexus. Nerve fibers rearrange themselves in the plexus and form two major nerves that branch in a complicated way (Fig. 3); the details of that branching need not concern us. Each muscle derives its innervation

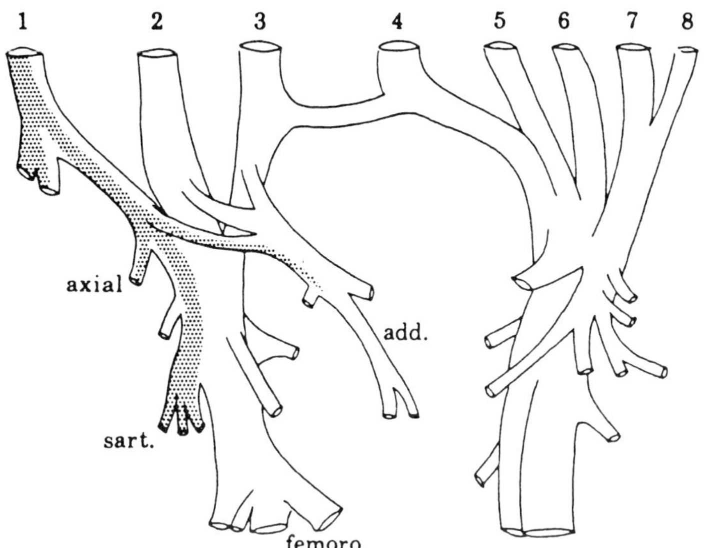

Figure 3.
Normal innervation of leg of chick embryo by segmental nerves 1-8. Nerves emerging from first segment of the plexus were labeled with HRP; add = edductor muscle; femoro = femorotibialis muscle; sart. = sartorius muscle. (From Ref. 37.)

from a cluster of neurons located in the lateral motor column; this is called its motor pool. Individual motor pools overlap and extend over several segments. Before explaining further, I should introduce the main actor in the pathfinding performance, the *growth cone*. This is the tip of the growing axon; it sends out filopodia that explore their microenvironment. As we will see the growth cones are endowed with exquisite capabilities for detecting subtle cues. Nerves grow into the limb bud at very early stages, and they reach their target muscles when these are just being formed by a process of segregation from two precursor muscle masses, the dorsal and ventral muscle mass. The leg of a seven-day-old embryo (Fig. 4), whose different parts and toes are barely outlined, already possesses an adult-type nerve branching pattern, and coordinated muscle contractions have begun (36). A detailed study of the early phases of innervation, using HRP and electrophysiologic techniques, has shown that nerves grow straight toward their target muscles. There is no indication of an initial random distribution or of aberrant pathways and subsequent error elimination (37,38). Nerves seem to know exactly how to find the direct route to their targets. How do they do this?

Some investigators have suggested a simple *spatiotemporal model*. They assume that axon bundles from neighboring motor pools retain their topo-

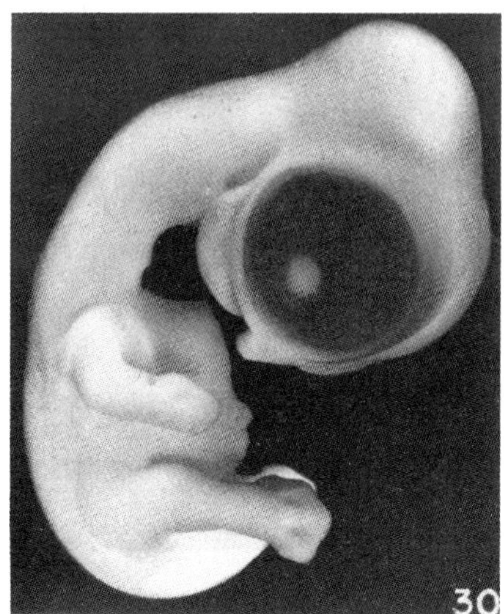

Figure 4.
Seven-day-old chick embryo (stage 30 of
the Hamburger-Hamilton stage series).

graphic relations to each other and that the precise and orderly arrangement of axon bundles within the mixed nerves, together with precise timing of outgrowth, is sufficient to guarantee their safe arrival at the appropriate muscles. This model, as well as mechanical contact guidance, has been subjected to experimental tests.

If the first three or four segments of the lumbar spinal cord, which innervate primarily thigh muscles, are removed before nerve outgrowth, the nerves from the intact posterior segments do not spread to the thigh but follow their normal pathways and terminate at the appropriate muscles. The thigh muscles remain nerveless and atrophy later (39). Likewise, if the anlage of the thigh located at the base of an early limb bud is removed, and the calf and foot part is implanted in the place of the whole limb bud, the nerves destined for the thigh remain unemployed; these nerves grow initially into the inappropriate distal limb parts but apparently lose out in a competition with the appropriate calf and foot nerves; they degenerate, and neurons in the thigh motor pools eventually die. On the other hand, the nerves for calf and foot muscles again follow their typical paths and reach their appropriate targets (40). The spatiotemporal model would have predicted that in both instances, the nerves growing out first would have taken possession of the nearest available muscles, and that the more remote muscles would have been innervated in spatiotemporal sequence. The fibers seem to have the capacity to recognize very specific cues. Both experiments indicate that the growth cones can discriminate between the pathways for thigh and those for calf and foot.

In another experiment, the first three or four segments of the lumbar spinal cord were lifted out and reimplanted in an inverted position (Fig. 5). The outgrowing growth cones were confronted with an unfamiliar environment; nevertheless, they managed to find the appropriate muscle by altering their course. The detour can begin in the plexus, or it can occur later, when the nerves have entered the limb. The fact that nerves can deviate from the normal course in their search for their target argues strongly against the mechanical guidance model and suggests that growth cones can recognize specific cues in their environment that guide them to a particular muscle (41). The growth cones can do this only within a limited radius, however. If longer pieces of the spinal cord are rotated, most fibers follow atypical pathways and terminate in foreign muscles (42). Other instances of nerves homing in on their specific targets by devious routes have been described.

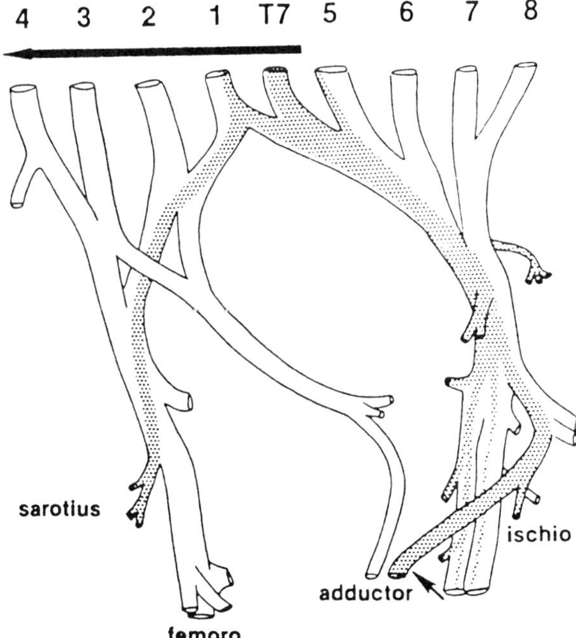

Figure 5.
Anterior-posterior rota-tion of the first 4-leg-in-nervating segments of the spinal cord. Thoracic segment 7 (T7) and le-ginnervating segment 1 were labeled with HRP. Note that on fiber tract of sgmental nerve 1 reaches the sartorius muscle by an atypical path across the plexus; another fibver tract of segment 1 nerve reaches the adductor muscle by a detour (ar-row), which is not present in Figure 3. (From Ref. 55).

Transplantations of supernumerary limbs in front of normal limbs have given additional information. One is impressed by the fact that the nerve patterns are remarkably precise replicas of the normal pattern. If the transplant is rotated 180° along its rostrocaudal axis, its nerve pattern is the mirror image of the normal pattern. This underscores the finding of Harrison, mentioned earlier, that the limb tissues play a decisive role in pattern formation. The difficulty growth cones have in recognizing cues over long distances explains the fact that most of the muscles in the transplants are innervated by inappropriate nerves. In other words, the synapses are mismatched. At closer inspection, however, it turns out that nerve distribution is by no means entirely random. When five different transplant muscles were injected with HRP and their motor pools identified, it was found that their innervation follows a strict rule: muscles derived from the dorsal muscle mass were consistently innervated from a motor pool that normally innervates another derivative of the dorsal muscle mass, and the same holds for muscles derived from the ventral muscle mass. This rule applies even to wing muscles when wing buds are transplanted in front of the leg bud. Some evidence suggests that the nerves make a still more subtle distinction: they can discriminate between muscles that are split off from the superficial subdivision and those that are split off from the deep subdivision of each muscle mass (43).

Several conclusions can be drawn from these and numerous other experiments dealing with limb innervation patterns.

1. The spatiotemporal model and contact guidance *alone* cannot account for all experimental data. The capacity of growth cones for recognizing very specific cues along their pathways and for deviating from their normal course, if necessary, to reach the appropriate targets, strongly suggests that *chemical markers* are the basis for making correct decisions at branching points. This idea goes back to R. Sperry, who formulated the concept of *chemoaffinity* in the 1940s; it was applied originally to synaptogenesis and was extended later to pathway selection (44).

2. It is clear that a combination of several mechanisms is involved in guiding nerves to their targets. These agents include mechanical tracks provided by limb tissues, selective adhesion, chemoaffinity, competition, and perhaps other agencies.

3. The decoding of chemical cues might occur in a sequential order. The outgrowing fibers would choose first between pathways leading to dorsal or ventral muscle-mass derivatives, then between superficial and deep subgroups, then between pathways to thigh or shank and foot, and finally between individual muscles. How the substrate markers originate remains to be explored.

4. Although guidance of nerve fiber bundles to the vicinity of their targets and synapse formation have a common denominator, chemoaffinity, the two components of neurogenesis should be distinguished from each other both conceptually and analytically. The limb transplantation experiment that I mentioned (43) illustrates this point. Whereas nerve and muscle are mismatched at the synapse, evidence indicates nevertheless that specific cues were recognized along the pathway and correct decisions were made between pathways leading to dorsal versus ventral muscle-mass derivatives, respectively. A persuasive argument for the necessity of distinguishing between mechanisms responsible for pathfinding and for synaptogenesis will be presented in the next section.

SYNAPTOGENESIS

The final step in neurogenesis, synapse formation, is the critical event that guarantees normal functional activity. I cannot discuss the molecular interactions by which the actual contacts are established; I will limit myself to the more general question of how the axon terminals recognize their specific targets when they have arrived in their vicinity. Because it is difficult to study these processes in embryos, the most widely adopted strategy has been the use of nerve regeneration. The strongest impetus in this field has come from experiments by R. Sperry on the projection of the retina to the optic tectum; this work began in the 1940s and led to the chemoaffinity hypothesis. In an early, classical experiment on frogs, the optic nerve was cut and the eye rotated 180°. After functional regeneration, it was observed that prey-catching movements were misdirected: objects placed above in the visual field were perceived as being located below, and objects to the right were perceived as being located at the left. Sperry concluded that axons from different parts of the retina had reconnected with the same regions on the optic tectum with which they had been connected before. In other words, the retinotectal projection map had been restored, notwithstanding the behavioral maladaptation that was never corrected. Sperry then made the bold extrapolation from the behavioral to the cellular level and proposed the *chemoaffinity hypothesis*, according to which two partners, the optic fibers and tectal neurons, recognize each other on the basis of matching biochemical labels (45). The hypothesis found support in an experiment in which one-half of the retina of an adult goldfish was removed and the projection pattern of the intact half was studied histologically. The finding that the fibers terminated at the appropriate half of the tectum meant that in some instances the inappropriate half was bypassed and remained devoid of nerves (44, 46).

From the beginning, the recognition mechanism was not perceived as a one-to-one relationship between individual neurons but was thought of in terms of graded chemical affinities in a coordinate system.

I cannot review the voluminous literature on retinotectal projection that has accumulated in the last two decades (47–50). But I think it is fair to state that although several mechanisms are involved in this process, as in pathfinding, chemoaffinity still is recognized as a significant component.

In some respects the peripheral nervous system is preferable to central connections for the analysis of selectivity in synapse formation. I will discuss briefly a set of regeneration experiments by D. Purves and associates on the superior cervical ganglion (SCG) of the adult guinea pig. The SCG innervates hair and iris muscles, blood vessels, and other structures in the head and neck. We are not interested in these modalities but rather in the relative position of these targets in the anterior, middle, or posterior parts of the head. It is important to realize that the SCG cells that innervate a particular region of the head are not organized in a topographic map, comparable to the retinotectal map, but are scattered throughout the ganglion. The SCG receives its input from preganglionic neurons located in the thoracic level of the spinal cord. Their axons exit through ventral roots and reach the SCG by way of the cervical sympathetic trunk (Fig. 6).

If the ventral roots are stimulated in rostrocaudal sequence and intracellular recordings are made from SCG neurons, one finds that *two rules of topographic contiguity* govern the innervation of SCG neurons. First, an SCG cell that innervates a target in the *anterior* head region receives input from a *rostral* subset of preganglionic neurons, such as thoracic segments 1–3 (T1–T3). On the other hand, a neuron that innervates a target in a relatively more *posterior* head region receives input from a more *caudal* subset of preganglionic neurons, for example, T4–T6. Second, each neuron receives input from a restricted but always contiguous set of preganglionic neurons, as for instance, from thoracic segments T2–T4, but never from a discontinuous subset such as T2, T3, T6. Considerable overlap exists in both instances, however.

What happens if the cervical sympathetic trunk is transected? One finds that in the regeneration process the same rules of topographic contiguity are strictly obeyed. For instance, a neuron that innervates an anterior head structure is reinnervated by a rostral contiguous subset of preganglionic fibers, such as T1–T3, and so on (52). Because the operation does not disturb the connections of SCG neurons with their targets in the head, one is justified in concluding that SCG neurons possess individual labels that identify their regional specificity, and that these labels are recognized by pregan-

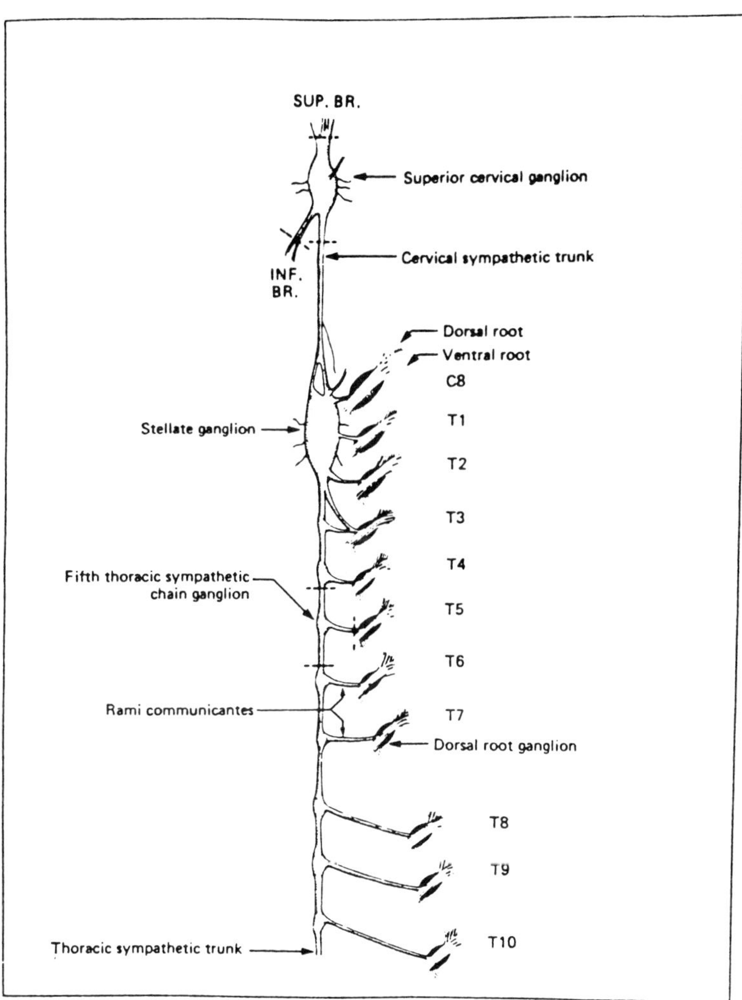

Figure 6. *Thoracic and cervical sympathetic system of guinea pig. The cervical sympathetic trunk was transected at the level of the lower dashed line. Sup. Br. = superior postaganglionic branch; Inf. Br. = inferior postganglionic branch; C8 = eighth cervical ganglion; Ti . . . T10 = thoracic segments 1-10. (From Ref. 53.)*

glionic fibers that enable them to make appropriate choices. Another experiment reinforces this notion. If the SCG is replaced by a sympathetic chain ganglion from the midthoracic level (T5) or the upper lumbar level, the rule of positional preferences is again upheld: these ganglia are innervated preferentially from relatively more caudal subsets than is the SCG (53).

Again, the most probable basis for this selectivity in the choice of targets is chemoaffinity. The case for chemoaffinity derives additional and, I think, crucial support because the data rule out the alternative explanation, that is, the spatiotemporal hypothesis of preganglionic fiber organization. Even if the regenerating fibers retain a strict, contiguous neighborhood relationship with each other, according to the rostrocaudal arrangement of their cells of origin, and if, in addition, one assumes a temporal sequence of rostral-to-caudal timing of outgrowth, this would be of no help when the regenerating sympathetic trunk arrives in the vicinity of the SCG. This is so, because as I mentioned earlier, there is no rostrocaudal map in the arrangement of SCG neurons; the rostral subset of preganglionic fibers has to seek out appropriate target cells scattered all over the ganglion, as do the more caudal subsets.

Purves states: "The evidence that . . . autonomic ganglion cells have some property that biases the innervation they receive offers the only direct support for the view that, amongst vertebrates, chemoaffinity might operate at the level of synapse formation" (54). The word "bias" deserves special emphasis; it underscores the point I made earlier, that selective affinities do not imply one-to-one relationships but merely preferences that are graded in some way.

CONCLUDING REMARKS

The flow chart of neurogenesis, from neural tube formation to synapse formation and functional activity, presents the picture of an elegantly smooth performance. This is the more remarkable, as the experimental analysis of over half a century has made it clear that each phase, such as proliferation, pathway search, and synapse formation, is governed by its own set of developmental mechanisms as diverse as are trophic maintenance and chemoaffinity. The secret of the success is obviously the very precise programming of every process in relation to all others. Nevertheless, I still think it a miracle that such an intricate network of developmental interactions engenders so few errors, and that a precision instrument emerges to serve us so well for so many years.

REFERENCES

1. Cowan WM: Selection and control in neurogenesis, in Schmitt FO, Worden, FG (ed): *Neurosciences Fourth Study Program.* 1979, pp 59–79.
2. Cowan WM: The development of the vertebrate central nervous system: An overview, in Garrod DR, Feldman JD (eds): *Development in the Nervous System.* Cambridge, Cambridge Univ Press, 1981, pp 3–33.
3. Detwiler SR: *Neuroembryology.* New York, MacMillan, 1936.
4. Duerken B: Ueber einseitigie Augenextirpation bei jungen Frosch larven. *Z f wiss zool* 105:192–242, 1931.
5. Hamburger V: Entwicklungs physiologische Bezichungen zwischen dem Extremitäten der Amphibien und ihrer Innervation. *Naturwiss.* 15:657–661; 677–681, 1927.
6. Hamburger V: Die Entwicklung experimentell erzeugter nervenloser und schwach innervierter Extremitäten von Anuren. *W Roux' Arch f Entw mech* 114:272–363, 1928.
7. Detwiler SR: On the hyperplasia of nerve centers resulting from excessive peripheral overloading. *Proc Natl Acad Sci USA* 6:96–101, 1920.
8. Hamburger V: The effects of wing bud extirpation on the development of the central nervous system in chick embryos. *J Exp Zool* 68:449–494, 1934.
9. Hamburger V, Levi-Montalcini R: Proliferation, differentiation and degeneration in the spinal ganglia of the chick embryo under normal and experimental conditions. *J Exp Zool* 111:457–501, 1949.
10. Hamburger V: Regression versus peripheral control of differentiation in motor hypoplasia. *Am J Anat* 102:365–410, 1958.
11. Levi-Montalcini R: NGF: An uncharted route, in Worden FG, Swazey JP, Adelman G (eds): *Neurosciences, Paths of Discovery.* Cambridge, Massachusetts, MIT Press, 1975, pp 245–256.
12. Levi-Montalcini R: One of Hans Spemann's pupils, in Cowan HWM (ed): *Studies in Developmental Neurobiology, Essays in Honor of Viktor Hamburger.* Oxford, Oxford Univ Press, 1981, pp 24–43.
13. Bueker ED: Implantation of tumors in the hind limb field of the embryonic chick and developmental response of the lumbosacral nervous system. *Anat Rec* 102:369–390, 1948.
14. Levi-Montalcini R, Hamburger V: Selective growth-stimulating effects of mouse sarcoma on the sensory and sympathetic nervous system of the chick embryo. *J Exp Zool* 116:321–362, 1951.
15. Levi-Montalcini R, Hamburger V: A diffusible agent of mouse sarcoma producing hyperplasia of sympathetic ganglia and hyperneurotization of the chick embryo. *J Exp Zool* 123:233–288, 1953.
16. Levi-Montalcini R, Meyer H, Hamburger V: In vitro experiments on the effects of mouse sarcoma 180 on the spinal and sympathetic ganglia of the chick embryo. *Cancer Res* 14:49–57, 1954.
17. Cohen S: A nerve growth promoting protein, in: McElroy WD, Glass B (eds): *Chemical Basis of Development.* Baltimore, Maryland, Johns Hopkins Press, 1958, pp 665–676.
18. Hamburger V, Yip J: Reduction of experimentally induced neuronal death in spinal ganglia of the chick embryo by nerve growth factor. *J Neurosci* 4:767–774, 1984.
19. Collins F: Neurite outgrowth induced by substrate associated material from nonneural cells. *Dev Biol* 79:247–252, 1980.
20. Ebendal T: Control of neurite extension by embryonic heart explants. *J Embryol Exp Morphol* 61:289–301, 1981.
21. Collins F, Garrett JE: Elongating fibers are guided by a pathway of material released from embryonic nonneural cells. *Proc Natl Acad Sci USA* 77:6226–6228, 1980.

22. Varon S, Adler R: Nerve growth factors and control of nerve growth. *Current Topics in Developmental Biology* 16:207–252, 1980.
23. Hamburger V: The mitotic patterns in the spinal cord of the chick embryo and their relation to histogenetic processes. *J Comp Neurol* 88:221–284, 1948.
24. Hamburger V, Wenger E, Oppenheim R: Motility in the chick embryo in the absence of sensory input. *J Exp Zool* 162:133–160, 1966.
25. Sidman R, Miale I: Histogenesis of the mouse cerebellum studied by autoradiography with tritiated thymidine. *Anat Rec* 133:429–430, 1959.
26. Sauer ME, Walker BE: Radiographic study of inter-kinetic nuclear migration in the neural tube. *Proc Soc Exp Biol Med* 101:557–560, 1959.
27. Rakic P: Neurons in rhesus monkey visual cortex: Systematic relation between time of origin and eventual disposition. Science 183:425–427, 1974.
28. Kahn AJ: Ganglion cell formation in the chick neural retina. *Brain Res* 63:285–290, 1973.
29. La Vail JH, Cowan WM: The development of the chick optic tectum. II. Autoradiographic studies. *Brain Res* 28:421–441, 1971.
30. Angevine JB: Time of origin in the hippocampal region. An autoradiographic study in the mouse. *Exp Neurol* (suppl) 2:1–70, 1965.
31. Levi-Montalcini R: The origin and development of the visceral system in the spinal cord of the chick embryo. *J Morphol* 68:253–284, 1950.
32. Noden D: Interactions directing the migration and cytodifferentiation of avian neural crest cells, in Garrod E (ed): *The Specificity of Embryonic Interactions.* London, Chapman and Hall, 1978, pp 5–49.
33. LeDouarin: Migration and differentiation of neural crest cells. *Current Topics in Developmental Biology* 16:31–85, 1980.
34. Harrison RG: Experiments in transplanting limbs and their bearing upon the problems of the development of nerves. *J Exp Zool* 4:239–281, 1907.
35. Weiss P: Nerve patterns: The mechanics of nerve growth. Third Growth Symposium, *Growth* 5(suppl) pp 163–203, 1941.
36. Bekoff A: Ontogeny of leg motor output in the chick embryo. A neural analysis. *Brain Res* 106:271–291, 1976.
37. Lance-Jones C, Landmesser L: Pathway selection by chick lumbosacral motoneurons during normal development. *Proc R. Soc London* 214(Biol):1–18, 1981.
38. Hollyday M: Motoneuron histogenesis and the development in limb innervation. *Current Topics in Developmental Biology* 15:181–215, 1980.
39. Lance-Jones C, Landmesser L: Motoneurone projection patterns in embryonic chick limbs following partial deletions of the spinal cord. *J Physiol London* 302:559–580, 1980.
40. Whitelaw V, Hollyday M: Thigh and calf discrimination in the motor innervation of the chick hind limb following deletions of limb segments. *J Neurosci* 3:1199–1215, 1983.
41. Lance-Jones C, Landmesser L: Motoneuron projection patterns in the chick hind limb following early partial reversals of the spinal cord. *J Physiol London* 302:581–602, 1980.
42. Lance-Jones C: Landmesser L: Pathway selection by embryonic chick motoneurons in an experimentally altered environment. *Proc Soc London* 214(Biol):19–52, 1981.
43. Hollyday M: Rules of motor innervation in chick embryos with supernumerary limbs. *J Comp Neurol* 202:439–465, 1981.
44. Sperry R: Chemoaffinity in the orderly growth of nerve fiber patterns and connections. *Proc Natl Acad Sci USA* 50:703–710, 1963.
45. Sperry R: Regulative factors in the orderly growth of neuronal circuits. *Growth* 10:63–87, 1951.

46. Attardi DG, Sperry R: Preferential selection of central pathways of regenerating optic fibers. *Exp Neurol* 7:46–64, 1963.
47. Gaze M: *The Formation of Nerve Connections*. New York, Academic Press, 1970.
48. Keating MJ: The formation of visual neuronal connections. An appraisal of the present status of the theory of "neuronal specificity," in Gottlieb G (ed): *Neural and Behavioral Specificity*. New York, Academic Press, 1976, pp 59–110.
49. Meyer RL, Sperry R: Retinotectal specificity: Chemoaffinity theory, in Gottlieb G (ed): *Neural and Behavioral Specificity*. New York, Academic Press, 1976, pp 111–149.
50. Constantine-Paton M: The retino-tectal hook-up. The process of neural mapping. In: Fortieth Growth Symp. Ed. S. Subtelny, P. Green, 1982, pp 317–349.
51. Nja A, Purves D: Specific innervation of guinea pig superior cervical ganglion cells by preganglionic fibers arising from different levels of the spinal cord. *J Physiol* 264:565–583, 1977.
52. Nja A, Purves D: Reinnervation of superior cervical ganglion cells by preganglionic fibers arising from different levels of the spinal cord. *J Physiol* 272:633–651, 1977.
53. Purves D, Thompson W, Yip J: Reinnervation of ganglia transplanted to the neck from different levels of the guinea-pig sympathetic chain. *J Physiol* 313:49–63, 1981.
54. Purves D: Selective formation of synapses in the peripheral nervous system and the chemoaffinity hypothesis of neuronal specificity, in Cowan WM (ed): *Studies in Developmental Neurobiology, Essays in Honor of Viktor Hamburger*. Oxford, Oxford Univ Press, 1981, pp 231–242.
55. Landmesser L: Pathway selection by embryonic neurons, in Cowan WM (ed): *Studies in Developmental Neurobiology, Essays in Honor of Viktor Hamburger*. Oxford, Oxford Univ Press, 1981, pp 53–73.

II. Development of Motility and Behavior

Reprinted from *The Quarterly Review of Biology* Vol. 38, No. 4, December 1963

SOME ASPECTS OF THE EMBRYOLOGY OF BEHAVIOR

By VIKTOR HAMBURGER

Department of Zoology, Washington University, St. Louis

Dedicated to my friend, Professor F. Baltzer, Berne, Switzerland,
on the occasion of his 80th birthday

ABSTRACT

In the past, the development of behavior in vertebrate embryos has been interpreted in terms of two opposing viewpoints. Coghill, on the basis of his observations on the salamander Ambystoma, has maintained that behavior represents an integrated pattern from beginning to end; partial patterns segregate from total patterns by a developmental process which he calls "individuation." The opposing view, advocated by Windle and other students of the behavior of mammalian embryos and fetuses, contends that local reflexes are the elementary units of behavior that combine secondarily to form integrated patterns.

The conceptual and factual foundations of both theories are reappraised, and their limitations and shortcomings are indicated. It becomes evident that Coghill's concepts, derived from lower forms, cannot be generalized and extended to higher forms. On the other hand, the reflexogenic theory is unacceptable on theoretical and factual grounds.

A reinvestigation of the embryology of behavior in the chick embryo has provided a new frame of reference that places the problem in a different perspective. An old observation of Preyer (1885) to the effect that overt motility is in operation several days before reflex circuits are formed has been followed up, and, in a series of observations and experiments, spontaneous, nonreflexogenic motility was established as a basic component of embryonic behavior. It can be shown that spontaneous motility patterns are not influenced by sensory input. Arguments are presented in favor of the idea that the spontaneous motility in the chick embryo is due to automatic, self-generated activity of neurons rather than to stimulation by agents in the circulation. One of the most significant characteristics of spontaneous motility is its rhythmicity: short phases of activity are followed by longer phases of inactivity. The analysis of recordings of the periodicity in normal embryos and in embryos with transected spinal cords has enabled us to dissociate a trigger mechanism for activation of motility from a device that determines the duration of the activity phases.

Spontaneous and reflexogenic motility are regarded as two independent basic constituents of embryonic behavior. The former is considered to be the primary component, not only because it precedes the reflex responses in lower forms but mainly because reflex motility remains latent in the embryo owing to the absence of adequate stimuli; therefore, it cannot contribute to the molding of behavior patterns in prehatching and prenatal stages.

INTRODUCTION

THE study of the ontogeny of behavior was conceived as a field in its own right by the ingenious psychologist and physiologist W. Preyer. His classical book, *Specielle Physiologie des Embryo* (1885), combined a survey of all previous work with a wealth of original observations and a deep insight into fundamental issues. This remarkable pioneer effort did not bear fruit immediately. After a latent period of 30 years in these studies, the anatomist G. E. Coghill took up the challenging problem, though from a new viewpoint — namely, that of the

342

neuroembryologist. The common title of a long series of his publications, "Correlated anatomical and physiological studies of the growth of the nervous system of amphibia" (1914-1936), indicates the focal point of his approach: he considered the development and elaboration of behavior patterns as the manifestation of an orderly sequence of steps of differentiation in the embryonic nervous system. His basic ideas and his philosophy are laid down in his well-known book *Anatomy and the Problem of Behavior* (1929a). His painstaking investigations, dedicated with single-minded devotion to the elucidation of the neuroembryological basis of behavior, and the theoretical concepts derived from them have served as a frame of reference for a generation of investigators who explored from this viewpoint a diversity of vertebrates, from the dogfish embryo to the mammalian and human fetus. His theoretical position is best stated in his own words:

> "The behavior pattern from the beginning expands throughout the growing normal animal as a perfectly integrated unit, whereas partial patterns arise within the total patterns, and by a process of individuation acquire secondarily varying degrees of independence . . . . Within the total, ever expanding integrated organism . . . partial patterns emerge and tend toward independence, but under normal conditions always remain under the supremacy of the individual as a whole" (1929b, p. 989).

In choosing a seemingly primitive form, Coghill had sought to establish basic principles that would be valid for vertebrates in general. However, later investigators found it increasingly difficult to fit their data into this mold. In particular, the idea of the primacy of integrated behavior was challenged, and the radically opposite stand was taken that local reflexes are the primary units of behavior which become secondarily integrated into coordinated action patterns. This view was largely based on observations on mammals; Windle (1940, 1944) became its most articulate spokesman. The controversy which ensued ended in a deadlock, despite some efforts to bridge the gap (e.g. Barron, 1950). The loss of interest in this field, resulting in another latent period during the last 2 or 3 decades, is in no small measure due to the unfortunate limitation of the 'dialogue' to these alternative viewpoints.

In a seemingly paradoxical way, the rise of experimental neuroembryology has contributed to the decline of this field. The analysis of basic mechanisms of neurogenesis has led to the overall concept that the complex structure of the nervous system is created through the instrument of intrinsic, inherited processes of differentiation and of embryonic interactions that occur within the developing nervous system as well as between nervous and non-nervous structures. The structural foundation of behavior is created in forward reference to functional activity but without benefit from it. Since the role of behavioral activity as a factor in neurogenesis is negligible, the neuroembryologists lost interest in the study of the ontogeny of behavior.

A revival of the study of behavior development requires a fresh, unbiased look at the facts, from which, it is to be hoped, a new frame of reference may emerge. We can now look with detachment on the old controversy. The viewpoint of reflexology which dominated the thinking of 30 years ago no longer prevails, and the general attitude in neurophysiology and ethology has shifted toward recognition of the importance of innate behavior patterns.

The following pages will not review the extensive field of the embryology of behavior. Rather, I shall attempt to clarify some major issues and in this way prepare the ground for a renewed study of this field that has significant implications for neurogenesis and neurology as well as for ethology and psychology. As a point of departure, I take our own new observations on the motility of the chick embryo. All these observations and experiments were done in collaboration with M. Balaban.

When we began to observe motility in the chick embryo, we were immediately impressed by two phenomena which, though known to several observers, had been generally neglected: one of them is *'spontaneous motility.'* ('Spontaneity' we define tentatively as nonreflexogenic motor activity.) The chick embryo shows overt motility several days before it is responsive to sensory stimulation, and it continues the same type of activity up to advanced stages. The second phenomenon is the *rhythmical* nature of spontaneous motility. The embryo shows definite periodicity in its overt performance, up to 12 days. Short phases of activity are fol-

166

lowed by longer phases of inactivity. The cycles are regular in their length. Activity becomes continuous from the 12th day on.

The first part of the discussion will be devoted to different aspects of spontaneous motility.

SOME CONCEPTS IN HISTORICAL PERSPECTIVE

In order to place our emphasis on spontaneous motility in the right perspective, a brief historical note seems to be justified.

In the 1930's and 1940's, when the study of the development of behavior flourished, behaviorism in the sense of the stimulus-response theory dominated the scene. In fact, students of behavior, with few notable exceptions, used 'behavior' and 'response' almost synonymously. Hooker begins his book on the development of behavior with the statement: "In essence, behavior is the sum total of the adjustments made by the organism to changes in its internal or external environment" (1952, p. 3). Consequently, the beginning of behavior in the embryo is dated from the inception of responses to stimuli. "With the advent of conduction from afferent to efferent neurons through synaptic centers reflex responses are manifested. At this point in development, behavior may be said to have its genesis" (Windle, 1940, p. 139). Carmichael, reviewing his studies of the guinea pig, stated: "True behavior, that is, response that results from stimulation and which is secondarily induced by nervous discharge can be elicited 10 to 14 hours before 'spontaneous' behavior appears" (1954, p. 90). It is understandable that, by adopting this frame of reference, spontaneous motility was either ignored or interpreted in terms of the stimulus–response concept. Barron, discussing probable mechanisms of spontaneous activity, stated: "The most probable explanation of the 'spontaneous' movements is that they are due to the activation of the central nervous and the muscular systems in the adult manner, i.e. by the arrival in the central nervous system of impulses via sensory nerves" (1941, p. 12). Perhaps a more widely accepted view was the assumption that agents in the circulation, such as excess of CO_2 or anoxia, elicit spontaneous motility, and the term 'endogenous stimuli' was

introduced. The model for this concept was the respiratory mechanism. In fact, Tracy suggested "that the respiratory center of the medulla in the adult is perhaps a specialized part of the motor system which develops or retains a higher susceptibility to asphyxiation conditions than the rest of the nervous system" (1926, p. 337); Visintini and Levi-Montalcini (1939), quite independently, have expressed the same idea in almost the same words. We shall come back to this point later on. The idea that there might be another alternative — namely, that spontaneous motility might be due to a self-generated discharge of motor neurons, without any stimulation from outside, was too far removed from current thinking to be considered, though the first report of generation of action potentials in isolated nerve centers, by Adrian, had been published in 1931.

The situation gives a thought-provoking illustration of the intricate interrelationship of theory and method and of the inherent danger of circular reasoning on that basis. In accordance with prevailing psychological theory, practically all investigations of the development of behavior concerned themselves with tactile stimulation experiments. Of course, such experiments can give answers only in terms of reflexogenous response and neural reflex pathways. Generalizations and theories of the ontogeny of behavior deduced from these experiments were of necessity inherently biased and limited by the method. We are now aware that responses to sensory input are only part of the repertory of the activities of organisms. Modern ethology postulates intrinsic action systems in the central nervous system and sees in the sensory information essentially a screening, channeling, and regulating device for adaptive use of the inherent potential for action. Against this background, Coghill's independence of thought stands in relief. On the basis of his own work, he not only took a clear stand against reflexology but he anticipated, and gave a developmental basis to, modern ideas. He made a distinction between what he called "form of behavior" and "conditioning of performance." "The general pattern of the primary nervous mechanism of walking is laid down before the animal can in the least respond to its environment. The form of the behavior pattern in

Amblystoma develops step by step according to the order of growth in particular parts of the nervous mechanism. . . . On the other hand, experience has much to do with determining where and to what extent the potentiality of behavior shall rise into action" (1929a, pp. 86–87). It is of interest to realize that his behavior studies were also based on tactile stimulation, but in his case this was an expedient rather than design. Salamander embryos do not exhibit much spontaneous activity; they have to be prodded into action. For Coghill, the stimulation procedure was a tool, divorced from theory.

"Myogenic," "neurogenic," "reflexogenic"

Although the temper of the time did not encourage the exploration of nonreflexogenic behavior, some phenomena pointing in this direction could not escape the attention of investigators. Two such types of motility were noticed. They were distinguished as 'myogenic' and 'neurogenic,' depending on whether muscle contraction was initiated within the muscle tissue itself or by nerve impulse. 'Myogenic' motility is well established for selachians, where it occurs spontaneously, preceding neurogenic spontaneous motility (Harris and Whiting, 1954). On the basis of experiments with curare, Coghill (1933) came to the conclusion that the earliest movements of the teleost, *Fundulus heteroclitus,* are myogenic. The experiments were never published in full, and the interpretation of the data is subject to several objections. Brinley (1951) reported that he could block all motility from the start in another species by high dosage of curare. Therefore, the question of myogenic motility in teleosts remains unsettled. In the chick embryo, motility seems to be neurogenic from the start (see p. 346). Whereas this meaning of the term 'neurogenic' is clear, the same term was unfortunately used later on also in antithesis to 'reflexogenic' action. Here again, the methodological bias obscured the issue. What was meant by neurogenic in this restricted sense was motility that was observed before reflexes could be elicited. This category included spontaneous motility such as that found in early chick embryos and contractions resulting from direct electrical stimulations of the nervous system. Such motility was considered to be an antecedent to 'true' — that is, reflexogenic — behavior, and the distinction between three successive phases of embryonic motility — namely, myogenic, neurogenic, and reflexogenic, found its way into the literature, where it still lingers on. Apart from the poor choice of the term (reflexogenic is also nerve-induced — i. e., neurogenic motility), this subdivision is based on a misconception: nonreflexogenic (so-called neurogenic) and reflexogenic nerve-stimulated motility are not two chronological phases but two different, basic types of motility. As we shall see in the following paragraph, nonreflexogenic motility in the chick embryo persists throughout the incubation period. Nonreflexogenic activity, be it spontaneous and overt or in response to direct stimulation of nerve tissue, often does precede the stage of development at which the first reflex responses can be elicited, but this is not always the case. In chick and teleost embryos, the time interval between the first appearance of nonreflexogenous and of reflex motility is of considerable length, but in mammals it is short or nonexistent. If one wishes to classify embryonic behavior in terms of mechanisms, the following tabulation might serve the purpose:

Motility: (1) *myogenic* (spontaneous or by direct stimulation of muscle tissue).

    (2) *neurogenic* (neuro-muscular mechanisms)

      (a) *self-generated* in motor or other neurons; 'automatic.'

      (b) *activated by endogenous stimuli* (such as agents in the circulation).

      (c) *reflexogenous,* activated by sensory systems.

REFLEXOGENOUS VERSUS SPONTANEOUS MOTILITY IN THE CHICK EMBRYO

The chick embryo is a favorable object for the analysis of nonreflexogenic behavior. As was mentioned in the introduction, overt spontaneous motility begins several days before reflex responses can be elicited. The motility is exhibited intermittently. Apart from its significance as a general biological phenomenon, rhythmicity proved to be valuable in the analysis of spontaneity. Furthermore, the exposed chick embryo, in contrast to the mammalian embryo and fetus after premature delivery, gives a normal performance for hours and days. The preparation of the embryo for observation is simple. A window is sawed in the shell, the shell membrane is carefully removed, and the embryo is placed in a temperature-controlled plastic box, which is kept under humidity conditions normal for chick development.

Activity starts at $3\frac{1}{2}$ days with a slight flexion of the head, caused by contraction of several head somites. It spreads from there in cephalocaudal direction, until undulating S-type movements extend from neck to tail. The extension of motility probably reflects the differentiation of the neuromuscular system in the cephalocaudal direction. In later stages, wings and legs, beak, tongue, eyeball, and eyelids join the trunk in spontaneous motility. However, the cephalocaudal sequence in the integration of the movements soon disappear; they give way to a generalized motility of all parts (see below, p. 353). One gets the impression that at a given stage all parts whose neuromuscular connections are functional actually participate in overt motility. Responses to tactile stimulation cannot be obtained until stage 31 or 32 (7 to $7\frac{1}{2}$ days; Hamburger and Hamilton, 1951) — that is, until $3\frac{1}{2}$ to 4 days after inception of the motility. Up to that stage, no amount of sensory stimulation can arouse the embryo into activity when it is in an inactive phase of the cycle. Preyer (1885) was the first to notice this long-time interval between the onset of motility and sensitivity. He realized the significance of this phenomenon for an understanding of the embryology of behavior. The structural basis for this behavior sequence has been described: the reflex circuit is closed at 6 to $6\frac{1}{2}$ days by the establishment of synaptic connections between collaterals from the dorsal sensory tract and internuncial neurons (Visintini and Levi-Montalcini, 1939). Obviously, there can be no doubt concerning the nonreflexogenic nature of the motility, up to the 8th day of incubation.

The possibility that the earliest motility in the chick embryo is *myogenic* has to be considered. As was mentioned, myogenic motility preceding neurogenic motility has been demonstrated for selachians and claimed for teleosts. Both Kuo (1939a) and Visintini and Levi-Montalcini (1939) have tested early chick embryos by electrical stimulation of nerve and muscle tissue and by curarization. They have come to the conclusion that no myogenic phase exists in the chick embryo and that its motility is neurogenic from the start. The latter investigators have also shown that neuromuscular connections do exist when motility begins. We have confirmed the observations concerning curarization: .03 mg curare in 0.1 ml saline injected into the amniotic fluid inhibit spontaneous motility from the earliest stages on. Pressure exerted on a myotome of a curarized embryo results in localized contractions which are different in pattern from spontaneous motility. Further evidence comes from an experiment (to be discussed on p. 359) in which segments of the thoracic spinal cord several somites in length were extirpated in 2-day embryos, and the motility of these embryos was observed at stage 32 ($7\frac{1}{2}$ days). Although the somite musculature was undisturbed and continuous, the parts in front of and behind the gap in the spinal cord showed independent motility cycles that were not synchronized with each other. On the basis of this concurring information we assume that all motility described in the following pages is neurogenic. However, a crucial experiment, such as total extirpation of the spinal cord, has not yet been performed. We are reinvestigating this problem. Pending the outcome, the possibility that the earliest movements are myogenic, or a combination of myogenic and neurogenic motility, cannot be ruled out completely.

The form of activity which the embryo performs spontaneously at 7 days continues without noticeable change except for the extension of motility to beak, tongue, and eyelids. Of particular significance in this respect is the continuation of the motility cycles, at first with the same periodicity as before and from 8 to 9 days on with increase of length of activity phases and decrease of length of inactivity phases, until spontaneous activity becomes continuous at 13 days. Since no evidence for any kind of external or self-stimulation of the embryo is apparent (see p. 347), the continuity of the motility pattern suggests that the same neural mechanism which is responsible for activity before the reflex circuits are closed remains in operation after the stage at which reflex responses can be elicited experimentally. Since this point is crucial for our basic position, it has to be scrutinized most carefully.

What arguments can be presented to exclude reflexogenous motility in stages after reflex arcs are established? Some authors (e.g., Kuo, 1939b) have contended that the rhythmical contractions of the amnion which begin at about the same stage as embryonic motility serve as a source of exteroceptive stimulation. But Preyer (1885) has already shown that amniotic and embryonic contractions are independent of each other, and this observation has been confirmed by others. We have observed that amniotic motility is not closely synchronized with embryonic motility; both can overlap, or the embryo can be rocked passively while it is in an inactivity phase without being aroused to activity. In older embryos, vigorous kicking of legs against the amnion can trigger its pulsations, but we have not observed the reverse.

Self-stimulation could be another source of reflexogenic motility. In later stages of incubation, the position of the embryo in the shell is such that the digits and toes come in contact with the head. Our observations over long periods have never given any indication that head movements are elicited when the legs in their vigorous movements touch the head, or vice versa.

The question of self-stimulation seemed to us of sufficient importance to warrant a series of experiments. The intermittent activity of the embryo offers an especially favorable opportunity for further investigation of this point. The question was raised whether tactile stimuli applied during the inactivity phase of a cycle could initiate extra cycles. The 11-day embryo (stage 37) was chosen because at this stage the activity phases are long, averaging about 40 seconds, and clearly distinguishable from reflex responses, such as withdrawal, which do not exceed 5 seconds. The skin of the wing or leg was stimulated by gentle strokes with a small loop made of baby hair, (an instrument used by experimental embryologists; see Hamburger, 1960). Stimulation was started 10 seconds after the termination of an activity phase. Seven successive strokes were applied at 4-second in-

tervals to the thigh or shank, or the upper or lower arm, or both. The responses to stimulation were invariably one or several withdrawals of wing or leg, the muscle contractions spreading frequently to adjacent parts of the trunk. The duration of these responses was not more than 5 seconds. These repetitive stimulations in no way changed the timing of the basic motility cycles. This was borne out by statistical evaluation of the data which are based on Kymograph recordings. Fig. 1 presents cumulative curves of activity and inactivity phases of stimulated and unstimulated embryos. It is evident that the basic motility pattern of the embryo is not modified by the stimulations. When stimulations were applied toward the end of an activity phase, they did not lengthen the activity phase.

The same experiment rules out proprioceptive self-stimulation as a source of embryonic motility. The withdrawal of the limb in response to tactile stimulation inevitably sets up proprioceptive stimuli, yet no sustained motility phase is thereby initiated.

We come to the conclusion that, in the chick embryo, during the incubation period at least up to 18 days, sensory input most likely plays no role as a trigger for the overt motility of the embryo, nor does it determine its pattern. Although the potentiality for reflex responses exists from $7\frac{1}{2}$ days on, it remains latent.

SPONTANEOUS MOTILITY IN OTHER FORMS

To what extent can the concepts developed from the investigation of the chick embryo be generalized? The available material is hardly

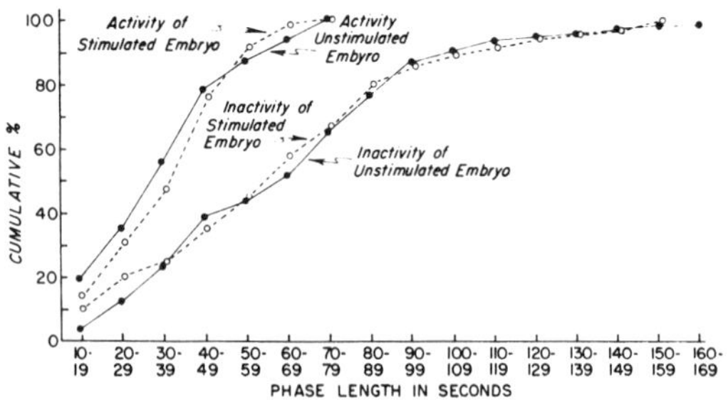

FIG. 1. EFFECTS OF TACTILE STIMULATION DURING INACTIVITY PHASES UPON SPONTANEOUS MOTILITY CYCLES

Cumulative percentage curves of length of activity and inactivity phases in normal embryos and in embryos subjected to tactile stimulation during inactivity (stage 37, see text).

sufficient to give an adequate answer. There are several reasons for this state of affairs. First, few investigators have focused their attention on the nonreflexogenic component; furthermore, spontaneous motility is infrequent in some forms, and the mammals present particular difficulties to the observation of fetal motility under entirely normal conditions.

Teleosts

Tracy (1926) has made a detailed study of the ontogeny of behavior in the toadfish *(Opsanus tau)*. His object eloquently persuaded him of the significance of spontaneous, nonreflexogenic motility for an understanding of the ontogeny of behavior. In the toadfish, there is a time interval of $2\frac{1}{2}$ weeks between the onset of motility and the beginning of responsiveness to tactile stimuli; the nonsensitive period extends beyond hatching to the beginning of free-swimming life. In this respect, the material is even more favorable than the chick embryo.

Motility begins at the 17- to 19-somite stage (about 9 days after fertilization) as a flexion of the anterior trunk. At hatching, about 13 days later, Tracy observed both slow and rapid spontaneous undulatory movements of the whole body. These movements, when performed in a more sustained fashion, lead to swimming over a short distance at about 25 days after fertilization. The first reflex responses can be elicited on day 27 by stimulation of the snout. Proprioceptive stimuli are also ineffective up to that stage. Swimming activity is thus perfected before it can be directed and adjusted by sensory input.

The spontaneous activity is intermittent, as in the chick embryo; however, the inactivity phases are much longer. Recordings of the periodicity were made from the beginning of motility through the stages when exteroceptive stimuli become effective, and no change in the pattern of motility was observed when the larva was left completely undisturbed. Tracy came to the same conclusions which we derived later from our parallel observations on the chick embryo. He stated: "There is no change in the character of the movements which the larva makes by itself which gives any ground for believing that the general development of the exteroceptive system as such inhibits or sub-

merges this spontaneous activity, or that the motor system does not continue to act as the result of internal stimuli" (Tracy, 1926, p. 270).

Tracy summarized his position as follows:

"There are two components of behavior, namely endogenous and exogenous activity. Endogenous activity constitutes the fundamental feature of the body motility (progression) and is conditioned by internal physiological adjustments in connection with metabolism. Exogenous activity is oriented activity; it appears to be essentially the modification of the endogenous activity which results either from the stimuli which the organism meets during its excursions in the environment, or from those aroused by changes in external energy relations."

(Tracy, 1926, p. 357). Although we have reservations as to the causation of spontaneous activity by internal physiological stimuli, his distinction of two separate components of behavior anticipates our own position.

Tracy went further than this: he made the interesting observation that a relation exists between the motility habit of adults and the motility pattern of the embryo. "The toadfish (young and adult) remains quiescent for considerable periods between bursts of movements, . . . whereas Fundulus and cunner are in continuous motion playing on the surface, darting in various directions" (Tracy, 1926, p. 275). Correspondingly, the spontaneous activity phases of the embryos and larvae of the latter two species have a higher frequency than those of the toadfish. "The presumption, therefore, exists that movements of endogenous origin continue through the whole life of the animal and determine its habitual activity, or, what, perhaps, may be called its motility pattern" (Tracy, 1926, p. 275).

Amphibia

In *Ambystoma*, and probably in other amphibians, spontaneous motility of the embryo within the jelly membranes is infrequent. To our knowledge, no special investigation has been devoted to this point. Coghill has shown that a reflex circuit is established very early through the Rohon-Beard cells, which are specialized sensory cells in the spinal cord. Hence, the time interval between the onset of motility and sensibility is very short. But even though Coghill had to apply stimulation in his own studies, the quotation given above (p. 343) leaves no doubt as to his position. The primacy of neuro-motor activity over reflex responses was one of the major generalizations derived from his work on *Ambystoma*. Herrick paraphrased his view: "The precocity of the motor systems in development is evidence of their autonomy, and there is experimental evidence that some measure of this autonomy and automaticity is preserved through life" (1949, p. 140). In this context, the significant conclusions of Weiss (1941a, b) should be mentioned, although they were derived from experiments on larval and adult salamanders and do not refer to the beginnings of motility. He has shown in a series of ingenious transplantation and extirpation experiments that the patterns of coordinated motor activity in locomotion result from self-differentiation of these centers, independent of sensory control.

Mammals

Generally speaking, the existence of a non-reflexogenic component of embryonic behavior can be demonstrated only if there is a sufficient time interval between the onset of motility and sensibility and if it can be shown that, once sensibility is present, it does not influence the spontaneous motility. The first condition is not fulfilled in the mammals. In contrast to amphibian and chick embryos, the mammalian embryo is far advanced in general body development when motility begins (Fig. 2). For instance, at this critical stage, limbs are absent or in the bud stage in submammalian forms, whereas contours of toes are distinct in mam-

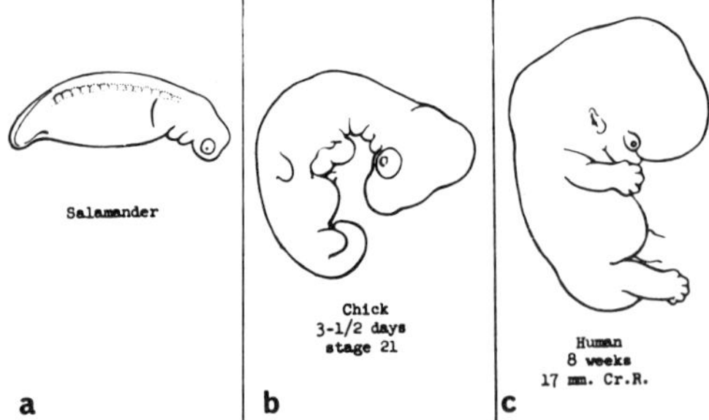

FIG. 2. BEGINNING OF MOTILITY
Comparison of developmental stages at which motility begins (a) in a salamander, (b) in a chick embryo (3½ days; stage 21), and (c) in a human embryo, (8 weeks; 17 mm crown-rump length).

mals. In all mammals, motility starts in the usual fashion as a flexion of the neck and extends through trunk and tail. In most mammals, limb movements are involved in early motility, and local limb reflexes can be elicited from the start. The earliest phases of behavior are thus telescoped.

While there is a wealth of information on stimulated movements in mammals, data on spontaneous motility are scanty and incidental to the reflex studies. It seems that, in most forms that have been studied, reflexes operate from the very beginning of motility, and the first spontaneous movements that were recorded occurred in later stages, or, at best, simultaneously with reflexes. (Angulo, 1932, for the rat; Bridgman and Carmichael, 1935, for the guinea pig; Hooker, 1952, for the human fetus; Windle, 1940, for the cat). In the sheep, Barcroft and Barron (1939) have observed spontaneous movements in very early stages preceding exteroceptive reflex responses, but the authors doubt the nonreflexogenic nature of the movements.

The study of motility in mammals, in general, is beset with great technical difficulties. Fetuses survive delivery for only a very short period owing to anoxia, and doubt has been raised concerning the validity of all but the most carefully controlled observations. One can therefore argue that the observed overt motility, whether spontaneous or reflexogenic, is elicited by anoxia or other abnormal conditions, and that it is not a normal phenomenon.

To my knowledge, the other approach — that is, the study of the role of stimuli in utero, has not been explored systematically. Since the reflex apparatus seems to be present from the start, the conditions for self-stimulation are present. Furthermore, amniotic contractions and maternal movements have to be taken into account, yet the extent of such stimulation and its role in the normal course of mammalian behavior development are dubious. Despite all the information that is on hand with respect to reflexogenous zones, changes in their extent, changes in response patterns, etc., we really do not know whether and to what extent reflex responses occur in the absence of the experimenter. In this connection, the statement of one of the strongest exponents of the reflex theory of the origin of behavior is of interest: "It should not be assumed that all responses which can be induced occur spontaneously within the uterus of the normal, intact individual. As a matter of fact, there is scanty evidence that any of them occur normally during the early part of the gestation period" (Windle, 1940, p. 164). Furthermore: "The relative quiescence of the normal fetus in utero is somewhat surprising when one considers all the activities of which the growing specimen is capable when removed from the uterus. The reasons seem to be: lack of adequate stimulation and high thresholds in the fetal central nervous system. . . . No significant excitation of the external receptors occurs" (Windle, 1940, p. 165). Barcroft and Barron (1939) have observed considerable spontaneous activity in the sheep during the first part of the gestation period, but this fetus is quiescent from the 60th day on. The absence of motility in this case is ascribed to inhibition by the brain stem. If adequate stimulations in utero are rare events, then they cannot play a significant role in the molding of behavior patterns of mammals.

In summary, the question of whether in the mammalian embryo and fetus there exists a motor action system as an independent component, apart from the reflex apparatus, is undecided. In the light of our experience with the chick embryo, the very existence of spontaneous motility in mammals speaks in favor of such a system, but crucial evidence for this viewpoint is still lacking. The claim that the spontaneous motility is reflexogenic is entirely without foundation.

A reinvestigation of spontaneous motility and of possible sources of stimulation in the undisturbed mammalian fetus is highly desirable.

CENTRAL AUTOMATISM VERSUS ENDOGENOUS
STIMULATION

We wish to carry the discussion of spontaneity one step further in an attempt to give this concept a more concise physiological meaning. So far, we have defined the term 'spontaneous' negatively, in the sense of nonreflexogenous. I have pointed out already (p. 345) that, apart from myogenic motility, two different mechanisms can account for spontaneous motility: (1) automatic, self-generated discharge of neurons, either continuously or intermittently, according to intrinsic phasic patterns — this type is referred to as 'central automatism' (e.g. by Bullock, 1961); or (2) the occurrence of chemical stimuli, impinging on neurons and activating them either continuously, or intermittently — this type has been referred to as 'endogenous stimulation.' The term 'endogenous' used alone as a single word is somewhat ambiguous: it has been used occasionally to designate the first category, but the combination 'endogenous stimulation' conveys perhaps with sufficient precision the idea that we are dealing with extra-neuronal agents whose stimulating effects are required for nerve cell action.

This categorization is, of course, a gross oversimplification; both categories lump together heterogeneous physiological mechanisms, and some further qualification is necessary. For instance, the categories are not mutually exclusive; they can be combined when a chemical agent modifies an intrinsic automatic rhythm. This seems to be the role of carbon dioxide in the respiratory mechanism. Furthermore, the conceptual separation of the two categories is easier than a physiological distinction. After all, nerve centers are not closed systems, and they are therefore subject to threshold conditions in their endogenous milieu. Hence there will be borderline situations in which a clear distinction between 'permissive condition' and 'trigger' is not possible. Nevertheless, at first approximation it may be useful to distinguish between discharge that results from intraneuronal metabolic changes going on in a relatively steady state of the internal milieu and, on the other hand, a resting or subthreshold state of neurons which requires for discharge a change in the steady state endogenous condition.

Traditionally, only the second mechanism has been invoked to explain spontaneous motility in embryos. The reasons for this have been discussed on previous pages. In particular, oxygen deprivation and CO_2 accumulation have been evoked frequently as triggers of embryonic motility. There is no point in reviewing and evaluating these investigations. We shall limit the discussion to the presentation of some data which lead us to believe that in the chick embryo we are dealing with central automatism.

Dr. Balaban and I have studied motility in embryos in which the spinal cord was transected

(Hamburger and Balaban, 1963). In 2-day embryos, spinal cord segments of the length of four to five somites were removed with glass needles or by electrocautery. The embryos were raised to $7\frac{1}{2}$ or 8 days of age (stages 31 or 32). Following cervical or thoracic transections, the isolated posterior parts retain their capacity for intermittent rhythmical activity, an observation which shows that there is no single pacemaker in the brain that triggers all activity. An influence of the brain on the periodicity of the trunk movements was observed (see p. 359), but this point is unimportant in the present context. However, it is of significance that the isolated parts are not in phase with each other; one part can be active and the other quiescent, or activity phases of isolated parts may partly overlap. This finding rules out the possibility that any agent or fluctuation in the endogenous milieu operates as a trigger of spontaneous motility. Furthermore, we have found that exposure to pure O_2 raises the activity somewhat in normal 8-day embryos, but not to the point of continuous activity. Exposure to a mixture of 95 parts O_2: 5 parts CO_2 has no influence on the periodicity; a mixture of 90 : 10 reduces the total amount of activity per time unit up to stage 28 and blocks it temporarily in later stages, but the rhythmicity is not eliminated in either instance. Hence, the periodic motility is maintained over a rather wide range of concentrations of the physiologically most important gases in the circulation (Hamburger 1964). It will be of great interest to continue these experiments and to find ways of modifying the rhythm.

In summary, some arguments speak in favor of the contention that in the chick embryo the spontaneous motility is due to self-generated automatic discharge of neurons. The issue will have to be settled by recording the electrical activity from neural tissue.

TWO BASIC COMPONENTS OF THE ONTOGENY OF BEHAVIOR

Our argumentation so far can be summarized as follows. Observations and experiments on the chick embryo demonstrate the existence of two behavior components that can be dissociated from each other on the basis of behavior characteristics. Beginning with the stage at which the first neuromuscular connections are established and continuing through the greater part of the incubation period, there exists a *motor action system* with the following characteristics: (1) it is overt and spontaneous — that is, it discharges independently of reflexogenous stimulation; (2) it performs in motility cycles of regular periodicity, up to 13 days, and from then on almost continuously; and (3) it involves generalized motility of many or all parts that are capable of motility at a given stage. Independently of this system, there develops the *reflex apparatus* which begins to attain functional maturity 3 to 4 days after the onset of spontaneous motility. It differs from the latter in the following points: (1) it remains latent in the normal, undisturbed chick embryo, due to the absence of adequate stimuli; and (2) the response following experimental stimulation is more or less localized and of brief duration. We assume that these two components are represented by different intracentral neural mechanisms.

The embryonic and larval spontaneous behavior in teleosts, according to Tracy's observations on several species, is remarkably similar to that of the chick embryo. The observations on mammals are inadequate and inconclusive, but they provide no argument against the existence of two behavior components.

We believe that the thesis of two components of embryonic behavior establishes a new frame of reference for the study of the ontogeny of behavior. The primacy of spontaneous motility is asserted, not only because it precedes the reflex responses in lower forms by several days or weeks but mainly because reflexogenic motility remains essentially latent in the embryo, due to the absence of adequate stimuli, and hence is ineffective in molding behavior.

To avoid misunderstanding, I should like to point out that the above-mentioned criteria which have led to the distinction of the two components are not essential constituents of these components. Take, for instance, the distinction between overt and latent motility. We take it for granted that the performance of overt motility requires a set of adequate or 'permissive' conditions. In the chick embryo, such conditions fortunately do exist: hence, the embryo provides direct information that otherwise could not be obtained. In mammals, on the other hand, the endogenous conditions are apparently less favorable for spontaneous motility (see above, p. 350), perhaps due to special conditions in the chemical milieu. Windle states that "in the cat, changing uterine tonus and impairment of oxygenation of fetal blood leads to activity in otherwise inactive fetuses" (1944, p. 249). In other words, relatively minor changes in the endogenous conditions can make the difference between overt activity and quiescence. The fact that spontaneous motility is reduced or absent for extensive periods, as seems to be the case in some amphibians and mammals, is no argument against the existence of an autonomous action system.

Nor does the distinction of two components hinge on the time difference between the maturation of the motor and sensory systems. This time interval permits us to study one component in the absence of the other, but the brevity or nonexistence of such a time interval is no argument against the existence of the two components. The following point is of much greater significance in this context: when the reflex system is structurally completed, it does not 'take over' or 'incorporate' the motor action system in the sense that the latter loses its identity as an operational unit. The motor action system has its own progressive differentiation and probably persists throughout life, as suggested by Coghill, Tracy, and others. When Tracy at one point speaks of the sensory system "capturing" the motor system, he refers to the obvious fact that the two components become amalgamated so that the reflex apparatus, in conjunction with other controlling and inhibitory neural mechanisms, can perform its role of channeling the motor action system into adaptive behavior patterns.

INTEGRATED BEHAVIOR VERSUS GENERALIZED
MOTILITY IN SUBMAMMALIAN EMBRYOS

So far, we have concerned ourselves with spontaneous motility as a component of embryonic behavior that can be distinguished from reflex response, both conceptually and factually. We turn now to a consideration of the form, or *pattern*, of embryonic motility. It is at this point that we encounter the most controversial issue in the past discussions of embryonic behavior. The focus of the controversy is Coghill's contention (quoted in the Introduction) that behavior is integrated from its beginning and that it does not lose this characteristic while it expands and becomes more intricate in the course of development. We can simplify the issue by disregarding, for the time being, the counterpropositions of his opponents and by asking the question 'Do observations on forms other than *Ambystoma* confirm or refute Coghill's idea'? Right at the beginning, it should be stated that my agreement with Coghill on the primacy of nonreflexogenous, intrinsic motility does not imply an endorsement of his holistic view concerning its pattern.

Very briefly, the observations on which Coghill has based his concepts are as follows. The *Ambystoma* embryo begins motility with a slight flexure of the neck (early flexure). At a slightly later stage, the movement extends to trunk and tail, resulting in a coil that starts in the neck. Next, two coils follow each other in such a way that the second begins in the neck before the first has reached the tail. The result is an S-flexure. The S-flexure is of paramount importance because it is the antecedent of integrated swimming. A sequence of several S-flexures results in locomotion for a short distance; more vigorous S-movements result in sustained swimming. The first adaptive form of behavior in *Ambystoma* is thus attained through an unbroken sequence of behavioral stages, each retaining the integrated nature of the preceding stage. Other behavioral patterns, such as walking, stalking, and catching prey, emerge in the same way as an elaboration of a total pattern. Local movements, such as limb withdrawal, postural reflexes, or eye movements, are first performed as part of the total pattern; they become emancipated secondarily from the total pattern ('individuation'). It was mentioned that the neurological basis for each step was ascertained. For further details, see Coghill (1929a).

In some teleosts, the behavior sequence that leads to swimming is very similar to that of *Ambystoma* (Tracy, 1926; Coghill, 1933). Behavior is somewhat modified by the large yolk sac. At hatching, single coils occur, as well as slow and rapid S-flexures, but no swimming as yet. In subsequent days, single coils become rare and rapid S-flexures ('flutter') become more frequent, and swimming emerges by a rapid sequence of vigorous flutters. It is difficult to judge from the scanty data on other teleosts whether this sequence holds true in all teleosts (see Whiting, 1955).

We turn now to our observations on the chick embryo. They are in essential agreement with the observations of Orr and Windle (1934). Those of Kuo (1932) are presented in a form which makes a comparison difficult. The initial phase of behavior seems to be in agreement with Coghill's notion. Motility starts, as in all vertebrates, with a slight bending of the head by a flexion of neck muscles. Contractions progress from there in cephalo-caudal direction. At $4\frac{1}{2}$ days (stage 25), typical S-flexures starting in the neck extend to the leg level, and shortly thereafter (stage 26) to the tail. The pattern of integrated cephalo-caudal waves prevails for a considerable period. When the wings and legs acquire motility at about $6\frac{1}{2}$ days (stage 29), their movements are at first synchronous with the trunk movements. Up to this point, the behavior of the chick embryo conforms to that of *Ambystoma*. However, in subsequent stages the motility pattern becomes increasingly complex. Whereas in early stages the waves begin at the neck level and extend from there in cephalo-caudal direction, one finds with increasing frequency that waves start in the trunk and either spread to posterior trunk and tail only or in both directions. Regularity is disrupted in another respect. Wing and leg movements are completely independent of each other; neither the left and right partners nor wings and legs are coordinated in any way. Limbs also get out of phase with neck and trunk movements; for instance, several limb movements can be performed during one S-wave. Furthermore, isolated movements confined to one or both legs or to one wing or to the tail occur occasionally during a short lull in the general activity. In other words, the initial pattern of cephalo-caudal waves gives way to a generalized motility of all parts that are capable of moving, with no sign of coordination. As was pointed out, these bursts of activity occur intermittently up to 13 days and from then on more or less continuously. From $6\frac{1}{2}$ to 7 days on, one cannot recognize integrated behavior

patterns by any stretch of imagination. At later stages, the lower jaw, the tongue, the eyeball, and the lower eyelid enter into the picture, one after another, and participate in the general activity without being integrated with each other or with the trunk or limbs. Only a film could give an adequate picture of the performance, which includes a jerky turning of the head, rapid vigorous kicks and flexions of the limbs, and slow opening and closing of the beak and lower eyelid, all these movements being performed in a wholly irregular and unpredictable sequence. Any part or combination of parts can be active, while other parts are temporarily quiescent, or all parts can move simultaneously but out of phase with each other.

Such a picture defies all of Coghill's concepts. Continuity of integration of behavior from start to hatching simply does not exist in the chick embryo. The initial sequence leading to cephalo-caudal waves is probably nothing else but the behavioral manifestation of the process of cephalo-caudal maturation of the neuromuscular connections of the trunk, including an early longitudinal interconnecting system. The very early appearance of the latter, and its connection with the motor neurons, has been described by Visintini and Levi-Montalcini (1939). Some authors refer to this sequence as 'swimming,' but little is gained by this designation.

The random character of motility indicates the absence of sensory stimulation as a patterning device. On the contrary, the spasmodic random bursts are suggestive of spontaneous, autonomous discharge. Apparently, different motor neuron groups fire independently of each other, and the impulses spread without reference to those pathways and circuits which integrate behavior from prehatching stages on.

Before we turn to mammals, we wish to point out that one can hardly expect in the chick embryo the same continuity of integrated behavior that is characteristic of the salamander embryo and larva. The structure of the embryonic nervous system is profoundly different in the two forms. In *Ambystoma* embryos, the starting point of the cephalo-caudal waves is guaranteed by the localization of special commissural cells, the 'floor plate cells,' at the cervical level. Sensory input feeds into these cells.

They do not exist in the chick embryo. The early integration of forelimb movements with trunk movements in *Ambystoma* has its basis in the structure of the axons of the primitive motor neurons,.which send collaterals to somite muscles and to limb muscles. In the chick embryo, the median and lateral motor columns, serving trunk and limb muscles, respectively, are separate units from the start. In *Ambystoma* the primitive motor neurons are actually multi-purpose cells which also provide longitudinal interconnecting pathways (see Coghill, 1929a, figs. 9–11). In the chick embryo, both systems are separate; other differences could be listed. These differences are readily understood in their biological context. In *Ambystoma*, neurogenesis and concomitant genesis of behavior are geared to an early readiness of the hatched larva to cope with life in an external milieu. The special temporary neural structures mentioned above serve this purpose. The swimming performance that is required after hatching emerges from earlier motility in the most direct and rapid way. On the other hand, the chick embryo, developing as it does in the sheltered milieu of the amniotic cavity, is not exposed to selective pressure of the same sort as is the salamander. It can afford spastic, unintegrated motility. If one seeks an adaptive significance of this performance, one can perhaps find it in the requirements for normal differentiation of skeleton and muscles. Embryonic limbs deprived of their innervation (Hamburger and Waugh, 1940), or kept immobilized by infusion of curare over prolonged periods (Drachman and Coulombre, 1962), frequently show ankylosis and other skeletal abnormalities. Furthermore, the discharge of impulses into muscles may be a necessary condition for the maintenance of their normal metabolism and structure.

The integrated embryonic behavior of *Ambystoma*, far from representing a generalized vertebrate pattern, is highly specialized. We have to inquire next whether that of the chick embryo is characteristic of amniotes in general.

BEHAVIOR PATTERNS IN MAMMALIAN EMBRYOS

We cannot undertake even a cursory summary of the extensive literature on the beginning of behavior in mammalian and human embryos.

Reviews may be found in Barron (1941), Carmichael (1954), Coghill (1940), Hooker (1952), and Windle (1940, 1944). Extensive data are available for the cat, rat, guinea pig, sheep and human embryos and fetuses. The discussion will be limited to a few basic issues.

Mammalian behavior was the central issue in the controversy which arose over Coghill's contention that behavior is integrated from the start and that local activity arises through individuation from a total pattern. Some investigators were impressed by the observation that localized responses — for instance, of the limbs — could be elicited very early, and in some instances even earlier than general body movements. On the basis of this kind of information, the reflexogenic theory of the origin of behavior was formulated in antithesis to Coghill's. It states that local reflexes are the units of embryonic behavior and that they are combined secondarily into integrated patterns. We shall see that in the subsequent controversy several issues were confused and that we are actually not dealing with a real antithesis of two mutually exclusive concepts. Despite the complexity of the situation, and despite species differences, a certain degree of clarification can be achieved and some generalizations can be established if one looks at the facts unbiased by the old controversy.

For the discussion it is important to remember that mammalian embryos at the inception of motility are more advanced in general body development than salamander or chick embryos (see Fig. 2) and that the sensory and motor innervation of the limbs is already functional in the mammal at that stage.

We inquire first whether the early phase of mammalian motility represents a sequence, similar to that found in *Ambystoma* and the chick embryo, leading to S-flexures and undulating cephalo-caudal waves. This question can be and should be divorced completely from that of early reflexes. It does not matter whether such a pattern, if it occurs, is a reflexogenic response or spontaneous action; the essential question is whether it does exist. A scrutiny of the literature shows that a case can be made for such a sequence. There seems to be general agreement that the first motion that can be observed in any mammal is a flexion of the neck and that, subsequently, contractions extend to trunk somites. Positive statements to that effect were made for the rat, sheep, guinea pig, cat, and human embryos. Angulo (1932) finds this sequence in the rat, beginning at 15½ days. It is definitely stated that first the forelimbs and then the hindlimbs are included in this total pattern. In this case, these movements were not observed as spontaneous movements but as results of stimulation of the snout, which is the first reflexogenous zone in the rat embryo. In the sheep, beginning on the 34th day of gestation, stimulation of the maxillary nerve elicits the same sequence, again including first the forelimbs and then the hindlimbs (Barcroft and Barron, 1939; Barron, 1941, p. 17). The earliest spontaneous movements of the sheep embryo, beginning at the same stage, are not described as a clearcut sequence, but neck, trunk, and forelimbs are involved from start, and it is stated that "generally, all the moving parts appear to be bound together or associated in their activity"; however, "spontaneous movements of quite localized parts of the embryo are quite often seen" (Barcroft and Barron, 1939, p. 480). The observations of Hooker (1952) on human embryos are in line with those mentioned so far. Beginning at 7½ weeks, stimulation of the maxillary and mandibular nerves results first in flexion of the neck and upper trunk. Later on, more posterior parts, including first the upper and then the lower extremities, are incorporated. "These early responses would appear to constitute a continually expanding total pattern" (Hooker, 1952, p. 68). The structural basis for the flexion of the neck, and the pathways involved, have been described in detail by Humphrey (1952). Carmichael and collaborators have given a detailed description of the early behavior of the guinea pig (see Carmichael, 1934, 1954). It is certain that the first movements, spontaneous or elicited by stimulation of the ear region, which is the first reflexogenous zone in this species, result in neck and upper trunk movement and synchronous but possibly independent forelimb movement. Subsequent stages show complex reflexogenous response patterns. Windle and Griffin (1931) have described spontaneous motility in cat embryos from 24 days to near birth. Again, movements begin at the neck and expand gradually;

on the 30th day, leg movements are observed in conjunction with rump movements, and the authors compare the pattern of rotation and flexion of head and trunk with the righting reflexes in postnatal life. They still find in 31-day to 33-day embryos total movements in the cephalo-caudal direction. This work was done before Windle started his observations on reflexes, which resulted in a complete shift of emphasis and in his reflex theory. However, the earlier observations are thereby not invalidated or repudiated by the fact that local reflexes can be elicited as early as the earliest spontaneous movements are performed.

The only objection that might be raised to this and all other studies is the ever-recurring skeptical question concerning the normalcy of the conditions under which the observations were made. Does the anesthesia of the mother that was applied in some of these experiments, or the possible hypoxemia of the embryo or fetus, distort the picture? In view of the concurrence of the observations relating to different species, this objection can hardly be considered valid. It is very probable that in mammals in general the early differentiation of the neuromuscular system proceeds in the same way as it does in submammalian embryos; it expands from the neck in the cephalo-caudal direction, and results in the performance of S-flexures of neck and trunk and including the limbs. Mammals differ from the chick in this early phase in one respect; in the mammalian embryo, some reflexogenous zones become functional very early, and, at least in some forms, the total pattern can be elicited from the start by appropriate stimulation. In the chick, S-flexures are performed spontaneously. The cephalo-caudal waves no longer prevail at the stage when the embryo becomes sensitive to stimulation (see p. 353).

In a very general way, then, one can claim a common denominator for the first phase of the ontogeny of behavior in all vertebrates, from teleost to mammal. However, one has to keep in mind that the underlying structural basis is different in details in different forms (see p. 353). It is of interest in the context of the previous discussion that this basic common feature concerns the neuromuscular system and not the sensory system.

Does mammalian behavior remain integrated in the subsequent stages? This question is more difficult to answer than in the chick, due to methodological limitations. We do not have available systematic observations on spontaneous motility, which in the chick embryo gave such a clear picture of the discontinuity of integrated behavior. Our only source of information is provided by stimulation experiments that by their very nature cannot give an unequivocal answer to our question. It is true that, in some mammalian embryos, local stimulation results in localized responses which do not spread to any extent and that such responses can be elicited very early. But do such stimulation experiments tell us anything concerning total integrated motility? All they can show is that the reflex circuit for a particular response was structurally complete when the experiment was performed. It is not too surprising to find that reflex responses representing localized function in the adult show the same restricted performance when they first appear in the embryo. But to use such results as an argument against the existence of integrated behavior is a non sequitur. Actually, not even those observers who favor Coghill's viewpoint claim that mammalian motility is totally integrated throughout the prenatal period. What they did observe was the emergence of *partial* patterns of different degrees of complexity, such as righting, grasping, sucking, and respiration. We have extensive information concerning this aspect of behavior development for a number of species, but, important as it is, we cannot discuss it in detail. We refer to the reviews of Carmichael (1934, 1954) and Hooker (1952).

Let us turn to the opposing view. Is there evidence for the contention that local reflexes are the basic units of mammalian embryonic behavior and that these units become integrated secondarily? Two arguments can be raised against this notion. First, we have shown at the beginning of this chapter that, in the initial phase, which is crucial for the theory, integrated behavior leading to S-flexures and including limb movements does exist. Second, spontaneous motility has been observed in some forms. Considering our experience with the chick embryo, this phenomenon strongly suggests the existence of a component of autono-mous, nonreflexogenous motility which would be contrary to the basic tenet of the reflex theory.

In summary, we must acknowledge that we have no clear conception of the pattern of behavior in intermediate and late stages of mammalian development. It is in all probability not of the integrated type found in the salamander and teleost. A reinvestigation of spontaneous motility might give valuable information on this point, as it did in the case of the chick embryo.

A RECONCILIATION OF DIFFERENT VIEWS ON
BEHAVIOR PATTERN

The discussion of the last section has shown clearly that an impasse is reached if embryonic behavior is viewed exclusively from the traditional opposing positions. Neither hypothesis can be generalized to provide an over-all theory that does justice to all facts. However, reduced to a more limited scope, the two viewpoints contain elements of truth that are not mutually exclusive.

The key to a uniform conception can be found in a consideration of the neurological correlates of the behavior pattern. If, for the sake of the argument, the complex situation is reduced to a greatly oversimplified scheme, then one can conceive of a dual hookup of the motor apparatus, on the one hand with local reflex systems, and on the other hand with an integrating system such as the fasciculus longitudinalis medialis that ties local action together with more general action. A similar idea was expressed by Barcroft and Barron (1939), though in a less general form. We shall present some examples from their observations on the sheep fetus to illustrate the point. They have found that, in early stages, stimulation of the maxillary branch of the trigeminal nerve results in a pattern of total motility that includes the limbs. In later stages, local stimulation of the limb skin results in strictly localized action of the limb. One performance of the leg does not preclude the other, just as in the adult sheep the leg can be part of an integrated walking pattern or can respond with withdrawal, depending on the appropriate stimulation. Another example concerns the movement of the mouth. It was observed that the jaw participates in the above-mentioned total activity elicited from

179

stimulation of the maxillary nerve. It does so at first occasionally, but "by the 40th day, the opening of the mouth is usually established as a definite part of the reaction" (Barcroft and Barron, 1939, p. 485). Somewhat later, the opening of the mouth can be elicited as a strictly local response by stimulation of another branch of the fifth nerve — namely, the mandibular nerve. In both instances, it depends on the position of the reflexogenous zone whether a particular action is performed as a local response or as part of an integrated pattern. The two performances are not mutually exclusive, and, in the consideration of embryonic behavior, it should be acknowledged that the demonstration of the existence of one does not reflect on the existence or nonexistence of the other.

Barcroft and Barron have pointed out the biological significance of the dual hookup for the case of limb function: "The peripheral distribution of the maxillary nerve to the anterior advancing pole of the animal determines that nerve from the outset of development as one whose sensory information is of importance to the animal as a whole, hence its ability to evoke a total response via the bulbo-spinal system with which it makes central connections at the lower end of the rhombencephalon" (Barcroft and Barron, p. 498). On the other hand, proprioceptive and exteroceptive stimuli evoked in muscles or skin of the limb and resulting in withdrawal give information that is "of first concern to that segmental region and only secondarily does it concern the total animal" (Barcroft and Barron, p. 499). In the examples that were cited, both types of action were reflexogenic in origin. However, the basic concept of the dual hookup is not tied to reflex response. Self-generated, automatic activation of motility can be imagined to feed likewise into either one or the other system, giving rise to either local or more generalized activity.

An instructive experiment of Holtzer and Kamrin (1956) shows for the salamander *Ambystoma* that the local hookup can differentiate independently of the other. The salamander forelimb is innervated by the third to fifth spinal cord segments and ganglia. These segments were isolated completely by extirpation of spinal cord pieces lying immediately anterior and posterior to them. The operation was done in premotile embryonic stages. After attainment of the larval stage, typical local withdrawal reflexes could be elicited in the isolated system, but not postural reflexes. The local reflex apparatus has reached functional maturity independently of other connections. In summary, the structural elements of *integrated* behavior that involve many parts exist and differentiate side by side with elements serving *localized* motility.

If one accepts this concept of a dual (or multiple) hookup, then some of the major controversial issues resolve themselves. Let us take a last critical look at the reflex theory from this viewpoint. When reduced to its essential core, it amounts to the assertion that the local reflex circuits are *invariably* established before the motor neurons involved in them connect up with longitudinal tracts or other integrating pathways. From our vantage point, the question of what comes first, local response or total integrated action, becomes irrelevant. The time schedule of neurogenesis is different in different forms, as has been discussed in previous sections. We recall, for instance, that in the chick embryo the collaterals from the fasciculus longitudinalis medialis establish synaptic connections with motor neurons a few days before the collaterals from the dorsal funiculus grow into the gray matter and thus close the reflex circuits, but, in some mammals, the two events seem to be nearly synchronous. Such differences are interesting from the viewpoint of neurogenesis, but they are of no particular significance for the basic issue under discussion.

Likewise, the troublesome issue of individuation loses much of its import. This concept is meaningful only if the behavioral process which it implies — namely, the emancipation of local action from a total action pattern — can be related to a neurogenetic differentiation process. To illustrate this point, we recall previous discussions of limb innervation. In the salamander, the early system of motor innervation by collaterals of multipurpose motor-integrator neurons is replaced in a later stage of development by the customary specialized motor neurons. Here, behavioral individuation has a structural correlate, but, in all higher forms, the lateral motor columns which are designated for limb innervation are assembled very early and administer to limb motility from its beginning. One could speak of individuation, in a certain sense, if these lateral motor neurons were *invariably* hooked up *first* with the longitudinal integrating tracts and only later with local reflex systems. But this sequence does not hold true in all cases, as we have seen, and, moreover, this point is not particularly important. Thus, the concept loses its general significance and should be abandoned.

PERIODICITY OF MOTILITY

We have stated in the introduction that the spontaneous motility of the chick embryo is performed in regular cycles. This periodicity has apparently escaped previous observers, with the exception of Visintini and Levi-Montalcini (1939), who gave some data on the time pattern. This rhythmicity has a number of interesting aspects, some of which we have begun to analyze. Some data have been published (Hamburger and Balaban, 1963; Hamburger, 1964).

From the beginning of motility at $3\frac{1}{2}$ days to the 9th day, the embryo performs in motility cycles that last, on the average, 60 seconds. Only about 5 to 15 seconds of each cycle are spent in bursts of activity; they are followed by about 30 to 50 seconds of complete inactivity. From about 9 days on, the activity phase gradually increases in length, and the inactivity phase decreases, though not proportionally (Fig. 3). From 13 days on, the cyclic nature of motility is lost, and motility is nearly continuous. Individual cycles show great variability; activity and inactivity phases vary independently of each other, but the statistical distributions give a consistent

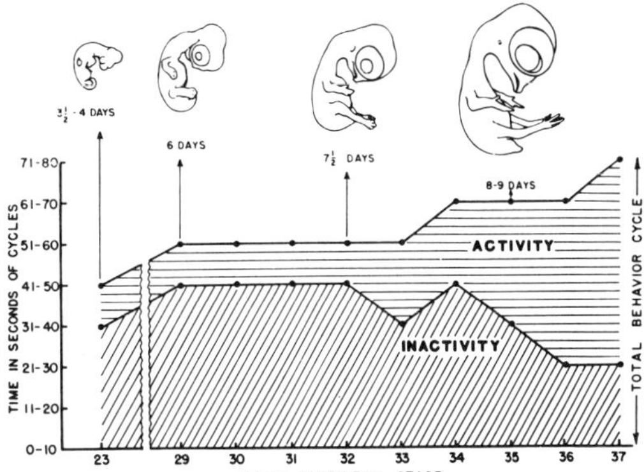

FIG. 3. AVERAGE DURATION OF ACTIVITY AND INACTIVITY PHASES IN TOTAL BEHAVIOR CYCLES, AT DIFFERENT DEVELOPMENTAL STAGES (Measured in seconds).

picture. We first recorded the cycles with a manually controlled Kymograph and then adapted a Sanborn polygraph for our purposes. Even the finest movements are recorded by placing electrodes either in direct contact with the embryo or in the fluid near the embryo (Fig. 4). The amnion is usually removed, and, in later stages, from 7 days on, the chorioallantoic membrane over the embryo is transected and displaced to the side. This can be done with a minimum of bleeding. Embryos thus exposed will perform normally for up to 24 hours when kept in a temperature-controlled small chamber.

We have inquired first whether one localized nerve center serves as the *pacemaker* that triggers the spontaneous activity. For this purpose, transections of the spinal cord were performed. In one set of experiments, a piece of spinal cord, several somites in length, was extirpated

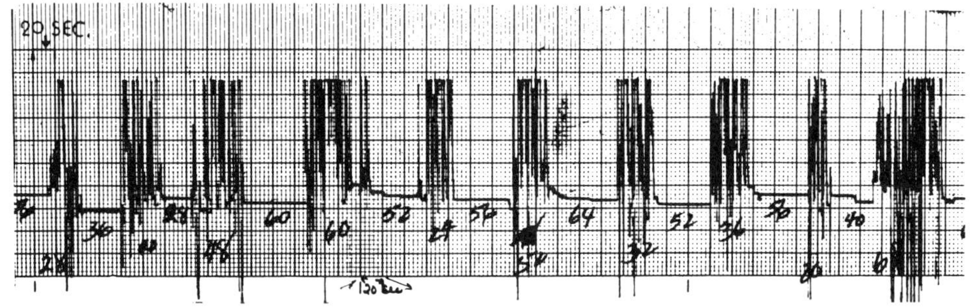

FIG. 4. CYCLES OF MOTILITY OF A NORMAL STAGE 37 EMBRYO

Part of a continuous automatic recording of the motility cycles of an embryo in stage 37, representing a total 15-minute period. The base line represents the inactivity phases. The numbers indicate the duration of a phase, in seconds.

in 2-day embryos (stage 12–13), at the cervical or brachial or thoracic levels, or at 2 levels simultaneously, and the embryos were then reared to 7½ and 10 days of age (Fig. 5). In another set of experiments, embryos of later stages, from 8 days on, were decapitated by placing a loop of fine thread around the neck and severing the head by a rapid pull on the ends of the loop. There is practically no loss of blood, and the trunk performs normally for as long as 24 hours afterwards, following a short period of recovery from the shock. The latter technique (developed by Dr. Balaban) has several advantages: it is extremely simple, there is no mortality, and it permits a recording of activity of the same embryo before and after transection. We have selected two stages for extensive analysis: stage 32 (7½ days) — that is, before reflexes can be elicited regularly — and stage 37 (11 days), after the establishment of functional reflex arcs and before the cyclic periodicity gives way to continuous activity.

In all instances, isolated parts retain the capacity for spontaneous rhythmical activity. A single pacemaker in the brain is thus ruled out. In spinal cord transections, the parts above and below the gap retain their typical cycles, but they are not synchronized. One part may be in an activity phase and the other inactive, or activity may overlap (Fig. 6). In cervical transections, an activity wave may start at the

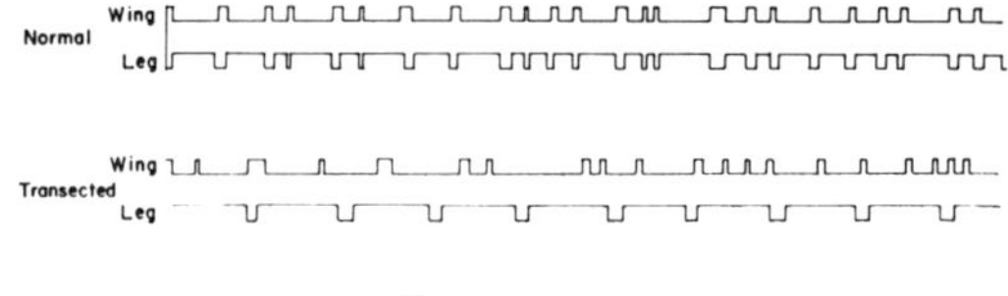

FIG. 6. ACTIVITY AND INACTIVITY PHASES OF A NORMAL EMBRYO AND AN EMBRYO WITH A THORACIC TRANSECTION (STAGE 32)

Running records. Activity phases for the wing are above the base line, those for the leg below the base line.

anterior cut end and move caudad, or it may start in a midtrunk region, as was found in normal embryos. In thoracic sections, motility may spread from the cut surfaces in opposite directions. The experiment shows that all parts of the spinal cord can initiate motility. We have to postulate that centers of activation are spread over the spinal cord. But does the brain have no influence on the motility of the trunk? We could not detect differences between normal embryos and isolated parts in the pattern of movements, but the recordings revealed an influence on the number of cycles in the spinal cord. The number per unit of time was reduced. The influence is not quite the same in stages 32 and 37, and the difference proved to be of importance for the analysis.

Brain influence at stage 32 (7½ days)

In isolated parts, the total time spent in activity during a given observation period is reduced by 25 per cent. This is particularly well demonstrated by transections of the cord at the thoracic level, where wing and leg periodicities can be recorded simultaneously. The anterior part with the wing shows normal timing, and the leg shows the reduced activity. Parts that are isolated by cervical cord transection show the same amount of reduction. This implies that the brain is the source of additional stimulation. Our data permit us to define the role of the brain more precisely. The data were grouped in classes of 5-second intervals for activity phases and 10-second intervals for inactivity phases. The histograms for normal activity show a preponderance of short activity phases of 1 to 10 seconds duration. These short phases are essentially lost after isolation from the brain (Fig. 7). Another detail is significant. The duration of

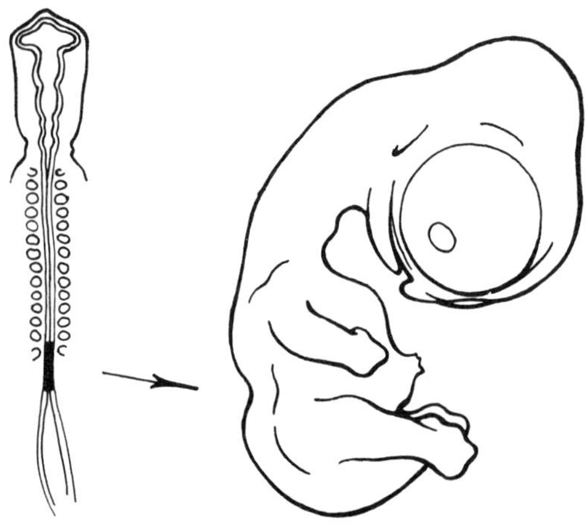

FIG 5. EFFECTS OF EXTIRPATION OF A SEGMENT OF THE THORACIC SPINAL CORD

Left, site of the segment extirpated at stage 11 or 12. Right, the same embryo in stage 30 (7 days), with a gap in the thoracic spinal cord.

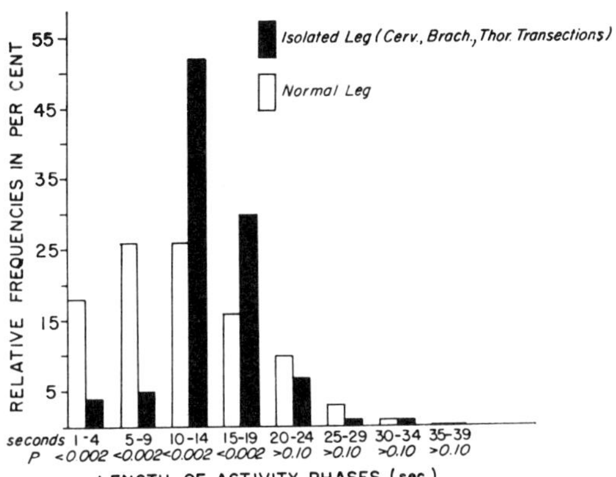

FIG. 7. COMPARISON OF RELATIVE FREQUENCIES (IN PERCENTAGES) OF DIFFERENT LENGTHS OF ACTIVITY PHASES IN NORMAL EMBRYOS AND EMBRYOS WITH SPINAL TRANSECTIONS (STAGE 32)

the activity phases of isolated parts has the same range of variation that we find in normal embryos. No especially long phases are added. There is merely a shift from both extremes toward the middle of the range, but not a change that would indicate a profound deviation from normal. The inactivity phases of isolated parts show also a shift to longer periods of inactivity, but, again, the duration remains within the normal range. The shift to longer inactivity phases is to be expected, since there are fewer activity phases per unit of time.

The effects of cranial on spinal levels can be explained tentatively in the following way. Triggering devices are distributed over the brain and the spinal cord. The spinal cord has its own intrinsic rhythm which, however, can be modified by activating stimuli from the brain. If the two are not in phase, the impulses propagated from the brain would occasionally cut into inactivity phases of the spinal cord and initiate extra activity, thus increasing the number of cycles per unit of time. The notion of extra stimulations of the spinal motor centers by the brain explains the shorter inactivity phases and also the greater amount of total activity of intact embryos compared to transected parts. However, it is not evident why these extra cycles should be predominantly of short duration.

This point raises the question of how the *duration* of an activity or inactivity phase is determined. We have found it helpful to look

at the motility cycles in terms of two devices, one that triggers, or turns the activity on, and one that turns it off. The two devices could either reside in the same neural system or in two separate systems. The first alternative would mean that both the on and the off mechanism which determine short phases reside in the brain and that these short activity phases are superimposed in their entirety on the longer ones in the spinal cord. Alternatively, one can postulate that certain centers that are distributed over brain and spinal cord are responsible only for the triggering of activity phases and that the mechanism for turning off resides in other centers, as for instance in the motor system. One can imagine that the latter, once triggered, would continue to fire until some physiological threshold is reached in the motor cells. The data for stage 32 do not permit a decision, but those for stage 37 favor the second alternative.

Brain influence at stage 37 (11 days).

Between the 8th and the 11th day, overall activity is stepped up considerably. The average length of the activity phases has increased approximately threefold, and there is no longer a predominance of short phases (Fig. 3, 8). Conversely, the inactivity phases have become shorter. What is the effect of decapitation or transection at this stage? This operation has no effect on the length of the activity phases (Fig. 9). On the other hand, one finds in the decapitated and transected embryos a significant increase in the length of the inactivity phases

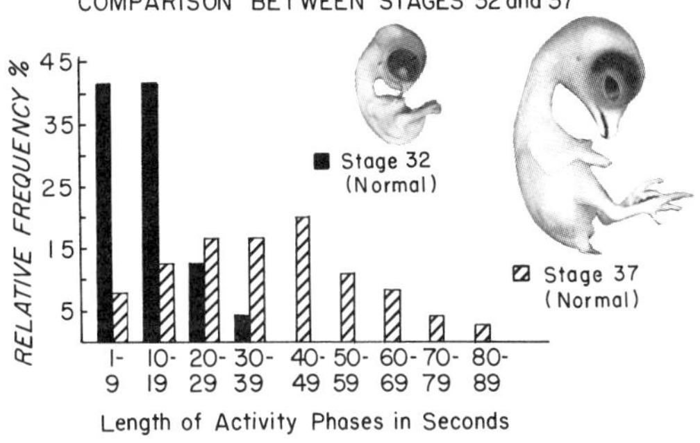

FIG. 8. COMPARISON OF RELATIVE FREQUENCIES (IN PERCENTAGES) OF DIFFERENT LENGTHS OF ACTIVITY PHASES IN STAGES 32 AND 37

Note the shift to longer activity phases in the older embryo.

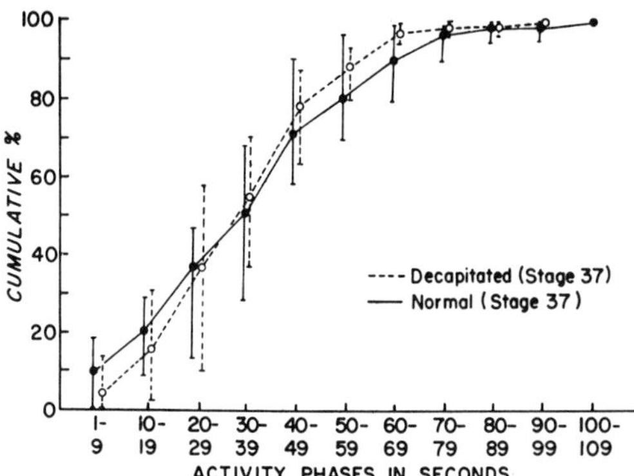

FIG. 9. COMPARISON OF DURATION OF ACTIVITY PHASES
BETWEEN NORMAL AND DECAPITATED EMBRYOS
(STAGE 37)
The graph shows the cumulative percentage distribution of activity phases of different durations. The vertical bars represent the ranges in the duration of activity phases of all embryos on the cumulative curve.

(Fig. 10). The cumulative curves for inactivity of normal embryos and isolated parts show no overlap despite a considerable range of variation within each class. In intact embryos, almost 50 per cent of all inactivity phases are shorter than 30 seconds. In decapitated and transected embryos, the 50 per cent level is attained at a duration of 60 to 69 seconds. In other words, the duration of the activity phases is not changed, but there are fewer per unit of time in isolated parts. It is of interest to note that the method of isolation makes no difference. The isolated parts of embryos in which the

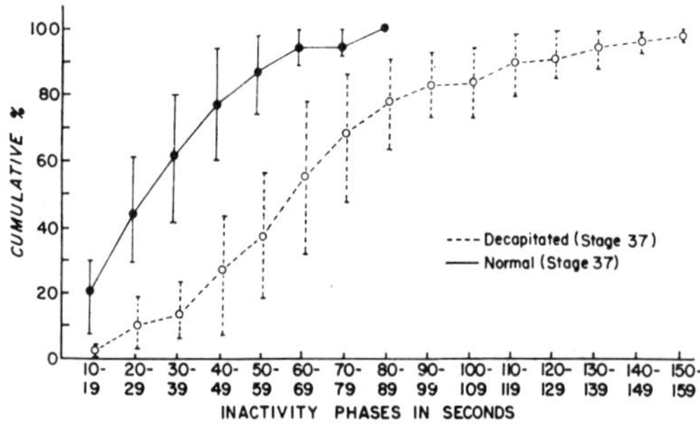

FIG. 10. COMPARISON OF DURATION OF INACTIVITY
PHASES BETWEEN NORMAL AND DECAPITATED EMBRYOS
(STAGE 37)
The graph shows the cumulative percentage distribution of inactivity phases of different durations. The vertical bars represent the ranges in the duration of inactivity phases of all embryos on the cumulative curve.

cervical spinal cord has been transected at 48 hours showed the same lengthening of inactivity phases at 11 days as the embryos which had been decapitated shortly before the recording. This comparison shows that the chronic isolation has not interfered with the normal development of the neural structures that are responsible for the performance of typical motility, nor can the shock or loss of blood caused by decapitation be made responsible for the longer phases of inactivity.

The observation that the duration of activity phases is the same in intact embryos and in isolated parts indicates that the *turning-off* device which controls the duration of activity operates independently of the brain. The experiment has therefore dissociated the turning-on device from the turning-off device. It gives substantial support to the idea that each resides in a different neural system. Our suggestion that the duration of activity is a function of some physiological property of the motor system or a system related to it is in agreement with the experimental data.

In another experiment, we have been able to dissociate the trigger of activity from the turning-off device in a different way. We have inhibited all motility by exposing 11-day embryos to a mixture of 10 per cent CO_2 and 90 per cent O_2 for 15 minutes and recorded the time pattern of recovery after return to normal atmosphere. The recovery may take several hours. We were not interested in the length of time that it took an embryo to recover fully, but we used the recovery process as a device for varying the percentage of activity during arbitrary units of time (15 minutes). The percentage of activity per unit of time is determined by both number and duration of activity phases. We inquired whether the change in percentage of activity could be accounted for by changes in either one of these two parameters or by a combination of both. The recovery of each animal contributed a number of time units of different percentages of activity, and the data were pooled according to the percentage of activity for these units. For this experiment, simple small perfusion chambers were used; they can be placed into a larger temperature-controlled box and connected with the polygraph. We have tabulated number of cycles, and mean length of activity and of inactivity phases in seconds, for the different percentages of activity (Table 1). We find that the number of cycles per unit of time was constant, whether an embryo was active 30 per cent of the time (that is, half-recovered) or 60 per cent of the time (that is, fully recovered). This means that the turning-on device, which gives the starting signal for each activity phase, operates at its normal rhythm from the beginning. Either the trigger recovers instantaneously or it is not affected by the CO_2 treatment. On the other hand, the length of the activity phases, once initiated, changes during recovery; it increases concomitantly with a corresponding decrease in the length of the inactivity phases. This implies that it is the turning-off mechanism which is affected by the CO_2 treatment and which undergoes recovery.

Here again, the triggering and the turning-off devices have been dissociated from each other, but in a different way from that in the previous experiment. In that experiment, the length of activity, for which the *turning-off* device is responsible, remained constant and in the present experiment, the periodicity of the *turning-on* signal remained constant. Our previously stated working hypothesis can be applied again. While the activating mechanism located in unspecified structures in brain and spinal cord shows no effect of CO_2, the motor neurons, or structures associated with them, are incapacitated

and gradually regain their capacity for sustained performance.

CONCLUDING REMARKS

In an endeavor to clarify the complex issues in the field of the development of behavior, we have separated in our presentation the problem of how motility is generated from the problem of what form or pattern it assumes in different species and in different stages of development. Although information on both issues is deplorably incomplete, some trends that might give a new frame of reference for future investigations can be delineated.

We have distinguished between two general modes of origin of motility: spontaneous and reflexogenic. In the chick embryo, the two can be shown to be separate and independent constituents of embryonic behavior, and our survey of data on other forms indicates that this basic concept can be generalized. The question of whether spontaneity is based on automatic, self-generated discharge of neurons or is stimulated by circulating or other systemic agents cannot be decided on the basis of available information. In our study of motility in the chick embryo we have found strong indications that we are dealing with self-generated nervous activity, but additional experimental work must be done to clarify the influence of 'endogenous stimuli' on motility.

A marked shift of emphasis from the impor-

TABLE 1

Recovery of Activity of Chick Embryos (Stage 37) from Inactivation by Exposure to 10% CO_2 and 90% O_2 for 15 Minutes

% ACTIVITY DURING 15 MIN. TIME UNITS	% 26–30	% 31–35	% 36–40	% 41–45	% 46–50	% 51–55	% NORMAL RANGE 56–60	% 61–65
Mean Number of Motility Cycles	9	10	10	11	10	10	11	10
Mean Length of Activity Phases, in Seconds	28	33	33	36	45	44	48	55
Mean Length of Inactivity Phases, in Seconds	77	65	54	47	51	40	36	34

tance of reflexogenic to that of spontaneous, nonreflexogenic motility is inherent in our presentation. We have postulated the primacy of the spontaneous over the reflexogenous component on the basis of several arguments. The motor components usually differentiate in advance of the sensory components, both structurally and in terms of overt behavior. In the chick and teleost embryos, which represent two very different levels of organization, overt motility starts long before the reflex mechanism is structurally completed. In both, the nonreflexogenic component can be shown to persist to larval or hatching stages, respectively. Furthermore, in the chick embryo, the reflex mechanism has no influence on the rhythmic spontaneous motility cycles of midincubation stages, and no evidence could be found for the contention that overt motility in prehatching stages is elicited by adequate reflex stimuli. There is no evidence that the situation is different in mammalian embryos and fetuses.

If we are correct in assuming that spontaneous, self-generated motility is the major issue in the development of behavior, then we can establish a direct link to an equally important aspect of adult behavior. Both in neurophysiology and ethology, the significance of the autonomous action of the nervous system, independently of sensory input, is increasingly recognized. The situation is epitomized by Bullock: "The fact of spontaneity introduces a tremendous contribution from the system itself into its behavior — on top of the system-characteristic determinants that are due to anatomical connections and transfer functions. We observe the capacity to discharge spontaneously under appropriate conditions in axons as well as in nerve cell bodies, with all phasic or triggering input removed. Besides the beat of the heart, there is reason to believe in spontaneity as the basis for many forms of peristalsis, of swimming, flying, breathing and perhaps even the elementary rhythm of walking" (1962, p. 103). And furthermore: ". . . Central patterning is the necessary and often the sufficient condition for determining the main characteristic features of almost all actions, whether stimulus-triggered or spontaneous" (1961, p. 56). Perhaps the spontaneous motility of the chick embryo represents the embryonic antecedent of this component of

the adult action system, even though it is not organized or integrated in the embryo. We shall come back to this problem presently.

So far, our own observations and experiments have been limited to investigation of overt motility performance. Clearly, neurophysiological investigations are indicated as a supplementary part of the analysis. Other investigators have made the beginning of direct recording from embryonic nerve tissue. These studies support our basic viewpoint in that they demonstrate the capacity of embryonic neurons for self-generated discharge.

Electrical activity has been recorded from nerve tissue of chick embryos both in vivo and in vitro. Peters, Vonderahe, and Huesman (1960) have found the beginning of electrical activity on the 13th to 14th day of incubation in the cerebral hemispheres of quiet embryos. Optic lobes and cerebellum show their first discharges about 2 days later. The amplitude increases with age. Electrophysiological experiments on the brains of mammalian fetuses are reviewed in Flexner (1955). The long-range tissue cultures of spinal cord of chick and human embryos and rat fetuses which were shown by Crain and Peterson (1963) to be electrically excitable and structurally healthy for months will probably be very useful for the further analysis of our problems.

Of great interest in our context are the recordings obtained by Cunningham (1962) and by Cunningham and Rylander (1961) from cultures in vitro of different brain regions of the chick embryo. Small pieces of telencephalon, cerebellum, and pons of 10-day to 15-day embryos were explanted and recordings were made over periods of up to 12 days. The recordings showed remarkably regular repetitive activity patterns which persisted for days with only slight changes. The investigators found differences between brain parts with respect to time patterns and amplitude of discharges. Alternations of activity and inactivity phases are indicated in some recordings; they are of a similar magnitude as those found by us in younger embryos.

One of the major unresolved problems concerns the emergence of patterned, integrated adaptive behavior from the unintegrated generalized motility of embryos of higher vertebrates. Whereas, in teleost and amphibian embryos, Coghill's scheme of a continuity of integration from the beginning of motility to free-swimming larval life seems to be materialized, we have described a completely unintegrated random activity of the chick embryo during the middle and later part of the incubation period, and the frequent reference of students of mammalian embryonic behavior to generalized 'mass action' indicates that the same situation exists in mammals. How is the transition made between this generalized motility and the integrated activities during and after hatching or parturition? What are the antecedents of patterned activity, both in structural and behavioral terms? How do exteroceptive and proprioceptive stimuli, which according to our view are essentially noneffective in the embryo, take control of activity patterns? Obviously, from the viewpoint of the behaviorist and of the neurophysiologist, this is the most serious gap in our understanding of the origin of behavior. We are now studying the prehatching and hatching behavior of the chick embryo from this viewpoint.

Finally, a brief comment on *rhythmicity*. We are entering here into another area that is of acute interest in present-day biology. Periodic phenomena in plants and animals are being investigated from a variety of viewpoints (see, for instance, Wolf, 1962). The chick embryo, and probably other embryos, such as those of teleosts, can perhaps make a contribution to this field. The periodicity of the motility of the chick embryo between $3\frac{1}{2}$ and 12 days has been discussed in the preceding section. One feature of this performance deserves special mention. Whereas most biological rhythms are either long-period phenomena, such as the circadian rhythm, or short periodicities, from heart-beat timing down to milliseconds in nervous discharges, the chick embryo performs in cycles of the order of seconds and minutes. Thus it takes an intermediate position between similar phenomena.

ACKNOWLEDGMENT

The research done in our laboratory is supported by Grant No. NB–03143 of the Public Health Service, National Institute of Neurological Diseases and Blindness.

LIST OF LITERATURE

ADRIAN, E. D. 1931. Potential changes in the isolated nervous system of *Dytiscus marginalis*. *J. Physiol.*, 72: 132–151.

ANGULO, A. W. 1932. The prenatal development of behavior in the albino rat. *J. Comp. Neurol.*, 55: 395–442.

BARCROFT, J., and D. H. BARRON. 1939. The development of behavior in foetal sheep. *J. Comp. Neurol.*, 70: 477–502.

BARRON, D. H. 1941. The functional development of some mammalian neuromuscular mechanisms. *Biol. Rev.*, 16: 1–33.

——. 1950. Genetic neurology and the behavior problem. In *Genetic Neurology*, P. Weiss, (ed.), pp. 223–231. University of Chicago Press, Chicago.

BRIDGMAN, C. S., and L. CARMICHAEL. 1935. An experimental study of the onset of behavior in the fetal guinea pig. *J. Genet. Psychol.*, 47: 247–267.

BRINLEY, F. S. 1951. Effects of curare on fish embryos. *Physiol. Zool.*, 24: 186–195.

BULLOCK, T. H. 1961. The origins of patterned nervous discharge. *Behaviour*, 17. 48–59.

——. 1962. Integration and rhythmicity in neural systems. *Am. Zool.* 2: 97–104.

CARMICHAEL, L. 1934. An experimental study in the prenatal guinea pig of the origin and development of reflexes and patterns of behavior in relation to the stimulation of specific receptor areas during the period of active fetal life. *Genet. Psychol. Monogr.*, 16: 337–491.

——. 1954. The onset and early development of behavior. In *Manual of Child Psychology*, L. Carmichael, (ed.), pp. 60–185. John Wiley & Sons, New York.

COGHILL, G. E. 1929a. *Anatomy and the Problem of Behaviour.* 113 pp. Cambridge, at the University Press.

——. 1929b. The early development of behavior in Amblystoma and in man. *Arch. Neurol. Psychol.*, 21: 989–1009.

——. 1933. Somatic myogenic action in embryos of *Fundulus heteroclitus*. *Proc. Soc. Exptl. Biol. Med.*, 31: 62-64.

——. 1940. Early embryonic somatic movements in birds and in mammals other than man. *Monogr. Soc. Res. Child Develop.*, Natl. Res. Council, Washington.

CRAIN, S. M., and E. R. PETERSON. 1963. Bioelectrical activity in long-term cultures of spinal cord tissues. *Science*, 141: 427–429.

CUNNINGHAM, A. W. B. 1962. Qualitative behavior of spontaneous potentials from explants of 15-day chick embryo telencephalon in vitro. *J. Gen. Physiol.,* 45: 1065–1076.

——, and B. J. RYLANDER. 1961. Behavior of spontaneous potentials from chick cerebellar explants during 120 hours in culture. *J. Neurophysiol.,* 24: 141–149.

DRACHMAN, D. B., and A. C. COULOMBRE. 1962. Experimental clubfoot and arthrogryposis multiplex congenita. *Lancet,* Sept. 15, 523–526.

FLEXNER, L. B. 1955. Enzymatic and functional patterns of the developing mammalian brain. In *Biochemistry of the Developing Nervous System,* H. Waelsch, (ed.), pp. 281–300. Academic Press, New York.

HAMBURGER, V. 1960. *A Manual of Experimental Embryology,* 221 pp. University of Chicago Press, Chicago.

——. 1964. Ontogeny of behavior and its structural basis. In *Comparative Neurochemistry,* D. Richter (ed.), p. 21–34. Pergamon Press, London.

——, and M. BALABAN. 1963. Observations and experiments on spontaneous rhythmical behavior in the chick embryo. *Develop. Biol.,* 7: 533–545.

——, and H. L. HAMILTON. 1951. A series of normal stages in the development of the chick embryo. *J. Morphol.,* 88: 49–92.

——, and M. WAUGH. 1940. The primary development of the skeleton in nerveless and poorly innervated limb transplants of chick embryos. *Physiol. Zool.,* 13: 367–380.

HARRIS, J. E., and H. P. WHITING. 1954. Structure and function in the locomotory system of the dogfish embryo. The myogenic state of movement. *J. Exptl. Biol.,* 31: 501–524.

HERRICK, C. J. 1949. *George Ellett Coghill, Naturalist and Philosopher,* 280 pp. University of Chicago Press, Chicago.

HOLTZER, H., and R. P. KAMRIN. 1956. Development of local coordination centers. I. Brachial centers in the salamander spinal cord. *J. Exptl. Zool.,* 132: 391–408.

HOOKER, D. 1952. *The Prenatal Origin of Behavior,* 143 pp. University of Kansas Press, Lawrence.

HUMPHREY, T. 1952. The spinal tract of the trigeminal nerve in human embryos between 7½ and 8½ weeks of menstrual age and its relation to early fetal behavior. *J. Comp. Neurol.,* 97: 143–210.

KUO, Z. Y. 1932. Ontogeny of embryonic behavior in Aves. I. The chronology and general nature of the behavior of the chick embryo. *J. Exptl. Zool.,* 61: 395–430.

——. 1939a. Studies in the physiology of the embryonic nervous system. I. Effect of curare on motor activity of the chick embryo. *J. Exptl. Zool.,* 82: 371–396.

——. 1939b. Studies in the physiology of the embryonic nervous system. II. Experimental evidence on the controversy over the reflex theory in development. *J. Comp. Neurol.,* 70: 437–459.

——. 1939c. Total pattern or local reflexes? *Psychol. Rev.,* 46: 93–122.

ORR, D. W., and W. F. WINDLE. 1934. The development of behavior in chick embryos: The appearance of somatic movements. *J. Comp. Neurol.,* 60: 271–285.

PETERS, J. J., A. R. VONDERAHE, and A. A. HUESMAN. 1960. Chronological development of electrical activity in the optic lobes, cerebellum and cerebrum of the chick embryo. *Physiol. Zool.,* 33: 225–231.

PREYER, W. 1885. *Specielle Physiologie des Embryo,* 644 pp. Grieben, Leipzig.

TRACY, H. C. 1926. The development of motility and behavior in the toadfish (*Opsanus tau*). *J. Comp. Neurol.,* 40: 253–369.

VISINTINI, F., and R. LEVI-MONTALCINI. 1939. Relazione tra differenziazione strutturale e funzionale dei centri e delle vie nervose nell'embrione di pollo. *Arch. Suisses Neurol. Psychiat.,* 43: 1–45.

WEISS, P. 1941a. Self-differentiation of the basic patterns of coordination. *Comp. Psychol. Monogr.,* 17: 1–96.

——. 1941b. Autonomous versus reflexogenous activity of the central nervous system. *Proc. Am. Phil. Soc.,* 84: 53–64.

WHITING, H. P. 1955. Functional development in the nervous system. In *Biochemistry of the Developing Nervous System,* H. Waelsch (ed.), pp. 85–103. Academic Press, New York.

WINDLE, W. F. 1940. *Physiology of the Fetus,* 249 pp. W. B. Saunders, Philadelphia.

——. 1944. Genesis of somatic motor function in mammalian embryos: a synthesizing article. *Physiol. Zool.,* 17: 247–260.

——, and A. M. GRIFFIN. 1931. Observations on embryonic and fetal movements of the cat. *J. Comp. Neurol.,* 52: 149–188.

WOLF, W. (ed.). 1962. Rhythmic functions in the living system. *Ann. N. Y. Acad. Sci.,* 98: 753–1326.

DEVELOPMENTAL BIOLOGY SUPPLEMENT 2, 251–271 (1968)

IV. EMERGENCE OF NERVOUS COORDINATION
Origins of Integrated Behavior

VIKTOR HAMBURGER

Department of Biology, Washington University, St. Louis, Missouri 63130

My topic, the emergence of order on the behavioral level, has two aspects. Since behavior emerges from the physiological activities of the nervous system, neurogenesis and the genesis of behavior are inseparable. The development of organization in the nervous system, in turn, has to be dealt with in terms of morphogenesis, cytogenesis, including ultrastructure, and physiological activity. Observation has to be supplemented by the analytical experiment. Such a multidisciplinary approach to our problem ranging from the behavioral to the ultrastructural level, is the great challenge for the future. It is beset with difficulties, not the least of which is the matter of communication between investigators in these different areas. A more immediate difficulty for my topic, apart from the fragmentary nature of the available material, is the fact that the relations between neurogenesis and the origin of behavior are by no means as straightforward and parallel as one might have expected. I shall attempt to deal with some of the intricacies of these relationships in a rather general way, but my old informant, the chick embryo, will supply most of the illustrations.

EXPERIMENTAL NEUROGENESIS

Experimental neurogenesis has elucidated some of the mechanisms by which the complex organization of the CNS, the patterns of central fiber tracts and of peripheral nerves, and the specific synaptic and terminal connections are established. I shall recall briefly some of the procedures that create order on the structural level. Mitotic activity in the CNS and in ganglia is programmed in space and time, and the mitotic patterns foreshadow the patterns of regional distribution of neuroblasts (Coghill, 1924; Hamburger, 1948; Källen, 1965; Watterson, 1965). Directional migration of neuroblasts, both within the CNS

251

and by neural crest and placodal derivatives, has been recognized as one of the most significant procedures by which the assembly of brain nuclei, stratifications, cell column formation and localization of peripheral ganglia is achieved (Levi-Montalcini, 1964). But the agencies that guide and direct the neuroblast migrations are obscure. The establishment of the complex patterns of intracentral fiber tracts and of peripheral nerve patterns can be attributed to subtle interactions between the "growth cone" and the substrate or matrix on which it is spun out. "Contact guidance" by oriented structural or ultrastructural elements in the matrix (Weiss, 1941a, 1955) is probably combined with specific biochemical matching properties of fibers and constituents of the matrix to provide the directional cues; but again, we know very little concerning their molecular basis. It would seem that this problem could be most profitably approached in tissue culture. A variety of experiments demonstrate a wide spectrum of degrees of selective affinities between different types of axons and their substrate, and along different segments of the pathways (Hamburger, 1962). The crucial step in the development of neural networks is the establishment of specific synaptic connections. On the basis of extensive experimental work, Sperry (1951, 1963, 1968) has developed the idea that selective chemo-affinity between the nerve ending and the neuron with which it synapses establishes the permanent contact between them. Such chemo-affinities are perhaps related to the above-mentioned matching properties between outgrowing fibers and the matrix, which we think are responsible for directional fiber growth, and perhaps also for directional migration of neuroblasts. As soon as order is established, problems of maintenance arise for the young neuroblasts. Their survival is threatened if they fail to establish, or lose, their interconnections with other neurons (transneuronal degeneration) or with the peripheral organs in which they terminate (Hamburger, 1956). For the growth of sympathetic and spinal ganglia, the nerve growth factor seems to be an essential metabolic requirement (Levi-Montalcini, 1966).

All these mechanisms taken together carry neurogenesis to an advanced state of neural organization. Later on, we shall scrutinize the question of whether or not sensory input is a necessary requirement for the completion of neurogenesis to the point where integrated behavior becomes possible. The mechanisms of neurogenesis include an extraordinary variety of ever-changing, yet carefully programmed interactions between parts of the nervous system, and between neural and nonneural structures. The baffling complexity of the neurogenetic process matches the complexity of the finished product. The unresolved problems arc clearly formulated. The field for the molecular neurobiologist is wide open.

BEGINNINGS OF FUNCTIONAL ACTIVITY

Since the development of functional activity is the central issue of this discussion, we shall inquire first into its beginning. The term "functional activity" has two meanings: the bioelectrical activity of neurons and neuron networks; and motility or behavior. Since recordings of potentials at the earliest stages of neurogenesis have not been made as yet, motility is the only criterion of early bioelectrical activity.

Motility begins remarkably early in embryos: in the salamander in intermediate tail bud stages; in the chick, in limb bud stages (3½–4 days), in the mouse at 14 days, and in the human at 7½–8 weeks menstrual age, when fingers and toes make their appearance. In all vertebrate embryos, the first sign of motility is rather uniformly a bending of the head; motility spreads from the neck muscles tailward, reflecting the cephalocaudal sequence of neuromuscular maturation. Sinusoid waves are characteristic of early stages, but in amniotes the pattern soon becomes irregular. In all forms in this first phase, spontaneous as well as evoked motility is total body movement involving all parts that are capable of motility. The movements may be coordinated or uncoordinated (see below).

There is a close correlation between structural and functional maturation: very young neuroblasts are capable of impulse transmission, and very primitive synaptic connections and neuromuscular contacts suffice to mediate overt motility. Coghill's pioneer work on *Ambystoma* (1929) has shown that each behavioral advance, for instance, that from head bending to coil and to S-flexure, follows immediately upon the completion of new synapses; this seems to be a general rule also in higher forms. For instance, in mammals, the earliest bending of the head can be elicited by tactile stimulation of the head surface. The area innervated by the trigeminal nerve is the first reflexogenous zone (Barron, 1941). Humphrey (1954) has shown that in 8–8½ week human embryos the descending fiber tract of the trigeminal nerve reaches the level of the 2nd–4th cervical segment, and dips into the gray matter, at exactly the time when the first neck muscle contractions occur. The fibers synapse with secondary commissural neurons that already have established connections with the contralateral motoneurons at a much earlier stage (Windle and Fitzgerald, 1937). The details of incipient *synapse formation* on motoneurons at the stage of onset of motility have been observed with the electron microscope in a 28-mm macaque embryo (Bodian, 1966). The stage corresponds to that of the aforementioned human embryo. The boutons are few in number and very primitive. Synaptic vesicles and the beginnings of junctional densities are present, but mitochondria are rare. All boutons are apposed to dendrites, and of one type only, in contrast to the variety found in the adult. Bodian comments that "onset of function follows very closely the minimal development of essential synaptic structures. . . . The suggestion is obvious that observed synaptic bulbs are excitatory, and that inhibitory synapses, implying more complex reflex patterns, are not yet developed" (1966, pp. 131–132).

The neuromuscular connections are equally primitive in the early phases of motility. When trunk motility begins in the 3½- to 4-day chick embryo, myofibrils just begin to differentiate in the trunk somites (Allen and Pepe, 1965) and cholinesterase (ChE) is diffusely distributed in the myofibrils (Mumenthaler and Engel, 1961). Distinct motor end plates do not appear until day 10 (Drachman, 1965). Likewise, distal leg muscles begin to contract at 7–7½ days when ChE is still diffuse; motor end plates are not well differentiated until day 13 or 14 (Drachman, 1965). It would be of great interest to study the ultrastructural details of these provisional neuromuscular contacts. Coghill (1929) had already fully realized that neuroblasts manage to combine growth and differentiation with functional activity, long before myelination begins.

The link between structural differentiation of synapses and the onset of bioelectrical activity is difficult to establish *in vivo*. This has been accomplished *in vitro*, in the long-term organ cultures of embryonic fetal rat spinal cord and brain which permit direct electrical recording (Crain, 1966; Crain and Peterson, 1967; Crain *et al.*, 1968b). Motility begins in the rat at 16 days (Angulo y Gonzalez, 1932). Spinal cord explants were made of 14- to 15-day embryos, i.e., prior to synapse formation. During the first 2 days in culture, only simple spikes can be obtained, indicating discharges of individual neuroblasts. After 2–3 days, long-lasting spike barrages and slow waves can be evoked. The increase in the complexity of bioelectrical activity suggests that polysynaptic networks are now in operation (Crain and Peterson, 1967). Parallel electron microscope studies on the same material by Bunge *et al.* (1967) have shown that, indeed, the neural tissue is practically free of synapses during the first 2 days in culture, and that primitive synapses appear with increasing frequency during the subsequent days. These experiments demonstrate the capacity of neuroblasts to produce action potentials before they synapse, and they confirm the finding that functional impulse transmission *in vivo* occurs immediately after the formation of a primitive synapse.

One can ask whether impulse propagation is a necessary prerequisite for the formation of individual synapses or of complex synaptic networks. The question was answered in the negative by experiments in which cultures of fetal rat spinal cord and fetal and newborn mouse neocortex were exposed to the blocking agent xylocaine during the critical period of synapse formation. (Xylocaine blocks all bioelectrical activity, not just synaptic transmission.) The block was started at a stage before the first synapses were formed and continued for 5–30 days. Within a few minutes after the removal of the blocking agent, evoked potentials of considerable complexity and long duration were obtained. Hence the chronic block of bioelectrical activity did not interfere with synapse formation in complex networks (Crain *et al.*, 1968a).

SPONTANEOUS MOTILITY IN EMBRYOS

The organ cultures of mammalian nerve tissue exhibit another characteristic feature of special interest to us—the capacity for spontaneous generation of bioelectrical activity. Our studies of the chick embryo have shown that this propensity of neural tissue for spontaneous discharges is the sole basis for its motility, up to 17 days. Since other forms display a similar type of embryonic behavior, this phenomenon would seem to be an important key in our understanding of the beginnings of behavior, in general.

Spontaneous neuronal activity has been defined by Bullock and Horridge (1965, p. 314) as "repetitive change of state of neurons without change of state of the effective environment, that is, activity without stimulation other than the sanding conditions." If such activities are transmitted to muscles, we speak of "spontaneous motility." We should distinguish further between "endogenous" and "spontaneous" discharges. *Endogenous* bioelectrical activity can be defined as resulting from the intrinsic metabolic processes of the neuron. *Spontaneous* activity is defined more broadly; it includes discharges that may be generated anywhere in the nervous system, and transmitted synaptically to other parts.

The characteristics of the motility of the chick embryo have been described repeatedly (Hamburger, 1963; Hamburger *et al.*, 1965), and I shall summarize them only briefly. As was mentioned, movements begin at 3½ days with the bending of the head, and extend subsequently to trunk, tail, and limbs. Beak clapping, eyeball and lid movements are added to the repertory as soon as the respective neuro-muscular connections are established. Activity builds up gradually from a few twitches per minute until, at 13 days, the embryo is in motion 80% of the observation time. Activity is performed in cycles, the activity phases lengthening in duration while the inactivity phases get shorter. It should be pointed out that the oscillations recorded in organ cultures of mammalian nerve tissue (Crain, 1966) and from the surface of optic tectum and cerebrum of old chick embryos *in vivo* (Peters *et al.*, 1960; Corner *et al.*, 1967) are of a different order of magnitude.

Lack of organization or integration is the main characteristic of this motility. The movements are mostly convulsive-type jerks and twitches and occasional head thrusts. They appear to be random movements in the sense that different parts are active independently of each other. During an activity phase, legs, wings, head, or beak may move synchronously but in an uncoordinated fashion, or any part or parts may be at rest while the others move. The combinations seem to be unpredictable. Our observations have failed to identify relationships that might be interpreted as antecedents to walking, pecking, drinking, or other posthatching activities (with the exception of occasional wing flutters); but a rigid statistical analysis is required to verify this point. We have called the random movements type I motility (Hamburger and Oppenheim, 1967). A modification of this type is designated as "startle" or type II motility. It is defined as a tremor of spasmodic movements passing rapidly through the body.

A distinction should be made between *integration* of movements of parts, as in alternating leg movements, and *coordination* of muscle groups within a part. Coordination, so defined, may well be present in type I motility, in the absence of integration. For instance, one might expect leg flexion to involve the excitation of synergistic flexor muscles and the inhibition of their antagonists. However, even this cannot be taken for granted. An EMG study of muscle reflexes in fetal sheep of 60–67 days (gestation time: 140–150 days) showed a myotatic response of the m. gastrocnemius to slight stretch, but the antagonistic m. tibialis anterior, instead of being inhibited, showed simultaneous excitation. Not until 30 days later did inhibition come into effect (Änggård et al., 1961). The experiment suggests that the central action systems have their own program of maturation. The diffuseness and jerkiness of the type I motility in the chick may be due, in part, to the lack of muscle coordination.

What is the evidence that the rhythmical embryonic motility up to 17 days is actually spontaneous (as defined above)? Alternative explanations are that the movements are triggered by changes in the biochemical milieu, or by sensory stimulation. It is unlikely that changes in the composition of agents carried in the circulation play a role. If, in 36-hour embryos, sections of the spinal cord are extirpated at different levels, the parts rostral and caudal to the gap show cyclic motility in later stages, but the parts are not synchronized; one part may be in an activity phase while the other is inactive.

Special attention was paid to the possible role of *sensory input* in type I activity. Such stimulation can be discounted for the period from the beginning of motility to day 7 or 7½, for the simple reason that the reflex circuits are not closed until that stage (Preyer, 1885; Visintini and Levi-Montalcini, 1939); hence afferent impulse transmission is not feasible. The period from 8 to 17 days is covered by the following experiments: A total deafferentation of both legs was achieved by a double operation performed on 2-day embryos: removal of the thoracic spinal cord to the extent of 5 somites, and extirpation of the dorsal half of the lumbosacral spinal cord, including the neural crest (Hamburger et al., 1966). The intact basal plate produced normal motor columns which supplied the legs with normal motor innervation. The legs were completely insensitive to extero- and proprioceptive stimulation. Leg motility was quantitatively within the normal range up to 15 days and qualitatively normal in more than half of the cases. The decline in motility observed between 15 and 17 days can be attributed to a deterioration of the neural tissue which was observed in all cases of reduced motility. Deafferentation of the head skin was achieved by Dr. Narayanan by bilateral extirpation of the neural crest primordia as well as the placodal primordia of the trigeminal ganglion (unpublished). Tactile stimulation tests proved that the deafferentation had been successful. Motility was quantitatively and qualitatively normal up to 15 days. Complete elimination of vestibular stimuli by bilateral extirpation of both otocysts in 3- to 4-day embryos by Dr. Decker (unpublished), likewise did not interfere with normal motility, up to 17 days. Incidentally, all these experiments rule out the claim that *self-stimulation,* for instance, by brushing of the legs against the head, plays an important part in the initiation and organization of motility patterns.

Our working hypothesis assumes that the overt motility up to 17 days is due to discharges that are generated spontaneously in neurons distributed throughout the CNS and that the discharges sweep through the entire system and activate all neuromuscular pathways indiscriminately. The brain contributes excitatory stimulation, since spinal embryos show a reduction in overall activity (Hamburger and Balaban, 1963; Hamburger *et al.,* 1965). Different brain parts participate differentially at different stages (Decker and Hamburger, 1967).

This hypothesis can be tested only by electrophysiological methods. It is essential now to find out what is going on in the nervous system during the activity phases and the inactivity phases, by recording electrical activity *in vivo.* Dr. Sharma, in collaboration with Dr. Sandel of our Biomedical Computer Laboratory has made a beginning. The previous recordings of Peters *et al.* (1960) and of Corner *et al.* (1967) were confined to EEG patterns of the brain and evoked potentials of embryos that were mostly older than 15 days, that is, some time after the brain influence on motility had been established.

Our findings on the chick embryo were confirmed in essential points by Corner and Bot (1967). (The lower total activity values reported by these observers can be explained by differences in the definition of an inactivity phase. We have defined it as a period lasting 10 seconds or longer. Corner and Bot have included the shorter rest periods in their calculation of duration of inactivity phases.) To what extent can these findings on the chick embryo be generalized? Rhythmical, unintegrated motility of the same type has been found in the lizard embryo (Hughes *et al.,* 1967) and in the turtle (Tuge, 1931; Decker, 1967). In the anuran *Eleutherodactylus,* a bufonid without a free-swimming larval stage, a phase of unintegrated motility precedes the coordinated postmetamorphic swimming and walking movements (Hughes, 1965).

The situation in *mammals* is not clear, because no detailed information on spontaneous motility is available. In the earlier work, in the 1930's, all interest was focused on evoked responses. One gathers from the few observations on record that the initial phase of head bending and sinusoid trunk flexions is followed by a period of so-called "total" or "mass" movements, in which all parts of the body are involved in unintegrated activity, much like that in the chick embryo (Windle, 1940). In the human embryo, this phase lasts for almost 2 weeks (Humphrey, 1964). Since in mammals the reflex circuits are in operation from the beginning of motility, these mass movements occur spontaneously as well as in response to stimulation of the early reflexogenous zones. Their spontaneous performance indicates that sensory input is not a necessary prerequisite; but the other alternative, that changes in the internal milieu are responsible, has not been ruled out.

In contrast to the amniotes, embryonic motility in amphibians (except *Eleutherodactylus*) and teleosts is integrated from the beginning (see below), and no phase of random motility has been observed. Yet, even in these forms, the early phases of development of behavior at least up to the swimming stage, seem to be based on nonreflexogenic spontaneous discharges. In the toadfish *Opsanus tau,* Tracy (1926) has found that responsiveness to tactile stimuli does not begin until after hatching, in the free-swimming stage, that is, 2½ weeks after the onset of motility; and Corner (1964) observed spontaneous, rhythmical swimming in anuran larvae. No relevant data are available for salamanders.

On the basis of all this material, we are inclined to generalize our notion that nonreflexogenic, endogenously generated activity of the embryonic nerve tissue, resulting either in random motility or in integrated motility, plays an important role in the development of behavior. The random type seems to be limited to those embryos that lead a prolonged sheltered life in the egg or uterus. The biological significance of random movements seems to be to guarantee the normal development and maintenance of joints, and the maintenance of muscles, since prolonged paralysis of the chick embryo results in ankylosis and muscle abnormalities (Drachman and Coulombre, 1962; Drachman and Sokoloff, 1966; Sullivan, 1966, 1967).

ORIGINS OF INTEGRATED BEHAVIOR

We turn next to the question: What are the origins of integration in behavior? As was indicated, the answer is different for lower and higher forms. Coghill (1929), in his studies of the salamander *Am-*

bystoma, has made a strong case for the continuity of integration from the first bending of the head through intermediate stages, such as coil and S-flexure to swimming, and from there to terrestrial locomotion, feeding, etc. Behavior development in this form is a "progressive expansion of a perfectly integrated total pattern" (p. 38), discrete local movements and reflexes arising by emancipation or "individuation" from the total pattern. In this case, the differentiation of neural patterns and of behavior patterns runs strictly parallel. The behavior development of teleosts seems to be very similar (Tracy, 1926).

Taking the chick embryo again as a representative of higher forms, we find a more complex situation. Unintegrated activity is the prevailing form of behavior up to 17 days. One finds occasional wing flutters in earlier stages, and Gottlieb and Kuo (1965) have described alternating leg movements in the 10-day duck embryo. But the general picture is that of unorganized motility. Day 17 is truly a turning point in the chick. From then on the types I and II movements decline, and a new type of integrated movement, which we designate as type III, makes its first appearance. These movements lead through a sequence of clearly definable intermediate steps to the attainment of the hatching position prerequisite for hatching. The hatching act itself (climax) is a modification of the prehatching type of motility. The whole process has been described in detail (Hamburger and Oppenheim, 1967), and I shall restrict myself to a few pertinent points. At the beginning of day 17, the embryo is oriented lengthwise in the shell, with the tarsal joints near the pointed end and the neck which is bent straight downward, near the membrane that separates the embryo from the air space at the blunt end. The beak is buried in the yolk sac between the legs. Two days later, most embryos are in the hatching position: The neck is twisted to the right in a tight coil. The right side of the head is tucked under the right wing which is apposed to the inner shell membrane. The beak is positioned obliquely against the shell; its tip is at a distance of a few millimeters from the shell. It has penetrated the inner membrane. Hatching is accomplished by sharp, powerful back thrusts of the upper beak against the shell. All other parts of the body are also involved, with the exception of the wings: A rapid wriggling movement passes from head to tail; the shoulder and tarsal joints are pressed against the shell. After these thrusts have been repeated several times, whereby the pipping hole is enlarged, a rotatory component is added involving the whole body and the legs. As a result, the beak thrusts which are now repeated at rather regular intervals, shift gradually along the outer circumference of the air chamber, in an anticlockwise direction, when viewed from the blunt pole. When the shell is opened approximately two-thirds around this circle, the cap is loosened sufficiently to be lifted off by a few vigorous body wriggles, wing flutters and stemming of the tarsal joints against the pointed end.

The complicated prehatching movements which result in the lifting of the head out of the yolk sac, the tucking of the head under the right wing, the shifting of the body to attain the hatching position, and pipping, have several features in common with the movements at climax which differentiate them from the type I movements. These characteristics are: the involvement of all parts of the body in an integrated fashion, and a distinct rotatory component. Furthermore, most of these movements, with the exception of the back thrusts of the head, are rather smooth, in contrast to the jerky type I movements. We have considered all prehatching and hatching movements as modifications of a basic pattern of integrated motility and designated them as type III motility (Hamburger and Oppenheim, 1967).

The question then arises as to the relation of this pattern to the type I motility. It does not seem possible to derive the former from the latter for the following reasons: First, they are very different in appearance, as was just mentioned. In particular, the rotatory component is not part of the repertory of the type I movements. Perhaps the most convincing evidence is the observation that unintegrated type I movements do not disappear after 17 days but are merely suspended during episodes of integrated movements, for instance, during tucking or pipping and climax. They are resumed during the intervals between such episodes, though at a reduced rate. According to Corner and Bot (1967) they continue even after hatching. Obviously, the unintegrated motility is not simply transformed into integrated motility.

Yet, in a different sense, there is a link between the two types of motility. The same muscle groups that flex and extend the legs during spontaneous motility, operate in walking and standing; the muscles involved in beak clapping before hatching are used in food pecking and drinking, etc. In other words, at the level of muscular units, components of type I motility are incorporated in the integrated hatching and posthatching action patterns.

Since, during the last 3–4 days before hatching, the type I movements are performed during the intervals between the type III movements, it is clear that the "final common paths" of Sherrington, that is the motoneuron connections with muscles, are activated alternately by massive electrical discharges that excite all motoneurons indiscriminately, and by highly selective discharge patterns characterized by a subtle interplay of excitation and inhibition of appropriate muscle groups. One is reminded of an orchestra, where the same players use the same instruments for tuning and for playing tunes. It would be of interest to find out how the prehatching and hatching (type III) movements are triggered off and the type I movements simultaneously inhibited. Are the former induced by changes in the biochemical milieu, such as O_2 or CO_2 or hormone concentrations in the circulation?

In summary, our observations on the chick embryo indicate that prehatching and hatching behavior, as well as the major posthatching activities, such as walking, pecking, righting, etc., do not emerge as the culmination of a gradual build-up from simpler antecedents; instead, they are activated rather suddenly and performed with a considerable degree of perfection the first time they are performed. Needless to say, practice and learning enter into the picture immediately after hatching.

At first sight, the notion that complex actions appear suddenly and without antecedents seems to violate the principle that all development, including that of behavior, is a continuous and gradual process. This, of course, is based on a misunderstanding. The continuity is found on the level of neurogenesis which proceeds gradually from a primitive structure to the most intricate organization of neuronal interconnections. This process of gradual elaboration of organization can be followed even on the behavioral level, by the simple expedient of eliciting responses through tactile stimulation, at different stages. This method was used extensively in the many studies that were made on mammalian embryos and fetuses during the 1930's and 1940's. By systematic stimulation experiments, the investigators followed, stage by stage, the gradual elaboration of reflexes (Carmichael, 1954; Hooker, 1952; Windle, 1940). In several instances it was possible to correlate rather closely the neurogenetic growth and differentiation processes with the progression in behavior (Humphrey, 1964). For instance, in the human embryo, the palmar surface of the hand becomes sensitive very early, at 10.5 weeks. The response is an incomplete closure of the fingers. Sensory nerve branches have reached the skin at that stage. At 13–15 weeks, the closure is complete and sustained for some time. At 17 weeks, a true grasp is observed, and at 27 weeks,

the fetus can almost support himself with the grasp of one hand
(Hooker, 1952).

THE ROLE OF SENSORY INPUT IN THE DEVELOPMENT
OF INTEGRATED BEHAVIOR

The stimulation experiments which we have just discussed reveal
the inventory of behavioral responses of the embryo; but they were
not used as an analytical tool to determine whether stimulation plays
a role in the molding of integrated behavior. We shall discuss briefly
a variety of other experiments that shed light on this problem.

Narcotization Experiments

As was mentioned before, behavior in urodeles is integrated from
the beginning of motility and beyond the swimming stage. In the frequently cited narcotization experiments on salamander larvae, from
the premotile stage through the stage of free swimming, it was found
that the performance of the embryos was normal, after the blocking
agent had been removed (Harrison, 1904; Carmichael, 1926; Matthews
and Detwiler, 1926). The experiments establish two points: neither
neuromuscular activity nor proprioceptive self-stimulation are necessary prerequisites for the attainment of the swimming activity. However, since the blocking agent, chloretone, operates on the motor end
plates, the bioelectrical activity in the nerves was not blocked. The
previously mentioned organ culture experiments on mammalian nerve
tissue are pertinent to this point (Crain et al., 1968a). In these experiments, the formation of synapses occurred while all bioelectrical
activity was suspended. If extrapolation to the chick nervous system
in vivo is permitted, then the propagation of bioelectrical activity in
the nerve may not be relevant. In previously mentioned experiments,
the spontaneous motility in chick embryos was paralyzed during the
middle phase of incubation (Drachman and Coulombre, 1962; Drachman and Sokoloff, 1966). The ensuing severe deformation of joints
usually prevents hatching. However, a few embryos did hatch, indicating that a 1- to 2-day paralysis does not interfere with the type III
movements. A systematic analysis of this problem would be of interest.

Autonomous Differentiation vs. Learning

In recent decades, the old theory that sensory information guides
the development of integrated activity by selecting adaptive patterns
of perception and motor activity from initial random performance, by

trial and error, experience and learning, has been thoroughly discredited, largely through the pioneer work of Weiss and Sperry. Nobody believes any more that walking or visual perception in its complexity are learned in embryonic or fetal stages. Ironically, the chick and other embryos start out actually with random movements, but they are not the raw material for locomotion or any other adaptive behavior. The widely accepted modern theory holds that the neural apparatus for integrated behavior differentiates autonomously, and that the appropriate interconnections are prewired in forward reference to functional activity, but without benefit from it. The evidence against the former and in support of the latter theory has been reviewed frequently in recent years (Weiss, 1955; Sperry, 1951, 1965; Sperry and Hibbard, 1968), and I shall not dwell on it. The case for the development of central action systems, independently of sensory input, has been strengthened by the demonstration that a number of complex activities in the adult are performed on the basis of patterned *spontaneous* neural activity, and that in these instances sensory input contributes at best a nonspecific tonic or modulating effect. This holds for the rhythmic fin movements in teleosts (von Holst, 1935), for the rhythmic flight patterns in the cicada (Wilson, 1961), for sexual behavior in insects (Roeder, 1963), and many other behavioral activities. One of the central issues of the modern theory concerns the mechanism by which the specificity of synaptic connections is guaranteed. Sperry's theory of selective chemoaffinities between the partners that enter into synaptic relationship was based originally on the retinotectal connections in regenerating optic nerves. It has found indirect strong support from the electrophysiological investigations of Gaze and his co-workers (Gaze, 1967; Jacobson, 1966); they have shown that regenerating optic fibers actually return to the tectal neurons with which they had been connected in the first place.

Although the bulk of the evidence for the present theory derives from regeneration experiments on adult teleosts and amphibians, several crucial experiments were done on embryos. For instance, Weiss (1941b) transplanted limb buds of the salamander *Ambystoma* in premotile stages from the left to the right flank, where they grew out in the wrong direction. He found that "from the very first stages of motility, the limbs moved in reverse" (p. 58). This implies that the spinal coordination center for locomotion differentiates through intrinsic developmental mechanisms in complete disregard of the resulting functional maladaptation which is never corrected. In another experiment, the legs were deafferented in frog tadpoles before leg function had started; yet coordinated locomotor function was not impaired. On the sensory side, we have the experiments of Székely (1954, 1968) and Jacobson (1966), in which embryonic eyes of urodeles were rotated at successive developmental stages. It was established that the regional specification of the retina cells on which the specification of their tectal connections is based, becomes fixed already before the optic nerve fibers have reached the brain. The same holds for the retina of the chick embryo (DeLong and Coulombre, 1965). Hence, sensory input has to yield to chemoaffinity as the crucial mechanism of synapse formation in development as in regeneration.

Sensory Guidance of Prehatching and Hatching Behavior

There are several other ways by which sensory input could influence embryonic behavior. For instance, it could provide receptor-specific information for control and orientation of integrated embryonic behavior, as it does in postnatal life. We have tested this point in the chick embryo. The previously discussed experiments have excluded the role of sensory input only for the nonintegrated spontaneous motility up to 15–17 days. Are the prehatching and hatching movements likewise driven exclusively by endogenously generated discharges, or do they require sensory guidance?

We have discussed before the (unpublished) deafferentation experiments of Dr. Narayanan, in which the trigeminal ganglia were removed bilaterally, and the bilateral otocyst extirpations of Dr. Decker (unpublished). In both instances, the spontaneous motility up to 17 days was unaffected, but none of the experimental embryos performed the type III movements with the rotatory component; hence none of them hatched. Most of them remained in the typical 16-day position, with the beak buried in the yolk sac, although several lived to day 20. The experiments are inconclusive. The failure to perform the prehatching and hatching movements could be due to the lack of orientation in space or to lack of orientation by tactile head stimuli, respectively. Alternatively, sensory input from these two sources could normally supply merely a tonic, facilitating influence on an otherwise endogenously driven system. A third possibility is an impairment of central nervous structures as the result of *transneuronal degeneration*. Levi-Montalcini (1949) has demonstrated degenerative changes in several cochlear nuclei and the absence of the nucleus tangentialis of the vestibular system following unilateral otocyst extirpation. We have not yet studied the trigeminal material in this respect.

Some other recent experiments indicate that tactile self-stimulation of the head or trunk by legs or wings, or proprioceptive feedback from the limbs, do not trigger prehatching or hatching motility or influence it in other ways. Bilateral extirpation of both leg buds was done by Miss M. Helfenstein. Absence of legs does not interfere with tucking and with the attainment of the hatching position, and 9 out of 15 embryos that were raised to advanced stages actually pipped. Obviously, the embryos were capable of performing the integrated type III movements. However, none hatched, probably because the rotatory movement of the body during climax requires the pressure of the tarsal joints against the shell and their alternating stepping movements. Dr. Narayanan extirpated the right wing bud. No changes in the type III movements were observed, and all embryos hatched. Of course, these experiments do not exclude all self-stimulation, and the question is not settled.

Facilitation of Posthatching Behavior by Self-stimulation

Developmental behaviorists contend that self-stimulation in prenatal stages may have significant formative effects on postnatal behavior, not necessarily in the sense of learning or conditioning, but by more subtle mechanisms, such as "facilitation" (Gottlieb and Kuo, 1965; Kuo, 1967). So far, there is no experimental evidence for this claim, but the following experiments of Gottlieb (1966) are suggestive in this respect. They deal with auditory cues for recognition of the species-specific maternal following-call by newly hatched chicks or ducklings. Ten to 35 hours after hatching, they were tested for their following response to replicas of hens of their own and other species that emitted selected types of calls. Prior to the experiment, that is, in the incubator and brooder, they had been exposed to their own chirping and to that of their siblings. Only one variant of the various experiments is pertinent to our discussion: One group was exposed to additional tape-recorded chirping, while in the brooder. The extra stimulation enhanced in several measurable parameters the following response to the species-specific maternal call, but to no other auditory cues. Here, then, the reinforcement of one type of auditory stimulation (chirping) facilitated selectively the response to an entirely different auditory cue. It is true that this experiment does not involve prehatching stimulation; nor is it claimed that the normally operating vocal self-stimulation before and shortly after hatching is a factor in the *formation* of the response to the maternal call. The experiment is presented merely as a model to show how sensory input in prehatching stages could conceivably influence posthatching behavior in subtle ways other than the conventional conditioning processes.

CONCLUDING REMARKS

One might have wished that the holistic dream of E. G. Coghill had been fulfilled: that behavior in all vertebrates is integrated from beginning to end. Instead, we are confronted with a puzzling diversity of phenomena that are difficult to fit in a coherent theory. A few solid building blocks have been assembled: the concept of selective chemo-affinity has proved its value as a fruitful heuristic hypothesis; the idea of autonomous neural differentiation that proceeds according to an intrinsically determined program has won over the rival idea that adaptive neural connections are the result of selection, by trial and error, from a randomly interconnected network; the role of sponta-neous motility in embryonic behavior has been recognized. But in all instances, probing in depth is the immediate challenge. Selective affinity is a general notion that needs a concrete underpinning on the molecular level. The exploration of the electrophysiological properties of the developing nervous system is at its very beginning; on the behavioral level, the speculations about storage of prenatal "ex-periences" and their influence on postnatal behavior are, up to now, without critical experimental foundations.

While the competence of the individual investigator determines the range and limits of his radius of action, compartmentalization in thought will not get us very far. I have tried to show that an overall view can be achieved only by pooling the resources of a variety of branches of neurobiology. Using as tools microsurgery on the em-bryonic nervous system, tissue culture, electron microscopy, cyto-chemical and biochemical microtechniques, modern electrophysiologi-cal approaches, and rigid experimental methods in developmental psychology, and adding a bit of ingenuity, we may achieve, eventually, a synthesis of our presently fragmented ideas of the way integrated behavior comes into existence.

The experiments from this laboratory were supported by grant No. 5721 of the NINDB of the PHS.

REFERENCES

ALLEN, E. R., and PEPE, R. A. (1965). Ultrastructure of developing muscle cells in the chick embryo. *Am. J. Anat.* **116**, 115–148.

ÄNGGÅRD, L., BERGSTROEM, R., and BERNHARD, C. G. (1961). Analysis of prenatal spinal reflex activity in sheep. *Acta Physiol. Scand.* **53**, 128–136.

ANGULO Y GONZALEZ, A. W. (1932). The prenatal development of behavior in the albino rat. *J. Comp. Neurol.* **55**, 395–442.

BARRON, D. H. (1941). The functional development of some mammalian neuro-muscular mechanisms. *Biol. Rev. Cambridge Phil. Soc.* **16**, 1–33.

BODIAN, D. (1966). Development of fine structure of spinal cord in monkey fetuses. I. The motoneuron neuropil at the time of onset of reflex activity. *Bull. Johns Hopkins Hosp.* **119**, 129–149.

BULLOCK, T. H., and HORRIDGE, G. A. (1965). "Structure and Function in the Nervous Systems of Invertebrates." Freeman, San Francisco.

BUNGE, M. B., BUNGE, R. P., and PETERSON, E. R. (1967). The onset of synapse formation in spinal cord cultures as studied by electron microscopy. *Brain Res.* **6**, 728–749.

CARMICHAEL, L. (1926). The development of behavior in vertebrates experimentally removed from the influence of external stimulation. *Psychol. Rev.* **33**, 51–58.

CARMICHAEL, L. (1954). The onset and early development of behavior. *In* "Manual of Child Psychology" (L. Carmichael, ed.). Wiley, New York.

COGHILL, E. G. (1924). Rates of proliferation and differentiation in the central nervous system of Amblystoma. *J. Comp. Neurol.* **37**, 71–109.

COGHILL, E. G. (1929). "Anatomy and the Problem of Behavior." Cambridge Univ. Press, London and New York.

CORNER, M. (1964). Rhythmicity in the early swimming of anuran larvae. *J. Embryol. Exptl. Morphol.* **12**, 665–671.

CORNER, M., and BOT, A. P. C. (1967). Developmental patterns in the central nervous system of birds. III. Somatic motility during the embryonic period and its relations to behavior after hatching. *Progr. Brain Res.* **26**, 214–236.

CORNER, M., SCHADÉ, J. P., SEDLAČEK, J., STOECKART, R., and BOT, A. P. C. (1967). Developmental patterns in the central nervous system of birds. I. Electrical activity in the cerebral hemisphere, optic lobe and cerebellum. *Progr. Brain Res.* **26**, 145–192.

CRAIN, S. M. (1966). Development of "organotypic" bioelectric activities in central nervous tissues during maturation in cultures. *Intern. Rev. Neurobiol.* **9**, 1–43.

CRAIN, S. M., and PETERSON, E. R. (1967). Onset and development of functional interneuronal connections in explants of rat spinal cord-ganglia during maturation in culture. *Brain Res.* **6**, 750–762.

CRAIN, S. M., BORNSTEIN, M. B., and PETERSON, E. R. (1968a). Maturation of cultured embryonic CNS tissues during chronic exposure to agents which prevent bioelectrical activity. *Brain Res.* **8**, 363–372.

CRAIN, S. M., PETERSON, E. R., and BORNSTEIN, M. B. (1968b). Formation of functional neuronal connections between explants of various mammalian central nervous tissues during development in vitro. *Ciba Found. Symp. Growth Nervous System* pp. 13–31. Little, Brown, Boston, Massachusetts.

DECKER, J. D. (1967). Motility of the turtle embryo, *Chelyda serpentina* (Linné). *Science* **157**, 952–954.

DECKER, J. D., and HAMBURGER, V. (1967). The influence of different brain regions on periodic motility of the chick embryo. *J. Exptl. Zool.* **165**, 371–384.

DELONG, R. G., and COULOMBRE, A. J. (1965). Development of the retinotectal topographic projection in the chick embryo. *Exptl. Neurol.* **13**, 351–363.

DRACHMAN, D. B. (1965). The developing motor end plate: curare tolerance in the chick embryo. *J. Physiol. (London)* **180**, 735–740.

DRACHMAN, D. B., and COULOMBRE, A. J. (1962). Experimental club foot and arthrogryposis multiplex congenita. *Lancet* **II**, 523–526.

DRACHMAN, D. B., and SOKOLOFF, L. (1966). The role of movement in embryonic joint development. *Develop. Biol.* **14**, 401–420.

GAZE, R. M. (1967). Growth and differentiation. *Ann. Rev. Physiol.* **29**, 59–86.

GOTTLIEB, G. (1966). Species identification by avian neonates: Contributary effect of prenatal auditory stimulation. *Animal Behavior* **14**, 282–290.

GOTTLIEB, G., and KUO, Z. Y. (1965). Development of behavior in the duck embryo. *J. Comp. Physiol. Psychol.* **59**, 183–188.

HAMBURGER, V. (1948). The mitotic patterns in the spinal cord of the chick embryo and their relation to histogenetic processes. *J. Comp. Neur.* **88**, 221–284.

HAMBURGER, V. (1956). Developmental correlations in neurogenesis. *14th Growth Symp.*, 191–212. Princeton Univ. Press, Princeton, New Jersey.

HAMBURGER, V. (1962). Specificity in neurogenesis. *J. Cellular Comp. Physiol.*, Suppl. 1, 81–92.

HAMBURGER, V. (1963). Some aspects of the embryology of behavior. *Quart. Rev. Biol.* **38**, 342–365.

HAMBURGER, V., and BALABAN, M. (1963). Observations and experiments on spontaneous rhythmical behavior in the chick embryo. *Develop. Biol.* **7**, 533–545.

HAMBURGER, V., and OPPENHEIM, R. (1967). Prehatching motility and hatching behavior in the chick. *J. Exptl. Zool.* **166**, 171–204.

HAMBURGER, V., BALABAN, M., OPPENHEIM, R., and WENGER, E. (1965). Periodic motility of normal and spinal chick embryos between 8 and 17 days of incubation. *J. Exptl. Zool.* **159**, 1–14.

HAMBURGER, V., WENGER, E., and OPPENHEIM, R. (1966). Motility in the chick embryo in the absence of sensory input. *J. Exptl. Zool.* **162**, 133–160.

HARRISON, R. G. (1904). An experimental study of the relation of the nervous system to the developing musculature in the embryo of the frog. *Am. J. Anat.* **3**, 197–220.

HOOKER, D. (1952). "The Prenatal Origin of Behavior." Univ. of Kansas Press, Lawrence, Kansas.

HUGHES, A. (1965). The development of behaviour in *Eleutherodactylus martinicensis. Proc. Zool. Soc. London* **144**, part 2, 153–161.

HUGHES, A., BRYANT, S., and BELLAIRS, A. (1967). Embryonic behaviour in the lizard, *Lacerta vivipara. J. Zool. London* **153**, 139–152.

HUMPHREY, T. (1954). The trigeminal nerve in relation to early human fetal activity. *Proc. Assoc. Res. Nervous and Mental Diseases* **33**, 127–154.

HUMPHREY, T. (1964). Some correlations between the appearance of human fetal reflexes and the development of the nervous system. *Progr. Brain Res.* **4**, 93–135.

JACOBSON, M. (1966). Starting points for research in the ontogeny of behavior. *Proc. 25th Symp. Soc. Develop. Biol., Haverford, 1966* pp. 339–383. Academic Press, New York.

KÄLLEN, B. (1965). Early morphogenesis and pattern formation in the central nervous system. *In* "Organogenesis" (R. L. DeHaan and H. Ursprung, eds.), pp. 107–128.

KUO, Z. Y. (1967). "The Dynamics of Behavior Development." Random House, New York.

LEVI-MONTALCINI, R. (1949). The development of the acoustico-vestibular centers in the chick embryo in the absence of the afferent root fibers and of descending fiber tracts. *J. Comp. Neurol.* **91**, 209–242.

LEVI-MONTALCINI, R. (1964). Events in the developing nervous system. *Progr. Brain Res.* **4**, 1–26.

LEVI-MONTALCINI, R. (1966). The nerve growth factor: its mode of action on sensory and sympathetic nerve cells. *Harvey Lectures Ser.* **60**, 217–259.

MATTHEWS, S. A., and DETWILER, S. R. (1926). The reactions of Amblystoma embryos following prolonged treatment with chloretone. *J. Exptl. Zool.* **45**, 279–292.

MUMENTHALER, M., and ENGEL, W. K. (1961). Cytological localization of cholinesterase in developing chick embryo muscle. *Acta Anat.* **47**, 274–299.

PETERS, J. J., VONDERAHE, A. R., and POWERS, T. H. (1960). Chronological development of electrical activity in the optic lobes, cerebellum and the cerebrum of the chick embryo. *Physiol. Zool.* **33**, 225–231.

PREYER, W. (1885). "Specielle Physiologie des Embryo." Grieben's Verlag, Leipzig.

ROEDER, K. D. (1963). "Nerve Cells and Insect Behavior." Harvard Univ. Press, Cambridge, Massachusetts.

SPERRY, R. W. (1951). Mechanisms of neural maturation. *In* "Handbook of Experimental Psychology" (S. S. Stevens, ed.), pp. 236–280.

SPERRY, R. W. (1963). Chemoaffinity in the orderly growth of nerve fiber patterns and connections. *Proc. Natl. Acad. Sci. U. S.* **50**, 703–710.

SPERRY, R. W. (1965). Embryogenesis of behavioral nerve nets. *In* "Organogenesis" (R. L. DeHaan and H. Ursprung, eds.), pp. 161–186.

SPERRY, R. W., and HIBBARD, E. (1968). Regulative factors in the orderly growth of retino-tectal connexions. *Ciba Found. Symp. Growth Nervous System*, pp. 41–52.

SULLIVAN, G. E. (1966). Prolonged paralysis of the chick embryo, with special reference to effects on the vertebral column. *Australian J. Zool.* **14**, 1–17.

SULLIVAN, G. E. (1967). Abnormalities of the muscular anatomy in the shoulder region of paralysed chick embryos. *Australian J. Zool.* **15**, 911–940.

SZÉKELY, G. (1954). Zur Ausbildung der lokalen funktionellen Spezifität der Retina. *Acta Biol. Acad. Sci. Hung.* **5**, 157–167.

Székely, G. (1968). Development of limb movements: embryological physiological and model studies. *Ciba Found. Symp. Growth Nervous System* pp. 77–93. Little, Brown, Boston, Massachusetts.

Tracy, H. C. (1926). The development of motility and behavior reactions in the toadfish (*Opsanus tau*). *J. Comp. Neurol.* **40**, 253–369.

Tuge, H. (1931). Early behavior of embryos of the turtle, *Terrapine carolina*. *Proc. Soc. Exptl. Biol. Med.* **29**, 52–53.

Visintini, F., and Levi-Montalcini, R. (1939). Relazione tra differenziazione strutturale e funzionale dei centri e delle vie nervose nell'embrione di pollo. *Arch. Suisse Neurol. Psychiat.* **43**, 1–45.

von Holst, E. (1935). Über den Prozess der zentralnervösen Koordination. *Pflügers Arch. Ges. Physiol.* **236**, 149–158.

Watterson, R. L. (1965). Structure and mitotic behavior of the early neural tube. *In* "Organogenesis" (R. L. DeHaan and H. Ursprung, eds.), pp. 129–159.

Weiss, P. (1941a). Nerve patterns: the mechanics of nerve growth. *Growth,* Suppl., **5**, 163–203.

Weiss, P. (1941b). Self-differentiation of the basic patterns of coordination. *Comp. Psychol. Monogr.* **17**, 1–96.

Weiss, P. (1955). Nervous system. *In* "Analysis of Development" (B. H. Willier, P. A. Weiss, and V. Hamburger, eds.), pp. 346–401. Saunders, Philadelphia, Pennsylvania.

Wilson, D. M. (1961). The central nervous control of flight in a locust. *J. Exptl. Biol.* **38**, 471–490.

Windle, W. F. (1940). "Physiology of the Fetus." Saunders, Philadelphia.

Windle, W. F., and Fitzgerald, J. E. (1937). Development of the spinal reflex mechanism in human embryos. *J. Comp. Neurol.* **67**, 493–509.

14 Embryonic Motility in Vertebrates

VIKTOR HAMBURGER

IN THE 1920s and 1930s, the theoretical ideas concerning embryonic behavior were polarized in two schools of thought: Coghill, on the basis of his pioneer studies of the motility of the salamander, *Ambystoma,* considered behavior to be integrated from beginning to end, from the first movements of the head to swimming, walking, feeding and so forth. He generalized this concept to cover all vertebrates, including man. His ideas have been very influential, up to this day. The opposing school, including most of those working on mammalian fetuses, led by W. F. Windle, held the view that local reflexes were the building units of behavior. They were thought to be integrated secondarily into complex action systems. Both viewpoints seem now untenable as generalized theories of behavior development.

Our own investigations have led to a different polarization of ideas. The earlier work in the 1920s and 1930s was dominated completely by the reflex-response concept. Behavior was said to begin at the stage at which the embryo or fetus became responsive to stimuli. As a corollary, the role of "experience" through sensory channels, during embryonic and fetal development, was considered by many as an essential element in the structuring of postnatal action patterns. More recent studies, primarily dealing with motility in the chick embryo, have revealed the importance of nonreflexogenic spontaneous motility, up to advanced stages. At the same time, the role of sensory input in the performance of the embryo has been relegated to a minor position. Evidence is accumulating that this type of motility is basic also in other forms.

Spontaneous motility—general

All vertebrate embryos perform movements when seemingly undisturbed and under adequate physiological conditions. In different forms, motility starts at different stages of development, and the movements exhibit changing frequencies and patterns in the course of development. What is the nature of these movements? In first approximation, we define "spontaneous" as nonreflexogenic. To establish spontaneous movements as a category distinct from

VIKTOR HAMBURGER Department of Biology, Washington University, St. Louis, Missouri

stimulated movements, it is necessary not only to exclude such obvious possible sources of stimulation as amnion and uterine contractions, but also possible hidden sources, such as self-stimulation by way of the proprioceptive system. The safest procedures are radical deafferentation experiments, some of which are described below. The first evidence for clearcut, nonreflexogenic motility, however, came to light not through experimentation but by the astute observation of the normal chick embryo by a great pioneer and innovator, the German physiologist, Wilhelm Preyer, who, almost singlehandedly, established the "Physiology of the Embryo" as a special branch of physiology. In his book *Specielle Physiologie des Embryo,* published in 1885, he reported his discovery that, although motility begins at about four days of incubation, responses to any kind of stimulation could not be elicited until after eight days (actually about seven days). He immediately recognized the importance of this prereflexogenic motility, and he also noticed the uncoordinated, aimless, seemingly nonadaptive nature of these movements, which he compared with the kicking and fidgeting movements of the infant. He called these spontaneous movements *"impulsive* movements," in distinction from *reflexive* and *instinctive* movements, two categories that are behaviorally adaptive and goal-directed. With uncanny premonition, he asserted that these impulsive movements are probably generated by some processes creating chemical energy in the motor cells that is then transformed into "actual energy," that is, motility, thus anticipating a more rigorous definition of "spontaneity."

Before I discuss spontaneous motility in detail, I should point out some limitations to an approach that relies on the unsolicited overt performance of the embryo. One never knows whether the embryo exhibits its full potential of motility. It is remarkable enough, and fortunate, that the salamander, chick, sheep, rhesus, and human embryos for which neurological data are available show spontaneous (or stimulated) motility very shortly after the necessary primitive neural connections are established; and one gets the impression that, in the chick embryo, new spontaneous movements of parts, such as limbs, beak, or eye, are added as soon as the prerequisite pathways and connections are established. But periods of silence have been reported during the middle period of gestation for such mammalian fetuses as those of cat and sheep, owing possibly either to the prevalence of inhibition from specific brain centers or to inadequate permissive physiological conditions. Very few studies are devoted to a critical analysis of the relation of motility patterns to O_2 or CO_2 tension in the blood or to other physiological parameters, and asphyxiation as a source of error has bedeviled many of the earlier studies of mammalian and human fetal motility. A thorough analysis of the permissive conditions for spontaneous motility in different species probably would be rewarding.

Analysis of spontaneous motility in the chick embryo

The spontaneous motility of the chick embryo has been described repeatedly (Orr and Windle, 1934; Hamburger et al., 1965; Hamburger, 1968b), and I confine myself to a brief characterization. Sawing a window in the shell directly above the embryo exposes it to direct observation; its position can be determined by means of candling. The movements consist of irregular, seemingly uncoordinated twistings of the trunk, jerky flexions, extensions, and kicking of the legs, gaping and later clapping of the beak with or without tongue movements, eye and eyelid movements, and occasional wing flapping in later stages. The movements are performed in unpredictable combinations. From their beginning at three and a half to four days up to about 13 days, a distinct periodicity is noticeable, activity phases alternating with inactivity phases (Figure 1). The activity phases become gradually longer and the inactivity phases shorter, so that between 14 and 17 days the rhythmicity becomes less clear. Even then, however, motility is not continuous but is frequently interrupted by short, quiet periods of a few seconds in duration. We have designated this type of motility as Type I, and occasional rapid, jerky, spasmodic movements passing through the whole body, or "startles," as Type II (Hamburger and Oppenheim, 1967). The total activity—that is, the time spent in activity during the standard observation period of 15 minutes—builds up gradually and attains a peak value of 75 to 80 per cent around day 13; this is maintained through day 17. Thereafter, Type-I motility decreases sharply to about 30 per cent at day 19. In these calculations, inactivity phase is defined as

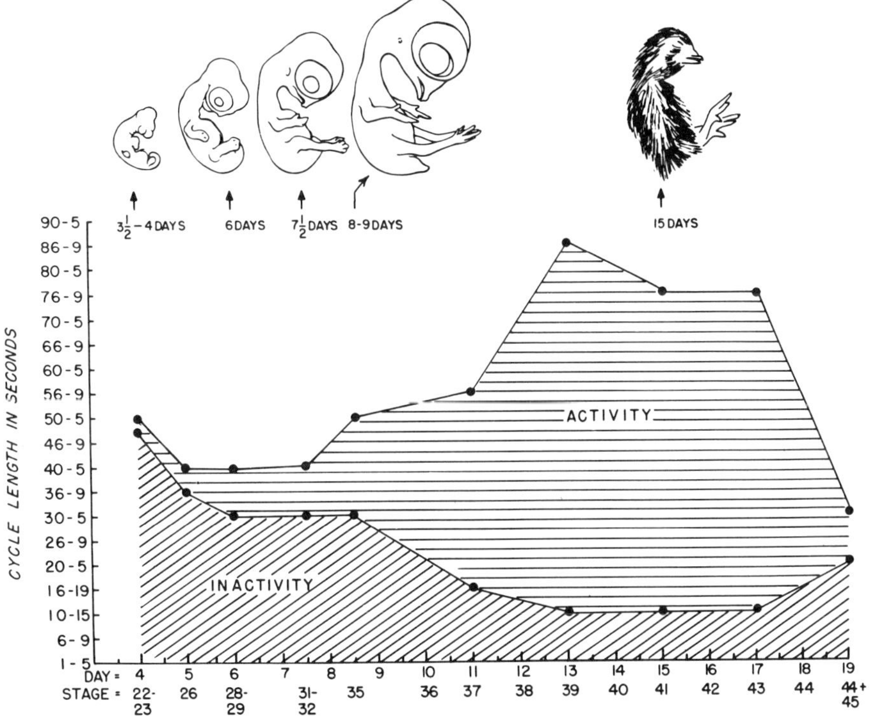

FIGURE 1 Mean duration of activity and inactivity phases and of length of cycles, in seconds, at different stages of the chick embryo. (From Hamburger et al., 1965, Figure 2.)

142 DEVELOPMENT OF THE NERVOUS SYSTEM

a period of quiescence, lasting 10 seconds or longer. If one includes shorter quiet periods in the calculation of inactivity phases, then peak total activity drops below 60 per cent.

Lack of integration of movements of different body parts and the cyclic nature of motility are characteristics indicative of nonreflexogenic activity, but rigorous deafferentation experiments are necessary to prove the point, at least for the embryonic period after seven days of incubation. As is pointed out above, the embryo is not sensitive to stimulation (Preyer, 1885), and the reflex circuits are not closed prior to that stage (Windle and Orr, 1934; Visintini and Levi-Montalcini, 1939). For later stages, deafferentation experiments are available. It is obvious that one cannot deafferent the whole embryo; one must do it piecemeal. The first such experiment was done on the legs (Hamburger et al., 1966). We made a gap several somites long in the thoracic spinal cord of two-day embryos to exclude sensory input from rostral levels. Simultaneously, we extirpated the dorsal part of the lumbosacral spinal cord, including the neural crest, thus eliminating the sensory ganglia and dorsal roots (Figure 2). These chronic preparations showed no response to exteroceptive or proprioceptive stimulation. The motor area developed normally up to 15 days (Figure 3), when progressive degeneration began. The motility of these operated embryos was compared not with that of normal embryos but with embryos in which only a thoracic gap had been made, because it had been shown that the separation of the spinal cord from the brain results in a reduction of body motility, the brain being a source of stimulation (Hamburger et al., 1965). In all stages, up to 15 days, spontaneous cyclic motility was comparable with that of control embryos with thoracic gap (Figure 4). The experiments demonstrated that sensory input is not necessary for the triggering and the maintenance of leg motility at the normal rate. It was suggested that the Type-I movements originated from spontaneous discharges of ventral internuncial or motor neurons.

In a similar operation, the deafferentation of all head structures supplied by the sensory trigeminal nerves was accomplished (Hamburger and Narayanan, 1969). The sensory nerves of the head emerge from the trigeminal ganglion, which is situated in front of the inner ear. The ganglion originates from two embryonic primordia, the neural crest of the preotic medulla and a local thickening of the epidermis, the so-called trigeminal placode. Both primordia were extirpated bilaterally in early embryonic stages. Stimulation tests of older embryos showed that tactile sensitivity was absent in all parts of the head skin in 31 of 35 operated embryos. In addition, the proprioceptive innervation of the jaw musculature was also greatly reduced or absent, in most cases, because the mesencephalic V nucleus which supplies the proprioceptive nerves was im-

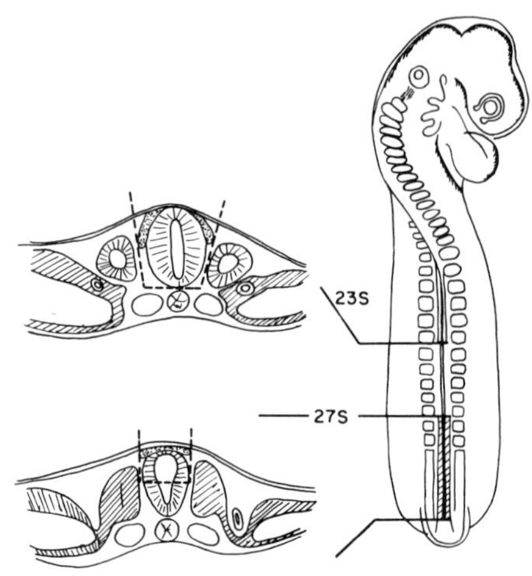

FIGURE 2 Schema of operation of deafferentation of the leg level in the two-day chick embryo. Upper left: Removal of the entire spinal cord along segments 23–27 inclusive. Lower left: Removal of the dorsal half of the spinal cord, including the neural crest, which gives rise to the sensory ganglia. (From Hamburger et al., 1966, Figure 1.)

paired, or absent, as a result of the operation. Nevertheless, recordings of Type-I motility showed that total activity, as well as the durations of activity and inactivity periods, was within normal range. Obviously, Type-I motility is independent of sensory input through the trigeminal system, up to 15 days. The decline of motility of the experimental embryos after that stage may be due to the absence of specific or nonspecific tonic sensory input. However, brain damage produced inadvertently in the majority of the embryos, and possible transneuronal degeneration, must be considered as alternative explanations. It should be noted that both the leg and the head deafferentation experiments also eliminate *self-stimulation* as a source of embryonic motility. In particular, the brushing of the legs against the head, which has been considered as a definite possibility for self-stimulation, actually does not serve this function. The same conclusion was reached in an experiment in which both leg buds were extirpated in four-day embryos. Again, total motility was not different from normal, up to 17 days (Helfenstein and Narayanan, 1970). Bilateral otocyst extirpation does not alter Type-I motility either, up to 17 days (Decker, in preparation). All these deafferentation experiments lead to the same conclusion: that Type-I motility is nonreflexogenic, up to 15 to 17 days of incubation. The obvious implication is that this type of motility is the result

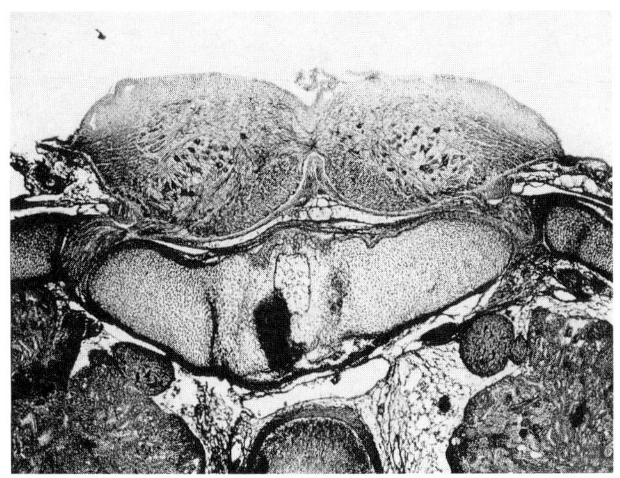

of discharges that are generated in neurons of the CNS.

The above-mentioned experiment, in which the lumbosacral spinal cord was isolated, and similar experiments, in which a chronic gap was made in the cervical cord, show that the different sectors of the spinal cord are autonomous in the generation of discharges. In both instances, the time pattern of total activity—that is, its gradual rise to a peak at 13 days and its periodicity—is unchanged. The performance, however, is quantitatively at a lower level. We infer that the brain also generates discharges which are transmitted to the cord. No inhibitory brain effect was detected, up to 17 days (Hamburger et al., 1965). A more detailed analysis of the contributions of different brain regions (Decker and Hamburger, 1967) showed that we are not dealing with a simple mass effect of brain tissue; rather, different parts of brain have different effects at different developmental stages. For instance, a definite influence of the cerebellum is demonstrable from day 15 on. The situation changes after 17 days (see below).

It has been pointed out that, as sensory input is not the source of embryonic motility, the most plausible hypothesis is the assumption of spontaneous discharges in the embryonic nervous system. It would probably be difficult to pinpoint by histological and cytological techniques the neuron types in which the activity generates. The above-mentioned experiments, in which the isolated ventral half of the cord showed its capacity to generate overt motility, suggest an involvement of ventral internuncial neurons or motor neurons, or of both. It is at this point of the analysis that the exploration of the electrical activity of the embryonic cord became mandatory.

Electrical activity of the embryonic spinal cord

In order to elucidate the neurophysiological basis of the motility patterns, electrophysiological investigations were started in 1967 by Dr. R. Oppenheim and Mr. R. Provine under the direction and with the generous aid of Dr. T. Sandel of the Psychobiology Laboratory of Washington University. The considerable technical difficulties in recording unit electrical activity from the spinal cord of chick embryos *in situ* were overcome eventually. In 1968, after the departure of Dr. Oppenheim, Dr. S. Sharma joined these efforts. In the following, I review briefly the results obtained by Dr. Sharma and Mr. Provine (see Provine et al., 1970; Sharma et al., 1970).

Numerous recordings have been made from normal embryos, particularly at the 17-day stage, to obtain information on the firing pattern within the lumbosacral spinal cord. Glass micropipettes, from 4 to 6μ in diameter at their tips, were filled with 3 molar KCl agar solution and inserted in the spinal cord at the level of dorsal root 25, anterior to the glycogen body. (It should be remembered that the lumbosacral plexus is formed by nerves 23 to 30.) By probing the cord from dorsal to ventral, one can distinguish three regions of activity (Figure 5):

A. Approximately the upper third of the cord shows relatively continuous activity. The interspike intervals are rather regular but vary from unit to unit. Most units recorded in this region show long-lasting periods of activity, with relatively short quiet periods, or none at all. We call this the "sensory region." Anatomically it corresponds to the dorsal column.

144   DEVELOPMENT OF THE NERVOUS SYSTEM

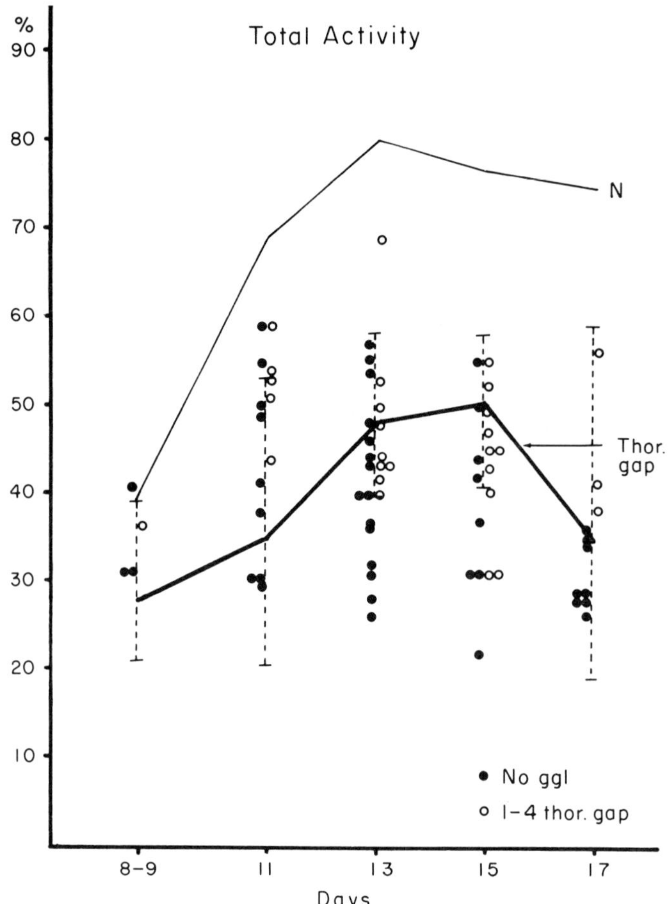

FIGURE 4   Per cent activity during 15-minute observation period. Upper solid line (N) means activity of normal embryos. Lower solid line (Thor. Gap) means activity of control embryos with thoracic gaps in the spinal cord. Each solid and open circle represents one recording from an experimental embryo. Dotted lines signify the range of controls. (From Hamburger et al., 1966, Figure 3.)

B. There follows a relatively silent region, which is about 100–200 μ in depth. This is the region immediately below the dorsal column.

C. Below this region, one obtains more discrete patterns of single-unit activity. Frequent "bursts" have been observed; these may result from the simultaneous discharge of several units. The bursts usually start abruptly and trail off into single units. Some units characteristically fire only in bursts. Others appear to fire continuously and may or may not fire synchronously during local burst activity deriving from other units. The deepest ventral region contains continuously firing units which possess relatively regular interspike intervals. All of region C contains the median and lateral motor columns and, in addition, a heterogeneous population of internuncial, commissural, and glial cells.

Thus, it is established for the first time that there exist in the spinal cord of the chick embryo patterns of single-unit electrical activity varying from intensive bursts to very low activity. Attempts are now being made to relate these patterns of neural activity to the behavioral periods of activity and inactivity which are discussed in earlier sections of this paper.

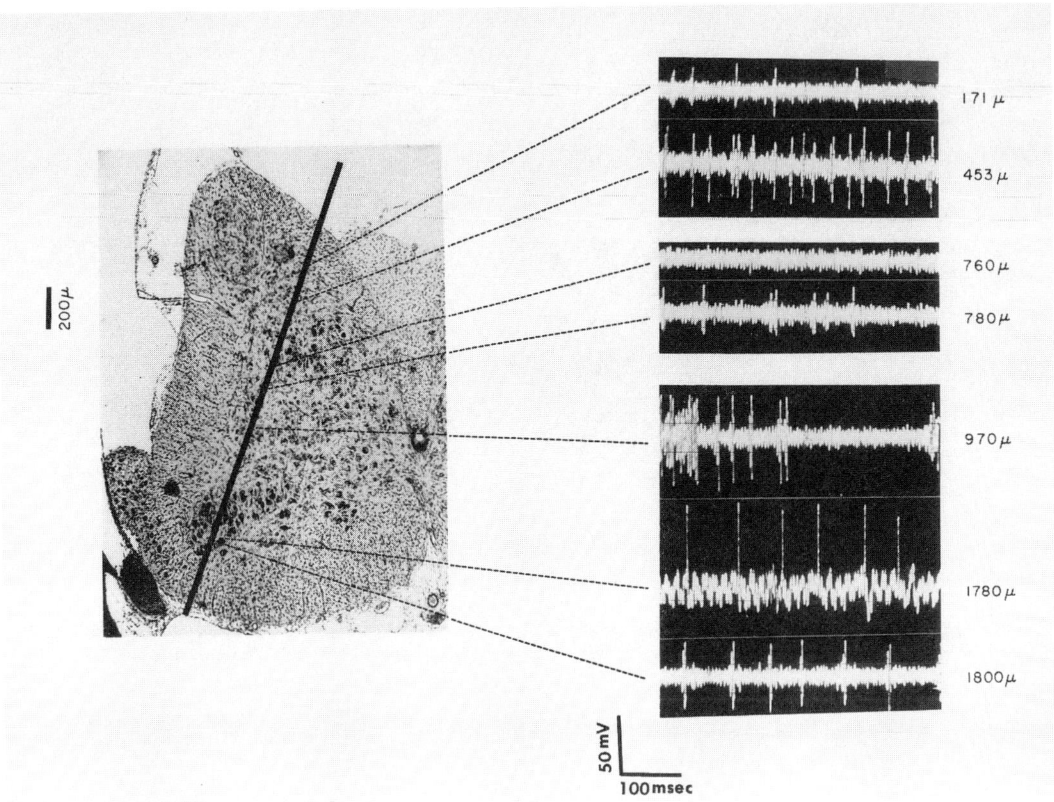

FIGURE 5 Electrode track plotted on the transverse section of the spinal cord of 17-day chick embryo at the level of dorsal root 25. Records at right show the activity picked up by the electrode tip at various points along the track. Upper two units represent the sensory region. Record at 760 μ represents zone B followed by the lower four units in motor column.

We discuss next some experimental results obtained by Sharma. They are relevant to the hypothesis formulated above—that we are dealing with spontaneous discharges of neurons, independently of sensory input.

1. Acute transection of the spinal cord was performed at the level of vertebrae one and two, in order to eliminate brain influence. A reduction of activity was observed at all levels of the cord. It is mentioned above that spinal transection reduced motility.

2. The same operation was performed and, in addition, dorsal roots 22 to 27 were transected on the ipsilateral side of the electrode. In this way, local sensory input to the recording site is blocked. As a result, the activity in sensory region A is lost, but activity in area C is not markedly affected. The contralateral side is unaffected.

3. After transection of the spinal cord, as under 1 above, 0.25 cc of 0.5 per cent xylocaine (Lidocaine-Ivenex) was injected into a thigh muscle on the ipsilateral side of the electrode. This drug blocks the afferent impulses by way of the dorsal roots without creating a trauma that might result from dorsal-root transection. All electrical activity is suspended within five minutes after injection, in region A, but recovery starts after 45 to 60 minutes. Activity in region C is not much affected, if at all. The contralateral side is not affected.

4. Acute transection of the spinal cord was performed immediately in front of the recording site (segment 21). In addition, all lumbosacral dorsal roots 23 to 30 were severed on both sides (Figure 6). This experiment tests directly our hypothesis that the ventral half of the spinal cord generates electrical activity in the absence of all sensory input. Some faint residual activity was recorded from the lower level of region A. Activity of units is present in region C throughout its depth (Figure 7).

146   DEVELOPMENT OF THE NERVOUS SYSTEM

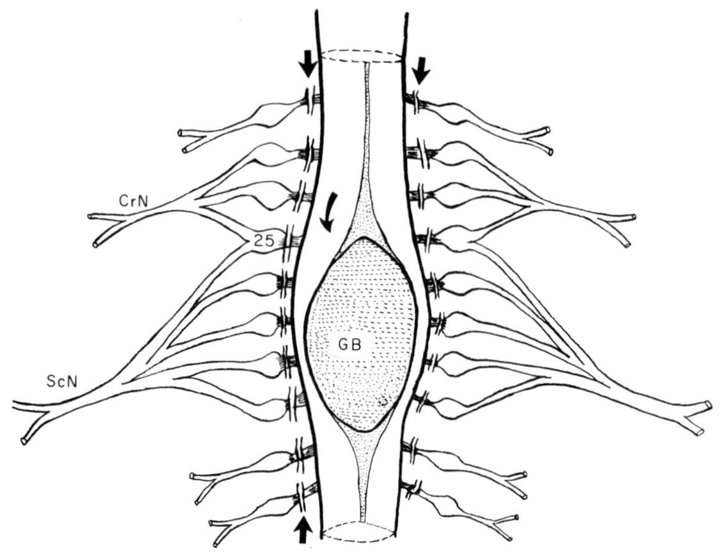

FIGURE 6   Diagram of the isolated lumbosacral spinal cord of 17-day chick embryo. Straight arrows point to the site of transection of the dorsal roots; curved arrow indicates the approximate electrode position. Spinal cord was cut at upper and lower extremities (marked by dots). CrN, cruralis nerve; GB, glycogen body; ScN, sciatic nerve.

It should be noted that, in all four experiments, the over-all level of spinal cord activity was reduced, as compared with that of normal embryos. It seems likely that this reduction can be attributed to the lack of input from higher centers. It is too early to speculate concerning the cause for the further reduction of electrical activity following de-afferentation.

One point is of particular importance: in no experimental case was the electrical activity of the ventral region of the spinal cord completely silenced. One cannot escape the conclusion that, although higher centers and sensory input contribute to the over-all firing level of this region, the latter contains elements that continue to initiate discharges of nerve cells in the absence of these sources of input.

Spontaneous motility in other vertebrate embryos

Spontaneous movements have been observed in all vertebrate embryos of which the behavior has been studied. Only an incomplete list can be given. Tracy (1926) found it in the teleost toadfish, *Opsanus tau,* in which it is cyclic, as in the chick embryo, although the inactivity phases are relatively longer. A prereflexogenic motility period of two and a half weeks precedes the stage at which the embryo becomes responsive to tactile stimulation. In the lizard *Lacerta vivipera* (Hughes et al., 1967), spontaneous motility is also intermittent during a substantial part of the embryonic period. There is a sharp rise in total activity during a period of about 12 days, but, in contrast to the chick, it does not maintain its high level; rather, it drops sharply during the subsequent 12 days. Several days intervene between the onset of motility and the sensitivity to tactile stimulation. The situation is very similar in the turtle embryo, *Chelydra serpentina* (Decker, 1967); cyclic motility reaches a short peak-activity period (at 30 days) and immediately drops gradually to a very low level. In both lizard and turtle the maximal total activity is 40 per cent or less, that is, considerably lower than in the chick embryo. "Turtle embryos like lizard embryos become sensitive to exteroceptive stimulation a few days after the onset of motility" (Decker, 1967, p. 954). Observations on spontaneous motility in mammalian fetuses were reported for the cat (Windle et al., 1933), the rat (Angulo y González, 1932), the sheep (Barcroft and Barron, 1939), the rhesus monkey (Bodian, 1966; Bodian et al., 1968), man (Hooker, 1952), and others. In no instance is there a reference to periodicity, probably largely owing to the preoccupation of the early observers with stimulated activity. In the 1930s, spontaneous activity was considered an odd phenomenon and of no particular interest. In a reinvestigation of fetal behavior in the rat, we have found typical spontaneous motility (Narayanan, Fox, and Hamburger, unpublished). It begins during the second half of day 16. Total motility (i.e., time spent in activity during the standard observation period of 15 minutes) builds up to a peak, as in all other forms. This is reached at day 18 and maintained through day 20—a day before parturition. The activity of the fetus is lower than in other forms; the highest level is between 20 and 28 per cent. The motility is intermittent, but an analysis of the Poisson distribution shows that we are dealing with a random periodicity: the movements seem to start and stop at random intervals.

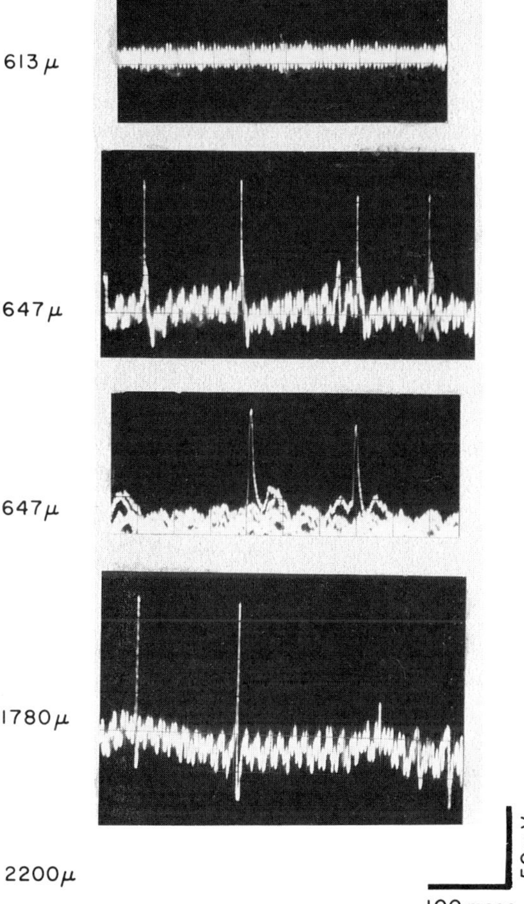

FIGURE 7 Electrical activity at various depths in the deafferentated lumbosacral spinal cord of 17-day chick embryo. The dorsal cord was almost inactive (up to 613 μ). Total depth in this probe was about 2200 μ.

Concerning the *nonreflexogenic* nature of spontaneous motility, deafferentation experiments of the kind reported above for the chick embryo are not available for any other form. A prereflexogenic period characteristic for the chick was found in the teleost, lizard, and turtle, but in none of the mammalian, fetuses. Hence, the only positive evidence we have for nonreflexogenic activity is for short periods in the development of motility of reptiles and for a longer period in the teleost, apart from the chick embryo. In this connection, it is of interest to note that amnion contractions, which are conspicuous in the chick embryo, are totally absent from the embryos of the turtle and of the rat. Hence, they are ruled out as stimulative agents in these two forms. They do occur in the lizard. The independence of embryonic motility from amnion contractions in the chick embryo was demonstrated by Oppenheim (1966). Uterine contractions occur in the rat, but they do not influence the fetal motility (Narayanan, unpublished).

The *patterns* and forms of behavior, that is, their qualitative aspect, have been described by different observers in different terms and from different viewpoints. It would be difficult and perhaps unrewarding to attempt to give a composite picture of the different types of head, trunk, and limb movements. The studies of many of the earlier investigators were guided by the question of whether behavior fitted in the scheme of Coghill, who, on the basis of his extensive work on the salamander, *Ambystoma,* had postulated that embryonic behavior is integrated from beginning to end. Although such may be true for urodeles and teleosts and perhaps other aquatic forms, which, for reasons of survival, must attain the capacity for swimming and other integrated performances as early as possible, the chick embryo presents an entirely different picture, as shown above. Its spontaneous movements are random, apparently based on massive discharges of large numbers of neurons that are not held in check or streamlined by higher inhibitory or integrating centers. As far as mammals are concerned, most investigators agree that, at least in the early phases, their movements are also mass movements, which involve all parts of the body, but that local movements of limbs, head, and so forth, make an early appearance. Our preliminary observations on rat fetuses give the impression, shared by others, that they are certainly not completely integrated but are more of the random type.

In summary: The contention of Preyer that spontaneous ("impulsive") motility is a special category of embryonic motility is amply confirmed; it is characteristic of most, or all, vertebrate embryos; it shows periodicity and a random pattern in many; and its nonreflexogenic nature is proved, at least for the chick embryo. I have pointed out elsewhere that its adaptive significance may be to insure the normal formation and maintenance of articulations and the intactness of the musculature. Our understanding of the neurophysiological basis of this type of motility is only beginning. Our hypothesis that we are dealing with discharges of motor neurons which fire indiscriminately and are probably driven by interneurons needs much further scrutiny. The periodicity cycles, which are on the scale of seconds or minutes, are, as yet, unexplained.

Integrated movements

The origins of integrated behavior have been reviewed on several occasions (Hamburger, 1963, 1968a, 1968b), and I limit myself here to a few general remarks. The pioneer work of Coghill on the salamander, *Ambystoma,* has clarified the issue for this form. In his "Correlated anatomical and physiological studies of the growth of the nervous system" (see Coghill, 1929), he has shown that behavior development in this urodele proceeds in a strictly programed way. The first bending of the head in the tail-bud stage is linked with the integrated swimming movements and with later walking and feeding action patterns by intermediate behavioral stages, which maintain the integrated involvement of all parts throughout behavior development. Local responses originate by a process of emancipation or "individuation" from the total pattern. The process is very similar in teleosts (Tracy, 1926). Coghill was successful in correlating the behavior sequence, step by step, with a corresponding sequence in the differentiation process of the central nervous system and with the formation of appropriate synaptic connections shortly before a new step in behavior development is attained.

Such an achievement was remarkable and provocative at the time, and an incentive for many other studies, but, although *Ambystoma* solves the problem of behavior integration in this simple and straightforward way, it was soon realized that the situation in amniotes is more complex and less transparent. As is stated above, the motility of the chick embryo, at least up to 17 days, is unintegrated in the sense that antecedents to organized posthatching behavior, such as walking, pecking, and so forth, cannot be clearly recognized. Wing flappings and alternate leg movements are rare. The same holds for reptilian embryos and, at least to a certain extent, for mammalian embryos. The first integrated activity in the chick embryo is the preparation for hatching and the hatching act itself, which occupies it from incubation days 17 to 20. The integrated head, trunk, and leg movements which are involved have been described in detail and designated as Type-III movements (Hamburger and Oppenheim, 1967; Hamburger, 1968b). It is not clear whether sensory input is necessary for this activity. Embryos with complete trigeminal or vestibular deafferentation fail to hatch. Yet, these experiments are inconclusive. The inability to perform these movements could be ascribed to the lack of specific sensory guidance or of tonic input. However, these chronic preparations show secondary transneuronal degeneration, and brain damage was produced inadvertently in most of the trigeminal operations. Hence, the behavioral deficiency could be the result of the impairment of central connectivity.

We have stated repeatedly that we do not find it possible to relate the Type-III motility to Type-I and Type-II motility. Not only are they different in pattern, but Type-I and occasionally Type-II movements continue during the prehatching period in the intervals between the episodes of integrated hatching movements. Nor do we see a direct continuity or relationship between any of the three embryonic motility types and the posthatching action patterns, such as walking, pecking, drinking, and so on. It seems that the neural apparatus for these activities is fully prepared during the embryonic period and that the action patterns are triggered by environmental or intrinsic signals.

We are not yet prepared to extend this notion to mammalian fetuses. It is conceivable that, particularly in forms that are very immature at birth and not immediately required to fend for themselves, the build-up of integrated patterns is more gradual. Alternating leg movements and trunk and leg movements that resemble righting (i.e., restoring the upright position) have been described, but that such antecedents are necessarily reflexogenic is not implied.

Stimulated embryonic movements

A wealth of information is available on responses to tactile and other stimulations in many forms, and particularly mammals, largely because research on embryonic motility was focused on this aspect in the 1930s and 1940s. The material on birds has been ably reviewed by Gottlieb (1968); the older literature, including that pertaining to mammals, is covered in the comprehensive review by Carmichael (1954). We limit ourselves to a few brief general remarks, mostly concerning birds and mammals.

The onset and progression of stimulated activity are as intimately tied to the differentiation of their neurological substrate as are those of spontaneous motility. A good example is the cutaneous sensitivity in the trigeminal area of the human fetus, which has been analyzed in detail by Humphrey (1964). The perioral region, which is the first reflexogenous zone, becomes sensitive at a menstrual age of seven and a half weeks, shortly after the cutaneous V fibers have reached the skin. The response, that is, contralateral neck flexion, coincides with the arrival of spinal tract V fibers at the level of the second and third cervical-cord segment. As soon as longitudinal fibers have grown caudally, the response is extended to include trunk and limb movements. It is of great interest that, for a considerable period (two weeks in the human fetus), local perioral stimulation elicits generalized total body movements. From the tenth week on, the generalized response subsides and gives way to local responses, such as mouth opening. In most vertebrate embryos, from amphibians to man, the trigeminal area is the first to become sensitive to tactile stimulation, but in the cat (Windle et al., 1933) and the rhesus monkey (Bodian, 1966; Bodian et al., 1968), the palmar surface of the forelimb becomes responsive at the same time. The sequence from generalized to local responses has been observed widely, but, in some forms, local reflexes occur simultaneously with, or even slightly earlier than, generalized movements. Obviously, the timetable for the differentiation of the central connections, and for the inhibitory mechanisms involved in restricted local movements, differs from form to form. Perhaps these differences have been overemphasized in the controversy over the primacy of total versus local responses as the basis of embryonic behavior.

Finally, I wish to raise the fundamental question of the relevance of these investigations of stimulated responses to the normal behavior development *in situ*. Windle, who has had extensive experience with the behavior of mammalian fetuses, has made the following comment: "It should not be assumed that all responses which can be induced occur spontaneously within the uterus of the normal intact individual. As a matter of fact, there is scanty evidence that any of them occur normally during the early part of the gestation period. . . . The fetus is adequately nourished and warmed in a medium lacking practically all the stimulating influences of the environment with which it will have to cope later. No significant excitation of the external receptors occurs" (Windle, 1940, pp. 164–165). The chick embryo is likewise sheltered in an environment that provides only a narrow range of stimulations—incomparably less than does the posthatching environment. Self-stimulation plays a minor role, if any, according to our experimental results (see above). Proprioceptive cues also have been excluded in several deafferentation experiments.

It follows that the stimulation experiments do not elucidate the actual overt performance of normal embryos *in situ*, at least up to fairly advanced stages. The stimulation experiments have revealed a great deal about the progressive differentiation of complex response patterns and the underlying neurological basis. These important results should have deserved a more extensive treatment in our presentation. However, if these reflexogenic response patterns hardly ever become overtly manifest during the major part of embryonic and fetal life, owing to the absence of appropriate stimuli, they cannot be considered as *the* elementary building stones in the genesis of behavior, as has been claimed frequently in the past. On the other hand, it is just as difficult to relate the motility patterns which the normal, undisturbed embryo and fetus actually perform *in situ* (mostly spontaneous in nature) to postnatal behavior. In the chick embryo, at least, we were unable to obtain clues from prehatching behavior for an understanding of posthatching behavior. The situation in mammals requires much further study from this viewpoint. One point seems clear: most embryos are *active* before they manifest their capabilities to respond to environmental cues. The precedence of action over reaction in embryonic development of behavior has interesting evolutionary and theoretical implications that we cannot follow up at this point.

Summary

Coghill's idea of a continuity of integration in behavior development from beginning to end cannot be generalized beyond his own material, the salamander, *Ambystoma,* and perhaps the teleosts. The antithesis: that local responses are the building blocks in behavior development is equally untenable as a general theory. We must admit frankly that the present state of our knowledge does not permit us to formulate broad generalizations. It is doubtful whether additional direct observations and simple stimulation procedures will carry us much farther. The neurological approach, particularly on the ultrastructural level; the neurophysiological approach combined with experimental procedures; and rigorously controlled behavior experiments would seem to give the best promise of further advances.

Acknowledgment

All investigations from this laboratory were supported by NIH Grant 5R01-NB 05721 to the author.

REFERENCES

Angulo y González, A. W., 1932. The prenatal development of behavior in the albino rat. *J. Comp. Neurol.* 55: 395–442.

Barcroft, J., and D. H. Barron, 1939. The development of behavior in foetal sheep. *J. Comp. Neurol.* 70: 477–502.

Bodian, D., 1966. Development of fine structure of spinal cord in monkey fetuses. I. The motoneuron neuropil at the time of onset of reflex activity. *Bull. Johns Hopkins Hosp.* 119: 129–149.

Bodian, D., E. C. Melby, and N. Taylor, 1968. Development of fine structure of spinal cord in monkey fetuses. II. Pre-reflex period to period of long intersegmental reflexes. *J. Comp. Neurol.* 133: 113–165.

Carmichael, L., 1954. The onset and early development of behavior. *In* Manual of Child Psychology (L. Carmichael, editor). John Wiley and Sons, New York, pp. 160–185.

150   DEVELOPMENT OF THE NERVOUS SYSTEM

Coghill, G. E., 1929. Anatomy and the Problem of Behaviour. Cambridge University Press, Cambridge, England.

Decker, J. D., 1967. Motility of the turtle embryo, Chelydra serpentina (Linné). *Science (Washington)* 157: 952–954.

Decker, J. D., and V. Hamburger, 1967. The influence of different brain regions on periodic motility of the chick embryo. *J. Exp. Zool.* 165: 371–384.

Gottlieb, G., 1968. Prenatal behavior of birds. *Quart. Rev. Biol.* 43: 148–174.

Hamburger, V., 1963. Some aspects of the embryology of behavior. *Quart. Rev. Biol.* 38: 342–365.

Hamburger, V., 1968a. Beginnings of co-ordinated movements in the chick embryo. *In* Growth of the Nervous System/A Ciba Foundation Symposium (G. E. W. Wolstenholme and M. O'Connor, editors). J. and A. Churchill, Ltd., London, pp. 99–105.

Hamburger, V., 1968b. Emergence of nervous coordination. Origins of integrated behavior. *In* The Emergence of Order in Developing Systems (M. Locke, editor). Academic Press, Inc., New York, pp. 251–271.

Hamburger, V., M. Balaban, R. Oppenheim, and E. Wenger, 1965. Periodic motility of normal and spinal chick embryos between 8 and 17 days of incubation. *J. Exp. Zool.* 159: 1–14.

Hamburger, V., and C. H. Narayanan, 1969. Effects of the deafferentation of the trigeminal area on the motility of the chick embryo. *J. Exp. Zool.* 170: 411–426.

Hamburger, V., and R. Oppenheim, 1967. Prehatching motility and hatching behavior in the chick. *J. Exp. Zool.* 166: 171–204.

Hamburger, V., E. Wenger, and R. Oppenheim, 1966. Motility in the chick embryo in the absence of sensory input. *J. Exp. Zool.* 162: 133–160.

Helfenstein, M., and C. H. Narayanan, 1970. Effects of bilateral limb bud extirpation on motility and prehatching behavior in chicks. *J. Exp. Zool.* 172: 233–244.

Hooker, D., 1952. The Prenatal Origin of Behavior. University of Kansas Press, Lawrence.

Hughes, A., S. V. Bryant, and A. d'A. Bellairs, 1967. Embryonic behaviour in the lizard, *Lacerta vivipara. J. Zool. (London)* 153: 139–152.

Humphrey, T., 1964. Some correlations between the appearance of human fetal reflexes and the development of the nervous system. *Progr. Brain Res.* 4: 93–135.

Oppenheim, R., 1966. Amniotic contraction and embryonic motility in the chick embryo. *Science (Washington)* 152: 528–529.

Orr, D. W., and W. F. Windle, 1934. The development of behavior in chick embryos: The appearance of somatic movements. *J. Comp. Neurol.* 60: 271–285.

Preyer, W., 1885. Specielle Physiologie des Embryo. Grieben's Verlag, Leipzig.

Provine, R. R., S. C. Sharma, T. T. Sandel, and V. Hamburger, 1970. Electrical activity in the spinal cord of the chick embryo, in situ. *Proc. Nat. Acad. Sci. U. S. A.* 65: 508–515.

Sharma, S. C., R. R. Provine, V. Hamburger, and T. Sandel, 1970. Unit activity in the isolated spinal cord of the chick embryo, in situ. *Proc. Nat. Acad. Sci. U. S. A.* (in press).

Tracy, H. C., 1926. The development of motility and behavior reactions in the toadfish (Opsanus tau). *J. Comp. Neurol.* 40: 253–369.

Visintini, F., and R. Levi-Montalcini, 1939. Relazione tra differenziazione strutturale e funzionale die centri e delle vie nervose nell'embrione di pollo. *Schweiz. Arch. Neurol. Psychiat.* 43: 381–393.

Windle, W. F., 1940. Physiology of the Fetus. Saunders, Philadelphia.

Windle, W. F., J. E. O'Donnell, and E. E. Glasshagle, 1933. The early development of spontaneous and reflex behavior in cat embryos and fetuses. *Physiol. Zool.* 6: 521–541.

Windle, W. F., and D. W. Orr, 1934. The development of behavior in chick embryos: spinal cord structure correlated with early somatic motility. *J. Comp. Neurol.* 60: 287–307.

ANATOMICAL AND PHYSIOLOGICAL BASIS OF EMBRYONIC MOTILITY IN BIRDS AND MAMMALS

VIKTOR HAMBURGER

Department of Biology
Washington University
St. Louis, Missouri

51

I. Introduction

In previous communications, I have paid tribute to Professor Wilhelm Preyer, to whom this volume is dedicated. The importance of his theoretical and observational contributions to the embryology of behavior has been acknowledged particularly in a recent essay (Hamburger, 1971). If his name does not appear in the following pages, this means that his influence has reached the stage of anonymity which is the fate of all great innovators.

Observations and experiments on embryonic motility in amniotes have been reviewed repeatedly in recent years (e.g., Gottlieb, 1971b; Hamburger, 1968, 1970, 1971), and I shall not go once more over familiar grounds. The task which I have undertaken is to scrutinize the relationships of processes going on at three different levels of organization: the level of motility and behavior, the level of bioelectrical activity, and the level of structural and ultrastructural organization. The discussion is limited to birds and mammals, but it includes spontaneous motility and evoked responses. I have selected a few key issues, but in these I have tried to go beyond generalities and to understand relatioships in depth and in specific detail.

The basic motility type of the chick embryo has three characteristics: it is spontaneous (nonreflexogenic); it is intermittent; and the parts of the embryo, such as head, trunk, limbs, beak, move in a seemingly uncoordinated fashion. This type of motility has been designated as Type I. Type II are startles, and Type III, coordinated hatching movements (Hamburger & Oppenheim, 1967). In a previous review (Hamburger, 1971) I have tried to establish spontaneity, based on autonomous discharges of neurons, as the constitutive element in embryonic motility, and I have made the point that neither periodicity nor the forms or patterns of overt motility are of equally universal character. In many of our investigations, durations of activity phases or number of movements per time unit had been used preferentially, for the simple reason that this parameter of Type I motility can be quantified. However, periodicity cannot be considered as a necessary prerequisite of spontaneity. Clear periodicity is obvious in chick embryos only up to about 13 days of incubation. Thereafter, motility and its correlate, burst activity, remain discontinuous, but the inactivity phases become very short and ill-defined.

The form, or pattern, of motility likewise is not universally uniform. Spontaneous motility can be either uncoordinated, as in reptiles, birds, and mammals, or coordinated, as in teleosts. In the latter, Tracy (1926) has found that embryonic and larval movements, including swimming, are spontaneous (nonreflexogenic) for about 2 weeks. The situation in the salamander is not quite clear. Spontaneity remains the one basic characteristic of overt, unelicited motility in vertebrate embryos.

II. Some Remarks on Lack of Coordination in Amniote Embryos

The unstructured performance of the chick embryo up to 17 days, and similar performances of reptilian and mammalian embryos, hardly deserve the designation of "behavior." I have usually referred to Type I and II ("startless") as motility, and only to the integrated prehatching and hatching performance (Type III) as behavior, since this term usually implies an integrated activity which subserves the whole organism (see also Gottlieb, 1971a, p. 6). In contrast, although Type I also involves the movements of several or all parts, the resulting motility appears to be merely the summation or combination of movements of individual parts, such as head, trunk, limbs, beak, eyelids. Such performance hardly occurs in postnatal behavior except perhaps in infants of immaturely born forms, including the human. The jerky uncontrolled movements of the chick have more resemblance to neuropathological convulsive conditions. The phenomenon of uncoordinated motility which, moreover, does not involve sensory input to any extent, may not seem to be of much interest to developmental psychobiologists. Yet, there remains the inescapable truth that this type of motility with all its monotony seems to be the basic type that amniote embryos and fetuses perform, notwithstanding occasional responses to stimulation. Hence, this activity is obviously of considerable importance for the embryo; it cannot be ignored in generalizations and theories concerning prenatal behavior and its relation to postnatal action patterns.

It has been objected that our claim that the combinations of movements in Type I motility are uncoordinated has not been subjected to a rigorous statistical analysis. It would be desirable, indeed, to make the recording method more objective and to do a correlation analysis of sequence of movements, perhaps using slow-motion moving pictures. However, even if such analysis would reveal a low degree of correlation between those particular types of movements that become integrated later, this would not necessitate a revision of my basic position, as I hope to be able to show when we analyze the electrophysiological patterns underlying Type I motility, and evoked responses.

In his Introduction, Dr. Gottlieb has stated that we have failed to recognize two particular instances of coordinated patterns in bird embryos. The one instance is nonrandom movement of the head, or, more specifically, head turning which is performed in the following context. Up to several days before hatching the beak is buried in the yolk sac and the neck is bent downward. The head is then lifted out of the yolk by head movements which are preferentially to the right. Eventually, the twist of the neck to the right becomes permanent and the head, which is now near the right wing, is subsequently tucked under the right wing. Oppenheim and I (Hamburger & Oppenheim, 1967) have described the preponderance of head movements to the right and considered this as the beginning of coordinated prehatching motility (Type III) and as preparatory to tucking. We have called this the pretucking phase. This interpretation is based (1) on a rotatory component that is involved and which is not observed in Type I, and (2) on the continuity of pretucking and tucking movements. We think that, in the chick, the head turning to the right is part of Type III and has nothing to do with Type I. Gottlieb and Kuo (1965) have apparently traced the preferential head turning to the right in the duck embryo to earlier stages than we did. The point deserves further consideration. Perhaps in the duck this nonrandom performance has to be considered as an aspect of Type I motility and as an exception to the rule.

The second instance are sinusoid waves performed by early chick embryos. They have been described by us and by others before us. In fact, S-waves are characteristic for very early motility stages in many vertebrate embryos, and they are, indeed, a type of coordinated movements. In fishes and amphibians, they are the direct precursors of swimming movements (Coghill, 1929; Tracy, 1926). But in the chick, this pattern breaks down soon after its inception at 4 days. At 5 days, only the first waves in a sequence usually go down all the way. Subsequent waves may end in the trunk or begin in the trunk and spread in both directions. At the same time, local contractions of trunk somites make their appearance (Hamburger & Balaban, 1963). In subsequent stages, the exceptions become the rule, and the S-waves disappear altogether.

Perhaps the neural mechanism available to the early embryo permits only the coordinated S-waves. A longitudinal fiber tract, which sends collaterals to the segmentally arranged motoneurons, and a ventral commissure are formed very early (Visintini & Levi-Montalcini, 1939), and perhaps only the rostrally located neurons of this fiber tract are at first sufficiently mature to initiate spontaneous discharges. The pattern breaks up, when sufficient numbers of neurons in the spinal cord become interconnected.

One might ask, why the chick embryo, and the amniote embryo in general, does not continue on its original path, following Coghill's scheme of continuously integrated behavior from beginning to end. Several answers to this question can be given; I shall come back to it later. One answer can be dismissed right away. The occurrence of S-waves in early embryos of higher forms could be considered as the recapitulation of an ancestral pattern, characteristic of aquatic forms, where it leads to swimming. While such a viewpoint is legitimate, it cannot be accepted as an adequate "explanation" of currently displayed behavior.

III. Spontaneous Motility in Rat Fetuses

Before we continue the discussion, I propose to broaden the basis by including mammals. Our recent studies of spontaneous motility in rat fetuses (Narayanan, Fox, & Hamburger, 1971) have shown basic similarities with the chick, and some differences. A considerable difference is in the stage of initiation of motility. Mammalian embryos are much more advanced in overall body development at the time when motility begins; their limbs are in possession of movable joints, whereas chick embryos have small limb buds, when the first neck flexions begin. In mammals reflex arcs are completed at the onset of motility, whereas chick embryos have a prereflexogenic period of 4 days. Motility in the rat embryo begins at $15\frac{1}{2}$ days; our observations cover the period of 16–20 days.

The females were immobilized by thoracic spinal transection to avoid anesthesia. Their lower body parts were submerged in a water bath at 37° C. Fetal movements were observed through the uterine walls, or after exposure of the fetus, with placenta and amnion intact. Such preparations show no changes in motility for several hours. Recordings of activity phases were made as in the chick, during 15-minute observation periods, both *in utero* and *ex utero*.

We have distinguished between total, regional, and local movements. By regional we mean, for instance, a combination of head and forelimb movements, with other parts at rest, or hindlimb, tail, and pelvis combinations. The movements are in general smoother than in the chick, though local movements of individual parts may be jerky. As in the chick, total activity builds up to a peak which is reached at 18 days and then declines. As in reptiles and birds, motility is intermittent; the activity phases occur at irregular intervals. The movements of parts are not correlated with each other, much like in Type I motility of birds. This is particularly striking in the lack of coordination of the movements of right and left forelimbs, which are the most active parts throughout.

Altogether, the similarities in the three major characteristics, namely spontaneity, periodicity, and lack of coordination, are so great that we have no doubt that we are dealing with the same basic type in birds and mammals. Although other investigators of motility in mammals have paid little attention to spontaneous motility, their casual observations on guinea pig, cat, sheep, and other fetuses, including the human, are in line with ours. What we call uncoordinated Type I motility is usually referred to as generalized or mass or total activity.

The nonreflexogenic nature of the fetal movements in the rat has not been established experimentally. We have excluded one possible source of extraneous stimulation: contractions of the uterus. Fetal movements are five times as frequent as uterine contractions, and their periodicities are un-related. The amnion is not contractile in the rat.

IV. The Continuity–Discontinuity Problem on the Behavioral Level

The shift from the coordinated S-waves to an uncoordinated type of motility, and then to the coordinated hatching behavior in bird embryos, raises a larger and more general issue: the continuity-discontinuity problem. In the following discussion, we shall keep in mind that we are dealing with phenomena on several levels, the behavioral, the electrophysiological, and the structural-ultrastructural. Much confusion can be avoided if this point is clearly recognized.

On the behavioral level, Oppenheim and I (Hamburger & Oppenheim, 1967) have asserted a discontinuity between Type I and the integrated pre-hatching and hatching behavior, Type III. We have given repeatedly the arguments for our stand that it is not possible to assume a smooth trans-formation of Type I into Type III. Briefly, they are the following (*1*) The two types are very different in their form, apart from the highly integrated character of Type III; a rotatory component is observed in most of the phases of prehatching and hatching motility; this is entirely missing in Type I. (*2*) The Type III movements are performed in episodes that are separated by intervals of different durations. The Type I movements do not disappear altogether in the prehatching period between 17 and 20 days. They are mere-ly suspended during Type III episodes, but resumed during the intervals, though at reduced frequency.

Oppenheim (1970) has examined the question of gradual transition versus rather sudden appearance of a behavioral act. He has analyzed the origin of one important component of Type III; namely, the vigorous back thrusts of head and beak which are instrumental in pipping and hatching. By record-ing head movements directed backward, lateral, or in other directions during prehatching stages in the duck, he has found that the frequency of back movements does not build up gradually; they increase rather suddenly beginning about 16 hours before initiation of hatching, compared to lateral and other movements. Oppenheim (1970) points out that individual back movements occur sporadically in earlier stages. In this sense, there is "not a de novo appearance of these component movements [p. 348]." The case is well stated when he points out that the novel aspect, apart from in-crease in frequency, is the incorporation of back thrusts in an integrated activity pattern, whereby they become effective in pipping and hatching. I think this kind of continuity whereby behaviorally uncommitted individual movements become integrated in a more complex behavioral act, at a higher organizational level, is paradigmatic for much of embryonic and fetal behavior. *In this way, aspects of continuity and discontinuity become compatible.* In a similar way, complex postnatal patterns, like walking or pecking, can be said to arise *"de novo"*; the newly hatched chick stands up and walks within $\frac{1}{2}$ hour; so does the newborn sheep or colt. Most of the component movements have been performed before, but not in the context of an integrated activity. To give another example: Embryonic swallowing is incorporated in postnatal feeding and drinking. Kuo (1932) has expressed a similar notion, except that in his view the component movements result from stimulation or self-stimulation.

We have observed another form of discontinuity within the framework of hatching behavior. I refer to the sudden onset of the climax, that is, the act of hatching, which involves a change in the temporal pattern of Type III movements. Kovach (1970) has also studied this and preceding processes in more detail, using a movement transducer for recording from intact eggs and eggs with windows. Between days 17 and 20, Type III rotatory movements are performed sporadically or in short repetitive sequences, at varying and sometimes hour-long intervals. Then, suddenly, at an unpredictable moment, at day 20+, they are enacted in regular cycles with short intervals of 10–30 seconds, for periods lasting $\frac{1}{2}$ to $1\frac{1}{2}$ hours, until the chick is hatched. In this kind of discontinuity, only the temporal pattern is changed, but not the form of behavior. The causation of this shift is not known; other contributors also discuss this question (Oppenheim article and Vince article, this volume). But it is clear that we are dealing with a systemic factor. Balaban and Hill (1969) have found that vocalization and raising of the upper eyelid, two parameters which are irrelevant for hatching, also increase suddenly and manifold exactly synchronously with the other climax symptoms.

There is yet another facet to hatching behavior, this time along the line of continuity. Kovach (1970) finds a great similarity between the rotatory Type III movements and the postnatal righting reflexes. In fact, he considers the latter as a resumption of Type III movements in postnatal life. Even if one recognizes differences in the release of the two acts and in their adaptive functions, the idea of a continuity of a specific embryonic behavior into postnatal life, with a change of function, is appealing. Time does not permit to pursue this interesting theme of prenatal antecedents to postnatal action patterns, though it is very much part of the continuity-discontinuity problem.

Before we turn the discussion to lower levels of organization, I shall try to clarify another point. In the Introduction, Dr. Gottlieb connects the continuity-discontinuity problem with the important question of *causal* or determinative relations between different behavioral stages. He states that "probabilistic theories. . . .hold (*1*) that behavior development is a gradual and continuous process wherein (2) certain features of late embryonic, fetal or early neonatal behavior can be traced to, and are in some sense (facilitative or determinative) dependent upon, earlier behaviors or stimulative events." (p. 17). He continues, "Certain other viewpoints (e.g., Hamburger, 1968), on the other hand, are explicit in their emphasis of discontinuities in embryonic behavior such that, for example, later movements are envisaged as arising *de novo*, in which case these movements could bear nothing other than a permissive relationship to earlier movements." The last part of this sentence imputes a *causal* relationship, whereas Oppenheim and I have claimed discontinuity only with respect to the *forms* of the motility patterns. We have not claimed that Type III could not be modified by experimental manipulation of Type I; there are no experimental data that shed light on this question. Nor have I claimed that our deafferentation experiments touch upon the more specific question of whether sensory input in earlier stages might have an effect (facilitative, determinative, or otherwise) on hatching behavior. We could not tackle this question because the deafferented embryos did not hatch. This problem is also unresolved. Our only claim is that *ongoing Type I motility up to day 17 is not dependent on sensory input* (e.g., Hamburger, 1970; Hamburger & Narayanan, 1969; Hamburger, Wenger, & Oppenheim, 1966). An earlier statement which went beyond this claim (Hamburger, 1963, 1964) stands corrected.

In summary, as far as the probabilistic claim of gradualness and continuity is concerned, it is fully realized in the behavior of anamniote embryos which follow Coghill's paradigm of continuously integrated behavior. In amniote embryos continuity is compatible with discontinuity, if one accepts the notion of incorporation of unspecific movements in newly established integrated patterns. And there is definitely continuity on the structural level, in neurogenesis, to which we shall turn next. The continuity–discontinuity problem is a complex one. Looked at close range, it dissolves itself into a number of specific problems. In this case, as in others, a dichotomous conceptualization, leading to an "either–or" position, does not do justice to the phenomena. In all such matters, the question is not "either–or" but rather: to what extent and in what respects does a specific sequence show one or the other alternative?

V. Incongruity of Neurogenesis and Development of Behavior

If complex integrated activities such as hatching behavior and many postnatal action systems make their appearance rather suddenly, *de novo*, in the sense indicated above, then one has to postulate that the neural machinery that makes such novelties in behavior possible must have been prepared in advance. The neural organization, including the circuitry that underlies all complex activities performed after birth, is created in a sequence of differentiation processess which are referred to as neurogenesis. This includes cellular and supercellular events and all aspects of fiber outgrowth and formation of central and peripheral nerve fiber patterns, and of course synaptogenesis. Whereas the early phases, such as proliferation, migration, axon and dendritic outgrowth, and some of the underlying mechanisms are fairly well known, we are much less well informed about the later phases, including the final fixation of synapses which is of crucial importance for function. Despite these gaps, the neuroembryologist is committed to the basic tenet of the continuity of all these process, in the sense that every step in progressive differentiation emerges from the preceding step, and that the increase in complexity is gradual and continuous.

There is nothing novel about all this. What has not been previously stated explicitly is the incongruity between neurogenesis and overt motility in amniote embryos. The unorganized movements in the chick embryo undergo little change, between 4 and 17 days, except that they become more frequent and more muscles get into the act. *In no way do they reflect the neurogenetic events which go on in the meantime, so to speak below the surface.* The gradual increase in the complexity of the circuitry finds no expression in the pattern of overt spontaneous motility; in fact, it is difficult to imagine how it could do so. Even if one admits that the resolution of our movement analysis is rather coarse, and that the structural and ultrastructural analysis is deplorably incomplete, I think that the lack of clear relationship between Type I motility and the observed or inferred neurogenetic processes is undeniable. Hence, the notion that neurogenesis fully "explains" or "determines" embryonic behavior development is not valid as a generalization. It does not apply to the chick and the rat embryos, though it may be valid for the salamander. In other words, even the most detailed knowledge of neural organization, including all significant synapses, in chick or rat embryos at a given stage would permit no prediction of the actual (Type I) movements performed at that stage. Nor would a progression in synaptogenesis from one stage to the other be reflected in the details of motility. However, in salamander embryos, the correspondence of progression in neurogenesis (especially synaptogenesis) and behavior is very close, indeed. It is one of Coghill's major achievements to have documented this in detail. But even in this form, there is a modicum of indeterminacy. For instance, in the "early flexure" stage, the head can move either to the left or to the right, though there is a high probability that it will move away from a unilateral stimulus. All one can say is that *the state of differentiation of the nervous system at a given stage delimits the range of behavioral potentialities.*

VI. Relation of Neurogenesis and Bioelectrical Phenomena

The electrophysiological events which are closely correlated with overt
motility (Provine, this volume) cannot be expected to give a more precise
reflection of ongoing neurogenesis than motility itself. The polyneuronal
bursts, though they do have their own structure and though they undergo
temporal changes (see Provine, this volume), appear as monotonous throughout the major embryonic period as the uncoordinated, jerky movements
which they generate. While the pathways for integrated activity are being

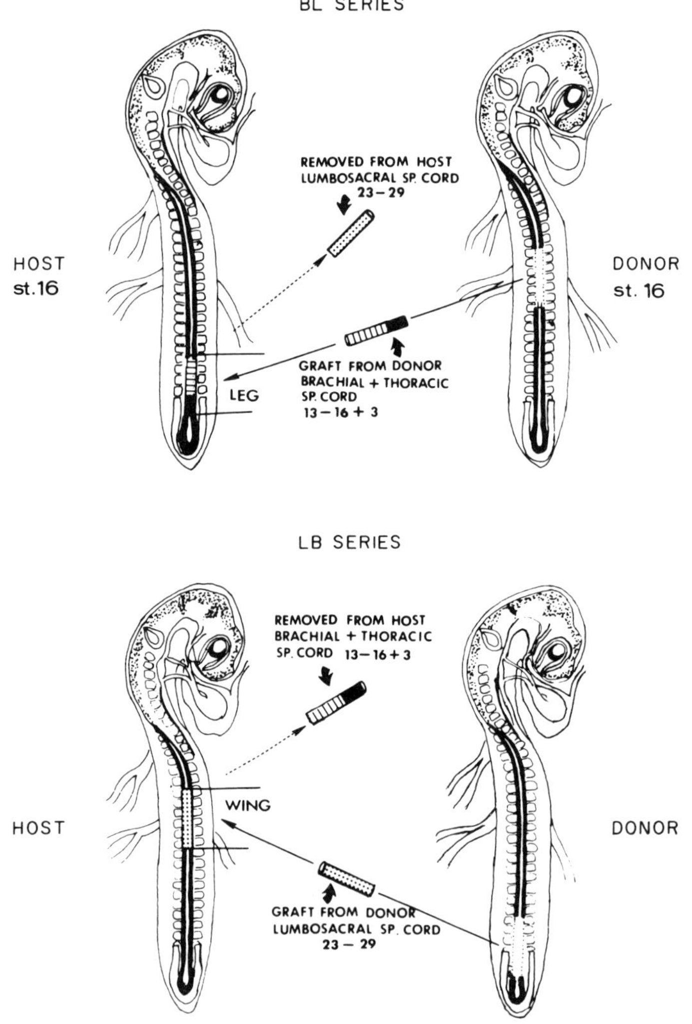

Fig. 1. Transplantation of brachial segments to lumbosacral level (BI series) and of
lumbosacral segments to brachial level (Lb series) in $2\frac{1}{2}$-day chick embryos. From Narayanan
& Hamburger (1971).

created, very little, if any, of this structural refinement is translated into the ongoing bioelectrical activity. On the contrary, the double-electrode findings of Provine seem to substantiate my earlier notion that the discharges spread indiscriminately through the ventral cord, ignoring, as it were, any specific pathways. They sweep through the system as if it were an unorganized network.

The incongruity of burst activity and neurogenesis thus leads to an impasse which has to be resolved in some way. It is conceivable that at our present level of analysis of electrical burst activity a more subtle, more highly organized pattern of discharges along specific channels is not revealed. The following experiment performed by Dr. Narayanan suggests that this is, indeed, the case, and that while Type I motility goes on, preferential pathways are actually being used for impulse transmission. In 48-hour chick embryos, the brachial spinal cord segments were transplanted in the place of the lumbosacral segments of another embryo, and vice versa; thus, one set (Bl) had two brachial, and the other set (Lb) had two lumbosacral cords. No differences between normal and experimental embryos were found with respect to total activity and periodicity. However, when another parameter was recorded, namely the number of movements performed per time unit, it was found that in both experiments wings and legs moved more frequently together, in concert, than in normal embryos. This holds for the entire observation period from 9 to 17 days (Narayanan & Hamburger, 1971).

The tight linkage of wing and leg movements when they are both innervated by homonymous spinal cord segments presupposes a corresponding coupling of the underlying bioelectrical discharges: two homonymous nerve centers are more frequently excited near-simultaneously than are the heteronymous centers in the normal embryo. One can explain this in different ways: either particular longitudinal fiber tracts have preferential chemo-affinities for brachial, or lumbosacral neuron sets, respectively, and connect with them preferentially, or bioelectrical messages are directed specifically to brachial, or to lumbosacral, levels, and decoded simultaneously by both homonymous segments in the experimental embryo. No matter which explanation one accepts, the results show that in this case the discharge pattern has special features; our assumption that *all* discharges sweep indiscriminately through the ventral cord has to be qualified.

It should be mentioned that coupling of wing and leg movements is even tighter after hatching. In hatched chicks of the Bl series, the four extremities almost always moved together. A similar observation was made by Straznicky (1963), who explained the phenomenon much as we did. Incidentally, the legs in our Bl series never displayed alternating stepping movements; they always performed simultaneous ab- and adduction, as in wing flapping. This is additional evidence that region-specific differences are built into the spinal cord at very early stages, and that the circuitry for wing- or leg-specific coordination cannot be modulated by the appendages.

Returning to our main theme, if one accepts the notion that not all discharges in the spinal cord spread indiscriminately, and that impulse transmission along discrete pathways can also occur, then the *distinction between coordinated and uncoordinated motility becomes less sharp.* One would no longer consider an occasional coordinated movement such as wing flapping or alternating movements of legs, when it is interspersed with uncoordinated movements, as an exception; one would rather expect this to happen once in a while. The conceptual dichotomy of coordinated versus uncoordinated movements which I have stressed in the context of embryonic motility is perhaps too rigid. Once more, dichotomous thinking should give way to a more flexible scheme. This will become more obvious when we turn to reflexogenesis.

VII. Responses to Stimulation (Concurrence of Generalized and Local Responses)

In our search for clues to discharge patterns in the embryonic spinal cord, we now turn to *evoked responses.* Postnatal reflexes such as withdrawal, stretch, or sucking are performed in a stereotyped way, and it is assumed that specific, discrete pathways are involved, even if not all details of the reflex circuits have been identified. The precursors of reflexes should provide evidence for the existence of discrete pathways in the embryo.

A great deal of information on evoked responses is available in the literature. Most experiments were conducted during the 30's and 40's, using principally mammalian fetuses (review in Carmichael, 1970). The data were interpreted from different viewpoints, as, for instance, in support or refutation of Coghill's or Windle's theories, or in terms of fetal antecedents of postnatal reflexes, or with respect to the sequential order in which different modalities make their appearance (Gottlieb, 1971b). I shall consider only cutaneous tactile stimulation in birds and mammals, and I shall limit myself to a few selected data that are pertinent to the issues under discussion.

The earliest responses in birds and mammals are elicited by perioral stimulation, but in some forms (e.g., rat, cat, rhesus monkey) the palmar surface becomes sensitive at the same time. The responses consist first of a bending of the head, and thereafter expand from there to trunk, tail, and limbs; or of forearm flexion in response to palmar stimulation. In some instances, a close correlation between the early movements and the underlying structural differentiation has been established. This holds for the human (Humphrey, 1964), the monkey (Bodian, Melby, & Taylor, 1968), cat (Windle, 1934), and others. However, very soon this relationship is lost sight of in amniotes. This is due, in part, to the lack of information concerning details of structural maturation and synaptogenesis. However, the main reason is that the responses, rather than progressing toward clearly defined reflexes and coordinated patterns, become unstructured and generalized. All observers agree that soon after the initial period of local responses, stimulation, for instance of the perioral region, evokes what has been called a *generalized*, total pattern, or mass response, that is, the activation of several or all parts of the musculature. In other words, we are dealing with a *diffusion* or spreading or irradiation of the motor responses (not to be confused with the extension of reflexogenous areas which occurs at the same time). The generalized responses are the exclusive or predominant pattern for considerable periods of prenatal life. For instance, in the chick, the generalized response is the only motility type observed from the beginning of sensitivity at $7\frac{1}{2}$ days to 11 days (Hamburger & Narayanan, 1969). In the guinea pig with a gestation period of 68 days, the period of exclusive or predominant total activity extends from about 31 to 40–45 days (Carmichael, 1934), in the rat from 16 to 18 days (Angulo y Gonzalez, 1932; Narayanan *et al.*, 1971), and in the human fetus from about 8 to 13 weeks (Hooker, 1952, 1958; Humphrey, 1964, 1970). The initial stages of head bending and forearm flexion have been interpreted either as local responses or in terms of incipient generalized activity; this is a minor point. It should be understood that generalized activity does not always involve all parts of the body; the diffusion of the response may extend only to certain regions. For instance, stimulation of the palm or of other parts of the leg usually elicits only segmental responses, whereas stimulation of the trigeminal area more frequently results in mass action. The term "generalized" motility is therefore perhaps preferable to "total pattern."

The generalized response reminds one of the spontaneous Type I motility with which it shares the uncoordinated form of movements. In fact, an observer would find it difficult to distinguish between them. The diffusion of evoked responses in mammals suggests that we are dealing with a spread of bioelectrical activity through the spinal cord, similar to that established for the chick.

However, this is not the full story. It was found that in all forms, sooner or later more restricted, local responses occur side by side with generalized responses. There is a general trend in that local responses become more frequent as development progresses, and in late stages specific stereotyped reflexes can be obtained as the sole response. This is not the occasion to discuss in detail the origin and elaboration of reflexes. In the context of our discussion, special interest is focused on the intermediate stages in which *generalized and local responses can be elicited side by side.* We shall give a few examples. In the chick (Table I), all responses to beak stimulation are

TABLE I

RESPONSES TO TACTILE STIMULATION OF TRIGEMINAL AREAS IN THE CHICK
(GENTLE STROKING)[a]

Age		Stimulated	Responses
(days)	(hours)	head areas	
7	15	Beak	Weak total body movements
8	5–9	Beak	Total body movements
9	5–8	Beak	Same, including head, legs
10	5–9	All areas	Same, including head, wings, legs
11	5–12	Beak	Local head withdrawal only
		Other areas	Total body movements, including legs, wings
12	5–9	Beak	Local head withdrawal only
		Other areas	Total body movements, including wings, legs
13	5–8	All except posterior head	Local head withdrawal
		Posterior head	Same with occasional wing, leg movements
14	5–7	Same	Same as 13, 5–8
15	5–7	All areas	Local head withdrawal, leg kicking
16	5–6	All areas	Same
17	5–8	All areas	Same
18	5–8	All areas	Same, occasional rotatory head movements

[a]Hamburger & Narayanan (1969).

generalized up to $11-11\frac{1}{2}$ days. At that stage, stimulation of the beak gives *local* head responses for the first time, whereas stimulation of posterior trigeminal head areas still gives the total response. At 15 days, one obtains consistently local head movement responses from all trigeminal areas, accompanied sometimes by kicking of the legs (Hamburger & Narayanan, 1969). In the guinea pig, the general trend is well illustrated by the stimulation of the lateral margin of the pinna. At 32 days, a generalized total pattern, including head, trunk, and extremities, is the only response. At 43 days, local contraction of the pinna is observed for the first time; it is accompanied by head, trunk, and limb movements. At 61 days, "a strictly localized twitch of the exact part of the pinna that had been touched [Carmichael, 1934, p. 438]" was recorded. In the human fetus, following perioral stimulation, the total pattern prevails up to 13 weeks. However, at $8\frac{1}{2}$ weeks, the stimulation of the edge of the lower lip already evokes an active mouth opening, along with other body movements; likewise, from $10\frac{1}{2}$ weeks on, stimulation of the upper eyelid elicits the contraction of the orbicularis oculi muscle, accompanied by mouth opening and movements of head and extremities (Hooker, 1958; Humphrey, 1970). *The local responses are so to speak embedded in the total pattern.* These few examples could easily be multiplied; they document the crucial point that generalized and localized responses occur side by side. The inference seems to be justified that exteroceptive sensory neurons which innervate the trigeminal and other cutaneous areas can discharge indiscriminately into neuron pools at different levels of the spinal cord, and at the same time along discrete pathways. The stimulation experiments thus lead to the same conclusion that was derived for spontaneous motility from the spinal cord transplantation experiment, namely, that *coordinated and uncoordinated behavioral activities are not mutually exclusive in embryos.* One can assume that while the reflex arcs are gradually refined by formation of the appropriate synaptic connections, they are preferentially used for channeling of impulses, though diffusion still continues up to late stages.

To summarize, as far as the structure ——>function relationship in evoked motility is concerned, in mammalian fetuses a close correlation has been established for the early stages of reflex responses (see p. 62). Again, at the end of the fetal period, the strictly localized responses which one obtains reflect the gradual perfection of the central reflex circuits. In the intermediate period, the picture is blurred by the phenomenon of diffusion of the motor responses, and no clear relationship can be established.

VIII. Inhibition

Generalized motility and the transition to coordinated activity cannot be properly understood without consideration of inhibition, inasmuch as all integrated activity depends on the subtle interplay of excitation and inhibition. Hence, the origin and elaboration of inhibition in the embryo is a matter of special interest. It is in the nature of inhibition that it cannot be demonstrated by observation of motility but only by experiment.

Data for the chick embryo are very limited. In our preparations of chronic gaps in the spinal cord, at the cervical or thoracic level, our observational methods have not given signs of inhibition up to 17 days; body regions caudal to the transection invariably show a quantitative reduction of motility (Hamburger, 1963; Hamburger, Balaban, Oppenheim, & Wenger, 1965). After 17 days, extirpation of the midbrain and of the otocysts raises the level of activity (Decker 1970; Decker & Hamburger, 1967). It is at this time, then, that the first inhibitory action of higher centers can be detected by these experiments. Yet, it is certain that inhibitory synapses become functional much earlier. A rise in activity was initiated by strychnine treatment at 14–15 days (R. Oppenheim and R. R. Provine, personal communications). Flat-vesicle synapses, which are considered as representing inhibitory synapses, were found in 11-day embryos (Foelix and Oppenheim, this volume). We assume that the inhibitory effects prior to day 17 are outweighed by excitation and therefore not detected by our methods. No further advance can be expected without application of electrophysiological and pharmacological methods.

Fortunately, we have much more precise information on the onset and development of inhibition in mammalian fetuses (review in Skoglund, 1969a, 1969b). We shall limit the discussion to the data on stretch reflex in sheep and guinea pig fetuses, though the extensive work on kittens is equally pertinent (see Purpura, 1971; Skoglund, 1969b). An essential feature of the Scandinavian studies is the use of electromyography.

TABLE II

STRETCH REFLEX IN THE FETUS

	Sheep (days)[a]	Guinea pig (days)[b]
Gestation period	150	65–68
First exteroceptive (plantar skin) response of leg	35–40	35
First sign of stretch reflex in gastrocnemius	60	45
Irradiation of excitation to antagonists	60–81	48–56
First observed antagonistic inhibition in leg	90 (no observations 82–89)	After 56

[a]Änggård et al. (1961).
[b]Bergström, Hellström, and Stenberg (1961), and Bergström et al. (1962).

In sheep, with a gestation period of 150 days, evoked responses to exteroceptive cutaneous stimulation can be obtained beginning at 35 days, whereas the proprioceptively mediated stretch reflex was first elicited at 60 days. This seems to be the stage at which maturation of muscle spindles occurs (see Table II). The most interesting finding, in our context, was that the excitatory response to stretch was not limited to the stretched muscle (in this case the gastrocnemius, an extensor), but it spread to an antagonist (the tibialis anterior, a flexor). This diffusion, or absence of antagonistic inhibition, goes on for an extended period, that is, up to 90 days. However, direct inhibition by another route, namely by stimulation of the ipsilateral leg skin, can be obtained earlier, i.e., from day 72 on (Änggård, Bergström, & Bernhard, 1961).

The corresponding data for the guinea pig are shown on Table II (Bergström, Hellström, & Stenberg, 1962). The gestation period, as mentioned, is 65–68 days. The first stretch reflex on a leg muscle was observed on day 45, and the diffusion of excitation to antagonists lasts to day 56. Plantar skin stimulation also results in a contraction of antagonistic muscles during the same period. After day 56, the plantar stimulation elicits only flexors (withdrawal), and the stretch only extensors (postural).

These experiments clarify several points: (*1*) Excitation precedes inhibition, at least in this particular system. (*2*) Lack of inhibitory mechanisms is responsible, at least in part, for the generalized diffusion, in evoked responses as well as in spontaneous motility. As stated by Änggård *et al.* (1961), "[the] pronounced lack of balance between the excitatory and inhibitory mechanisms . . . may explain the diffuse and widespread character of the reflex activity [p. 134]." (*3*) The transition from generalized to precisely localized response seems to be paralleled by the elaboration of the inhibitory mechanisms.

IX. Influence of Function on Structure

Gottlieb (1970 and Introduction, this volume) distinguishes between two different theoretical viewpoints. According to one view (predeterministic), the relationship is unidirectional (structure———>function); according to the other (probabilistic), the relationship is bi-directional (structure $\rightleftharpoons$ function), that is, there is a feedback of function on neurogenesis. Function is defined broadly, to include bioelectrical activity, movements, sensory input through stimulation, and other experiences.

In the previous sections, I have stressed the difficulties which one encounters when one tries to come to grips with the specifics of the unidirectional structure ——> function relations in amniote embryos, both with regard to spontaneous and to evoked motility.* With respect to the reverse relation, the scarcity of reliable data leaves most questions unresolved. Since Dr. Gottlieb has dealt with this matter in the Introduction, I shall limit myself to a few additional comments. For the purpose of analysis, the different aspects of function, that is, bioelectrical activity, motility, and sensory input, are considered separately.

A. Does Impulse Transmission Play a Role in Structural Differentiation?

The theories of neurobiotaxis (Ariëns Kappers, 1932) and of stimulogenous fibrillation (Bok, 1915) have postulated such influences. For instance, impulse transmission along particular fiber tracts was supposed to induce directional outgrowth of dendrites from adjacent neuroblasts toward the tract, and of axons in the opposite direction. These theories were not based on accurate observations nor on experimental evidence and are now discarded for good reasons (see Jacobson, 1970; Sperry, 1951). But this does not settle the problem. The very early start of motility in all vertebrate embryos is evidence that immature neuroblasts generate impulses, and

---

*It is clear from previous (and also following) sections that my viewpoint is not that of "predetermined epigenesis" as defined by Gottlieb elsewhere in this serial publication.

Provine (this volume) has recorded activity from the spinal cord of chick embryos as early as 4 days. Hence, there is ample opportunity for electro-physiological influences during a considerable part of neurogenesis. It is conceivable that electrical activity of a neuroblast or neuron plays a role in the regulation of its metabolism or, more specifically, in its differentiation processes; but the experimental design for tests *in vivo* will be difficult. Under the more favorable conditions of tissue culture, Crain, Bornstein, and Peterson (1968) have found that synapses can form in mammalian embryonic nerve tissue when electrical activity is blocked with Xylocaine (see Crain, Volume 2 of this serial publication). This observation does not support such a notion, but it is perhaps too early to generalize from this experiment.

B. Does Motility as such Play a Role in Neurogenesis?

A negative answer was obtained in a particular instance. In the frequently quoted experiments of Harrison (1904), Carmichael (1926), and Matthews and Detwiler (1926), salamander and frog embryos were kept in chloretone narcosis during the critical preswimming stages; they showed normal swimming after removal from the narcotic. The inference is that the sequence of synapse formations described by Coghill (1929), by which the swimming mechanism matures, can proceed normally in the absence of function. Similar experiments by Fromme (1941) in which he found impairment of swimming after removal from the narcotic are sometimes quoted as evidence to the contrary; however, this claim is not warranted. He found only that "the earliest appearance of swimming behavior does not compare favorably with that of the control group [p. 238]." He found normal swimming after full recovery and was probably dealing with a transient impairment. He used frog embryos, which have a much lower tolerance for anesthetics than salamander embryos. The fact that these time-honored but poorly controlled experiments have not been superseded to this day testifies to our ignorance in this matter. The performance of similar experiments on chick embryos is marred by troublesome side effects: paralysis by curare for 24–48 hours results in ankylosis and muscle atrophy (Drachman & Coulombre, 1962). Perhaps the deleterious effects can be avoided by applying intermittent treatment over longer periods which would reduce motility drastically, though not abolish it. There is another way of circumventing the difficulty. Since according to Provine (this volume), motility is closely correlated with burst activity, the monitoring of burst activity might give information on development of motility and, by inference, on synaptogenesis in embryos with ankylosis and myopathy.

But one should realize that even if such efforts were successful, any con-clusions from this type of experiment concerning neurogenesis would be by inference only, and it seems altogether hopeless to expect precise information on function ——→ structure relationships by this approach. It would be preferable to stay at one level of organization and ask either, whether experimental modification of bioelectrical activity at one stage influences such activity at a later stage, or whether manipulation of motility, including its suppression, during embryonic development affects later behavior. The assumption that this is the case is the basic tenet of Kuo's (1932) and Schneirla's (1965) theories, but the crucial experiments still have to be done.

We turn next to the question which, for a long time, has been in the center of theoretical considerations and controversy.

C. The Role of Sensory Input in Neurogenesis

Different aspects of this problem have been approached from different viewpoints and by a variety of experimental designs, such as: deafferentation in embryos, sensory deprivation or increase of stimulation postnatally, enrichment and impoverishment of the environment. I shall add only a few comments to those of Dr. Gottlieb in the Introduction.

The nervous system of birds and mammals is already at an advanced state of differentiation when the sensory-motor connections become functional. Our deafferentation experiments show that neurogenetic differentiation responsible for spontaneous (Type 1) motility in the chick embryo does not require sensory input (see Hamburger, 1968, 1970). This applies to the major part of the embryonic period up to 17 days. But, as was stated above, the question of whether or not subsequent pre- and postnatal coordinated action patterns and the concomitant neurogenetic events are dependent on sensory input at any stage remains unresolved.

As a broad generalization, one would assume that the central circuitry for those action patterns which are performed in a rather stereotyped way, such as locomotion, would be established according to an intrinsically programmed blueprint. On the other hand, those higher brain centers which subserve activities requiring a high degree of adjustment to environmental exigencies, such as the mammalian cortex as the seat of learning and memory, would retain a certain degree of plasticity up to postnatal stages.

The first point is illustrated by our spinal cord transplantations (p. 60). They show what other experiments had demonstrated before (e.g., Weiss, 1955), that the program for the circuitry underlying coordinated wing and leg movements, respectively, is built into the brachial and lumbosacral spinal cord regions at very early stages, and, furthermore, that the connectivities are not modified by sensory input from an atypical periphery. Concerning *plasticity*, two questions arise: Is it limited to functional specification of neurons or does it extend to the structural level? And, if the latter is the case, can sensory input control or modify structural differentiation, as for instance, synaptic linkage? Since higher centers in mammals do not complete their maturation until after birth, opportunities exist for a variety of sensory experiences to impinge on functional as well as structural specification of neurons and synapses. Of course, the major fiber tracts have long been established by then, and the fiber terminals have reached their assigned sites. For instance, the highly specific topographic projection of the retina fibers onto the optic tectum in chick embryos is completed long before visual perception begins (La Vail & Cowan, 1971). The mechanism by which this is accomplished is probably selective chemoaffinity between the nerve ending and the neuron with which it establishes contact (Sperry, 1963, 1965; see, however, Székely, Volume 2 of this serial publication). But this organizing mechanism may still permit a considerable degree of randomness on a smaller scale, such as modifications or shifts of synaptic links. It certainly does not exclude the possibility that in some systems the fine details of synaptic structure and linkage could be regulated by sensory experiences. In brain centers that retain plasticity after birth, connections of units may remain diffuse and uncommitted, and visual and other stimulation may play a determinative role in the specification of a synapse, or little-used synapses may disappear.

Experimental studies of the modifiability of neural structures by sensory stimulation are of rather recent date and not yet very numerous. A few examples may be cited. Several investigations have used visual deprivation in newborn mammals. The visual system has many advantages, but it is often difficult to decide whether one deals with regressive effects in an already differentiated structure or with the arrest of ongoing differentiation. Valverde (1967) has shown that in mice raised in darkness from birth, the number of dendritic spines on the apical dendrites of pyramidal cells in the visual cortex is markedly reduced, and it is very probable that the lack of sensory input causes the failure of normal spine formation. In a later investigation, Valverde (1971) has identified two populations of pyramidal cells: one in which dendritic spine growth is independent, and another one in which it is dependent on stimulation by light. Szentágothai and Hámori (1969) have described a special kind of dendritic spines in the lateral geniculate body of the dog which penetrate into deep invaginations of the surface of optic terminals, where they synapse. In the newborn dog these spines are not yet developed. If the eyes are sutured after birth, the spines fail to differentiate. They are completely absent at 2 months, when the adult pattern is established in normal puppies. In both instances, we are dealing with the arrest of a differentiation process, resulting from absence of sensory input. Finally, we present an example of enhancement of differentiation by stimulative experience. Rosenzweig and collaborators (see Rosenzweig, 1971) have shown that when young rats, after weaning, are exposed to an enriched or impoverished environment, respectively, a number of anatomical and biochemical parameters show significant changes. The investigations have now been extended to the synaptic level (Møllgaard, Diamond, Bennett, Rosenzweig, & Lindner, 1971). The axodendritic synapses in layer III of the occipital cortex (mostly visual) were used for quantitative comparison. In animals raised in the enriched environment, the mean length of synapses was 52% greater and the number of synapses per unit area of neuropil was 35% less than in animals raised in the impoverished environment. The total synaptic area was 40% greater in the former than in the latter. While the relation of these findings to behavior is not immediately obvious, it is of critical importance that the principle of regulation of at least some aspects of synaptogenesis by sensory input has been validated. The far-reaching consequences for behavioral performance in general, and learning and memory in particular, are obvious.

D. Performance Effect of Prenatal Sensory Experience on Postnatal Behavioral Performance

Finally, I shall comment briefly on this question, although this topic does not relate directly to the function ——→ structure problem. Claims that such a role is of great significance have been made by Kuo (1967) and Schneirla (1965), but experimental tests have not been forthcoming, with the exception of the important experiments of Gottlieb (1971a) which will be reviewed briefly (see also Impekoven and Gold, this volume). The experiments are limited, so far, to acoustic stimulation of birds. As is known, birds begin to vocalize several days before hatching, and, as Gottlieb has shown, these late embryos are able not only to respond to acoustic stimulation, but to discriminate between the species-specific maternal call and other calls. The role of prenatal acoustic experience in postnatal responses was tested in deprivation experiments. Since acoustic isolation before hatching still leaves the embryo exposed to its own vocalization, it was necessary to devocalize the embryos. A simple but very effective procedure was devised which consists of coating the tympaniform membranes of the syrinx with nonflexible collodion, preventing the vibration of the membranes. The operation was done about 2 days before hatching, the embryos being partly pulled out of the shell. All experiments were done on mallard duck embryos, and their postnatal discriminative ability was tested using maternal mallard, duckling (sibling), maternal pintail, and maternal chicken calls. The individuals that had been reared prenatally in complete acoustic deprivation actually did show postnatally several deficiencies in their responses, as for instance a time lag in the development of the usual discriminatory abilities. The most significant deficiency is the inability of the experimental ducklings to discriminate between the mallard and the chicken maternal call, whereas they are capable of distinguishing between the duckling (sib) call and the pintail maternal call. Experiments are underway to identify the acoustic characteristic (rate, fundamental frequency, etc.) which makes the discrimination between the mallard and chicken maternal calls difficult. This is apparently the first instance in which experimental evidence has been provided for the theoretically important notion that stimulative events occurring normally before birth play a role in the perfection of species-specific perception after birth.

X. Concluding Remarks

The picture that I have presented of stucture-function relationships in amniote embryos will not satisfy a mind in search for clearly definable connections and broad generalizations. The only claim that my picture can make is that it is a fairly realistic portrait of the present situation. Part of its imperfection is due to our very limited knowledge in these matters. But there are other reasons. One is the fundamental difficulty inherent in every reductionist effort to "explain" phenomena at one level in terms of events occurring at lower levels. This becomes evident when one realizes that the different units with which one operates at different levels are incommensurate. On the behavioral level, a representative unit is the reflex; on the level of motility, the units are contractions of muscles and muscle groups; on the level of bioelectrical activity, we are dealing with single-unit and burst discharges; and on the structural level with differentiating neurons, neuron pools, connectivities, and synapses. What kind of relationship would one accept as an "explanation" of events at one level in terms of events at lower levels? In our search we have detected only two relationships that could make such a claim: that between bioelectrical bursts and activity phases in the chick embryo, and the relation between structural differentiation of reflex arcs and evoked responses, at the beginning of sensory competence and then again toward the end of fetal life, when local reflexes can be elicited. But why do we find incongruity of patterns in all other respects? I think this cannot be attributed to conceptual difficulties alone. A major difficulty is inherent in the phenomena themselves, or, to be more specific, in the prevalence of uncoordinated motility in amniote embryos and fetuses. The situation would look quite different if the embryos would make the gradual assemblage of embryonic motility units into integrated postnatal behavior patterns overtly manifest; or, in other words, if prenatal antecedents to postnatal action patterns were clearly recognizable. In this case, specific questions could be asked concerning relationships prevailing at specific stages of development. But the reality is different. We are, then, back at the question raised earlier: Why is there the prevalence of uncoordinated motility in amniotes? One could argue that autonomous discharge of electrical impulses is an elementary property of immature nerve tissue which generates automatically the observed motility type, as long as excitation prevails over inhibition, and in the absence of selection pressure against this form of overt motility, when the embryo and fetus is protected in the egg and uterus. Or, as I have indicated, the electrical discharges may play a facilitative or even indispensable role in the functional or structural maturation of neurons. The maintenance of articulations and musculature would then be a fringe benefit of the fetal exercise, as has been suggested by Eisenberg (1971). Be this as it may, the incongruities that I have spoken of are real.

Where do we go from here? This will depend largely on the preferences of the individual investigator. It seems to me that further studies of the forms and sequences of overt motility would not be particularly rewarding. At this moment, the best strategy would seem to be to confine oneself to one level of organization at a time, building on the beginnings that have been made, but with the problems posed by behavior development constantly on one's mind. On the behavioral level, the pioneer experiments of Gottlieb and others, which have been limited so far to the perinatal period, can be expanded to earlier stages, to different sensory modalities and different species. They can answer questions in which the developmental psychologist is particularly interested, without necessarily referring to lower levels of organization. The electrophysiological analysis is in its infancy; and the ultrastructural analysis, particularly of synaptogenesis, is likewise at its beginnings. Needless to say, in both areas answers to many questions will be available in the near future. If the focus of such investigations is on behaviorally critical stages, such as onset of sensory competence for different modalities, or initiation of specific integrated behavior patterns, then one can be confident that a more unified picture will emerge very soon.

References

Änggård, L., Bergström, R., & Bernhard, L. G. Analysis of prenatal spinal reflex activity in sheep. *Acta Physiologica Scandinavica*, 1961, **53**, 128–136.

Angulo y Gonzalez, A. W. The prenatal development of behavior in the albino rat. *Journal of Comparative Neurology*, 1932, **55**, 395–442.

Ariëns Kappers, C. U. Principles of development of the nervous system (Neurobiotaxis). In W. Penfield (Ed.), *Cytology and cellular pathology of the nervous system*. Vol. I. New York: Harper (Hoeber), 1932.

Balaban, M., & Hill, J. Perihatching behaviour patterns of chick embryos (*Gallus domesticus*). *Animal Behavior*, 1969, **17**, 430–439.

Bergström, R. M., Hellström, P. E., & Stenberg, D. Prenatal stretch reflex activity in the guinea pig. *Annales Chirurgiae et Gynaecologiae Fenniae*, 1961, **50**, 458–466.

Bergström, R. M., Hellström, P. E., & Stenberg, D. Studies in reflex irradiation in the foetal guinea-pig. *Annales Chirurgiae et Gynaecologiae Fenniae*, 1962, **51**, 171–178.

Bodian, D., Melby, E. C. & Taylor, N. Development of fine structure of spinal cord in monkey fetuses. II. Prereflex period to period of long intersegmental reflexes. *Journal of Comparative Neurology*, 1968, **133**, 113–166.

Bok, S. T. Die Entwicklung der Hirnnerven und ihrer zentralen Bahnen. Die stimulogene Fibrillation. *Folia Neurobiologica*, 1915, **9**, 475–565.

Carmichael, L. The development of behavior in vertebrates experimentally removed from the influence of external stimulation. *Psychological Review*, 1926, **33**, 51–58.

Carmichael, L. An experimental study in the prenatal guinea pig of the origin and development of reflexes and patterns of behavior in relation to the stimulation of specific receptor areas during the period of active fetal life. *Genetic Psychology Monographs*, 1934, **16**, 337–491.

Carmichael, L. The onset and early development of behavior. In A. Mussen (Ed.), *Carmichael's manual of child psychology*. Vol. I. New York: Wiley, 1970. Pp. 447–563.

Coghill, E. G. *Anatomy and the problem of behavior*. London & New York: Cambridge University Press, 1929.

Crain, S., Bornstein, M. B., & Peterson, E. R. Maturation of cultured embryonic CNS tissues during chronic exposure to agents which prevent bioelectric activity. *Brain Research*, 1968, **8**, 363–372.

Decker, J. D. The influence of early extirpation of the otocysts on development of behavior of the chick. *Journal of Experimental Zoology*, 1970, **174**, 349–364.

Decker, J. D., & Hamburger, V. The influence of different brain regions on periodic motility in the chick embryo. *Journal of Experimental Zoology*, 1967, **165**, 371–384.

Drachman, D. B., & Coulombre, A. J. Experimental club foot and arthrogryposis multiplex congenita. *Lancet*, 1962, **ii**, 523–526.

Eisenberg, L. Persistent problems in the study of the biopsychology of development. In E. Tobach, L. R. Aronson, & E. Shaw (Eds.), *The biopsychology of development*. New York: Academic Press, 1971. Pp. 515–529.

Fromme, A. An experimental study of the factors of maturation and practice in the behavioral development of the embryo of the frog, *Rana pipiens*. *Genetic Psychology Monographs*, 1941, **24**, 219–256.

Gottlieb, G. Conceptions of prenatal behavior. In L. R. Aronson, E. Tobach, D. S. Lehrman, & J. S. Rosenblatt (Eds.), *Development and evolution of behavior*. San Francisco: Friedman, 1970. Pp. 111–137.

Gottlieb, G. *Development of species identification in birds: An inquiry into the prenatal determinants of perception*. Chicago: University of Chicago Press, 1971. (a)

Gottlieb, G. Ontogenesis of sensory function in birds and mammals. In E. Tobach, L. R. Aronson, & E. Shaw (Eds.), *The biopsychology of development*. New York: Academic Press, 1971. Pp. 67–128. (b)

Gottlieb, G., & Kuo, Z.-Y. Development of behavior in the duck embryo. *Journal of Comparative and Physiological Psychology*, 1965, **59**, 183–188.

Hamburger, V. Some aspects of the embryology of behavior. *Quarterly Review of Biology*, 1963, **38**, 342–365.

Hamburger, V. Ontogeny of behaviour and its structural basis. In H. Waelsch (Ed.), *Comparative neurochemistry*. Oxford: Pergamon, 1964. Pp. 21–34.

Hamburger, V. Emergence of nervous coordination. Origins of integrated behavior. *Developmental Biology, Supplement*, 1968, **2**, 251–271.

Hamburger, V. Embryonic motility in vertebrates. In F. O. Schmitt (Ed.), *The neurosciences: Second study program*. New York: Rockefeller University Press, 1970. Pp. 141–151.

Hamburger, V. Development of embryonic motility. In E. Tobach, L. R. Aronson, & E. Shaw (Eds.), *The biopsychology of development*. New York: Academic Press, 1971. Pp. 45–66.

Hamburger, V., & Balaban, M. Observations and experiments on spontaneous rhythmical behavior in the chick embryo. *Developmental Biology*, 1963, 7, 533–545.

Hamburger, V., & Narayanan, C. H. Effects of the deafferentation of the trigeminal area on the motility of the chick embryo. *Journal of Experimental Zoology*, 1969, 170, 411–426.

Hamburger, V., & Oppenheim, R. Prehatching motility and hatching behavior in the chick. *Journal of Experimental Zoology*, 1967, 166, 171–204.

Hamburger, V., Balaban, M., Oppenheim, R., & Wenger, E. Periodic motility of normal and spinal chick embryos between 8 and 17 days of incubation. *Journal of Experimental Zoology*, 1965, 159, 1–14.

Hamburger, V., Wenger, E., & Oppenheim, R. Motility in the chick embryo in the absence of sensory input. *Journal of Experimental Zoology*, 1966, 162, 133–160.

Harrison, R. G. An experimental study of the relation of the nervous system of the developing musculature in the embryo of the frog. *American Journal of Anatomy*, 1904, 3, 197–220.

Hooker, D. *The prenatal origin of behavior*. Lawrence, Kans. University of Kansas Press, 1952.

Hooker, D. *Evidence of prenatal function of the central nervous system in man*. New York: American Museum of Natural History, 1958.

Humphrey, T. Some correlations between the appearance of human fetal reflexes and the development of the nervous system. *Progress in Brain Research*, 1964, 4, 93–135.

Humphrey, T. Reflex activity in the oral and facial area of the human fetus. In J. F. Bosma (Ed.), *Second symposium on oral sensation and perception*. Springfield, Ill.: Thomas, 1970. Pp. 195–233.

Jacobson, M. *Developmental neurobiology*. New York: Holt, 1970.

Kovach, J. Development and mechanisms of behavior in the chick embryo during the last five days of incubation. *Journal of Comparative and Physiological Psychology*, 1970, 73, 392–406.

Kuo, Z.-Y. Ontogeny of embryonic behavior in aves: V. The reflex concept in the light of embryonic behavior in birds. *Psychological Review*, 1932, 39, 499–515.

Kuo, Z.-Y. *The dynamics of behavior development*. New York: Random House, 1967.

La Vail, J. F., & Cowan, W. M. The development of the chick optic tectum. I. Normal morphology and cytoarchitectonic development. *Brain Research*, 1971, 28, 391–419.

Matthews, S. A., & Detwiler, S. R. The reactions of Amblystoma embryos following prolonged treatment with Chloretone. *Journal of Experimental Zoology*, 1926, 45, 279–292.

Møllgaard, K., Diamond, M. C., Bennett, E. L., Rosenzweig, M. R., & Lindner, B. Quantitative synaptic changes with differential experience in rat brain. *International Journal of Neuroscience*, 1971, 2, 113–128.

Narayanan, C. H., Fox, M. W., & Hamburger, V. Prenatal development of spontaneous and evoked activity in the rat (*Rattus norwegicus albinus*). *Behaviour*, 1971, 40, 100–134.

Narayanan, C. H., & Hamburger, V. Motility in chick embryos with substitution of lumbosacral by brachial and brachial by lumbosacral spinal cord segments. *Journal of Experimental Zoology*, 1971, 178, 415–432.

Oppenheim, R. Some aspects of embryonic behaviour in the duck (Anas platyrhynchos). *Animal Behaviour*, 1970, 18, 335–352.

Purpura, D. Synaptogenesis in mammalian cortex: problems and perspectives. In M. B. Sterman, D. J. McGinty, & A. M. Adinolfi (Eds.), *Brain development and behavior*. New York: Academic Press, 1971. Pp. 23–41.

Rosenzweig, M. R. Effects of environment on development of brain and of behaviour. In E. Tobach, L. R. Aronson, & E. Shaw (Eds.), *The biopsychology of development*. Academic Press, 1971. Pp. 303–342.

Schneirla, T. C. Aspects of stimulation and organization in approach-withdrawal processes underlying vertebrate behavioral development. In D. S. Lehrman, R. A. Hinde, & E. Shaw (Eds.), *Advances in the study of behavior.* Vol. 1. New York: Academic Press, 1965. Pp. 1–74.

Skoglund, S. Growth and differentiation, with special emphasis on the central nervous system. *Annual Review of Physiology,* 1969, **31**, 19–42. (a)

Skoglund, S. Reflex maturation. In M. A. B. Brazier (Ed.), *The interneuron.* Berkeley: University of California Press, 1969. (b)

Sperry, R. Mechanisms of neural maturation. In R. Stevens (Ed.), *Handbook of experimental psychology.* New York: Wiley, 1951. Pp. 236–280.

Sperry, R. Chemoaffinity in the orderly growth of nerve fiber patterns and connections. *Proceedings of the National Academy of Sciences, U.S.,* 1963, **50**, 703–710.

Sperry, R. Embryogenesis of behavioral nerve nets. In R. L. DeHaan & R. Ursprung (Eds.), *Organogenesis.* New York: Holt, 1965. Pp. 161–186.

Straznicky, K. Function of heterotopic spinal cord segments investigated in the chick. *Acta Biologica (Budapest),* 1963, **14**, 145–155.

Szentágothai, J., & Hámori, J. Growth and differentiation of synaptic structures under circumstances of deprivation of function and of distant connections. In S. H. Barondes (Ed.), *Symposia of the international society for cell biology,* 1969, Vol. 8. *Cellular dynamics of the neuron.* New York and London: Academic Press.

Tracy, H. C. The development of motility and behavior reactions in the toadfish (*Opsanus tau*). *Journal of Comparative Neurology,* 1926, **40**, 253–269.

Valverde, F. Apical dendrite spines of the visual cortex and light deprivation in the mouse. *Experimental Brain Research,* 1967, **3**, 337–352.

Valverde, F. Rate and extent of recovery from dark rearing in the visual cortex of the mouse. *Brain Research,* 1971, **33**, 1–11.

Visintini, F., & Levi-Montalcini, R. Relazione tra differenziazione strutturale e funzionale dei centri e delle vie nervose nell'embrione di pollo. *Schweizer Archiv für Neurologie und Psychiatrie,* 1939, **43**, 1–45.

Weiss, P. Nervous system (neurogenesis). In B. H. Willier, P. A. Weiss, & V. Hamburger (Eds.), *Analysis of development.* Philadelphia: Saunders, 1955.

Windle, W. F. Correlation between the development of local reflexes and reflex arcs in the spinal cord of cat embryos. *Journal of Comparative Neurology,* 1934, **59**, 487–505.

III. History of Neurogenesis

CHANGING CONCEPTS IN DEVELOPMENTAL NEUROBIOLOGY*

VIKTOR HAMBURGER†

I should like to take this opportunity to meditate on half a century of explorations in the field of experimental neurogenesis which I have witnessed or participated in. I shall do this in a rather personal and informal way. This, I am sure, would have the blessing of George Bishop, who had little use for formalities.

To me it is still a miracle to watch the nervous system transform itself within a few weeks from a simple tube, composed of a few hundred seemingly undifferentiated embryonic cells, into the most complex organ system that has evolved in nature. So many interlocking production lines must operate with the highest precision, so many tightly programmed schedules must be met, so much could go wrong that one marvels that we all function as well as we do. But unraveling and understanding these intricacies is another matter. The neuroembryologist is overwhelmed by problems on the cellular, supercellular, and ultrastructural level. How do the literally hundreds of neuron strains, each with its own structural and biochemical identity, originate? How do they organize themselves into the supercellular units, the strata, columns, and nuclei in their precise topographic relationships? How is the circuitry established? How does an axon know on which dendritic spine to settle down? As if all this were not enough, we have to cope with the propensity of the neurons to establish intimate relations with any number of peripheral structures. This confronts us with problems of directional axon outgrowth and specific sensory and motor connections. Moreover, these relationships involve mutual dependencies of such stringency that they often decide on the life or death of the partners.

By posing the problems the way I do, I reveal my bias, which is that of the experimental neuroembryologist brought up in a school where the dynamics of developmental relations and mutual interactions between

*Nineteenth George H. Bishop Lecture in Experimental Neurology, given on April 19, 1974, at the Washington University School of Medicine, Saint Louis.
†Department of Biology, Washington University, Saint Louis 63130. Research has been generously supported by the NINDS, NIH, Bethesda, Maryland.

FIG. 1.—George H. Bishop (1889–1973)

embryonic primordia were the primary concern. My bias was acquired in the laboratory of H. Spemann, which in the twenties and thirties (while I was there first as a Ph.D. candidate and later as a *Privatdozent*) was at the zenith of its activity. The "organizer" story was unraveling, and although I did not participate directly in this adventure, I became imbued with the spirit and the canons of experimental embryology, which in essence is the inquiry into the immediate causes or factors that determine the fate of cells and organ primordia. Today one hardly speaks of causality; we analyze mechanisms. And the term "determination," which was then the key concept, is now translated into computer language as "program-

Fig. 2.—Viktor Hamburger

ming," with little net gain in basic insight. The art of microsurgery on amphibian embryos was brought to high perfection by the masters of experimental embryology, H. Spemann and R. Harrison. The aesthetic appeal added to the satisfaction of being engaged, as Spemann put it, in a direct dialogue with the living embryo, a pleasure which most modern molecular embryologists have to forgo.

The focus of interest was on embryonic induction, which is a special category of developmental interactions. As it happens, the primordium of the nervous system, the neural plate, owes its existence to such an interaction, which occurs during a very early stage, the gastrulation phase. During that process, the mesoderm invaginates and its median portion applies itself closely to the overlying outer layer, the ectoderm. The mesoderm mantle then induces the formation of the neural plate in the overlying ectoderm; that is, it initiates neural differentiation by a chemical interaction. The induction was demonstrated by H. Spemann and Hilde Mangold in the classical organizer experiment on salamander embryos [1], which earned Spemann the Nobel Prize. If by appropriate transplantation a piece of mesoderm is brought into contact with a part of ectoderm that would normally form body wall, a secondary neural plate is induced in that region. Transplant and induced structure could be distinguished by choosing a salamander species with unpigmented eggs as donor and one with pigmented eggs as host. If the induced structure is allowed to continue its differentiation, the secondary nervous system, together with the transplant and additionally induced organs, forms a whole secondary embryo on the flank of the host embryo; hence, the transplanted mesoderm was designated the "organizer." By the way, the inductive act creates not just neuralization in general but the regional patterning into three brain divisions and the spinal cord. This seems to be accomplished by two macromolecules distributed in the mesoderm mantle along two concentration gradients in opposite directions along the main axis [2]. I have always considered it a particularly friendly gesture of the embryo toward the neurologist to single out the creation of the nervous system as the most prominent and celebrated, Nobel Prize–winning event in all embryonic development.

At the suggestion of Spemann, I began to work on developmental relationships in later phases of neurogenesis, and thus moved into the orbit of R. Harrison, the founder of experimental neuroembryology, who at that time was "the Chief" (as everybody called him) at Yale. His historic contribution was the definitive demonstration that the axon is a pseudopodial outgrowth of the neuroblast, which he observed under the microscope in fragments of explanted neural tube of frog embryos. The experiment settled a highly controversial issue and at the same time gave us the tissue culture method. He also confirmed an old observation of Ramón y Cajal, dating back to 1890, who had described in silver preparations the formation of a club-shaped "growth cone" at the end of the advancing tip of the axon. Harrison states in his 1907 paper: "These observations show beyond question that the nerve fiber develops by the outflowing axoplasm from the central cells. This protoplasm retains its ameboid activity at the distal end, the result being that it is drawn out into a long thread which becomes the axis cylinder" [3]. As you see, axoplasmic flow is an old story. Its present popularity dates from the discovery of P. Weiss that it is not limited to embryonic and regenerating fibers but is a continuous process [4]. Both Cajal and Harrison recognized an essential corollary to the nerve outgrowth theory—the postulate that the growth cone must be endowed with some kind of sensitivity to perceive cues and signals that guide it to its destination.

The Detwiler Experiment

One of the earliest experiments in this field, done in the early twenties at the suggestion of Harrison by his oldest and most active student and co-worker, S. Detwiler [5], illustrates the way in which the experimental embryologist tackles problems such as directional fiber outgrowth. To answer the simple question of how outgrowing axons would behave in a foreign environment, he transplanted forelimb primordia in the tail bud stage of the salamander embryo one to five segments caudad. The general result was that trunk nerves alone or in combination with limb nerves would form a reasonably typical limb nerve pattern. Two conclusions are warranted: (1) that the limb tissues make a major contribution to nerve pattern formation by providing tracks and (2) that, at least in this instance, the responsiveness of growth cones to environmental cues is rather broad and unspecific. The ingenuity of this particular experimental design becomes apparent when one realizes that it led to two other major discoveries. Detwiler found that a limb transplanted to the flank would participate in coordinated walking movements only if at least one of its nerves originated in the brachial segments of the cord. If innervated exclusively by trunk nerves, it would perform irregular twitches at best. This implies a basic difference between the limb-innervating and the thoracic segments of the cord; only the former have the capacity to build the circuitry for coordinated locomotion. This difference exists already in very early prefunctional stages, when neuronal differentiation has hardly begun.

Detwiler made another observation. He found that those ganglia which were deprived of their peripheral field of innervation became hypoplastic and thoracic ganglia that were overloaded by the implanted limb became hyperplastic. Thus, the "trophic" dependency of developing neural structures on nonnervous structures was revealed. This deceptively simple experiment, which required only a pair of iridectomy scissors, some glassware, and a few embryos from a nearby pond, thus identified three of the major issues in neurogenesis which I shall deal with.

Trophic Effects

I shall take up first the trophic effects. I indicated at the beginning that trophic relationships between neuroblasts and their target organs may become obligatory to the point that the absence of one partner may spell death to the other. The fate of denervated muscle in the adult is well known. In amphibian and chick embryos, one can obtain nerveless limbs by preventing nerve ingrowth from the start. The general morphogenesis is relatively normal, but the limbs are smaller than normal and atrophic. Their musculature differentiates to the point of cross-striation but then breaks down [6]. In insects, lack of innervation has more severe effects. In moths, the larval musculature is broken down during pupation and the adult musculature is built anew from remnants of larval muscle fibers. If a thoracic ganglion of the pupa is extirpated, adult muscle differentiation is completely inhibited; indeed, it does not even get started [7]. A spectacular illustration of trophic nerve influence was found in amphibian limb regeneration. A denervated amputation stump in a salamander larva not only fails to regenerate but actually regresses rapidly down to the base of the arm [8]. We still refer to this remarkable extracurricular activity of the axon by the century-old term "trophic," which indicates that we have made little advance in the clarification of the mechanism involved. But it is obvious that impulse propagation is only one aspect of neuron function. To quote George Bishop,

What is a nerve fiber for, anyway? and what, in fact, is a nerve cell? The intriguing mechanism, by means of which it generates and conducts an impulse, has chiefly preoccupied three generations of neurophysiologists with the performance of too many and ingenious experiments. . . . Conduction of an impulse is in fact somewhat incidental to another essential functioning of a neuron, however useful as a sign that the neuron has functioned. Where does one come out, if he looks at the neuron as a secretory organ? This proposal is not new, but nobody has done much lobbying for it. To wit, the prime function of a neuron is to produce and apply to other tissues a chemical activator. [9, p. 14]

The reverse relationship has been analyzed in considerable detail. If a limb bud of the chick embryo is extirpated at 2½ days, this is not noticed by the primary motor and sensory centers for a while. Differentiation goes on until the time when the axons would normally have contacted muscle fibers and sense organs in the growing limb. But then, at around 6–7 days, a sudden dramatic disintegration of the motor column sets in which wipes out the entire population of 20,000 cells within 3 days [10]. The ganglia are also reduced greatly. Obviously, unspecified conditions at the periphery control the maintenance of the centers. There exists an interesting parallel between this breakdown process and a similar phenomenon that occurs in normal development. The motor column, like other neuron populations, normally engages in overproduction of neuroblasts, which is followed by a depletion that amounts to 40 percent in the case of the chick motor column [11]. The fact that it occurs at approximately the same time as the breakdown after limb ablation suggests a similar mechanism in both instances. Probably only those neuroblasts survive whose axons manage to make contacts at the periphery. This would explain selective survival in the normal situation and total loss in the experimental situation.

The hypoplasia story has had a much happier ending for me than for the motor neuroblasts. I confirmed in the chick the old observation of Detwiler that an increase of the peripheral field by implantation of a supernumerary limb results in a distinct hyperplasia of the spinal ganglia and a less conspicuous effect on the motor column. I proposed a feedback hypothesis which I thought was quite clever in that it would explain both hyper- and hypoplasia by the same mechanism. I suggested that the demand for nerve fibers at the periphery might regulate the production of neuroblasts at the center. The hypothesis turned out to be essentially wrong; but my error was a blessing in disguise. It resulted in two related events: the appearance of Dr. Rita Levi-Montalcini on the scene, in 1948; and the discovery of the nerve growth factor (NGF) a few years later. Rita had done the same limb extirpation experiments in Italy during the war. She had obtained the same results, but her interpretation was different and, as usual, the correct one: absence of the limb results in the retrograde degeneration of the frustrated neuroblasts. I invited her to Saint Louis, and we settled the argument. We then tried to get to the heart of the matter of hyperplasia. We repeated a rather bold experiment of one of my former students, Bueker, who had implanted a piece of mouse tumor in chick embryo in the hope that this fast-growing tissue would incite a more impressive hyperplasia than the supernumerary limbs. He did find invasion of the tumor by nerves but was not encouraged by the results. At this point, Rita's flair for detecting subtle clues which the embryo manages to hide from the eyes of other observers asserted itself, and in rapid sequence the NGF yielded its secrets. This story has been told by her [12], but I would like to reminisce on two milestones in the early days. One day she showed me a chick embryo with a large intracmbryonic tumor. The tumor had produced the typical conspicuous hyperplasia of spinal and sympathetic ganglia, and it had been invaded massively by nerve fibers from these ganglia. But she had made another, novel observation. Some prevertebral sympathetic ganglia, quite remote from the tumor, were also hyperplastic; their axons ended in adjacent viscera rather than in the tumor. This was the first hint that a diffusible agent was involved. It did not take long to prove the point. Tumor tissue was transplanted to the chorioallantoic membrane, far from the embryo, where it grows well. It exerted the familiar effect: hyperplasia of ganglia, by remote control via the circulation [13].

The second memorable event was the arrival of a sketch from Rita from Rio de Janeiro, where she was working with a tissue-culturing friend. The sketch showed the first ganglion in tissue culture with a halo of fibers induced by a piece of tumor which had been placed at some distance from it and had released NGF. To this day, the halo provides the indispensable bioassay for NGF—and honors for Rita.

To conclude this "trophic" chapter: I am not sure of the actual role of NGF in the normal development of adrenergic neurons, but I am sure that this discovery is a major breakthrough in developmental neurobiology. It is the one instance in which a highly target-specific nerve growth regulator has been tracked down to its molecular structure. With the sequencing of the active NGF protein by Ruth Angeletti and R. Bradshaw in 1971 [14], the road is open for an understanding of its mechanism of action. But the fact that this discovery, which grew out of a seemingly peripheral problem (peripheral in every sense of the word), has blazed so many new trails is its greatest contribution to neuroembryology.

Central Circuitry

Of the other two problems raised by the Detwiler experiment, that of the construction of central circuitry has been advanced most successfully. In the more precise formulation as the problem of formation of specific synapses, it has moved to the center of the stage. Before taking up this topic, I shall mention briefly two experiments which support the contention of Detwiler that the central circuitry for locomotion is built into the system prior to function and without benefit from it. Narayanan and I found very early specification of the brachial and lumbosacral segments of the spinal cord for their respective functional assignments. We used a different experimental design, which, incidentally, also goes back to Detwiler. In 1½-day chick embryos, we switched the brachial to the lumbosacral level, and vice versa. A few embryos with legs innervated by a brachial plexus were hatched. The legs moved only synchronously, as in wing flapping, but they never showed alternating stepping movements. Obviously, the program for the construction of the central circuitry is built into the limb segments of the spinal cord even before the onset of neuroblast differentiation, and it cannot be modulated by the appendages [15]. Straznicky [16] had reported similar findings.

The same point was demonstrated by Bentley and Hoy in a more sophisticated way. They studied the embryonic origin of the neural circuitry which commands the motor output to wing muscles in the flight pattern of the cricket [17]. Crickets undergo approximately 10 molts before emergence of the adult. Individual flight muscles are innervated by only one to three neurons. Motor output was monitored by electromyograms taken simultaneously from four muscles: hind-wing depressor, hind-wing elevator, fore-wing depressor, and fore-wing elevator. The tests were made by suspending the larvae in a wind tunnel. In the last instar before final molt, the complete flight pattern is estab-

256

lished; long trains of spikes are performed, equivalent to dozens of wing strokes, although the wing primordia are immobile and tightly packed in the larval envelope. The gradual emergence of flight muscle coordination, which reflects the gradual establishment of neuronal circuitry, was traced back to the fourth to the last instar, when the first irregular spikes occur in the hind-wing depressors, which are the muscles that eventually lead the flight pattern. There could be no better demonstration of the nonparticipation of functional activity in the buildup of circuitry for complex, stereotyped motor patterns.

Synaptogenesis

Before I turn to the modern aspects of synapse specificity, I would like to indicate the conceptual metamorphosis which the problem has undergone. In the twenties and thirties there prevailed among developmental psychologists a belief in the powerful role of learning, experience, and sensory input in the molding of behavior and a downgrading of instincts and built-in action systems. These views carried over into neurogenesis. It was seriously proposed that neuronal connections were at first indiscriminate and random and that functionally adaptive pathways and synapses would be reinforced by practice and would survive while all others would vanish. This extreme view of the creative role of function was founded on shaky evidence and challenged by neuroembryologists such as Coghill, P. Weiss [18], and others. Their observational and experimental data supported the diametrically opposed view that the maturation of neural connectivities precedes and conditions functional activity and behavior, not vice versa. The neuroembryologists have won the argument, but, as we shall see, the final, or rather the present-day, verdict is by no means quite that simplistic.

It was in this context, as a challenge to the viewpoint of functional adaptation, that Roger Sperry initiated his analysis of retinotectal relations [19]. The choice of the retinotectal system was fortunate and was one major reason for the rapid advances in our insight into the mechanism by which neuronal connections are formed in synaptogenesis. It is of interest that his major discoveries were based on purely behavioral observations on lower vertebrates, frogs and salamanders, which he subjected to minor surgery. His basic experimental design capitalizes on the capacity of the optic and other nerves in the frog to regenerate to the point of full functional recovery. In the first crucial experiment, optic nerve transection was combined with the rotation of the eyeball by 180°. The optic fibers were scrambled at the cut surface to reduce the chance of their return to their old channels. After functional regeneration, the visual experience of the frog was found to be abnormal: objects were seen upside-down and backward instead of forward, and this maladaptation was never corrected. If only the dorsoventral axis of the eye was rotated (by shifting the left eye to the right orbit, over the head), then an object presented above was seen as if it were located below; if only the anteroposterior axis was inverted (by shifting the left eye to the right orbit around the nose), then the forward-backward direction only was misjudged. In other words, the errors were not random, but systematic. The explanation for abnormal responses was based on the well-known fact that there is a precise topographic projection of the different quadrants of the retina onto the optic tectum. Sperry then postulated that the retinal ganglion cell axons had reestablished synapses at exactly the same tectal region with which they had been connected before. We speak of "position specificity." Numerous experiments on other sensory systems supported this conclusion. Sperry then made the bold and ingenious inference from the behavioral data to a biochemical mechanism of synapse formation. He proposed the chemoaffinity hypothesis, according to which synapse formation is based on matching or complementary biochemical affinities between the axonal growth cone and the neuron which it contacts. In the course of time, the wide gap between data and inference was almost filled by him and others.

The first step was taken by M. Gaze in England [20]. He confirmed the precise retinotectal projection in the frog by electrophysiological recording from the tectum. Then he applied the mapping technique to inverted eyes. He showed that, indeed, retinal ganglion axons returned to their specific locations on the tectum. Attardi and Sperry [21] came a step closer to histological verification of the synaptic site in an experiment in which optic nerve transection was combined with ablation of half-retinas. The residual fibers returned to their specific sites, leaving other tectal areas vacant, or even bypassing the inappropriate areas. But none of these approaches could trace the axon terminals to the actual synaptic sites; and they all dealt with the reestablishment of previously existing connections. In both respects, the recent experiments of Cowan and his co-workers on the retinotectal projection in chick embryos [22] provide the finishing touch. The half-retina experiments were repeated on 2½-day chick embryos, and the ingrowth of the fibers from the residual retina to their correct positions was ascertained histologically and by autoradiography. The crucial point is that the actual synaptic sites could be identified by two different methods. The first was based on rapid axoplasmic flow: labeled protein precursors were injected into the miniature eyes at prehatching stages. The sites of projection of the ganglion cells were determined from the grain condensations found in those strata of the optic tectum which are known to be the sites of optic fiber termination. In the other method, which is more precise, eyes with reduced retinas were extirpated some time after hatching, and a few days later the sites of degenerating synapses were identified by the appearance of ringlike boutons. Since we are dealing here with initial embryonic synaptogenesis, the "position specificity" of optic fiber connections is established beyond doubt.

But the final step, from position specificity to chemoaffinity, remains to be taken. So far, chemoaffinity is an abstract concept devoid of substantive content, not an explanation. It is a challenge to the molecular neuroembryologist to give it a concrete meaning. Synapse formation is a special case of cell recognition. The latter topic is now under intense investigation in cell dissociation and reaggregation studies. It is obvious that the solution to the problem of synapse specificity will be in terms of the biophysical and biochemical properties of membranes. The beginnings that have been made with regard to retinotectal adhesive recognition are promising. Preferential adhesion of dissociated retina cells to appropriate tectal sites has been reported [23]. In another experimental design, isolated plasma membranes from chick embryo retina cells were found to inhibit differentially the reaggregation of retina cells and, to a lesser degree, that of tectal cells [24]. Of particular interest is the demonstration that cell membrane recognition changes profoundly with stage of development.

This is not the only unresolved problem. The pursuit of the question concerning the precision of neuronal specification has led to difficulties which present a serious challenge to the chemoaffinity hypothesis. Nobody thinks in terms of a rigid one-to-one relationship between individual retinal and tectal cells. But what are the limits of "recognition"? Sperry built into his hypothesis a certain degree of plasticity in the form of two overlapping gradient fields, one perpendicular to the other, and each representing a major axis in the retina and tectum, respectively. A neuron group would be identified by its relative position in this coordinate system [25]. Gradient fields have been invoked in experimental embryology whenever the problem of regulation becomes acute, because the relative position of the subunits remains unchanged when the fields are experimentally reduced or expanded or even distorted. An unexpected degree of regulation in the retinotectal projection was revealed when the size of the tectum rather than that of the retina was reduced. For instance, if half of the tectum or a median strip is removed, the entire retina is projected onto the residual tectum surface in a compressed, orderly fashion [26, 27]. Further difficulties arise in mammals, in which the retina projects onto several different centers [28, 29]. It remains to be seen whether these new data can be accommodated in Sperry's double-gradient model, or whether the chemoaffinity hypothesis will have to undergo a major revision.

Another series of recent investigations makes it appear as if we were on the way to returning full circle to the viewpoint of functional adaptation. Recent experiments from different laboratories give clear evidence that cells in the visual cortex of newborn mammals can be specified as to their functional role by exclusive exposure to particular visual patterns, such as vertical bars [30–32]. I do not believe that these findings are in conflict with those on prefunctional specification. It would seem that built-in fixity and functional plasticity both have their place in synaptogenesis, the former as the basis of stereotyped behavior and fundamental organization and the latter in higher centers, where the capacity for postnatal adjustments and refinements would be of adaptive value.

Finally, one has to ask, To what extent are the findings on the visual system paradigmatic for other systems? A clear answer cannot be given at this moment. Extensive work on neuromuscular specificity, which my limited time does not permit me to review, has not given unequivocal results, and the issue is highly controversial. We are not much better off in the matter of the central organization of localized skin reflexes. In both instances, two alternative possibilities have been debated: that the peripheral structures specify the nerves which contact them, or that prespecified nerves selectively establish contact with the proper end organ. Recently, skepticism has been expressed concerning both viewpoints [33]. Indeed, we are far from any generalization in this field. However, this should not minimize the great heuristic value of the chemoaffinity concept, which has served as a guide and frame of reference for all subsequent experimental analysis.

Peripheral Nerve Patterns

In contrast to synaptogenesis, the problem of how nerves reach their goals has not advanced much. I state this with regret, because this problem has interested me for many years and I worked on it in my earlier days. It is not even generally realized that pathfinding is a separate problem, distinct from synapse formation. Fibers must arrive before they can make connections. There is one way to combine both aspects in a unified theory, namely, by assuming that fibers are attracted at a distance by their targets. This was indeed the prevailing view when tropisms were fashionable, around the turn of the century. Ramón y Cajal was one of the first to suggest chemotropism as the guiding force; others postulated galvanotropism. Yet there is an inherent improbability in the tropism theories: How can overlapping local gradients in an embryo remain sufficiently stable for a long enough period to be effective? How would one account for the branching of dorsal root fibers in opposite directions? Then there is the fact that the major branching pattern, for instance in the embryonic limb, is laid down before the individual muscles are formed. Apart from such considerations, all efforts to test tropism theories in vitro or in vivo have given negative results. But perhaps the last word has not been said in this matter.

Harrison has always stressed the importance of the solid substrate on which the axon is spun out, and I mentioned earlier that the growing limb provides tracks for nerves. Weiss once thought that the micellar organization of the substrate might provide the necessary directional cues; this is expressed in the contact guidance concept [34]. But mechanical guidance alone cannot be the answer. Since growth cones are compelled to make choices at every branching point and have to decide whether to enter a proximal or a distal muscle or a particular sense organ, one has to postulate additional cues which would provide more specific guidance than mechanical tracks can give. The obvious suggestion is that substrates are biochemically specified. Thus, we are back at chemoaffinity, this time between the growth cone and the substrate on which it advances [25, 34].

This brief survey of facts and fancies in pathfinding and synapse formation should end on an optimistic note. If the prevailing view of the inherent structural, functional, and biochemical specificity of neuron strains is correct, and if one also accepts the corollary that interneuronal and neuron and target recognition is based on matching biochemical membrane specifications, then we are in sight of one of the most formidable problems in neurogenesis: How does neuronal strain specificity originate in the embryo? What constellation of intrinsic and extrinsic agencies is instrumental in conferring identity on a neuron? Is this done in one step or in a sequence of steps? Do the partners acquire their matching specifications independently, or does one consult with the other? Here is a great challenge for a new generation to invent the concepts and tools necessary for the pursuit of this goal—indeed an optimistic outlook.

Embryonic Motility

Near the start of my career, I wrote an article for *Naturwissenschaften* in which I outlined a research program for the rest of my life [35]. I have followed it—to some degree. But I had omitted one point—the embryology of behavior—for reasons that are now not clear to me. Perhaps, as an embryologist, I looked at the developing nervous system primarily as a playground for collecting developmental correlations, and I was apparently not fully aware of the *raison d'être* of the nervous system. This omission has been corrected in the last decade with the help of a number of able collaborators, M. Balaban, J. Decker, C. H. Narayanan, R. Oppenheim, R. Provine, S. Sharma, and E. Wenger [36–38]. In the course of this work, again, a fundamental shift in conceptualization was forced on us by the phenomena. It will not surprise you to learn that the change of outlook is in the same direction as for previously discussed areas, namely, from a behavioristic, stimulus-response notion to the primacy of built-in, autonomously generated spontaneous motor activity. For developmental psychologists of the thirties and forties, like Kuo and Schneirla, who dominated the field, the role of sensory input, learning, and similar experience in the emergence of behavior was self-evident, and the extensive observations of Kuo on chick embryo were interpreted in this way, without experimental validation. A competing theory of that period, that is, Coghill's notion that behavior is integrated from the first head movement of the embryo to the adult stage [39], may be valid for his material, the salamander, but it is definitely not applicable to birds and mammals.

My contention that embryonic behavior results from autonomous, nonreflexogenic activity of the central nervous system is based on three lines of evidence: observation of some peculiar features of overt motility on chick and rat embryos, deafferentation experiments, and electrophysiological recordings. The peculiar features I refer to are (1) the spontaneity of embryonic movements, not caused by any obvious stimulation; (2) the periodicity of motility, activity phases alternating with inactivity phases, which is evident in the chick embryo from the beginning of motility at 3½ to 13 days; and (3) the lack of coordination of parts, such as right and left leg or head movements. During an activity phase, different parts move in seemingly random and unpredictable combinations (I do not speak of muscle coordination) and in a jerky, convulsive-like fashion. These features suggest that an integrating influence of sensory information is not yet present. This peculiar motility pattern is in contrast to a highly integrated, quite different behavior type which starts at 17 days and is instrumental in hatching. The fetal motility of the rat between 16 days and parturition shows the same characteristics, that is, spontaneity, periodicity and lack of coordination. I believe that this type of motility is characteristic of amniote embryos in general.

Experimental evidence for the nonreflexogenic nature of the embryonic motility type was obtained by deafferentation experiments in which the primordia of sensory ganglia to the leg, the primordia of the trigeminal ganglia, and those of the accousticovestibular system were extirpated in early chick embryos. I shall discuss only the leg deafferentation, which was accomplished by a double operation; the removal of a thoracic section of the cord to prevent input from rostral regions; and the extirpation of the dorsal half of the lumbosacral cord, which includes the primordia of sensory ganglia that feed into the lumbosacral plexus. Periodicity and general characteristics of the motility were normal, up to 15–17 days, when the half-cord began to deteriorate. Normalcy in these features was also found in the other deafferentation experiments.

Since the activity is not blocked by chronic gaps in the cervical cord, we concluded that the motility results from autonomously generated discharges of the spinal cord. This assumption was substantiated by direct recordings from curarized embryo, in vivo done mostly by Provine [40]. He found, after preliminary exploration of the lumbosacral cord, a close correlation between patterns of polyneuronal bursts generated in the ventral (motor) part of the spinal cord and the overt motility patterns. This holds for all stages, from the beginning of motility, at 4 days, to hatching. The conformity of bursts and activity phases was definitely established when floating electrodes were used to record from uncurarized, freely moving embryos. One observer recorded the motility phases, and an oscillograph recorded the bursts. There was precise synchrony throughout development. Finally, the proof that the bursts are the cause of motility, and not movement or other artifacts, came from recordings that were made before, during, and after a 15-minute curarization period. It was found that bursts continue while the embryo is immobilized.

This peculiar embryonic spontaneous motility pattern has no parallel in postnatal behavior; it persists, perhaps, in the phasic movements of REM sleep. Its adaptive significance is probably in preventing muscle atrophy and fusion of joints; both occur when embryos are immobilized for longer periods.

The strange performance of the chick embryo and of the mammalian fetus leaves us in a puzzling situation. It looks as if integrated activities such as walking or pecking of food which are performed with reasonable perfection soon after hatching or birth by all precocious animals have no antecedents in prenatal motility. This apparent paradox can be resolved if one assumes that the neuronal circuitry for these activities is prepared during embryonic development but does not find expression in the overt motility of the embryo. In fact, it can be shown that, at least at the level of muscle coordination within the leg, more order exists than is apparent from the jerks and twitches that the embryonic leg performs or from the burst pattern. Anne Bekoff, who is presently monitoring the motor output in embryos in situ by electromyography, finds antagonistic inhibition in leg muscles as early as 7 days of incubation, that is, shortly after onset of leg motility [41].

Concluding Remarks

When I was invited to give this lecture and asked for a title, I tried to think of something timely and "relevant," and it occurred to me that "Law and Order in Neurogenesis" would be a fitting title. But, on second thought, I discarded this idea, not only because I have too much respect for the nervous system to associate it with dubious company, but also for another reason. Although there is definitely order, I was at a loss to think of a single "law" or even a broad generalization. Apparently, experimental neuroembryology, after passing the half-century mark, is still too young to show such signs of maturity. The old-timer admires the vigor and vitality of its second youth and the rapid and imaginative progress that has occurred in neuroembryology in recent years. In the same spirit, the younger generation that is now taking over should be grateful to us old-timers for not being smart enough to solve all the problems.

REFERENCES

1. H. Spemann and H. Mangold. W. Roux' Arch. Entwicklungs-Mechanik, **100**:599, 1924. English translation *in:* B. H. Willier and J. Oppenheimer (eds.). Foundations of experimental embryology. Englewood Cliffs, N.J.: Prentice-Hall, 1964.
2. H. Tiedemann. *In:* R. Weber (ed.). The biochemistry of animal development, vol. **2**. New York: Academic Press, 1967.
3. R. G. Harrison. Anat. Rec., **1**:116, 1907.
4. P. Weiss and H. B. Hiscoe. J. Exp. Zool., **107**:315, 1948.
5. S. R. Detwiler. J. Exp. Zool., **31**:117, 1920.
6. V. Hamburger. W. Roux' Arch. Entwicklungs-Mechanik, **114**:272, 1928.
7. H. Nüesch. Annu. Rev. Entomol., **13**:27, 1968.
8. O. E. Schotté and E. G. Butler. J. Exp. Zool., **87**:279, 1941.
9. G. Bishop, Annu. Rev. Physiol., **27**:3, 1965.
10. V. Hamburger. Am. J. Anat., **102**:365, 1958.
11. ———. J. Comp. Neurol., in press.
12. R. Levi-Montalcini. *In:* The Harvey Lectures. New York: Academic Press, 1966.
13. R. Levi-Montalcini and V. Hamburger. J. Exp. Zool., **123**:233, 1953.
14. R. H. Angeletti and R. A. Bradshaw. Proc. Natl. Acad. Sci. USA, **68**:2417, 1971.
15. C. H. Narayanan and V. Hamburger. J. Exp. Zool., **178**:415, 1971.
16. K. Straznicky. Acta Biol. Acad. Sci. Hung., **14**:145, 1963.
17. D. R. Bentley and R. R. Hoy. Science, **170**:1409, 1970.
18. P. Weiss. Comp. Psychol. Monogr., **17**:1, 1941.
19. R. W. Sperry. *In:* S. S. Stevens (ed.). Handbook of experimental psychology. New York: Wiley, 1951.
20. M. Gaze. The formation of nerve connections. New York: Academic Press, 1970.
21. D. G. Attardi and R. W. Sperry. Exp. Neurol., **7**:46, 1963.
22. W. J. Crossland, W. M. Cowan, L. A. Rogers, and J. P. Kelly. J. Comp. Neurol., **155**:127, 1974.
23. A. J. Barbera, R. B. Marchase, and S. Roth. Proc. Natl. Acad. Sci. USA, **70**:2482, 1973.
24. D. I. Gottlieb, R. Merrell, and L. Glaser. Proc. Natl. Acad. Sci. USA, **71**:1800, 1974.
25. R. W. Sperry. *In:* R. L. deHaan and H. Ursprung (eds.). Organogenesis. New York: Holt, Rinehart & Winston, 1965.
26. S. C. Sharma. Proc. Natl. Acad. Sci. USA, **69**:2637, 1972.

27. M. Yoon. Exp. Neurol., **33**:395, 1971.
28. R. W. Guillery and J. H. Kaas. J. Comp. Neurol., **143**:73, 1971.
29. R. M. Gaze and M. J. Keating. Nature, **237**:375, 1972.
30. H. B. Barlow and J. D. Pettigrew. J. Physiol., **218**:98, 1971.
31. C. Blakemore and G. F. Cooper. Nature, **228**:477, 1970.
32. H. V. B Hirsch and D. N. Spinelli. Exp. Brain Res., **13**:509, 1971.
33. C. Székely. *In:* G. Gottlieb (ed.), Behavioral embryology, vol. **2**. New York: Academic Press, 1974.
34. P. Weiss. *In:* Third Growth Symp., Growth Suppl. to vol. **5**, 1941.
35. V. Hamburger. Naturwissenschaften, **15**:657, 1927.
36. ————. *In:* F. O. Schmitt (ed.). The neurosciences: second study program. New York: Rockefeller Univ. Press, 1970.
37. ————. *In:* E. Tobach, L. R. Aronson, and E. Shaw (eds.). The biopsychology of development. New York: Academic Press, 1971.
38. ————. *In:* G. Gottlieb (ed.). Behavioral embryology. New York: Academic Press, 1973.
39. E. G. Coghill. Anatomy and the problem of behaviour. London: Cambridge Univ. Press, 1929.
40. R. Provine. *In:* G. Gottlieb (ed.). Behavioral embryology, vol. **1**. New York: Academic Press, 1973.
41. A. Bekoff. Soc. Neurosci., Fourth Annu. Mtg., 1974, p. 136 (abstr.).

Perspectives in Biology and Medicine 23: 600–616, 1980

S. RAMÓN Y CAJAL, R. G. HARRISON, AND THE BEGINNINGS OF NEUROEMBRYOLOGY

*VIKTOR HAMBURGER**

Santiago Ramón y Cajal (1852–1934)

It is rare that the birth date of a branch of science can be determined rather precisely. The beginnings of modern developmental neurobiology can be traced to the eighties of the last century and to two eminent men: the German embryologist and anatomist Wilhelm His (1831–1904) and the Spanish neurologist Santiago Ramón y Cajal (1852–1934). Of course, the development of the nervous system had been studied before, but the foundations of our present view were laid by these men during the years 1886–1890.

Since our focus will be on Ramón y Cajal, he should be introduced briefly to nonneurologists. I rank him among the leading biologists of the last century, a peer to Darwin, Carl Ernst von Baer, Pasteur, Johannes Müller, von Helmholtz. He is the founder of modern neurology, which is also the basis of neurophysiology, neuropathology, and physiological psychology. Almost singlehandedly, he unraveled the design of the central nervous system of the vertebrates and man and traced its structure to the most intricate details. Some of his drawings, all of which bear the stamp of his originality, may still be found in modern textbooks, testifying to the amazing accuracy of his observations. His monumental *Histologie du système nerveux de l'homme et des vertébrés* (1904) is still a standard work.

The combination of extraordinary conceptual insight and observational power which characterizes his genius were displayed right at the beginning of his work, around 1887 and 1888, in a breakthrough which liberated neurology from a fallacy that had hindered all progress and at the same time set it on the right track. The then-prevailing conception of the structure of the nervous system was embodied in the reticular theory. It envisaged the nervous system as a syncytial network of nerve

*Professor emeritus, Department of Biology, Washington University, St. Louis, Missouri 63130.

fibers which were continuous with each other; the cell bodies were considered as trophic elements, at the intersection of the web. The fatal flaw of this theory is obvious: it obviates the establishment of specific pathways and connections, which are the necessary prerequisites of integrated function. Cajal revolutionized the concept of the nervous system by asserting—and demonstrating—that nerve fibers are not continuous but contiguous, that they possess terminal structures which contact other nerve cells but do not fuse with them. The contacts are now called "synapses." The hypothesis of contiguity had been proposed independently, then unknown to Cajal, by two German investigators, A. Forel and W. His. But, as Cajal points out, their hypothesis, based largely on inferences, does not take us much farther than the reticular theory as long as the possibility of diffuseness of contacts is not ruled out. He states: "To settle the question [of contiguity vs. continuity] definitely, it was necessary to demonstrate clearly, precisely, and indisputably the final ramifications of the central nerve fibers, which no one had seen, and to determine which parts of the cells made the imagined contacts" [1, pp. 337–338]. The momentous discovery of the synapse was made in 1888. During an investigation of the structure of the cerebellum of birds, he observed that terminal branches of the axons of the so-called stellate cells "applied closely to the bodies of the cells of Purkinje about which they form a kind of complicated nests or baskets" [1, p. 330]. Other synapses of different types were observed in rapid succession, and synaptic contact was recognized as a basic phenomenon. Ironically, Cajal's success in demonstrating synapses was based on the method of chrome-silver impregnation of nerve fibers which had been introduced by the Italian neurologist C. Golgi, the major proponent of the reticular theory. The same method, later improved by Cajal himself, enabled him to identify specific nerve centers and specific connections of nerve centers on a large scale.

The idea that individual nerve cells, or neurons, are the basic units of the structure of the nervous system and that axons and dendrites are parts of the neuron became known as the *neuron theory*. For many years it was pitted againt the reticular theory.

But why did Cajal turn to embryos? This was done by deliberate design. His motive is told best in his own words.

... the great enigma in the organization of the brain was the way in which the nervous ramifications ended and in which the neurons were mutually connected. Repeating a simile already used, it was a case of finding out how the roots and branches of these trees in the gray matter terminate, in that forest so dense that, by a refinement of complexity, there are no spaces in it, so that the trunks, branches, and leaves touch everywhere.

Two methods come to mind for investigating adequately the true form of the elements in this inextricable thicket. The most natural and simple apparently,

267

but really the most difficult, consists of exploring the full-grown forest intrepidly, clearing the ground of shrubs and parasitic plants, and eventually isolating each species of tree, as well from its parasites as from its relatives. Such was the approach employed in neurology by most authors. ... Such tactics, however, are inappropriate for the elucidation of the problem proposed, by reason of the enormous length and extraordinary luxuriance of the nervous ramifications, which inevitably appear mutilated and almost indecipherable in each section.

The second path open to reason is what, in biological terms, is designated the ontogenetic or embryological method. Since the full grown forest turns out to be impenetrable and indefinable, why not revert to the study of the young wood, in the nursery stage, as we might say? Such was the very simple idea which inspired my repeated trials of the silver method upon embryos of birds and mammals. If the stage of development is well chosen, or, more specifically, if the method is applied before the appearance of the myelin sheaths upon the axons (these forming an almost insuperable obstacle to the reaction), the nerve cells, which are still relatively small, stand out complete in each section; the terminal ramifications of the axis cylinder are depicted with the utmost clearness and perfectly free; the pericellular nests, that is the interneuronal articulations, appear simple, gradually acquiring intricacy and extension; in sum, the fundamental plan of the histological composition of the gray matter rises before our eyes with admirable clarity and precision. As a crowning piece of good fortune, the chrome silver reaction which is so incomplete and uncertain in the adult, gives in embryos splendid colourations, singularly extensive and constant. ...

Realizing that I had discovered a rich field, I proceeded to take advantage of it, dedicating myself to work, no longer merely with earnesteness, but with fury. In proportion as new facts appeared in my preparations, ideas boiled up and jostled each other in my mind. A fever for publication devoured me. [1, pp. 323–25]

His "fever for publications" produced 12 papers and monographs in 1889 and 16 in 1890, his most productive years. Very soon, what began as a "strategic subterfuge" became an endeavour in its own right, with intriguing problems of its own.

In 1890, Cajal was successful in obtaining splendid silver impregnations of the spinal cord of early (2½-day) chick embryos [2]. They showed the early stages of differentiation of an embryonic neuron, or neuroblast; that is, cell bodies with a short outgrowth which was identified as the incipient axon. It terminated in a club-shaped thickening with short spikes. The latter were recognized later as filamentous pseudopodia. Cajal designated the terminal structure as the "growth cone." As was stated, the neuron theory asserted that the axon is part of the neuron. The discovery of the mode of origin of the axon was the categorical affirmation of this aspect of the neuron theory. Cajal observed that neuroblasts are polarized in the sense that the site of the outgrowth of dendrites is opposite to that of the axons, and he established the general rule that dendrites differentiate later than the axon. He found the clearest demonstration of neuroblast polarity in the earliest differentiation stages of spinal ganglion neuroblasts: they are at first bipolar, with two outgrowths at opposite ends of the cell. These extensions fuse later at their bases to form the single sensory fiber. The recognition of structural polarity later on became the basis of the theory of physiological polarity. But of all his observations in the field of neurogenesis, Cajal was most intrigued by the growth cone. We shall return to this point later.

To appreciate the fundamental importance of these discoveries, one has to place them in their historical setting. The axon outgrowth theory had two formidable rivals: *the cell chain theory* of Schwann postulated that nerve fibers are produced by chains of Schwann cells which connect the nervous system with the peripheral organs. The nerve fibers are considered as products of these cells which fuse with each other and with the neuroblasts. More widely accepted was the *plasmodesm theory* of Hensen and Held. It was based on the ubiquity of protoplasmic bridges, or plasmodesms, resulting from incomplete cell divisions. In the original version of Hensen [3], some plasmodesms would be transformed into nerve fibers by functional validation. In the more sophisticated version of Held [4], an approach to the His-Cajal notion of axon outgrowth is evident. According to Held, the axon is built by two components. One is *neurofibrillar* material spun out by neuroblasts (demonstrable in Held's silver-impregnated material and distinguished by him from the protoplasmic outgrowth described by His and Cajal). The neurofibrils penetrate into plasmodesms, and these intraplasmatic neurofibrils are then transformed into nerve fibers by utilization and incorporation of plasmodesm material. Ramón y Cajal fought all his life battles on two fronts: for contiguity and against continuity in the structure of the nervous system; and for protoplasmic outgrowth, against cell chains and plasmodesms, in the origin of the nerve fiber.

During the crucial years in his career, from 1887 to 1892, Cajal was professor of histology in Barcelona. In this provincial place, he was remote from the mainstream of scientific research and not aware of the work of Wilhelm His, one of the leading German anatomists and embryologists of that time. In 1886 and 1889 [5, 6], His had given a very detailed account of the development of the spinal cord, first in human embryos and then in other vertebrate embryos. He had described the transformation of the neural epithelium into mantle and marginal velum. The neuroblasts were derived from mitotic cells at the inner lining of the central canal. Erroneously, he considered these proliferating "germinal cells" as a special strain of neuroblast precursors, and he derived ependymal layer and glia from the neural epithelium. We know now that the "germinal cells" are merely the mitotic phase of neuroepithelial cells which give rise to both neurons and glia. He coined the terms "neuroblast" and "dendrites." He was the first to describe the transformation of the postmitotic cell into a neuroblast and the forma-

269

tion of axon and dendrites as protoplasmic outgrowths from the neuroblast. Earlier than Cajal, he described the migration of neuroblasts to the periphery, where they form the mantle. He also noticed the originally bipolar configuration of the spinal ganglion cell.

In 1886 he stated the neuron theory very concisely: "I consider it as an established principle that each nerve fiber emerges as an outgrowth from a single cell. This is its genetic, trophic and functional center. All other connections of fibers are either indirect or secondary" [5, p. 513]. This statement includes implicitly the concept of contiguity as against a network.

Cajal did not learn of these findings until 1890, when His sent him copies of his work. From then on, Cajal gives His full credit for his discoveries, but he states explicitly that his own discoveries were made independently of those of His. He adds: "This coincidence in thought on the part of the leading workers in the field, without any oral or written collaboration, constitutes the best moral encouragement and the strongest guarantee of the validity of the adopted interpretation" [7, p. 6].

Yet it is the merit of Cajal to have realized fully the dynamic implications of the outgrowth theory. In this he went far beyond His. I shall turn to this aspect and omit any further references to his substantial contributions to the development of many structures, such as retina, cerebellum, spinal cord, optic tectum [8]. The silver-impregnation method had permitted Cajal the observation of the growth cone which was not discernible in the material of His, treated with ordinary stains. Cajal describes it as a swelling with spiny extensions, sometimes triangular or lamellar, and ramified. In his treatise on histology he gives the following interpretation: "From the functional point of view the growth cone may be regarded as a sort of club or battering ram, endowed with exquisite chemical sensitivity, with rapid ameboid movements, and with certain impulsive force, thanks to which it is able to proceed forward and overcome obstacles met in its way, forcing cellular interstices until it arrives at its destination" [9, p. 599]. In this quotation, two points deserve attention: the uniquely dynamic interpretation of the static microscope slide picture; and the clear visualization of problems of pathfinding which are implicit in the outgrowth theory. Sherrington has an interesting comment on the first point:

A trait very noticeable in him was that in describing what the microscope showed he spoke habitually as though it were a living scene. This was perhaps the more striking because not only were his preparations all dead and fixed, but they were to appearance roughly made and rudely treated—no cover-glass and as many as half a dozen tiny scraps of tissue set in one large blob of balsam and left to dry, the curved and sometimes slightly wrinkled surface of the balsam creating a difficulty for microphotography. He was an accomplished photographer but, so

604 | *Viktor Hamburger · S. Ramón y Cajal and R. G. Harrison*

far as I know, he never practiced microphotography. Such scanty illustrations as he vouchsafed for the preparations he demonstrated were a few slight, rapid sketches of points taken here and there—depicted, however, by a master's hand.

The intense anthropomorphism of his descriptions of what the preparations showed was at first startling to accept. He treated the microscopic scene as though it were alive and were inhabited by beings which felt and did and hoped and tried even as we do. It was personification of natural forces as unlimited as that of Goethe's *Faust,* Part 2. A nerve-cell by its emergent fibre "groped to find another"! We must, if we would enter adequately into Cajal's thought in this field, suppose his entrance, through his microscope, into a world populated by tiny beings actuated by motives and strivings and satisfactions not very remotely different from our own. He would envisage the sperm-cells as activated by a sort of passionate urge in their rivalry for penetration into the ovum-cell. Listening to him I asked myself how far this capacity for anthropomorphizing might not contribute to his success as an investigator. I never met anyone else in whom it was so marked. [10, pp. xiii–xiv]

Indeed, the climbing fibers climbed and the synapses were "protoplasmic kisses, . . . the final ecstasy of an epic love story" [1, p. 373]. Cajal's dynamic view was all-pervasive. For instance, it led him to postulate the polarization of impulse conduction, based solely on morphological data.

As to the second point, his immense intellectual analytical power equals his power of observation. In fact, both are two facets of his creative genius. Whatever he observed took on a meaning transcending the microscope picture. There are few problems on our present-day mind on which he did not reflect at one occasion or another. I shall elaborate on one example, his theory of neurotropism, which deals with the problem of how nerves find their way to their targets. This problem does not exist in the cell chain and plasmodesm theories. Cajal became aware of it when he discovered the growth cone. In his monograph on the retina he mentions for the first time a solution that had occurred to him, in terms of a chemical attraction of the growth cone by substances produced by the target structures (chemotropism):

How does the mechanical development of the nerve fibers occur, and wherein lies that marvelous power which enables the nerve fibers from very distant cells to make contact directly with certain other nerve cells or the mesoderm or ectoderm without going astray or taking a roundabout course?

His has concerned himself with this important question and is of the following opinion: The axis cylinder of the neuroblasts, whether in the medulla or in the mesoderm, always follows the path of least resistance. That resistance is offered by bone, cartilage, connective tissue, etc. which are found along the route of growing nerves. This accounts for the major part of the phenomenon.

Without wanting to deny the importance of such a mechanical influence, especially in the growth of the nerve fibers from the retina to the brain and vice versa, I believe that one could also think of processes like the phenomenon called Pfeffer's chemotaxis, whose influences on the leukocytes was established by Massart and Bordet, Gabritschewsky, Buchner, and Metchnikoff. . . .

271

If a chemotaxic sensitivity in the neuroblasts is assumed, then it must be supposed that these cells are capable of amoeboid movement and are responsive to certain substances secreted by cells of the epithelium or mesoderm. The processes of the neuroblasts become oriented by chemical stimulation, and move toward the secretion products of certain cells. [11, p. 146]

The discoveries of plant physiologists concerning tropisms (chemotropism, geotropisms, etc.) and taxies figured prominently in contemporary thought. Taxies refer to directed movements of cells and organisms, and tropisms to directed outgrowth of parts, such as roots. Cajal refers specifically to the German plant physiologist W. Pfeffer, who among many other discoveries had described the attraction of sperm in mosses by malate produced by the ovary. Chemotaxis was suggested to Cajal for the first time when he observed the migration of granule cells from the superficial layer in the embryonic cerebellum to deeper layers [8, p. 291]. But the notion of chemotropism came to fruition only a decade later in the context of nerve regeneration.

A brief discourse on the status of nerve regeneration at the time when Cajal became active in this field (around 1905) is necessary. In the middle of the nineteenth century, Waller had discovered the degeneration of the distal stump, if regeneration after transection is prevented; and the central stump was recognized as the necessary "trophic center." It was also known that in the case of regeneration, Schwann cells in the distal stump proliferate and penetrate the scar between proximal and distal stump. Cajal encountered here again the unsettled controversy between those who, beginning with Waller and Ranvier, postulated that the regenerated nerve is an outgrowth from the proximal stump, and the adherents of the Schwann cell chain and plasmodesm theories who brought forth the same arguments as in nerve fiber origin in the embryo. In fact, the first decade of this century is marked by a remarkable revival of the old erroneous theories, combined with renewed attacks on the axon outgrowth theory, even in the face of Harrison's tissue culture experiment of 1907 (see, for instance, the treatise of Held [4]). Cajal, applying his silver-impregnation method, had no difficulty in finding growth cones both in the proximal stump and in later stages in the distal stump, thus bringing regeneration in line with axon production in the embryo. I omit again numerous other original findings by Cajal on nerve regeneration. I may mention the observation that, after nerve constriction, strings of beads are formed in the region proximal to the ligature. They were then rediscovered by Weiss and Hiscoe [12] and interpreted correctly as indication of axoplasmic flow.

In the meantime, the chemotropism theory had gotten a foothold in neurogenesis; and since, at the turn of the century, there were no methods available for testing it in the embryo, experimentation was carried out on regenerating nerves. Forssman [13], one of the experimenters in this field, believed that degenerating axons and Schwann sheath produced a chemotropic agent. He coined the term "neurotropism," which was adopted by Cajal, though he believed that Schwann cells (Buengner bands) rather than degenerating material generate the tropic agent. A representative example of this type of experiment is the following, done by Cajal (fig. 1): The sciatic nerve of a kitten was split longitudinally. One half was transected once, the other half was transected at the same level and also at a more proximal level. In the latter half, the nerve sector between the two cuts degenerated. Six days after the operation, a strong bridge of nerve fibers connected the distalmost cut ends. They originated in the half that had been cut only at that level, and entered the degenerating tubes of the other half, where they grew in proximal direction. This and numerous similar experiments by Cajal, Tello, Forssman, and others were interpreted as evidence for neurotropism (see [14]). However, as was pointed out later by Weiss and others, they can be explained in a different way, that is, in terms of original random outgrowth of fibers in all directions, and survival of those which happened to grow in the direction of the other stump. In fact, at the right distal stump in figure 1, fibers do grow out in all directions.

At this point, I wish to follow the theoretical reflections of Cajal, which, though purely speculative, have a bearing on very recent developments. The simplistic statement of neurotropism in 1892, quoted above, was superseded by a very sophisticated version in 1913 [15]. He recognized three basic conditions for successful regeneration: "The nervous reunion of the peripheral stump and restoration, without physio-

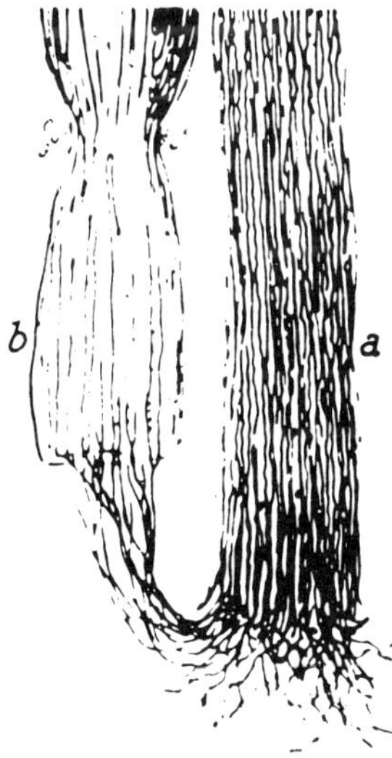

FIG. 1.—Nerve regeneration in sciatic nerve of a kitten. For details see text (from [15, p. 317]).

273

logical errors, of the terminal nerve structures, are the combined effect of three conditions: the neurotropic action of the sheath of Schwann and terminal structures; the mechanical guidance of the sprouts along the old sheaths; and, finally, the superproduction of fibres, in order to insure the arrival of some of them at the peripheral motor or sensory regions.—Of all these conditions the most essential, especially as regards the reconstruction of the terminal apparatus, is the trophism or neurotropism of the peripheral stump, motor plates, and sensory structures" [15, p. 371]. The strange juxtaposition of "trophism and neurotropism" will be commented on presently.

Furthermore, he distinguishes between general neurotropic action on the part of the peripheral stump, guiding nerve fibers toward the target, and specific action guaranteeing the appropriate connection with specific target structures: "The orienting chemical stimuli are probably, so far as their selective power is concerned, both generic and specific.—The attractive substance elaborated by the embryonic connective cells and by the cells of Schwann of the peripheral stump have a generic character, acting without distinction on all sprouts; while the attractive substances given out by the spindles of Kühne, motor plates, cutaneous sensory structures, etc., have a specific character, acting only on certain functional categories of regenerated axons" [15, p. 371]. Apart from neurotropism, his distinction between pathfinding toward the target and specificity of synaptic connections is now generally accepted. His notion of an overproduction of fibers, to insure the safe arrival of some of them at the target, in the earlier quotation has a very modern ring. In fact, he suggests that in synaptogenesis unsuccessful fibers and branches atrophy and unsuccessful neurons disappear, thus anticipating the phenomenon of naturally occurring neuronal death.

There is no doubt that, at first, many imperfect connections are formed, and that many duplications and errors of distribution occur. But these incongruences are progressively corrected, up to a certain point, by two parallel methods of rectification. One of these occurs in the periphery, and is the atrophy through disuse of superfluous and parasitic ramifications, in combination with the growth of congruent sprouts. The other occurs in the ganglia and spinal centres; by this there would be a selection, due to the atrophy of certain collaterals and the progressive disappearance of disconnected or useless neurones, of the sensory-motor fibres capable of being useful. [15, p. 279]

He even anticipates our present idea of a process of competition (for a synaptic site, or for a trophic agent) which figures prominently in our search for the explanation of naturally occurring neuronal death. ". . . It is only those expansions which are able to establish useful relations with afferent nerve fibres which survive in this contest for space and connections. In nervous regeneration this process of hyperformation is repeated" [15, p. 278].

I was particularly intrigued by the refinement of the original notion of attraction at a distance. In his later view, the distal stump (and, more specifically, the Buengner bands) would release an agent whose function is to stimulate metabolism and assimilation in the sprouting axon growth cones. What I regard as a novel conceptualization is the combination of the idea of a trophic action with *tropism*—that is, directional growth—to which I had called attention in an earlier quotation. The following paragraph clarifies what he has in mind: "The neurotropic stimulus acts as a ferment or enzyme, provoking protoplasmic assimilation. . . . While in the present state of knowledge we cannot penetrate the mechanism of the neurotropic action, an analysis of all the facts of nervous reunion known to us suggest the hypothesis that the orienting agent of the sprouts does not operate through attraction, as many have supposed, but by creating a region that is favourable, eminently trophic, and stimulative of the assimilation and growth of the newly-formed axons" [15, p. 372]. In other words, he envisages the production, by the target, of a trophic agent which stimulates growth in the growth cone and then, so to speak, nurses the axon along toward the target. I shall come back to this point presently.

Ingenious as it was, the neurotropism theory has not fared well in recent decades. It is true that not very extensive efforts have been made to test it and that practically all experiments, both in vivo and in vitro, to that effect have given negative results. Admittedly, the regeneration experiments of Cajal, Forssman, and others are not conclusive, as I have pointed out above. But the negative results of Weiss and others are not a final verdict either. When such efforts are unsuccessful, one can always raise the question of whether the experimental design was sufficiently subtle. Anyway, neurotropism has been pronounced dead as recently as 1976 [16].

As it happened, the deceased was resurrected in the same book by R. Levi-Montalcini [17]. She had discovered a case of neurotropism in the central nervous system. In order to test the claim of Swedish investigators that transected axons of monoaminergic neurons in the brain stem of young rodents can be stimulated to sprouting by Nerve Growth Factor (NGF), a nerve-growth-stimulating protein, she injected NGF into the medulla of newborn rats, near the locus coeruleus. She observed a conspicuous enlargement of the sympathetic chain ganglia on the side of injection and a massive invasion of sympathetic fibers through dorsal roots to the site of injection (fig. 2). Histofluorescence treatment demonstrated the passage of these fibers in the dorsal funiculus. They did not innervate any particular structure and disappeared when NGF injection was discontinued.

Of particular interest is her interpretation of this phenomenon which, mutatis mutandis, comes remarkably close to what Cajal had envisaged to be a tropic-trophic mechanism. I quote from a recent publication: "The entrance of sympathetic nerves into the CNS of neonatal rodents injected intracerebrally with NGF should however not be regarded as evidence of an attraction at a distance produced by the high NGF concentration gradient in the neural tube of the experimental mice and rats." (One remembers Cajal's statement that "the orienting agent does not operate through attraction as many have supposed.") "The direct access of NGF to the sympathetic ganglia through the motor and sensory roots is clearly indicated by the hypertrophic and hyperplastic effects elicited by the intracerebral NGF treatment. The same roots which served as transport channels and are presumably imbued with NGF provide in turn most convenient routes for the sympathetic fibers which engage in these paths and . . . gain in this way entrance into the

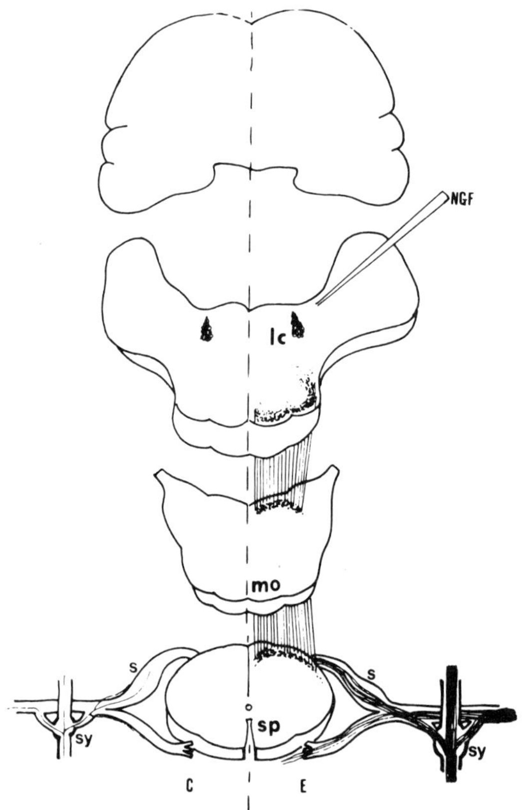

Fig. 2.—Chemotropic attraction of sympathetic fiber bundles to NGF injected intracerebrally into neonatal rats (from [17, p. 244]). *C* = control side, *E* = experimental side, *lc* = locus coeruleus; *mo* = medulla oblongata, *s* = sensory ganglia; *sp* = spinal cord; *sy* = sympathetic ganglia.

610 | *Viktor Hamburger · S. Ramón y Cajal and R. G. Harrison*

CNS. . . . The axonal tip of the fibers moves along gradients of diffusion of *trophic and tropic* factors released by end organs. . . . Neurotropism would assist, rather than determining the course of nerve fibers toward their correct destination" [18, pp. 79–80]. The linkage of the words "trophic and tropic" by Cajal in 1913 and Levi-Montalcini in 1978 is startling; one is reminded again of what Cajal had to say of independent discoveries. What was envisaged by Cajal, by pure reflection, namely that a growth promoting agent, released by the target, imbues the intervening tissue and guides the growing tip of the axon along to its source, seems now to be demonstrated in a controlled experiment. NGF has long been established as a trophic agent. If a tropic function can be added, then the revival of Cajal's idea that trophic and tropic may be two sides of the same coin could become an important new model in modern developmental neurobiology. Indeed, it is very difficult to be original in neurogenesis with Cajal looking over one's shoulder.

Ross Granville Harrison (1870–1959)

Harrison's major contribution to biology in general and to neuroembryology in particular is the invention of the tissue culture method. It will be remembered that in his classical experiment he isolated pieces of the neural tube of a frog embryo and reared it in frog's lymph, in a depression slide. What links this experiment directly with Ramón y Cajal is the fact that it was motivated by the same concerns which preoccupied Cajal: to find a direct test for the axon outgrowth theory. Of equal importance is another major contribution of Harrison: Almost single-handedly, he introduced the analytical experiment, that is, microsurgery on the embryo, as a tool for the exploration of neurogenetic problems. These two achievements would have been sufficient to rank him among the leading experimental biologists of the first half of the century. If one adds the solution of another fundamental problem, the origin of axial polarization and bilateral symmetry in vertebrates by ingenious experiments, then it is hard to understand that he did not share the Nobel Prize with Spemann, in 1935, as had been expected. In fact, he was proposed twice: first in 1914, but the prizes were suspended on account of the war, and then again in 1933. According to the official account of the Nobel Committee, in a special committee in 1933 "opinions diverged, and in view of the rather limited value of the method and the age of the discovery, an award could not be recommended" [19, p. 259]. What was actually of limited value was the judgment of the committee, and not Harrison's achievements.

I knew Harrison well. He was a frequent summer visitor in the laboratory of his friend H. Spemann, in Freiburg, where I took my Ph.D. and then advanced to an instructorship. After a vacation in the Swiss moun-

277

tains, he would occasionally spend a few weeks in Freiburg. Since the interest of Spemann in the nervous system ended with the closure of the neural tube and that of Harrison began at that stage, I got more help in my neuroembryological work from him than from Spemann.

Communication with Harrison was easy. He spoke German fluently, he had spent several pre- and postdoctoral years in Bonn, where he obtained an M.D. in 1899; he had published some of his early papers in German, he had a German wife—in short, he was at home in Germany. He was informal, unassuming, soft-spoken, and reserved; he had a good sense of humor. And he had an amiable human trait, the capacity for procrastination.

Harrison had been a graduate student in one of the best graduate schools for zoology of that time, at the Johns Hopkins University. W. K. Brooks was outstanding in embryology and remarkable for the number of prominent men who were his students. Among Harrison's fellow graduate students were T. H. Morgan and E. G. Conklin. He obtained his Ph.D. in 1894 and became then a staff member in the Anatomy Department under F. P. Mall, who was well known as a human embryologist. In 1907 he was called to Yale, and soon its Zoology Department became one of the most prominent in the country. Yale and Freiburg shared the reputation of being the leading centers of experimental embryology. For his students, most of whom became his friends, Harrison was "the Chief." He was influential in raising the standard of excellence at Yale, both in the sciences and in the medical school. For many years, he was the managing editor of the *Journal of Experimental Zoology*, the most prestigious in its field. He was not particularly enthusiastic about teaching or administration. Most administrative chores were handled by his student, later colleague, and successor, J. S. Nicholas. Harrison's due place was in the research laboratory.

One of Harrison's first experiments was based on an experiment of a young German anatomist, G. Born. In 1894, Born had discovered by chance that parts of frog embryos when cut apart could be healed together again. Taking advantage of this extraordinary healing power, he had been able to fuse parts of embryos of different genera, such as frog and toad. This method of "xenoplastic" combinations, made possible by the absence of immunological barriers in embryos, was used widely by Harrison and Spemann. Harrison was the first to apply this kind of experimentation on the embryo to neurogenetic problems. In one of his earliest experiments, he employed the method of Born to demonstrate the mode of origin of the lateral line sense organs of aquatic vertebrates. These are sensors for water perturbations; they are evenly spaced in several rows in the head and along the trunk and tail. Each sense organ consists of sensory hair cells and supporting cells; those of the head are innervated by a branch of the facial nerve and those of the trunk and tail by a branch of the vagus nerve. Harrison ingeniously took advantage of species differences in pigmentation. He fused the darkly pigmented head of an early embryo of *Rana sylvatica* with the body of a lightly pigmented *R. palustris* embryo. He observed in the living composite tadpole the step-by-step deposition of dark spots, identified as embryonic lateral line sense organs, from tissue that emerged from the dark head and moved in several rows down the yellowish trunk and tail. He had thus uncovered a peculiar, unique long-range migration of cell clusters that followed prespecified paths, as was shown by variants of the experiment [20].

The major contributions of Harrison to experimental neurogenesis were motivated by the controversy between the axon outgrowth theory of His-Cajal and the cell chain and plasmodesm theories of the origin of the nerve fiber. In the first decade of this century, when Harrison became active, the plasmodesm theory had regained ground, and even support for Schwann's cell chain theory had not subsided completely. Harrison had become convinced of the correctness of the outgrowth theory in his earlier work on neuroblast differentiation in the salmon embryo and, like Cajal, he set out to put the competing theories to a test. It was clear to him that the best histological techniques could not solve the problem, so he took the crucial step of applying the powerful tool of experimentation to its solution.

First, he took on the relatively easier task. He addressed the question, Is nerve fiber formation dependent on Schwann cells? Assuming that the Schwann cells originate in the neural crest, he removed the dorsal part of the neural tube and the adjacent neural crest, in early tail bud stages of frog embryos. He found that in the tadpole normal ventral roots and motor nerve fibers had developed which were naked and devoid of any cellular companions. Hence, the independence of nerve fiber formation from Schwann cells was proven. Furthermore, the then controversial question of the origin of Schwann cells was settled in favor of the neural crest, at least for frog embryos. These experiments date back to 1904 and 1906 [21].

Harrison turned next to the problem of the role of protoplasmic bridges in nerve fiber formation. We remember the claim of Hensen that the substance of plasmodesms is actually incorporated in the formation of nerve fibers. As was mentioned, this theory had been revived by Held and others, and the opinion of leaders in the field was divided. Experiments by the German anatomist H. Braus, in which limb transplantations were used for the first time to address neurological questions, had been interpreted by the author in support of the plasmodesm theory [22]. In the spring of 1906, Harrison repeated the experiments; and his findings, which I shall not describe in detail, led him to a rejection of the claim of Braus [23]. Two points deserve mentioning. First, this was Harrison's first experience in the transplantation of limb primordia, an experiment which was to preoccupy him in his later work devoted to problems of regulation and determination of laterality. Second, Braus had found that the nerves in the limb transplants formed a normal pattern. This was to be expected if plasmodesms in the limb are transformed into nerves. But in Harrison's reinterpretation of limb innervation in terms of the axon outgrowth theory, the fact that ingrowing nerves from any source form a typical limb pattern can be interpreted only in one way: "that the structures contained within the limb must have a very important directive action upon the developing nerve fibers, in that they determine their mode of distribution" [23, p. 276]. The contribution of the limb structures to pathfinding and patterning of their innervation is basic to an understanding of directional nerve outgrowth.

Yet, Harrison realized clearly "that in all of the first experiments the nerve fibers had developed in surroundings composed of living organized tissues, and that the possibility of the latter contributing organized material to the nerve elements stood in the way of rigorous proof of the view that the nerve fiber was entirely the product of the nerve center. The really crucial experiment remained to be performed, and that was to test the power of the nerve centers to form nerve fibers within some foreign medium which could not by any possibility be suspected of contributing organized protoplasma to them" [24, p. 790]. At another point, he said: "In order to reach a final settlement of this question, it thus became necessary to devise a method by which to test the ability of a nerve fiber to grow outside the body of the embryo, where it would be independent of protoplasmic bridges" [25, p. 402].

The decisive step had been taken in 1907. Pieces of neural tube of frog embryos, prior to nerve outgrowth, were grown in a hanging drop of frog's lymph. The outgrowth of individual fibers and their growth cones was observed under the microscope; the rate of growth was determined, and the important fact was established that nerves require a solid substrate for extension. Thus the plasmodesms, or, for that matter, any microscopic or submicroscopic materials in the embryo, are assigned their proper role: they serve for guidance but do not contribute materially to the formation of the nerve fiber.

It is clear that the design of the tissue culture method was the logical final step on the long road toward the solution of the problem of the origin of the axon. The immediate purpose was the crucial test of the plasmodesm theory. But, as the title of the detailed report of the tissue culture experiments in 1910 [24] indicates, the emphasis is shifted immediately from the critical to a positive aspect. The phenomenon of "the outgrowth of the nerve fiber as a mode of protoplasmic movement" is placed in the center of the scene. This, we remember, was a key element in Cajal's appraisal of the growth cone. Harrison states "the primary factor, protoplasmic movement, must be regarded as definitely established and it will have to form the basis of any adequate theory of nerve development. The chief claim to progress that the present work has is that it has taken this factor out of the realm of inference and placed it upon the secure foundation of direct observation." And he goes a step further and fits this discovery in a broader frame of reference: "the first manifestations of activity observable in the differentiating nerve cell are of the same fundamental nature as those found not only in other embryonic cells but also in the protoplasm of the widest variety of organisms" [24, p. 840]. Thus, the tradition of Cajal's dynamic view of the growth cone was continued and it became a reality. The later discovery of axoplasmic transport [12] follows the same tradition.

It may seem strange that while the tissue culture method opened up a new field of knowledge and became an indispensable tool in a very broad range of biological endeavors, Harrison himself never made use of it again. The answer suggests itself to those who knew Harrison. The method was designed by him to solve a specific problem—which it did. The time was not ripe for analysis of protoplasmic movement in depth. He became intrigued by other fundamental problems and turned to their solution. His primary concerns were the basic theoretical issues in embryology and *not* the exploitation of what he called a technique. He made decisive contributions to the analysis of a key phenomenon in animal development, the "morphogenetic field" [26] and, as was stated earlier, he solved one of the most difficult problems in embryology, the origin of bilateral symmetry, which is a basic morphological attribute of vertebrates [27]. This led him to the consideration of the polarization of the three main axes, rostro-caudal, dorso-ventral, medio-lateral, in terms of molecular repeat patterns of protein molecules. He actually went to Leeds, to the laboratory of the great biophysicist V. T. Astbury, and they published jointly a paper on X-ray diffraction pictures of embryonic materials [28], a premature step in the direction of molecular embryology. The fact that this enterprise was doomed to failure at that time is less important than the insight it gives in the train of thought of a truly great scientist who was far ahead of his time. His later achievements fully justify the abandonment of his gifted brainchild that was born in 1907 and is still very much alive and thriving.

REFERENCES

1. S. RAMÓN Y CAJAL. Recollections of my life. Trans. E. HORNE-CRAIGIE. Philadelphia: American Philosophical Society, 1937.
2. S. RAMÓN Y CAJAL. Anat. Anz., 5:609, 1890.
3. V. HENSEN. 1903. Die Entwicklungsmechanik der Nervenbahnen im Embryo der Säugetiere. Kiel and Leipzig: Lipsius & Tischer, 1903.

4. HELD. Die Entwicklung des Nervengewebes bei den Wirbeltieren. Leipzig: Barth, 1909.

5. W. HIS. Abhandl. Kgl. Sächs. Gesellsch. D. Wiss., **13**:479, 1886.

6. W. HIS. Arch. Anat. Entwicklungsgesch., **10**:249, 1889.

7. S. RAMÓN Y CAJAL. Neuron theory or reticular theory? Trans. M. U. PURKISS and C. A. Fox. Madrid: Instituto "Ramòn y Cajal," 1954.

8. S. RAMÓN Y CAJAL. Studies on vertebrate neurogenesis. Trans. L. GUTH. 1929. Reprint. Springfield, Ill.: Thomas, 1960.

9. S. RAMÓN Y CAJAL. Histologie du système nerveux de l'homme et des vertébrés. Madrid: Instituto "Ramòn y Cajal," 1909.

10. C. SHERRINGTON. *In:* D. F. CANON. Explorer of the human brain: the life of S. Ramón y Cajal, p. xiii. New York: Schuman, 1949.

11. S. RAMÓN Y CAJAL. The structure of the retina. Trans. THORPE and GLICK. Springfield, Ill.: Thomas, 1972.

12. P. WEISS and H. B. HISCOE. J. Exp. Zool., **107**:315, 1948.

13. J. FORSSMAN. Beitr. Z. Pathol. Anat. Allg. Pathol., **24**:56, 1898; **27**:407, 1900.

14. J. F. TELLO. Gegenwärtige Anschauungen über den Neurotropismus. Vorträge und Aufsätze über Entwicklungsmechanik der Organismen 33. Ed. W. ROUX. Berlin: Springer Verlag, 1923.

15. S. RAMÓN Y CAJAL. Degeneration and regeneration of the nervous system. Trans. R. MAY. 1928. Reprint. New York: Hafner, 1968.

16. P. WEISS. *In:* M. A. CORNER and E. F. SCHWAB (eds.). Brain research, p. 11. New York: Elsevier, 1976.

17. R. LEVI-MONTALCINI. *in:* CORNER and SCHWAB (eds.), [16], p. 235.

18. M. G. M. CHEN, J. S. CHEN, and R. LEVI-MONTALCINI. Arch. Ital. Biol., **116**:53, 1978.

19. NOBEL COMMITTEE (eds.). Nobel, the man and his prizes. New York: Elsevier, 1962.

20. R. G. HARRISON. Arch. Mikr. Anat., **63**:65, 1903.

21. R. G. HARRISON. J. Comp. Neurol., **37**:123, 1924.

22. H. BRAUS. Anat. Anz., **25**:433, 1905.

23. R. G. HARRISON. J. Exp. Zool., **4**:239, 1907.

24. R. G. HARRISON. J. Exp. Zool., **9**:787, 1910.

25. R. G. HARRISON. Anat. Rec., **2**:385, 1908.

26. R. G. HARRISON. J. Exp. Zool., **25**:413, 1918.

27. R. G. HARRISON. J. Exp. Zool., **32**:1, 1921.

28. R. G. HARRISON, W. T. ASTBURY, and K. M. RUDALL. J. Exp. Zool., **85**:339, 1940.

Reprinted from Trends in NeuroSciences–*July 1981*, Vol. 4, No 7, pp 151–5

Perspectives

Historical-landmarks in neurogenesis

Viktor Hamburger

The nervous system has the distinction of being the first organ primordium formed in the early embryo. After the fertilized egg has undergone its initial development i.e. cleavage and gastrulation (the latter resulting in the formation of the three germ layers: ecto-, meso-, and endo-derm), a pear-shaped ectodermal thickening appears on the dorsal side. This is the neural or medullary plate, the *anlage* of the nervous system. Its margins fold up, and the neural folds converge in the midline to form the hollow neural tube. From the beginning, its anterior part, the future brain, is broader than the posterior part, the spinal cord. The origin of the nervous system was described first by Carl Ernst von Baer, the founder of modern embryology, in his classical work *Development History of Animals* (1828) with the significant subtitle 'Observation and Reflection'. Considering that he had only a hand lens and sharpened needles at his disposal—his work antedates the microtome and serial sections—his accurate description of the formation of most organs of Vertebrates was a monumental achievement.

Spemann and the 'organiser' experiment

Toward the end of the century, when descriptive embryology had achieved its major goals, a new branch, analytical or experimental embryology (*Entwicklungsmechanik*), was inaugurated by W. Roux, H. Driesch and others. They addressed fundamental questions concerning the factors which determine organ differentiation. Embryonic 'induction' was recognized as a major mechanism and one of the first instances (the induction of the crystalline lens by the underlying optic vesicle) was discovered by H. Spemann. In a second classical experiment, Spemann demonstrated that the formation of the neural plate is induced by the subjacent mesoderm mantle—the precursor of the notochord, somites, kidneys and other structures. This was the famous 'organizer' experiment—it was published by H. Spemann and H. Mangold in 1924[13]. To understand its significance, a brief account of gastrulation in amphibians is necessary. (See also Fig. 1.) During this stage of embryogenesis the ventral half of the hollow, spherical gastrula invaginates around the so-called blastopore. Its upper lip first moves forward and inward to form the anterior mesoderm underlying the brain. The parts which subsequently invaginate around an extending sickle-shaped blastopore form the mesoderm underlying the trunk and tail region, and parts invaginating around the ventral blastoporal lip form the endoderm, the precursor of the intestine.

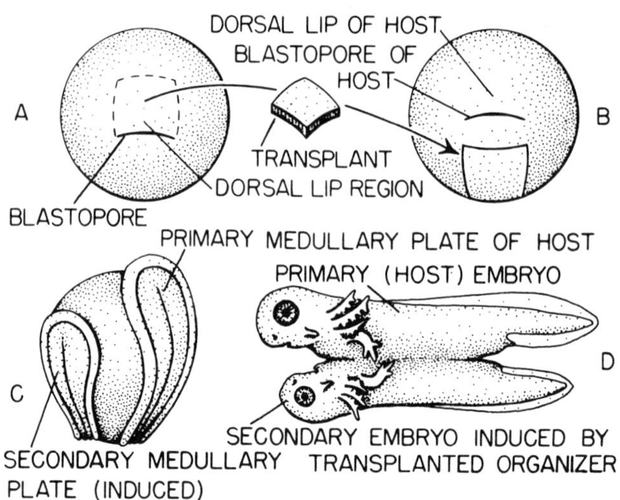

Figure 1

The organizer experiment consisted of the transplantation of the upper blastoporal lip of a salamander embryo to a relatively indifferent position in another gastrula. To distinguish donor from host tissue, piecs from a species with unpigmented eggs were transplanted to a pigmented host. The transplant invaginated in the new position, and on the following day, a pigmented secondary neural tube was found, with an elongated piece of unpigmented tissue underlying it. Obviously, the secondary neural plate had been induced by the subjacent transplant. A day or two later, the secondary structures had formed a more or less complete secondary embryo, a spectacular effect which earned the upper blastoporal lip the name of 'organizer'.

I had the good fortune of witnessing this momentous event as a graduate student in Spemann's laboratory. H. Holtfreter, who contributed more than anybody else to the further analysis of the organizer, Hilde Mangold (née Proescholdt), the co-discoverer, and I worked on our Ph.D. theses in the same graduate laboratory at Freiburg University. I remember our excitment, when the first secondary embryo appeared on the scene. Spemann received the Nobel prize 10 years later. Unfortunately, Hilde Mangold died tragically in an accident a few months after the publication of the organizer paper.

Wilhelm His and S. Ramón y Cajal

It is a long way from the undifferentiated neural tube to the fully differentiated brain. The foundations for our understanding of the developmental processes involved were laid by the German anatomist W. His (1831–1904) and the Spanish neurologist, S. Ramón y Cajal (1852–1934). While Cajal is widely known, few people know of His, yet he was one of the most farsighted anatomists and embryologists of his time. For instance, long before W. Roux founded experimental embryology, His was sure that the time was ripe for raising questions concerning the mechanical forces underlying developmental

H. Spemann.

S. Ramón y Cajal.

processes—for instance, the folding of plates in the development of the nervous system or intestine. He worked with models but did not join the experimental embryologists.

His and Cajal share the credit for laying the groundwork for both modern neurology and neuroembryology. First, they established the 'neuron' as the structural unit of the nervous system, and second, they discovered the mode of origin of the axon as an outgrowth of the neuroblast. These discoveries were made independently between 1884 and 1890. The full scope of their achievements can be assessed by turning to the contemporary authoritative views on these topics. The nervous system was held to be a network of fibers with cell bodies at the intersections and the dendrites as trophic elements. The fatal flaw of this reticular theory was stated by Cajal in his colorful way: 'The axon branches are lost in that unfathomable physiological sea into which pour the streams arriving from the sense organs, and from which the motor or centrifugal conductors were supposed to spring like rivers originating in mountain lakes. This was admirably convenient, since it did away with all need for the analytical effort to determine the course of the nerve impulse through the gray matter. The reticular hypothesis, by dint of explaining everything explains absolutely nothing and hinders and makes superfluous future inquiries regarding the intimate organization of the centers' (Ref. 11, p. 336). The crucial discovery that axons are not continuous but contiguous and the discovery of what we now call the 'synapse', was made in 1888 in the cerebellum of birds, where the free endings of basket cell terminals of mossy and climbing fibers on the Purkinje cells were demonstrated in Golgi preparations. A few years later, the complex structure of the entire cerebellum had been elucidated. Ironically, these discoveries were made possible by the silver-impregnation method developed by the Italian anatomist Golgi who himself remained a stout adherent of the reticular theory. In fact, in 1906, when both Cajal and Golgi received the Nobel prize, Golgi used the occasion to attack Cajal and the neuron theory once more, in a not very tactful way.

W. His.

Unknown to Cajal, His had formulated the neuron theory and the axon outgrowth theory a few years earlier in Leipzig, largely on the basis of embryological studies. He stated: 'The axon emerges as an outgrowth from a single cell which is its genetic, trophic and functional center' (Ref. 6, p. 513). In 1890, Cajal independently observed the outgrowth of axons from neuroblasts in chick embryos, and he discovered in his silver impregnations the 'growth cone' which was not visible with ordinary stains. The outgrowth theory was diametrically opposed to the two then current theories of axon origin; the theory of Schwann considered them as products of the Schwann cells which became connected secondarily with the CNS and ganglia. According to Hensen and Held, nerve fibers are formed by ever-present protoplasmic bridges between cell chains, or 'plasmodesms'. In Held's version a concession to the outgrowth theory was made, in that the axons were considered to be the joint product of plasmodesms and neuron-produced fibrils entering the plasmodesms. The controversy went on for decades, and in one of his last publications in the early 1930s, Cajal still found it necessary to marshal all evidence for the neuron theory and the axon outgrowth theory.

Within two decades Cajal accomplished the unbelievable feat of unravelling the cytoarchitecture and the major pathways of the brain and spinal cord of higher Vertebrates and Man, laid down in two big volumes which appeard in 1909[10]. At the same time he vigorously pursued his neuroembryological work, which took a high place in his priorities[12]. In fact, he says at one point that of all his discoveries the growth cone was the most cherished. Why was this? The particular genius of Cajal, like that of C.E. von Baer, was the combination of very powerful observation and conceptual insight— observation and reflection. Cajal's imagination translated the static microscope picture of a growth cone into a vivid dynamic process. The growth cone is described as 'a sort of club or battering ram, endowed with exquisite chemical sensitivity, with rapid ameboid movements, and with impulsive force thanks to which it is able to proceed forward and overcome obstacles met in its way, forcing cellular interstices, until it arrives at its destination' (Ref. 10, p. 599). Implicit in this visionary conception is the problem of 'directional' axonal outgrowth, and it is only a short step from there to a concrete theory of axon guidance. The idea of 'chemotropism' or 'neurotropism', terms which were then used for chemical guidance, took shape in his mind shortly after the discovery of the growth cone.

Harrison and the origin of axons

It was no acident that the controversy over the origin of the axon was finally settled not by a histologist but by an experimental embryologist. In fact, around the turn of the century, a stalemate was inevitable. In 1908, after a voyage to Germany, Cajal reported that he had seen the excellent silver preparations of the German anatomist H. Held and found to his surprise that they were nearly identical with his own. Yet both men stuck to their diametrically opposed interpretations of the same microscope sections.

Harrison had favored the nerve outgrowth theory since his early descriptive work on the development of the nervous system of the salmon in 1901. But even before that time he had been drawn to experimental embryology. He had been intrigued by an experiment of the German anatomist Gustav Born, who had cut early frog embryos apart in order to study regeneration and found that parts which he had left in a dish would fuse together spontaneously. He used this chance observation to produce the most unlikely combinations and monstrosities but died shortly thereafter before he could exploit his discovery. Harrison realized the potentials of the grafting method and used it in one of the first experiments designed to solve neuroembryological problems, namely the problem of the origin of the lateral line organs of aquatic Vertebrates. These are sensors for detection of water perturbation spaced in three rows on the head and in two rows along the trunk and tail. Those of the head are innervated by a special branch of the facial nerve and those of the trunk and tail by a component of the vagus nerve. Harrison had the ingenious idea of fusing the head of a pigmented frog embryo, *Rana palustris*, with

R. G. Harrison.

R. Sperry.

the posterior end of a light yellowish *Rana sylvatica* embryo. He observed, in the living embryo, the sequential deposition of dark spots along the light trunk and tail and traced the origin of these sense organs to a post-otic placode or epidermal thickening. Thus, he discovered a unique mode of formation of a neural structure by extensive directional cell migration.

In another experiment he used the extirpation method to test the hypothesis of the origin of axons from a chain of Schwann cells. He removed the dorsal part of the neural tube including the neural crest and found that normal motor axons would emerge from the intact ventral half of the spinal cord. They were devoid of Schwann cells. Thus, the hypothesis of the origin of axons from Schwann cells was refuted. At the same time the controversial question of whether Schwann cells are derived from mesenchyme or from neural crest was settled in favor of the latter. By applying the concepts and methods of experimental embryology to neurogenesis, Harrison became the founder of experimental neuroembryology.

Harrison realized that the controversy over the origin of the axon could not be settled by histology but only by experiment, and the Schwann cell exclusion experiment was the first step. But to test the plasmodesm theory, a much more radical procedure was necessary. In his own words: 'The really crucial experiment remained to be performed, and that was to test the power of the nerve centers to form nerve fibers within some foreign medium, which could not by any possibility be suspected of contributing organized protoplasm to them' (Ref. 5, p. 790). Thus, the famous tissue culture experiment was conceived, in which neural tube of a frog embryo was grown in frog's lymph in a depression slide, and the outgrowth of axons from neuroblasts was observed under the microscope. A brief report appeared in 1907[3] and a detailed paper in 1910[5]. Hence, the tissue culture method, which is now an indispensable tool in all biology originated, so-to-speak as a spin-off from the controversy over the origin of axons. Harrison also saw the growth cones and the microscopic filaments which are spun out in all directions and partly withdrawn, and he made the important discovery that the fibers require a solid substrate on which to grow.

Harrison was an occasional summer visitor at the laboratory of his friend Spemann in Freiburg, when he returned from a vacation in Switzerland. I had the good fortune of getting advice from him in my early experimental work on the role of innervation in amphibian limb development, a topic which he had been interested in earlier. Among the leading experimentalists of his time he was one of the most clear-minded and perceptive, and at the same time unassuming and genteel, and also one of the wisest scientists, who was fully aware of the limitations imposed on scientific inquiry by the nature of its methodology and by human nature.

Harrison made a major contribution to the problem of nerve pattern formation. For the second time he turned an experiment initiated by a German anatomist into a gold mine. The anatomist H. Braus was the first to use limb bud transplantations in the context of neuroembryology. The results of his experiments on frog embryos were interpreted by him as evidence in favor of Hensen's plasmodesm theory. Harrison repeated these experiments in 1906, and proved convincingly that the nerves did not originate *in situ* but as outgrowths from the spinal cord. In limb transplants to the flank of frog embryos he observed that the foreign nerves formed a nearly normal nerve pattern, and he concluded 'that the structures contained within the limib must have a very important directive action upon the developing nerve fibers, in that they determine their mode of distribution' (Ref. 4, p. 276). In other words, nerve pattern formation results from an interaction of actively growing and searching axonal growth cones and the substrate on which they grow. Needless to say that this axiom as well as the experimental design of limb transplantation have served us well to this day in the analysis of several basic problems of neurogenesis.

Sperry and the chemoaffinity hypothesis

The discovery of normal nerve patterns in transplanted limbs encouraged the notion that nerves can reach their targets by following mechanical preneural pathways which seem to be relatively nonspecific since they can be retraced by thoracic or even cranial nerves. This idea goes back to His, who had spoken of 'paths of least resistance'; later, this same notion was referred to as 'stereotropism'. During the 1930s, Weiss showed, in very impressive tissue culture experiments, that nerve fibers can be oriented by creating channels in the substrate over which they grow. He coined the term 'contact guidance', and it seemed plausible to extrapolate from the *in vitro* behavior of nerves to the *in vivo* situation. Moreover this notion fitted in with other contemporary views, namely, that synaptic terminations are indiscriminate, that any motor nerve can activate any muscle, and that the integration of behavior is taken care of either by functional validation of reflex and other pathways, or by impulse specificity or other physiological mechanisms. Cajal's idea of chemo- or neuro-tropism was forgotten or disparaged.

It is against this historical background that the impact of the 'chemoaffinity' hypothesis has to be assessed. It was proposed by Roger Sperry, a student of Weiss, around 1940,

Reversing the major trend of current thinking, he postulated that synaptic conections are highly selective and that the precision of neuronal circuitry can be accounted for only by chemical affinities between nerve terminals and their target cells. The classical regeneration experiments in amphibians on which this hypothesis was based are well known and I mention only one of many[14]. In an adult frog, the optic nerve was cut and the eye rotated 180°. After functional regeneration and enucleation of the other eye, a moving object was presented up and backward in the visual field. The animal lashed its tongue downward and forward, and the maladaptive behavior was never corrected. The aberrant behavior followed strict rules: If only one axis of the eye was inverted, as for instance, the dorso-ventral axis, then an object presented above was located below; after anterior—posterior eye rotation, an object presented in back was located in front in the visual field. The conclusion was drawn that the original pattern of retina fiber projection onto the optic tectum is restored precisely.

The original version of the chemoaffinity hypothesis has been elaborated; and the hypothesis now has three components: specific chemical identity of like neurons belonging to the same strain, such as a motor pool; selective matching affinity of growth cones with chemical cues in their local micro-environment which they encounter along their pathway; and selective affinity of nerve terminals and their targets in synapse formation.

The chemoaffinity hypothesis owes much of its acceptance to the wise choice of the retinotectal system as its first experimental testing ground. It is an ideal model because we are dealing with the projection of a quasi-two-dimensional retina field on a quasi-two-dimensional tectal field. Thus, it is possible to refer the relative position of a neuron group in one field to that of the target neurons in the other field along the two axes of a co-ordinate system, and Sperry introduced early the notion of gradient fields. Second, the choice of regenerating systems in preference to embryonic systems was fortunate, because at that time techniques like retrograde and anterograde horseradish peroxidase transport which make tests in the embryo possible were not available. Third, the further analysis of the specificity of the retinotectal projection system owes very much to the introduction of electrophysiological methods by M. Gaze and his collaborators. But the greatest achievement was Sperry's bold and ingenious extrapolation from relatively simple behavioral performance to ehemoaffinity of the structural elements.

I shall not deal with later validations of the chemoaffinity hypothesis by electrophysiological, histological and other methods, but will return briefly to limb nerve patterns. The basic new idea is the assumption that Harrison's preneural pathways become chemically specified by agents produced by the target structures and that the growth cones recognize these cues in their micro-environment. Recent experimental findings of M. Hollyday[7] and L. Landmesser[8] and their associates on the establishment of motor axon connections with limb muscles in chick embryos can hardly be explained in any other way. Fibers from a specific motor pool find their proper target muscle even if limbs or segments of the lumbar spinal cord are rotated or limb buds displaced in other ways. It is clear that this precision in 'homing in' on a specific target, often bypassing more proximate targets, or requiring detours, cannot be explained by mechanical guidance alone, and chemoaffinity seems to be the most likely answer.

Lillie and trophic interrelations

Perhaps the discovery of the nerve growth factor (NGF) can be considered as a significant event in more recent times. I shall not discuss this topic, but it is of some interest to trace its historical origins. This gives me an opportunity to pay tribute to Frank R. Lillie, longtime Chairman of the Department of Zoology at the University of Chicago, and a leading embryologist and influential statesman. As the Director of the Marine Biological Station in Woods Hole, Massachusetts, he made this institution world famous and the credit for establishing the chick embryo in research and teaching goes to him. His book *Development of the Chick Embryo*, which appeared first in 1908 and was brought up-to-date by H. Hamilton, is still a classic.

It was in his laboratory that the chapter on 'trophic relations' between nerve centers and their target structures was initiated. At his suggestion, M. Shorey did limb bud extirpations in the chick embryo, to study the reprecussions of the removal of target structures on the nerve centers which innervate them. She discovered the hypoplasia in both the motor column and the sensory ganglia. Her work was not continued at that time but as fate would have it, it was taken up by me 25 years later, in the same laboratory. In 1932, I received a fellowship from the Rockefeller Foundation to spend a year in Lillie's laboratory with the intent of applying the microsurgical technique of Spemann to the chick embryo. At that time, the only experimental technique applicable to the chick was the transplantation onto the chorioallantoic membrane. Lillie and I agreed that the repetition of the Shorey experiment would provide a suitable testing ground. I confirmed and extended her results and found hyperplasia in spinal ganglia following limb transplantation. During the war, R. Levi-Montalcini repeated the limb extirpation experiment, but her interpretation of the results was different from mine. After the war, she joined me in St. Louis and we studied the effects of limb extirpation and transplantation on spinal ganglia in detail[2]. Her contention that the hypoplasia is due to the degeneration of differentiated neurons was confirmed. We were much impressed by the hyperplasia which indicated a growth potential that is not realized in the normal course of development, and continued the analysis of this phenomenon by repeating an original experiment designed by my former student, E. Bueker. With a similar idea in mind, he had implanted mouse sarcoma into the coelom of 3-day embryos and observed the invasion of sensory fibers into the tumor and the hyperplasia of the spinal ganglia that were involved. The story of the discovery of the NGF which began with this experiment has been told repeatedly[1,9]. Looking back, while the discoveries of Ramón y Cajal and W. His are the cornerstones of neuro-embryology, the other chapters which have been outlined are still open-ended.

Reading list

1 Hamburger, V. (1980) *Annu. Rev. Neurosci.* 3, 269–278
2 Hamburger, V. and Levi-Montalcini, R. (1949) *J. Exp. Zool.* 111, 457–502.
3 Harrison, R. G. (1907) *Anat. Rec.* 1, 116–118

4 Harrison, R. G. (1907) *J. Exp. Zool.* 4, 239–281

5 Harrison, R. G. (1910) *J. Exp. Zool.* 9, 787–846

6 His, W. (1886) *Abhandl. d. Math.—Phys. Klasse der Koenigl. Saechs. Ges. d. Wiss.* Vol. 13

7 Hollyday, M. (1980) in *Current Topics in Developmental Biology* (Moscona, A. and Monroy, A. eds), Vol. 15, pp. 181–215, Academic Press, New York and San Francisco

8 Landmesser, L. (1980) *Annu. Rev. Neurosci.* 3, 279–302

9 Levi-Montalcini, R. (1975) in *Neurosciences: Paths of Discovery* (Worden, F. G., Swazey, J. P. and Adelman, G. eds), pp. 245–265, M.I.T. Press, Cambridge, Massachusetts and London

10 Ramón y Cajal, S. (1909) *Histologie du Système Nerveux de l'Homme et des Vertébrés* (reprinted in 1952 by Istituto Ramón y Cajal, Madrid)

11 Ramón y Cajal, S. (1937) *Recollections of My Life* (Horne Craigie, E., transl.) M.I.T. Press, Cambridge, Massachusetts and London

12 Ramón y Cajal, S. (1960) *Studies in Vertebrate Neurogenesis* (Guth, L., transl.) C. Thomas, Springfield

13 Spemann, H. (1938) *Embryonic Development and Induction*, Yale University Press, New Haven and London

14 Sperry, R. (1951) in *Tenth Symposium on Development and Growth*, pp. 63–78, Growth, Worchester, Massachusetts

---

Viktor Hamburger is Professor Emeritus of Biology at the Department of Biology, Washington University, Campus Box 1137, St. Louis, MO 63130, U.S.A.

The Journal of Neuroscience, October 1988, 8(10): 3535–3540

Feature Article

Readers will notice a new addition to this issue. The following article by Viktor Hamburger is the first of a series of general interest articles that the Editors plan to include in the Journal pages. Because of the backlog of primary research reports (see Society for Neuroscience Newsletter, Vol. 19, No. 2 (March/April), 1988, pp. 6–7), feature articles will appear only occasionally at first, As the backlog and the resulting publication delays are diminished, however, we plan to make such features a regular part of the Journal. Our intention is to present brief essays on subjects of broad importance to neuroscientists, including historical accounts, tributes to prominent figures, reports of important advances, and other noteworthy issues in our field.

The Editors welcome the response of subscribers to the introduction of this feature section. Further, we are happy to receive specific suggestions from subscribers for future articles.

Dale Purves, Editor-in-Chief

Ontogeny of Neuroembryology

V. Hamburger

E.V. Mallinckrodt Distinguished Service Professor Emeritus,
Washington University, St. Louis, Missouri 63130

This essay commemorates the 100th anniversary of the birth of neuroembryology. One cannot, of course, ascribe the beginning of a branch of science to a single year, but the years between 1885 and 1890 saw major publications by the German anatomist Wilhelm His (1831–1904) and the Spanish histologist S. Ramón y Cajal (1852–1934), both of whom laid the foundation to our present understanding of the structure and embryonic origin of the nervous system.

Modern developmental neurobiology emerged from the convergence of two traditions that had their roots in quite different and separate fields of inquiry, and with different conceptual and methological frames of reference. The one, *the histogenetic tradition*, was descriptive and became sophisticated through refined technology. The other, *experimental neuroembryology*, was causal-analytical and experimental, and was originally a modest side branch of general experimental embryology.

This essay is based on a lecture delivered in November 1987 at the 13th Annual Meeting of the Society for Neuroscience.

Correspondence should be addressed to Professor V. Hamburger, Washington University, Department of Biology, St. Louis, MO 63130.

The Histogenetic Tradition

The neurohistologists of the 1860s and 1870s, among them O. Deiters, had already worked out a clear picture of the neuron. It had been obtained by teasing out a motor neuron from the adult spinal cord and clearly shows perikaryon, axon, and dendrites (Fig. 1). Nevertheless, Deiters and his contemporaries subscribed to the *network* or *reticular theory* of the structure of the nervous system. The impulse-conducting elements were pictured not as autonomous units but as part of a network of nerve fibers in which the cell bodies and dendrites were of subordinate importance. Many thought of them as nutritive elements. An earlier version of the reticular theory dated back to Theodor Schwann, one of the founders of the cell theory. He had postulated that the cells bearing his name form cell chains whose protoplasmic connections are transformed into nerve fibers. The more refined versions of the reticular theory of the 1870s and 1880s, associated with the names of Golgi, Hensen, Gerlach, had one important point in common: nerve fibers were supposed to be the product of a preneural protoplasmic network, referred to as *plasmodesms*, which, from the outset, connects the central nervous system with its targets. According to some, the plasmodesms originated by incomplete separation of postmitotic cells; according to others, the plasmodesms were formed secondarily as bridges between cells. The major problem of how the plasmodesms were transformed into nerve fibers remained unresolved.

It is against this background that Wilhelm His's conceptual breakthrough to the neuron theory has to be judged. He was a native Swiss who had become Professor of Anatomy and Physiology at the university of his home town, Basel, at the young age of 26. (At that time the two disciplines were still combined at most universities.) His's title was somewhat deceptive; his institute consisted of two rooms: one his office and laboratory, the other a classroom for his 8 to 12 students that also housed the anatomical collection. In time the department grew rapidly. Later, His, who had a remarkably broad range of interests, became the Director of the Anatomy Department of the prestigious University of Leipzig and one of the leading figures of his generation. Only a very independent mind of his stature could accomplish a complete break with the tradition.

In the early 1880s he began to concentrate on the development of the nervous system. When he looked at the spinal cords of a series of early human embryos he recognized at once that they are not composed of a syncytium but of a layer of individual epithelial cells. He described correctly the neural tube, the precursor of the central nervous system, as a flat epithelium, which became columnar and then loosened up to form what he appropriately called the spongy layer, the *spongiosa*. He identified it as the precursor of the ependymal layer, that is, a glial structure. The mitotic cells at the inner lining of the tube, which he called "*germinal cells*," had been observed before him, but he was the first to recognize that they were the precursors of nerve cells. He observed that after their terminal mitosis they became pear-shaped and formed a protoplasmic outgrowth at their distal ends which he identified as the incipient axon. These young *neuroblasts*, as he called them, supposedly migrated across the spongiosa and assembled at the outer margin of the neural tube where they formed what he called the *mantle layer*. In this particular case, the neuroblasts were motor neurons. The tips of their axons pierced

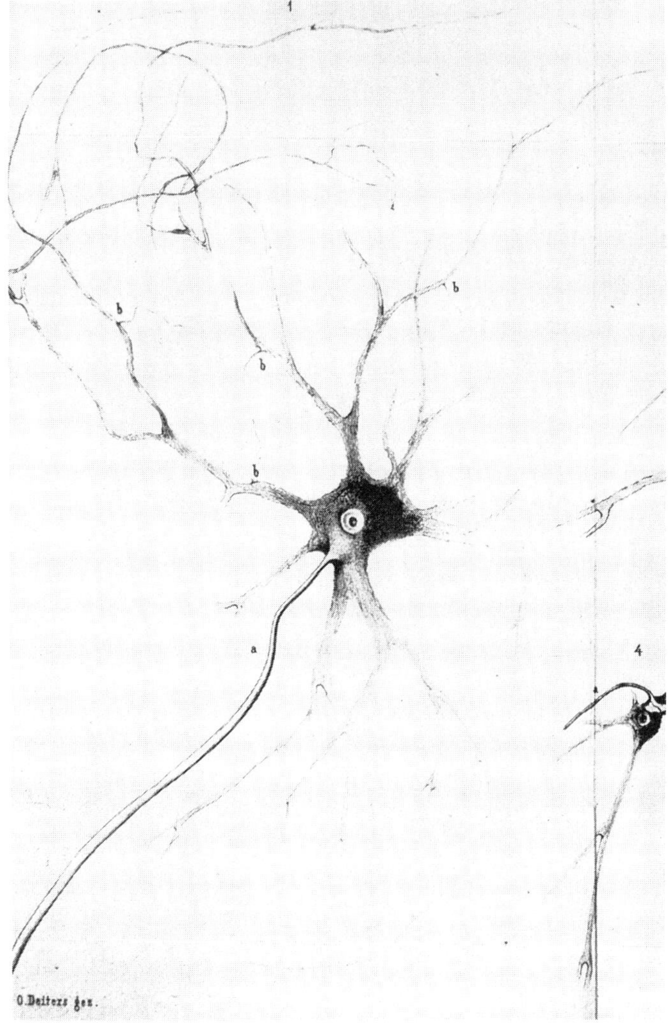

Figure 1. Drawing of a neuron by O. Deiters (1865).

through the external limiting membrane, formed a bundle, the motor nerve, and grew toward their target, the somites. They were perhaps supported and guided by the plasmodesms, but were neither nourished nor transformed by them. These observations formed the foundation of the concept of the autonomous neuron on which the *neuron theory* is based. At the end of his classical monograph of 1886 he generalized his findings: "I consider as a definitive principle the theorem that every nerve fiber originates as the outgrowth of a single cell. The latter is its genetic, nutritive and functional center. All other connections are either indirect or they originated secondarily" (1886, p. 513). By genetic he meant embryonic. Among His's other discoveries: the neural crest and the derivation of spinal and sympathetic ganglia from this structure, and the observation that dendrites (which were named by him) always differentiate later than axons.

Implicit in His's theorem is the idea that there is contiguity but not continuity between nerve cells and other nerve cells or their targets. However, his histological methods did not permit him to demonstrate cell-to-cell contact as a fact. It is at this point that Ramón y Cajal enters the stage. He was a genius in science and an extraordinary human being. One can glean from his delightful autobiography (1937) the magnetism of his personality —a personality that was certainly more colorful than that of Herr Geheimrat His. But his most outstanding trait was his iron will, which he had inherited from his father, and his singleness of purpose. Cajal, who was 20 years younger than His, was born in a small, desolate mountain village in northern Spain where he received very little formal education. Although this lack of education was largely his own fault, he managed to enter professional life. Up to 1887 he had done rather undistinguished work in histology at the University of Valencia, but in 1887 he moved to Barcelona and, according to his own testimony, this year was of decisive importance. On the occasion of a visit to Madrid, a colleague showed him microscope slides of nerve tissue treated with Golgi's silver impregnation method. The method had been available since 1873, and Golgi had made some important discoveries using it, but it was otherwise neglected. The incredible clarity with which the nerve cells and fibers appeared against a faint background made a profound impression on Cajal. It struck him immediately that here was the key to the unraveling of the structure of the nervous system, and the realization of this idea filled the rest of his long life.

His efforts were at first disappointing. The complexity of the central nervous system seemed to be an insurmountable barrier, despite the selectivity of the Golgi method. He then had the ingenious idea to turn to the embryo. "Since the full-grown forest turns out to be impenetrable, why not revert to the study of the young wood in the nursery stage? This was the very simple idea which inspired my repeated trials of the silver method on embryos of birds and mammals. If it is applied before the appearance of the myelin sheaths upon the axons, the nerve cells stand out complete ... the terminal ramifications of the axis cylinder are depicted with the utmost clearness and perfectly free. The interneuronal articulations appear simple, gradually acquiring intricacy and extension: in sum, the fundamental plan of the histological composition of the gray matter rises before our eyes with admirable clarity and precision" (1937, pp. 324–325).

Cajal looked first at the embryonic cerebellum. He obtained the first convincing evidence for contact—as against fusion—when he observed that the terminal ramifications of the axons of the stellate cells in the molecular layer form basket-like endings around the bodies of the Purkinje cells. In the same preparations he observed the behavior of the climbing fibers. "When they reach the level of the first branches of the dendritic trunks of the Purkinje cells, they break up into twining parallel networks which ascend along the protoplasmic branches, to the contours of which they apply themselves like ivy or lianas to the trunks of trees" (1937, p. 332). It is still incomprehensible to me how he managed within a year or two to unravel the development and structure of the cerebellum in its finest details.

But are we really taking about neuroembryology? Was not the embryo merely recruited by the histologist to provide evidence for the neuron theory, as it had been recruited by the Darwinists to provide evidence for evolution? Does Cajal deserve admission to the guild of neuroembryologists? I would say not, if we consider only his early work on the cerebellum. But the embryonic nervous system captivated his interest in its own right, and he began to study the embryonic spinal cord. At that time, in his seclusion in Barcelona, he was cut off from the mainstream of anatomical research and was not aware of His's investigations. As a result he rediscovered in chick and mammalian embryos the early history of the neuroblast and the outgrowth of the axon. In 1890. His sent him his publication and Cajal acknowledged later the priority of His.

In 1890 Cajal made what he described as one of his most cherished discoveries: the *growth cone.* "In my sections of the 3-day chick embryo, this ending appeared as a concentration of protoplasm of conical form, endowed with ameboid movements. It could be compared to a living battering ram, soft and flexible, which advances, pushing aside mechanically the obstacles which it finds in its way, until it reaches the area of its peripheral distribution. This curious terminal club I christened the growth cone" (1937, p. 369). To this day, the growth cone has remained one of the major challenges to neuroembryologists. Of the many other embryological discoveries of Cajal I mention only one: the mass migration of embryonic neurons. He observed the details of the differentiation of the granule cells in the cerebellum. He saw how the postmitotic cells on the surface became unipolar, then bipolar, how the 2 processes fused and became T-shaped, as in DRG, how the cells migrated to the depth, across the layer of Purkinje cells, and eventually settled down in the granular layer. In the meantime these cells had acquired dendrites. His had already pointed out in 1890 that "the capacity of embryonic nerve cells to migrate seems to be a principle of decisive importance" (1890, p. 115). Indeed, mass migration of embryonic neurons is widespread and is a phenomenon unique in embryonic development.

A penetrating mind of Cajal's stature could not fail to become aware of the central issue in cell migration and axon outgrowth: Which are the forces that give their movements direction? As early as 1892 he opted for *chemotropism,* that is, attraction at a distance by chemical signals emanating from the target. In this speculation he was far ahead of his time, but while he asked the right question, his answer, as described below, turned out to be incorrect.

The way Cajal looked at the growth cone is as interesting as the discovery itself. In the quotation above he reveals one of his prominent traits, his immensely dynamic interpretation of what he saw under the microscope. He "saw" the ameboid movements of the growth cone and the force that pushes aside obstacles, just as he "saw" the climbing fibers climb. As Sherrington (1949) remarked on the occasion of Cajal's Croonian Lecture in London in 1894:

> A trait very noticeable in him was that in describing what the microscope showed he spoke habitually as though it were a living scene.... The intense anthropomorphism of his descriptions of what the preparations showed was at first starting to accept. He treated the microscope scene as though it were alive and were inhabited by begins which felt and did and hoped and tried even as we do. A nerve-cell by its emergent fibre "groped to find another"! We must, if we would enter adequately into Cajal's thought in this field, suppose his entrance, through his microscope, into a

world populated by tiny beings actuated by motives and strivings and satisfactions not very remotely different from our own. Listening to him I asked myself how far this capacity for anthropomorphizing might not contribute to his success as an investigator. I never met anyone else in whom it was so marked.

I have dealt with the histogenetic tradition very selectively, focusing on the two leading figures and disregarding the contributions of other important investigators, such as Kölliker, Retzius, van Gehuchten, Lenhossek, and Bielschowsky. However, I should like to add one point. Early in this century, the reticular theory made a remarkable comeback. One of the leading revivalists was Hans Held, who became the successor of His in Leipzig. In 1909 he wrote a weighty tome of almost 400 pages, the gist of which can be summed up in one sentence: What emerges from the embryonic neuroblast is not a protoplasmic outgrowth but a bundle of neurofibrils that become nerve fibers by amalgamating with Hensen's plasmodesm network. This was a futile effort to salvage the reticular theory by combining it with the outgrowth theory. Nevertheless, the reticularist ideas still had adherents in the 1940s. When I attended a conference of neuroembryologists in Chicago in 1949, convened by Paul Weiss, the Dutch histologist Jan Boeke treated us to an animated defense of reticularist ideas. The controversy was finally settled by electron microscopists in the 1950s.

The final victory of the neuron theory was based as much on superior technique and observation as on its rationale. From the physiological perspective, the neuron theory made sense, but the reticular theory was seriously flawed. Cajal points out that the network theory" ... takes it for granted that the final axonal branches ... are lost or disappear in the network, in that sort of unfathomable physiological sea into which, on the one hand, were supposed to pour the streams arriving from the sense organs and from which, on the other hand, the motor or centrifugal conductors were supposed to spring like rivers originating in mountain lakes. This was admirably convenient, since it did away with all need for the analytical effort involved in determining in each case the course through the gray matter followed by the nervous impulse.... The reticular hypothesis, by pretending to explain everything easily and simply, explains absolutely nothing"' (1937, pp. 336–337). Like Cajal. His was fully aware that only the neuron theory can account for integrated functional activity which requires specific connections between specific neuronal assemblies.

Experimental Neuroembryology

From the histogenetic tradition, we turn to experimental neuroembryology. It is based on an entirely different tradition, that is, a causal-analytical approach and problem-solving by the analytical experiment. Experimental embryology was conceived and pioneered by the German anatomist Wilhelm Roux in the 1880s, at the same time that His and Cajal started the histogenetic tradition. Roux did his first experiments on frog embryos in 1888. His choice of amphibian embryos was ideally suited for his purpose, so much so that H. Spemann (1869–1941) and R. Harrison (1870–1959), who soon assumed the leadership in the new field, never used any other embryos. Here, then, was another difference which separated the two traditions.

One can consider the organizer experiment of H. Spemann and Hilde Mangold of 1924 as the beginning of experimental neuroembryology because the outcome explains the origin of the neural plate (the precursor of the central and peripheral nervous system) as the result of induction by the organizer. In this experiment, a small piece of the so-called upper lip of the blastopore of a salamander gastrula was transplanted to the flank of another gastrula. The transplant invaginated into the interior and induced in the overlying ectoderm a secondary neural plate and, within a few days, an entire secondary embryo. Embryos of different species differing in the pigmentation of tissues were used, the pigmentation serving as a permanent cell marker. In this way it was established beyond doubt that the secondary neural plate had been induced by the subjacent organizer in tissue that would normally have formed epidermis.

Yet Spemann, like Cajal, can hardly be considered as a neuroembryologist, since his interest in the development of the nervous system ended with the neural tube.

In fact, Ross Harrison was the founder of experimental neuroembryology, although he actually got his start in the histogenetic tradition. He had a Ph.D. from Johns Hopkins University and a German M.D. from the Anatomy Department of the University of Bonn. On the basis of his first investigation of neuron development in the salmon embryo, in 1901, he opted for the axon outgrowth theory as opposed to the reticular theory. While he witnessed the rather acrimonious fight between outgrowth theorists and reticularists, which had flared up again after 1900, he—and apparently he alone—perceived that, in principle, the problem could not be resolved by the histological methods available at that time and that only an analytical experiment could decide the issue. This marked another conceptual breakthrough. He took the bold step of growing embryonic nerve tissue in complete isolation: "The really crucial experiments remained to be performed, and that was to test the power of the nerve centers to form nerve fibers within some foreign medium, which could not by any possibility be suspected of contributing organized protoplasm to them" (1910, p. 790). He succeeded in 1907 in growing pieces of the spinal cord, from early frog embryos, in clotted frog lymph in hanging drop cultures. He was the first to observe axon outgrowth and the formation of growth cones and filaments in the living cell, and he extended these observations over a period of hours and days. He made the important observation that the nerve fibers would not grow out in liquid medium but rather attached to the cover glass, or to fibrin fibers or spider webs which he provided.

One might have expected that this ingenious experiment would have been hailed by the outgrowth theorists and might even have converted some reticularists. Far from it! Nothing can show the gulf between the two traditions better than the cool reception that the tissue culture experiment received in both camps of the histologists. In his last book, which appeared in 1933. Cajal collected once more all the evidence for the neuron theory and against the reticular theory; but he devoted only a few sentences to the tissue culture experiment. From Cajal's vantage point. Harrison had nothing new to say. His opponent, the reticularist Held, in his book of 1909, voiced for the first time a theme which has been repeated ever since: that the behavior of neurons *in virto* does not necessarily reflect their behavior *in vivo*. He insisted that "the histogenetic investigation of the embryo shows more than the experiment of Harrison. It shows that the intraembryonic nervous system is not formed in the manner of an outgrowth from the neuroblast but that a substance which is present already along its future path and which connects different cells and organ primordia is utilized in the formation of the definitive nerve. For this reason, Harrison's experiment cannot decide according to which principle Nature develops a nervous system in the embryo" (1909, p. 261). Fortunately, posterity has treated Harrison's achievements more kindly.

Harrison was aware, of course, of the problem of how growth cones are guided to their targets. It is interesting to consider the difference between the approach of the experimental embryologist to this problem and Cajal's speculation on chemotropism, which was not amenable to an experimental test. Harrison made a major methodological contribution by choosing the limb innervation pattern as the testing ground for the analysis. This paradigm has served us well to this day. He transplanted limb buds of frog embryos to the flank and made two observations: that the innervation is provided by the region to which the limb is transplanted, and that the foreign nerves form a normal limb pattern. Harrison concluded from his experiment: "The structures contained in the limb must have a very important directive action upon the developing nerve fibers in that they determine their mode of branching" (1907, p. 276). Of course, he was aware that such a general statement leaves open the problem of the specificity of nerve connections. "One of the most baffling questions ... is the selectivity of the fibers in establishing their proper terminations—motor neurons with muscle fiber and sensory neurons with the epithelium of the skin or with muscle spindles.... It seems necessary to assume some specific reaction between each kind of end organ and its nerve, and Cajal and Tello have pointed out that this could scarcely be other than of a chemical nature" (1935, p. 184). In the meantime, two other mechanisms for guidance had been suggested by others: *stereotropism*, or mechanical guidance, and *galvanotropism*, or orientation in an electrical field. Eventually all theories proposing an action at a distance were discarded and the view was adopted that the growth cone is guided by signals encoded in the structures with which it is in direct contact.

As early as 1904 Harrison opened up the broad field of *trophic relations* between nerves and their target structures. For a long time pathologists had been aware of muscle atrophy resulting from denervation. Harrison inquired whether the initial differentiation of muscles is dependent on nerve supply. He removed the trunk segment of the spinal cord of frog embryos prior to nerve outgrowth and found that the trunk musculature differentiated normally; it showed fiber formation and cross-striation and responded to electrical stimulation; however atrophy and degeneration began soon thereafter. Harrison did not continue the analysis of trophic relations. For reasons of his own, he turned (around 1910) from neuroembryology to other basic problems of development, but his most prominent student, S. Detwiler, continued the tradition.

By chance I became involved in the problem of the trophic role of innervation in the development of target structures. My Ph.D. thesis was supposed to put to the test the rather improbable claim of a German experimental embryologist that eye extirpation in early frog larvae would create a chain reaction of neural deficiencies from eye, to midbrain, to the spinal cord, and to the motor centers that would then result in neurogenic limb abnormalities. The repetition of the experiment gave ambiguous results and I decided to do the crucial experiment: to create nerveless limbs by removing the limb-innervating segment of the spinal cord before nerve outgrowth (1928). I found that limb development was entirely normal. As in the Harrison experiment, the musculature developed normally but atrophied and degenerated later. The pattern of skeletal elements and even joints had been formed normally in these paralyzed limbs. The fact that limb development in amphibians does not require nerve supply is in strange contrast to the dependence of amphibian limb regeneration on innervation. My result was definitive and did not suggest further experiments. I was then prepared to abandon this field and did actually turn to my other interest in developmental genetics.

In the meantime, Detwiler had encountered by chance the one problem that had escaped Harrison's attention but which eventually became one of the most exciting in neurogenesis: the trophic role of the targets in the differentiation of the nerve centers that innervate them. Harrison had suggested to Detwiler the transplantation of forelimb primordia of salamander embryos to different positions on the flank to find out whether limbs innervated by foreign nerves would be capable of motility. Detwiler found that coordinated movements were performed only if at least one limb nerve was derived from the brachial plexus. When he studied his material, he made the seminal discovery that the brachial ganglia, which were deprived of their target, were hypoplastic whereas the thoracic ganglia, which were overloaded, were hyperplastic. Strangely enough, he did not observe changes in the motor centers. The findings were first reported in 1919 and 1920 and were followed by a series of experiments which, however, did not advance the analysis significantly (see Detwiler, 1936).

A decade later fate brought me back into the fold and I landed in Detwiler's territory. In 1932 I joined the laboratory of Dr. Frank Lillie at the University of Chicago as a Rockefeller Fellow. I was supposed to apply Spemann's microsurgery with glass needles on the chick embryo which had been placed on the map by Lillie's classic book on the development of the chick, first published in 1908. By some intuition he had the idea that the limb might have an influence on the development of the nervous system. In 1909 his student, M. Shorey, had destroyed the wing bud by electrocautery and found that, indeed, wing bud removal resulted in the hypoplasia of both spinal ganglia and motor columns. However, there was no follow-up to the experiment and it was almost forgotten. It was only natural that I should start my explorations by repeating this relatively simple experiment. As it happened, my success within a few months in limb extirpation and transplanation shifted the emphasis from amphibian to chick embryos. Their more highly differentiated nervous system was more favorable for in-depth analysis. I could, therefore, add an important point to the findings of Shorey: I established by semi-quantitative methods that the hypoplasia in the motor column was proportional to muscle loss, and that the hypoplasia in the spinal ganglia was roughly proportional to skin loss. In other words, the different centers responded independently of each other. I interpreted this to mean that each center receives a signal from its own target, and I subsequently suggested in 1934 that "the stimuli going from the peripheral fields to their nerve centers are probably transmitted centripetally by the nerve fibers" (1934, p. 491). Thus, I had an inkling of the retrograde axonal transport of a signal from the target. The transplantation of supernumerary limbs resulted in a distinct hyperplasia of the spinal ganglia and a slight increase in the number of cells in the motor column, but otherwise the transplantation experiments shed no further light on the problem of trophic relations (1939).

I come now to a critical issue: How to explain all these findings in terms of a mechanism by which the targets regulate the differentiation of the centers that innervate them. Detwiler and I had two explanations: either the target regulates the proliferation in the nerve centers of it regulates cell numbers in a more complicated fashion. We proposed a highly speculative recruitment hypothesis that involved pioneer fibers which would explore the target area, and a pool of hypothetical uncommitted cells in the nerve centers. The pioneer fibers would send signals back to the centers indicating the size of the target area, and the appropriate number of cells would then be recruited from the pool of undifferentiated cells. Both explanations had the advantage that they could explain hypo- and hyperplasia by the same mechanism. A disadvantage was that they were wrong.

The correct answer was provided by Rita Levi-Montalcini. She had repeated my limb extirpation experiment in the 1940s, together with her mentor, Guiseppe Levi, who was Professor of Anatomy in Turin and a distinguished neurohistologist. They confirmed my results but provided an entirely different explanation. They had made cell counts in spinal ganglia at different stages and suggested that the so-called hypoplasia comes about, not by interference with proliferation or differentiation, but by the gradual loss of fully differentiated neurons—an entirely novel concept. However, this notion did not explain the hyperplasia resulting from limb transplantation. I suggested to Rita that we collaborate and pursue the matter further. Her arrival in St. Louis in 1947, and the repetition of both limb extirpation and transplantation experiments, turned out to be the start of a new chapter in neuroembryology.

I think that our collaboration profited greatly from our different backgrounds. Rita was more familiar with the intricacies of the nervous system; I was more familiar with the subtle ways of the embryo. The combination of the experimental method with the very powerful silver-impregnation method in which Rita had expertise was indispensable for further progress. The idea that regressive changes could be an integral part of development was not in the conceptual repertory of the experimental embryologist: Rita, however, was not encumbered by this mindset. But I would hesitate to identify Rita with the histogenetic tradition, or, for that matter, with any tradition. I know from my long association with her that her intuition and ingenuity are uniquely her own. Yet, the discoveries of every one of us have roots somewhere in the past. Therefore, one can assert that, in historical perspective, the discovery of NGF by Rita was founded on the confluence of the histogenetic and experimental neuroembryological traditions.

In a broader sense, both the histogenetic-descriptive and the analytical-experimental approaches are now part of history. It is true that the silver-impregnation method and experimentation on embryos are still widely employed tools. And the fundamental questions that were then formulated rather precisely still form one frame of reference for modern developmental neurobiology. Yet, the reductionist turn to the cellular, subcellular, and molecular levels has changed our perspective profoundly. We can now hope for sophisticated solutions of some of these problems, solutions which could not have been anticipated a few decades or even a few years ago. The brilliant successes of the new era have tempted some members of the younger generation to believe that all essential ideas and methods were born in the 1950s. The older generation does well to remind them once in a while that they too stand on the shoulders of their predecessors.

Bibliography

Boeke, J. (1950) Nerve regeneration. In *Genetic Neurology*, P. Weiss, ed., pp. 78–91, University of Chicago Press, Chicago.

Deiters, O. (1865) *Untersuchungen über das Gehrin und Rückenmark des Menschen und der Säugetiere*, Hrsg. M. Schultze, Vieweg, Braunschweig.

Detwiler, S. R. (1919) The effects of transplanting limbs upon the formation of nerve plexuses and the development of peripheral neurons. Proc. Natl. Acad. Sci. USA *5*: 324–331.

Detwiler, S. R. (1920) On the hyperplasia of nerve centers resulting from excessive peripheral loading. Proc. Natl. Acad. Sci. USA *6*: 96–101.

Detwiler, S. R. (1936) *Neuroembryology*, Macmillan, New York.

Golgi, C. (1886) *Sulla fina Anatomia degli Organi centrali del Sistema nervosa*. U. Hoepli, Milano.

Hamburger, V. (1925) Ueber den Einfluss des Nervensystems auf die Entwicklung der Extremitäten von *Rana fusca*. Roux' Arch. f. Entw. Mech. *105*: 149–201.

Hamburger, V. (1928) Die Entwicklung experimentell erzeugter nervenloser und schwach innervierter Extremitäten von Anuren, Roux' Arch. f. Entw. Mech. *114*: 272–363.

Hamburger, V. (1934) The effects of wing bud extirpation on the development of the central nervous system in chick embryos. J. Exp. Zool. *68*: 449–494.

Hamburger, V. (1939) Motor and sensory hyperplasia following limb bud transplantations in chick embryos. Physiol. Zool. *12*: 268–284.

Harrison, R. G. (1901) Ueber die Histogenese des peripheren Nervensystems bei *Salmo salar*. Arch. Mikr. Anat. *57*: 354–444.

Harrison, R. G. (1904) An experimental study of the relation of the nervous system to the developing musculature in the embryo of the frog. Am. J. Anat. *3*: 197–220.

Harrison, R. G. (1906) Further experiments on the development of peripheral nerves. Am. J. Anat. *5*: 121–131.

Harrison, R. G. (1907a) Experiments in transplanting limbs and their bearing upon the problems of the development of nerves. J. Exp. Zool. *4*: 239–281.

Harrison, R. G. (1907b) Observations on the living developing nerve fiber. Anat. Rec. *1*: 116–118.

Harrison, R. G. (1910) The outgrowth of the nerve fiber as a mode of protoplasmic movement. J. Exp. Zool. *9*: 787–846.

Harrison, R. G. (1935) On the origin and development of the nervous system by the methods of experimental embryology (Croonian Lecture). Proc. R. Soc. Lond. [Biol.] *118*: 155–196.

Held, H. (1909) *Die Entwicklung des Nervengewebes bei den Wirbeltieren*. J. A. Barth, Leipzig.

His, W. (1886) Zur Geschichte des menschlichen Rückenmarkes und der Nervenwurzeln, Abhandl. Königl. Sächs. Akad. d. Wiss. Math.- Phys. Klasse *13*: 479–514.

His, W. (1887) Die Entwicklung der ersten Nervenbahnen beim menschlichen Embryo. *Arch. f. Anat. Entw. Gesch. (Anat. Abt.)* 368–378.

His, W. (1889) Die Neuroblasten und deren Entstehung im embryonalen Mark. Abhandl. Königl. Sächs. Akad. d. Wiss. Math.-Phys. Klasse *15*: 312–372.

His, W. (1890) Histogenese und Zusammenhang der Nervenelemente. Arch. f. Anat. Entw. Gesch. Suppl. 95–117.

Levi-Montalcini, R., and G. Levi (1944) Correlazione nello sviluppo tra varie parti del systema nervoso, Comment. Pontif. Acad. Sci *8*: 527–568.

Ramón y Cajal, S. Note: All embryological publications are collected in: S. Ramón y Cajal (1960) *Studies on Vertebrate Neurogenesis*, transl. L. Guth, Charles Thomas, Springfield, IL.

Ramón y Cajal, S. (1890a) The time of appearance of nerve cell processes in the chick spinal cord. Anat. Anz. 5: 631–639.

Ramón y Cajal, S. (1890b) The nerve fibers of the cerebellar granular layer and the development of the granular layers. Int. Monatsschr. f. Anat. u. Physiol. 7: 12–31.

Ramón y Cajal, S. (1890c) Some bipolar cerebellar elements and new details concerning the development of the cerebellar fibers. Int. Monatsschr. f. Anat. u. Physiol. 7: 447–468.

Ramón y Cajal, S. (1893) La rétine des vertébrés. La Cellule 9: 119–258.

Ramón y Cajal, S. (1937) *Recollections of my Life*, transl. E. Horne Craigie, *Mem. Amer. Philos. Soc.*, Vol. 8 (reprinted, M.I.T. Press).

Ramón y Cajal, S. (1954) *Neuron Theory or Reticular Theory?*, transl. M. U. Purkiss and L. A. Fox, *Consejo Sup. de Investig. Cient. Istituto Ramón y Cajal. Madrid* (first published in: *Arch. de Neurobiol.*, 1933).

Ramón y Cajal, S. (1960) *Studies on Vertebrate Neurogenesis*, transl. L. Guth, Charles Thomas, Springfield, IL.

Sherrington, Ch. (1949) A member of Dr. Cajal. In *Explorer of the Human Brain*. D. F. Cannon, pp. 9–15, Henry Schuman, New York.

Shorey, M. L. (1909) The effect of the destruction of peripheral areas on the differentiation of the neuroblasts. J. Exp. Zool. 7: 25–63.

IV. Developmental Genetics and Evolution

Reprinted from BIOLOGICAL SYMPOSIA, Vol. VI, 1942.

THE DEVELOPMENTAL MECHANICS OF HEREDITARY ABNORMALITIES IN THE CHICK

VIKTOR HAMBURGER

WASHINGTON UNIVERSITY, ST. LOUIS

1. INTRODUCTION

THE problem of the mode of gene action in development can be approached in different ways. The geneticist will choose the genotype as the variable; he will select those cases in which the effect of a large number of genotypic variants on a single phenotypic character, for instance pigmentation, can be studied. Dr. Wright will presently discuss one of the most illuminating examples of this type. The embryologist will follow a different approach. He begins the analysis at the other end, at the phenotypic manifestations; he will attempt to trace backward, step by step, the chain of events from the established structural expressions towards the initial gene action. In doing so, he will apply his way of thinking and his methods. He is aware of the extremely complex interplay of actions and reactions between different parts of the developing embryo. He recognizes that any change in one part, at one moment, is reflected in many others, and the analysis of the mechanisms of such embryonic correlations is his domain. A gene-controlled modification of a developmental process will not only find expression in the primordium where it originates but in other parts as well. The subject of the following discussion is to illustrate by a single example, the Creeper fowl, some of the intricacies of indirect gene action. The chick, in general, is favorable for such studies because it is one of the few materials which are suitable for genetic as well as experimental embryological work and because mutants of embryological interest are being recorded in increasing numbers. The Creeper factor was chosen because, in this instance,

311

a single Mendelian factor is responsible for a multitude of well circumscribed effects on the organism, and because invaluable information on this material was already available through the investigations of Dr. Landauer and his associates. I wish to acknowledge his generous cooperation in our project.

Time will not permit me to correlate our findings with pertinent results obtained on other Vertebrates and on Invertebrates. I prefer to stress some theoretical implications of the problems involved, and this can be done best by using one case as an example.

2. THE MATERIAL

The Creeper factor is a dominant lethal. In the heterozygous condition it manifests itself mainly in the leg skeleton. All long bones are shortened. This is evident as early as on the 11th day of incubation (Pl. I, Fig. 4). The tibio-tarsus is bent and the fibula is relatively long. It extends to the distal end of the tibio-tarsus with which it is fused. Chondrogenesis and osteogenesis are abnormal. All symptoms are strikingly similar to a human congenital malformation known as *chondrodystrophy* (Landauer, 1931). In the homozygous condition the factor is lethal. In the stock which we used most embryos die at the end of the third day of incubation or shortly thereafter. They exhibit the following symptoms (Pl. I, Fig. 2): a general retardation of growth and differentiation, the head being most seriously affected; a delayed flexure and rotation of the head; asymmetries of the head which are very pronounced in eyes and otocysts, the structures of the left side being more retarded than those of the right side; the absence of the vitelline circulation; blood pools around the embryo, etc. (Landauer, 1932; Rudnick and Hamburger, 1940; Cairns, 1941). A few homozygous embryos (2% or more) survive the critical third day but die later on and never hatch. They exhibit marked abnormalities which are much more widespread than those of the heterozygous condition (Pl. I, Figs. 5 and 7)

(Landauer, 1933). The head is abnormal in shape. The eyelids are rudimentary. The eyes show a number of deficiencies, outstanding among them an abnormality which is known as "typical coloboma," to be described later. The limbs are extremely short, so that the toes seem to be attached to the body. This condition is known in human pathology as *"phocomelia."* The wing and leg skeleton fails to undergo enchondral ossification and shows other gross abnormalities like fusion or even elimination of elements. All phalanges of a digit or toe are fused into one tapering rod (Pl. II, Fig. 11). Altogether, we are dealing with three distinctly different phenotypes: the heterozygotes which will be referred to as *"chondrodystrophic"* embryos; the homozygotes which die on the 3rd or 4th day and are called *"prothanic"* embryos; and the few surviving homozygous embryos known as *"phocomelic"* embryos.

3. Multiple Effects Mediated by Basic Physiological Integrators

The logical beginning of an inquiry into the mode of gene action from an embryological point of view is an attempt to trace the gene-conditioned structural modifications back to the point where they first deviate visibly from the normal course of development. Such studies in themselves may bring about a clarification of complex situations. This is illustrated in the reinvestigation of the cause of death of the prothanic embryos by Cairns (1941).

That they do not die of a lethal effect on all organ primordia was first demonstrated by David (1936). He succeeded in rearing prothanic tissue (heart, limb, etc.) beyond the critical 3rd day by means of tissue culture and chorio-allantoic grafting. Cairns found that at 48 hours of incubation (14–22 somites) the prothanic embryos cannot be distinguished from normal embryos, whereas 6 hours later (23–27 somites) most of the symptoms mentioned before have made their appearance, including the delay in the flexure and the asymmetry of the head. A point of particular interest was added to the previous observations of Landauer (1932) and Rudnick and Hamburger (1940). Most of the deviations from normal can be considered as a drastic retardation or even cessation of growth, morphogenesis and histogenesis of the different organs. The thyroid primordium, visceral pouches, the left eye, etc., remain in a stage corresponding to a 20-somite or younger embryo. In contrast to these, the circulatory system presents an anomaly which can only in part be considered as a cessation of development. The vitelline circulation fails to develop. In addition, the majority of the prothanic embryos exhibit a unique feature: a conspicuous anastomosis between the dorsal aorta and the anterior cardinal vein in the posterior head level (Pl. I, Fig. 6). This abnormality is not an entirely new creation but rather an elaboration of an earlier condition. Evans (1909) has shown that the origin of the anterior cardinal is by way of capillaries which grow out from the dorsal aorta. In the normal embryo these connections disappear at the 15–20 somite stage. In the prothanic embryo, the capillaries, instead of regressing, enlarge and create a short-circuit by which the blood is immediately returned from the aortic arches and the dorsal aorta to the heart instead of being distributed to the vitelline circulation. Some embryos which were probably prothanic embryos showed a weak yolk sac circulation, and it is to be assumed that the surviving phocomelic embryos are those in which this defect is slight enough to permit them to survive this critical period. It is reasonable to assume that the failure of the yolk sac circulation, which at these stages is the respiratory and nutritive source of the embryo, accounts for all other symptoms, namely the cessation of growth and differentiation, and that subsequent death is caused by asphyxia. The head asymmetries are readily explained by assuming that the organs adjacent to the shell have a slightly better access to oxygen and are therefore affected later.

If this interpretation of the early lethal effect is correct, then it should be possible to produce experimentally all symptoms of the prothanic embryo in a genetically normal embryo by inhibiting its yolk sac circulation. The following experiment was devised by Cairns (*op. cit.*). Normal embryos of 44 hrs. (10–18 somites), *i.e.*, immediately before onset of the yolk sac circulation, were cauterized or cut with an iridectomy scissors near the

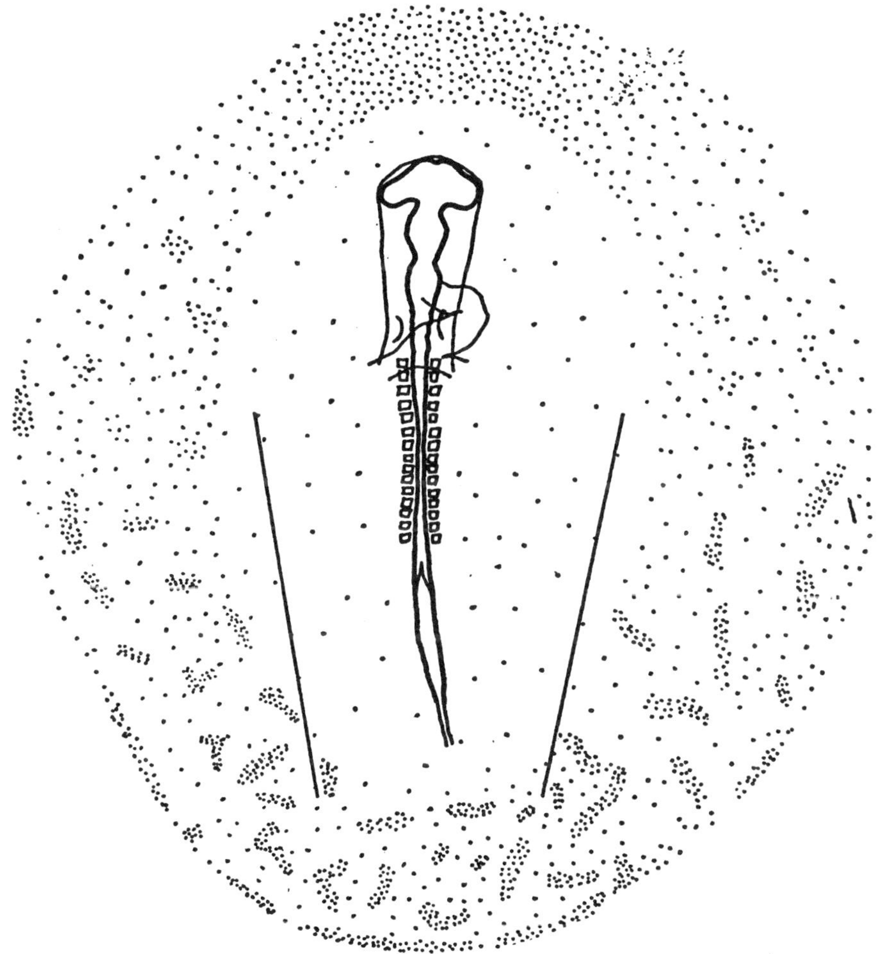

Fig. 1. Blocking of the vitelline circulation by cutting along the straight lines (from Cairns, 1941).

base of the vitelline arteries (Fig. 1). The operation was successful in 29 out of 93 cases, the others either healed completely or died. Seventeen embryos showed in every detail a striking resemblance to prothanic Creeper embryos. Growth and differentiation are retarded to the same degree; the head asymmetry is clearly present, and most convincing of all, the anastomosis between dorsal aorta and anterior cardinal vein is retained as in prothanic embryos (Pl. I, Fig. 3).

Obviously, experiments of this type can never furnish crucial evidence in favor of a hypothetical mode of gene action. They give merely indirect support to a working hypothesis. What the experiment does show is this: the multitude of widespread effects found in the dying prothanic embryos (including asymmetries and anastomoses) can be brought about by interference at one "focal" point, at one single organ system which, due to its key position, mediates widespread secondary effects.[1] In this respect, the case is similar to the well known dwarf gene in mice (Smith and McDowell, 1931) which interferes with the elaboration of a growth controlling hormone in the hypophysis, and thus produces indirectly a certain type of dwarfism. Equally striking is the case of the *Frizzle Fowl* which was analyzed by Landauer and his associates (Landauer and Upham, 1936; Landauer, 1937). The gene interferes with the keratinization of the feathers and thus upsets indirectly the thermoregulation. The loss of heat, in turn, causes a great variety of structural and functional disturbances. This case will be discussed by Landauer in another symposium.

All three cases illustrate one mode of indirect gene action: an interference with the development of one vitally important structure from which the effects spread through channels of physiological integrating mechanisms.

[1] This statement refers only to those features of the homozygous embryos which are directly related to their early death. It is very doubtful whether the structural deficiencies of the surviving phocomelic embryos have anything to do with this effect of the Creeper factor on circulation (see below). The prothanic embryo is to be considered as a potential phocomelic embryo in which one of several effects of the homozygous Creeper factor, namely that on the vitelline circulation, is so severe that it kills the embryo before other effects can manifest themselves.

4. EMBRYONIC CORRELATIONS AS MEDIATORS OF GENE EFFECTS

The phase of visible differentiation of an organ primordium is preceded by a phase during which its developmental tendencies are gradually established. The process of stabilization is usually called "determination." We know of many instances in which the stabilization in a primordium is brought about by interaction between neighboring areas. If such an interaction is by direct contact with an adjacent structure we call it "embryonic induction." However, we are aware that this term includes a variety of types of correlations.

"Determination" is not a single act but a succession of stabilizing processes which fix irreversibly, first the general, and then, gradually, the detailed characters of an organ or structure. For instance, in amphibian and chick embryos, an area is set apart very early which has eye-forming properties in a general way as demonstrated in transplantation experiments. But this area has, at first, no definite boundaries, nor are any details of the eye fixed irreversibly. This is shown best by regulation experiments on the "eye field." The prospective area of each eye can, under proper experimental conditions, give rise to two complete eyes. This shows that details were not rigidly fixed at the stage of operation. The decision as to what part of this "eye field" is to form retina, what part pigment epithelium, etc., is made later. The same stepwise determination of details probably occurs within the retina, and so forth. Likewise, brain and spinal cord are determined early as neural structures in a general way, but many details, *e.g.*, the number of neurons to be differentiated in the different parts of the central nervous system, or the pathways of the peripheral nerves, are fixed much later and by agents which, we know, are entirely different from those responsible for the initial neural determination by the archenteron roof.

As far as the role of hereditary factors in development

is concerned, it seems to me of fundamental importance to find out for each phenotypic characteristic of an organ whether or not it is irreversibly fixed at the moment when the organ as a whole is determined. For instance, are the skeletal abnormalities of Creeper or phocomelic limbs irreversibly fixed at the moment when the prospective limb area acquires its general limb forming properties? Is coloboma determined at the time of determination of the eye field? If so, then this would mean that the Creeper factor, or more precisely, the genotypic configuration for which Cp stands as a symbol, is called into action in the cells of the limb or eye material itself, without further stipulation from agents extrinsic to the primordium. In other words, structural determination and gene activation would be two aspects of the same process (see Hamburger, 1936). Only in such an instance can one speak of *"local gene action"* in the strict sense of the word. This situation is exemplified in many instances of induction in Amphibian development. For instance, the balancers of *Triturus taeniatus* and *Triturus cristatus* differ in size and shape. If indifferent gastrula ectoderm of *taeniatus,* say from the prospective belly region, is exposed to a "balancer inductor" of *cristatus,* then a balancer is induced which, from the beginning, shows *taeniatus* characters (Rotmann, 1935).

The alternative is *"indirect gene action"* in the widest sense of the word. A primordium starts out as potentially normal, and characteristics which are considered as phenotypic gene expressions are imposed on it during later phases of its development by mediation of inductions or similar intraembryonic correlations. For example, we know that in Amphibians the growth of the visual centers in the brain is controlled by the eye. Extirpation of the eye primordium results in a reduction of the optic lobes. The same happens if the eye development is suppressed by a hereditary factor. It is not necessary to assume a local gene action in the brain tissue. However, a crucial proof can be obtained only by transplantation experiments. Indirect gene action on the brain would be proven if the substitution of a microphthalmic eye by a normal eye would result in an optic lobe of normal size. Likewise, it is probable that the skull size is determined by the brain size. In cases of hereditary head abnormalities, such as the otocephalic monsters in the guinea pig (Wright and Wagner, 1934), the braincase fits neatly about the undersized brain. It is reasonable to assume that the effect on the skull is an indirect one.

In every such instance, transplantation experiments are required to decide whether a given phenotypic character is brought about by "local gene action" or by "indirect gene action." In favorable instances of indirect gene action, such experiments may even give a clue as to the mechanism of its mediation. With this aim, limb and eye transplantations were made on embryos from Creeper stock.

a. *Limb Transplantations*

The first question was this: at what stage are the chondrodystrophic and phocomelic skeleton abnormalities determined?

Limb primordia were transplanted from homozygous and heterozygous Creeper embryos during the 3rd day of incubation (19–30 somites, Fig. 2). At that stage, they range from barely visible condensations of mesenchyme to clearly demarcated limb buds. All homozygous donor embryos were in stages which manifested the prothanic symptoms. The heterozygous (Cpcp) embryos cannot be identified at the stage of operation. They were raised for 9–10 days (Pl. I, Fig. 9) and their skeletons then stained with methylene blue. The results were clear cut: all transplants from heterozygous embryos showed all symptoms characteristic of chondrodystrophy (Pl. I, Fig. 8, Pl. II, Fig. 10). The transplants from homozygous donors developed all characteristics of phocomelia: extreme shortening, absence of ossification, occasional fu-

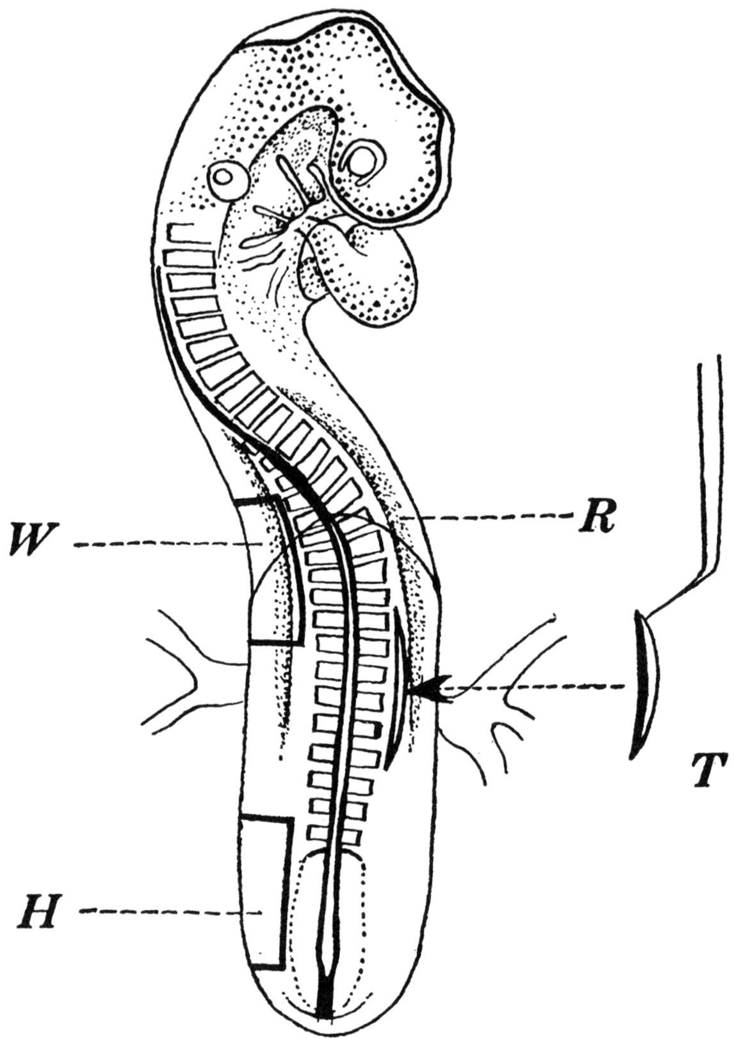

FIG. 2. Transplantation of limb primordia (27 somite embryo, from Hamburger, *J. Exp. Zool.*, 77, 1938). H = Hind limb area, R = Wolffian ridge, W = Wing area.

sions of tibia and fibula; and, invariably, fusion of all phalanges (Pl. II, Fig. 11, Hamburger, 1941).[2]

The experiments show, first, that prothanic limb primordia are viable but potentially phocomelic, which was to be expected; second, that from the stage of operation

[2] Coelomic and flank transplants from normal stock also show abnormalities but they are of a different type, *e.g.*, ankylosis, hypodactyly, etc. (Hamburger and Waugh, 1940). They may be found superimposed on chondrodystrophic or phocomelic symptoms; but in an individual transplant it is always possible to distinguish them from the chondrodystrophic abnormalities.

on, both chondrodystrophic and phocomelic skeletal conditions are fixed in the limb primordium. Since the transplantations were done before the onset of circulation, it is certain that hormones and other agents carried in the blood stream are not instrumental in causing the phenotypic effect on the limbs. This is of some clinical interest, as the chondrodystrophy and phocomelia of the chick resemble closely those found in mammals and man. It remains to be decided whether we are dealing with an instance of local gene action or if the deviations from normal were imposed on the limb by extrinsic agents during the period between its "determination" and the stage of operation.

b. *Eye Transplantations*

The homozygous phocomelic embryos show a severe coloboma, that is, a failure of the choroid fissure to close, combined with eversions of the retina tissue around the edges of the open fissure which thus occupies part of the outer layer of the cup (Pl. II, Fig. 14). Coloboma is thus easily recognizable in cleared specimens as an unpigmented area (Pl. I, Fig. 5). Both sclera and choroid coat are greatly impaired; the sclera forms only small fragments of scleral cartilage near the entrance of the optic nerve; the chorio-capillaris is poorly developed (Landauer, 1933).

Again, the question arises: from what stage on are these symptoms irreversibly fixed in the optic primordium? In an attempt to answer this question optic vesicles prior to cup formation (9–21 somites) were transplanted from embryos from Creeper stock to the flank of slightly older normal embryos (Fig. 3, Gayer, 1942). The genotypes of the donors were not identifiable at this stage. Therefore they were operated on *in situ* and then raised until their genotype became manifest. Eighteen of twenty successfully transplanted eyes from prothanic embryos exhibited colobomata. Some showed a definite improvement of the condition found in the phocomelic head; others showed very severe coloboma (Pl. II, Fig. 15). In the two exceptional cases, no choroid fissure was formed at all, and the optic fibers were caught inside of the cup. At first sight, these results seem to indicate that coloboma is fixed in the eye primordium at the stage of operation. However, such a conclusion would have been premature. Very unexpect-

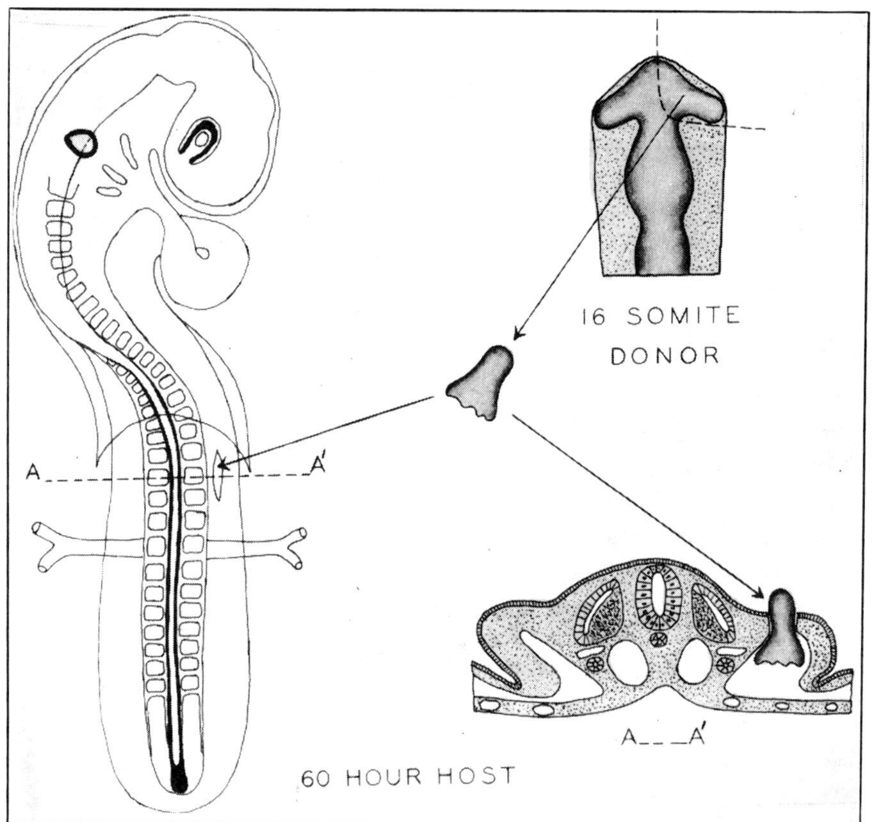

FIG. 3. Flank graft of eye primordia from 16 somite donor onto 60-hour host (from Gayer, 1942).

edly, all but 2 control transplants (53 eyes from normal and heterozygous embryos) exhibited exactly the same typical coloboma with about the same range of variability. These controls indicate that the colobomatous condition can be imposed on a genetically normal eye by a modification of factors extrinsic to the primordium, and that the eye is susceptible to these agents as late as in the

21 somite stage, that is shortly before cup invagination begins. It is therefore quite possible that the eye of a prothanic embryo is potentially normal up to that stage and that the Cp-factor operates on a mesodermal structure around the eye which in turn causes coloboma. The flank transplantations of homozygous eyes are therefore inconclusive. Their colobomatous condition may be due either to flank factors or to intrinsic factors. A decision can be expected only if a more favorable site could be found in which normal eyes would develop normally. It then occurred to us that the orthotopic position would naturally be the optimal site. The right optic vesicles and the adjacent right halves of the forebrain were exchanged between normal and homozygous embryos and between normal embryos *inter se* (Gayer, unpublished). Both donors and hosts ranged between 5 and 17 somites. The majority of all control eyes developed normally without coloboma (7 colobomata in 45 cases. See Table I). In the operations from Creeper stock 53 do-

TABLE I

ORTHOTOPIC EYE TRANSPLANTATIONS
(donors and hosts: 2–17 somites)

Donor genotype	Total operated	Hosts died	Transplants failed	Transplants without coloboma	Transplants with coloboma
Normal stock	56	29	8	15	4
Creeper stock:					
Cp+ and ++	53	24	3	23	3
CpCp	16	11	0	5	0

nors were identified as normal or heterozygous, and 16 as homozygous, which is precisely the expected 3:1 ratio (Table I). All 5 successful transplants from homozygous donors were perfectly normal without a trace of coloboma or other abnormalities (Pl. II, Fig. 13). The CpCp genotype of the donors was definitely diagnosed on sections. We have thus crucial evidence that this type of hereditary coloboma is not caused by local gene action operating within the optic vesicle. This is a striking and unexpected case of *"indirect gene action."* It is quite likely that the homozygous eye is potentially normal up to the 17-somite stage, and this despite the fact that each of its cells contains the Creeper factor in double dose. Since we have in the flank graft method a means of producing coloboma experimentally, it may be possible to detect the mechanism by which it is brought about.

The *scleral cartilage* is almost entirely missing in phocomelic embryos. Little is known of the developmental mechanics of this structure. Weiss and Amprino (1940) have shown that it is determined on the fourth day, and theirs as well as other experiments make it likely that the optic cup is instrumental in its formation. The question arises: is its absence in phocomelic heads due to some deficiency of the eye cup in its capacity as its "inductor" or to a failure of the mesoderm to respond to a normal inductive agent? The results are clear cut. Homozygous eye transplants developing in normal heads (orthotopic series) were surrounded by absolutely normal scleral cartilage; and even two homozygous flank grafts had an almost complete scleral coat although they were colobomatous. This proves that the homozygous eye, even if colobomatous, can induce scleral cartilage.[3] Its failure in phocomelic heads must be due to a lack of responsiveness of the head mesoderm. It is quite possible that this head mesoderm abnormality and the one which causes coloboma have the same root.

c. *Functional Hypertrophy*

We mention briefly another mechanism by which indirect gene effects are mediated through interplay of intrinsic factors: functional hypertrophy. The Creeper case presents a very striking example of this type, as

[3] The amount of head mesenchyme transplanted with the optic vesicle was not controlled. It is likely that the scleral cartilage both in flank and in orthotopic grafts was, at least in part, contributed by the host. However, this point has no bearing on our conclusions.

Landauer (1939a) has shown. Whereas in phocomelic embryos most organs show a distinct progressive reduction of weight, the relative heart weight and the relative and absolute spleen weight are above normal. Landauer correlates this with the absence of bone marrow in the extremities which results in a progressive anemia. The anemia is probably the immediate cause of death. ''The enormous enlargement of the spleen and the gradually developing enlargement in relative size of the heart are functional adjustments to the anemia'' (*op. cit.*, p. 22).

5. GROWTH

Finally, we have to consider the effect of the Creeper factor on growth. We owe to Landauer (1934, 1939a) a very extensive study of the growth of Creeper organs during the incubation period as well as after hatching. One of the most striking effects of the Creeper factor in a single dose is its local effect on the appendages, the hind limbs being much more severely affected than the wings. Length measurements of the different long bones showed that within the leg and wing there is likewise a differential growth reduction of the different elements. A careful analysis of all data enabled Landauer to demonstrate that the reduction of growth rates follows two rules: First, there exists a proximo-distal gradient. In the leg, for instance, the distalmost element, the tarso-metatarsus, is more severely affected than the tibio-tarsus, and this bone, in turn, more severely than the femur. This sequence corresponds to the sequence in the time of appearance of the different elements in the early limb bud, as well as to gradient differences in inherent growth rates. Lerner (1937) has found that the tarso-metatarsus has the highest growth rate, and the femur has the lowest rate. Second, the element which is ultimately the longest, namely the tibio-tarsus in the leg, shows the greatest relative reduction if its decrease in per cent of total length of all long bones is taken as the base. The femur which is ultimately the shortest element shows the slightest relative effect, and the phalanges are not affected at all. In a later publication, Landauer (1933a) has shown that the same two rules hold for the much more severe growth disturbances found in the phocomelic embryos. Here, the toes are also affected.

The signal importance of these findings was clearly recognized by the author. The disproportionate size relations of Creeper and phocomelic legs and wings are nothing but distortions of the normal growth pattern; their different elements are differentially susceptible to the growth reducing effect of the Creeper factor. A multitude of effects can thus be subordinated to one single causal agent. All facts are satisfactorily accounted for if one assumes that one specific growth controlling reaction or substance is interfered with and that "the localized effects are due merely to quantitative differences in physiological needs of certain parts at certain periods" (Landauer, 1941a).

This interpretation finds strong support in data obtained by Landauer (1939b, 1941b) on two other mutations of the fowl: the lethal mutation of the Cornish fowl which is, in many respects, similar to the Creeper fowl, and a semilethal mutation, affecting length of the upper beak and of the long bones. Considering only the long bones of the legs it was found that in both instances their growth reduction follows the same rules as in Creeper legs. The end results are fundamentally identical though modified in details. Three different genes, causing probably deficiencies in three slightly different growth controlling agents, modify the same fundamental growth pattern.

If one considers the fact that the Cp-factor in double dose causes a growth retardation in almost every organ (Landauer, 1939a), one is inclined to assume that the Creeper factor, as far as its growth controlling effect is concerned, acts locally in every part of the organism. Yet numerous experiences in experimental embryology caution us against premature generalizations. The final size of an organ is, of course, the resultant of its intrinsic gene controlled growth rate and of extrinsic growth controlling and growth limiting agents, some of which may likewise be gene controlled. Not only hormones and nutritive agents carried in the blood stream but contact interactions and other embryonic correlations as well play a role. To mention only a few instances of the latter type: The growth of the skull is controlled by that of the brain. The growth of the visual centers in the brain is controlled by the eye. The initial size of the eye is controlled by the size of the underlying mesodermal inductor. In later phases, eye growth is controlled by the growth of its own lens, as shown by the classical experiments of Harrison (1929) in which optic cups of the large *Ambystoma tigrinum* were combined with lens epithelium of the small *Ambystoma maculatum*, and vice versa. In general, then, we may expect the same hierarchy of secondary gene effects which we discussed for structures to reappear in the realm of growth processes. Only rigidly controlled transplantation experiments can decide to what extent intrinsic and extrinsic gene effects coöperate synergistically to determine the final size of an organ. Again, our observations on limb and eye transplants of Creeper embryos may serve as an illustration.

In the transplanted homozygous and heterozygous Creeper *limb* primordia we found the following situation: the length of transplanted chondrodystrophic femora was measured at 11–12 days of incubation and compared with that of non-transplanted chondrodystrophic and with normal host femora. At this stage, the size of the non-operated Creeper femur is 20% below normal (Landauer, 1934). Previous measurements on normal limb transplants (Hamburger and Waugh, 1940) had shown that transplantation itself results in a growth reduction of genetically normal femora which is, incidentally, of the same order of magnitude, namely 20%. Transplanted Creeper femora were found to be 40% reduced in size which is clearly a cumulative effect of the Creeper condition and of the transplantation *per se*. Less accurate measurements on homozygous transplants indicate that the same holds for homozygous femora. This means that the growth rate of the femur which is representative for the whole leg is firmly established in the early leg bud. It implies that the specific agent for growth restriction in Creeper legs is not carried in the blood stream. Local gene action in the limb primordium is indicated but not proved until transplantation of still younger primordia has been done.

Again the situation with respect to the *eye* is different. It will be remembered that phocomelic embryos are microphthalmic. The orthotopic transplants from homozygous embryos grew to fully normal size on a normal head despite the fact that they are built up of cells which carry the Creeper factor in double dose (Pl. II, Fig. 13). Clearly, the microphthalmia of the phocomelic embryo, like its coloboma, is due to conditions extrinsic to the eye, perhaps insufficient blood supply, and not to "local gene control of growth." The claim that the Creeper factor has a universal, *direct* effect on the growth rates of all structures is not warranted.

6. The Relation between Growth and Structural Differentiation

In the Creepers, as well as in other hereditary monsters, growth reductions and structural deficiencies are so intimately associated that a causal relation between the two has been suspected. It has often been postulated that genes in general exert their influence on development by controlling growth rates and that all structural manifestations are secondary effects. In the present case, the histopathological features of chondrodystrophy and of phocomelia, as well as the early lethal gene action could be regarded as caused by a preceding effect on growth rates. This scheme of gene action is based on the hypothesis that growth determines histological differentiation. This assumption in itself is not well substantiated by facts. It may hold for some cases; but in other instances, growth can be dissociated from differentiation, or the two were found to be mutually exclusive.

The statement that growth determines differentiation is meaningful only if one considers growth as an entity which is defined as an increase in volume per time unit or in a similar way. However, one can apply a more dynamic concept of growth and consider it as the resultant of an interplay between proliferating cells and a multitude of specific growth controlling agents. One may then find that one or the other specific growth controlling agent affects also histological differentiation; that a deficiency in one specific substance, or the interference with one specific reaction impairs the growth and also the differentiation of the same primordium. In this alternative scheme, differentiation is not subordinated causally to growth, but both are subordinated to a third agent. Applied to gene action this would mean that genes affect differentiation not by way of growth but by controlling specific physiological mechanisms which are essential for both differentiation and growth.

What is the situation with respect to the Creeper factor? Several experiments have been interpreted to support the first scheme. Fell and Landauer (1935) reared limb primordia of normal chick embryos in vitro in a normal and a special growth-restricting medium and found morphological and histological abnormalities which resembled strikingly the phocomelic condition. Likewise, David (1936) in chorio-allantoic grafts of normal 72-hr. limb buds found occasionally chondrodystrophy-like deficiencies in the cartilage. It seems to me that the results of both experiments can be fitted into either scheme and are therefore crucial for neither one. The abnormal cartilage condition may be due to a general growth restriction, as the authors assumed at that time, or they may be due to the deficiency of the culture medium in one specific substance which is particularly necessary for chondrogenesis, and also for growth. Another argument in favor of the first scheme requires reinvestigation. According to it, a growth reduction should always precede the histological abnormalities. According to the second scheme both may, or may not, appear simultaneously. In his first communication, Landauer (1932) assumed that the prothanic homozygous Creeper embryos show first a general growth inhibition and that other structural deficiencies appear subsequent to this. However, the measurements were based on doubtful identifications

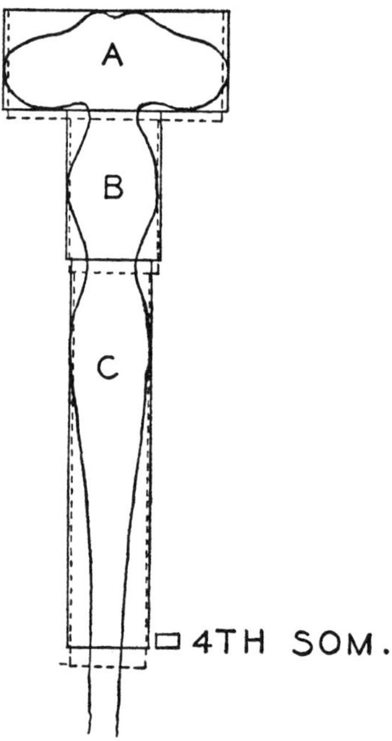

FIG. 4. Measurements of widths and lengths of forebrain, midbrain and hind brain of 7–15 somite embryos from Creeper stock. Solid lines: normal and heterozygous embryos. Dotted lines: homozygous embryos.

of homozygous embryos (Rudnick and Hamburger, 1940). Cairns (unpublished) has made preliminary measurements of the length and width of the different parts of the brains of vital stained embryos from Creeper stock, at 40 to 44 hours of incubation (7–15 somites), that is at a stage immediately before structural differences become noticeable (Fig. 4). The embryos were then raised for several days until a definite identification was possible. No statistically significant length difference between prothanic and non-prothanic embryos was found (Table II).

TABLE II

BRAIN MEASUREMENTS OF CpCp AND NORMAL CHICK EMBRYOS
40–44 hours incubation, 7–15 somites

	Normal embryos (44)		Homozygous CpCp embryos (17)	
	Width	Length	Width	Length
Forebrain (A, Fig. 4)	36.5 ± 3.44	15.8 ± 1.92	34.8 ± 6.05	17.2 ± 3.11
Midbrain (B, Fig. 4)	15.4 ± 2.26	23.8 ± 2.79	14.6 ± 2.64	24.3 ± 3.10
Hindbrain (C, Fig. 4)	13.3 ± 1.82	61.8 ± 5.20	12.7 ± 1.85	62.9 ± 6.98

The figures given are the dimensions in millimeters of camera lucida drawings of living embryos *in situ*.

The errors indicated are standard deviations.

Altogether, there is at present no convincing indirect or direct evidence to show that the structural Creeper symptoms are preceded or determined by a primary gene effect on growth. It is equally plausible to assume that the Creeper factor produces a specific alteration of a basic metabolic process or substance which is required both for growth and differentiation. The restricted localized responses in growth and in histogenesis would be due to locally different susceptibilities and to the manifold intraembryonic interactions discussed before. This interpretation of the action of the Creeper factor seems to me preferable as a working hypothesis, and I find myself in full agreement with Landauer's recent communications (1941a, 1941b). This concept, of course, does not negate the assumption that genes act by controlling rates of chemical processes. On the contrary, this is the most likely mechanism of their action. It merely implies that their control of *growth* rates is not the cause of all their other manifestations.

7. CONCLUSIONS

At present, we know of at least four channels through which the Creeper factor exerts its widespread influence: an effect on the early yolk sac circulation; an effect on chondrogenesis (which will automatically result in a dis-

TABLE III

ACTION OF THE HOMOZYGOUS Cp-FACTOR

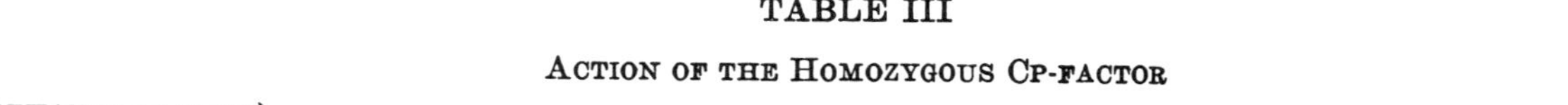

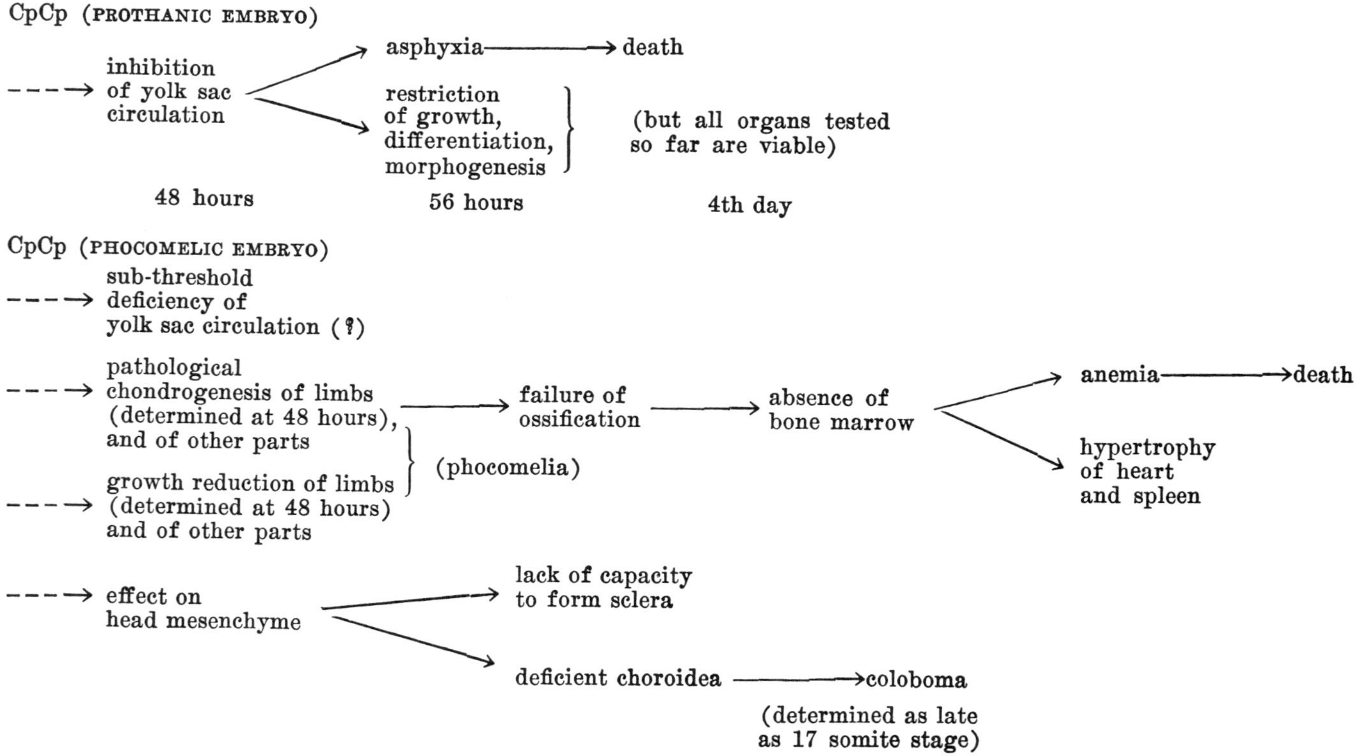

turbance of ossification; see Fell and Landauer, 1935); an effect on the head mesoderm which is indirectly responsible for the eye abnormalities; and an effect on growth (Table III). It is not possible yet to reduce this multiple action to one single basic gene action. However, such an attempt would not be within the scope of embryological analysis. What the embryologist can do is to simplify a complex situation, to uncover the developmental mechanics of indirect gene action during the phase of determination and thus to prepare the material for the final analysis. The first phase, from primary gene action to the beginning of determination remains for physiological and biochemical analysis. The complete story of the mode of gene action must be written jointly by geneticists, embryologists and physiologists.

LITERATURE CITED

Cairns, J. M.
    1941. *J. Exp. Zool.,* 88: 481–500.
David, P. R.
    1936. *Arch. f. Entw'mech.,* 135: 521–551.
Evans, H. M.
    1909. *Anat. Rec.,* 3: 498–518.
Fell, H. B., and W. Landauer.
    1935. *Proc. Roy. Soc. Lond., B.,* 118: 133–154.
Gayer, K.
    1942. *J. Exp. Zool.,* 89: 103–133.
Hamburger, V.
    1936. *J. Exp. Zool.,* 73: 319–364.
    1941. *Physiol. Zool.,* 14: 355–364.
Hamburger, V., and M. Waugh.
    1940. *Physiol. Zool.,* 13: 367–380.
Harrison, R. G.
    1929. *Arch. f. Entw'mech.,* 120: 1–55.
Landauer, W.
    1931. *Zeitschr. f. Mikr. Anat. Forsch.,* 25: 115–180.
    1932. *J. Genet.,* 25: 367–394.
    1933. *Zeitschr. f. Mikr. Anat. Forsch.,* 32: 359–412.
    1934. Storrs Agr. Exp. Sta. Bull. 193.
    1935. *J. Genet.,* 31: 237–242.
    1937. *Am. J. Med. Sci.,* 194: 667–674.
    1939a. Storrs Agr. Exp. Sta. Bull. 232.
    1939b. Storrs Agr. Exp. Sta. Bull. 233.
    1941a. Proc. 7th Int. Genet. Cong. Edinburgh, 1939.
    1941b. *Genetics,* 26: 426–439.

Landauer, W., and E. Upham.
1936. Storrs Agr. Exp. Sta. Bull. 210.
Lerner, I. M.
1937. *Hilgardia*, 10: 511–560.
Rotmann, E.
1935. *Arch. f. Entw'mech.*, 133: 193–224.
Rudnick, D., and V. Hamburger.
1940. *Genetics*, 25: 215–224.
Smith, P. E., and E. C. MacDowell.
1931. *Anat. Rec.*, 50: 85–93.
Weiss, P., and R. Amprino.
1940. *Growth*, 6: 245–258.
Wright, S., and K. Wagner.
1934. *Am. J. Anat.*, 54: 383–447.

PLATE I

FIG. 1. Normal 72-hour chick embryo.

FIG. 2. Prothanic (homozygous) Creeper embryo.

FIG. 3. Copy of a prothanic embryo experimentally produced by blocking of the vitelline circulation (see text; from Cairns, 1941).

FIG. 4. Heterozygous Creeper embryo, 11 days old. Note the curvature of the tibio-tarsus and the long fibula.

FIG. 5. Homozygous (phocomelic) Creeper embryo, 11 days old.

FIG. 6. Anastomosis (AN) between dorsal aorta and anterior cardinal vein in a homozygous (prothanic) Creeper embryo.

FIG. 7. Phocomelic embryo, 11 days old.

FIG. 8. Flank transplant of the right leg primordium of a heterozygous Creeper embryo (donor: Fig. 9) to the flank of a normal embryo.

FIG. 9. Donor of transplant shown in Fig. 8. Note the absence of the right leg.

PLATE II

FIG. 10. Middle: transplant of Fig. 8 stained in methylene blue. Left: the left leg of the donor embryo, Fig. 9. Right: the right leg of the host embryo, Fig. 8, for comparison (from Hamburger, 1941).

FIG. 11. Right: transplant of right leg of homozygous (prothanic) Creeper embryo onto a normal embryo. Left: leg of a phocomelic embryo for comparison (from Hamburger, 1941).

FIG. 12. Flank transplant of an eye from a homozygous (prothanic) Creeper embryo, optic vesicle stage, onto a normal embryo. C = colomba (from Gayer, 1942).

FIG. 13. Orthotopic transplant from a homozygous (prothanic) Creeper embryo onto a normal embryo. Upper figures: right and left view of the host embryo. Lower figure: ventral view of the head of the host cleared in oil of wintergreen. Note equal size and normal choroid fissures.

FIG. 14. Cross section through the right eye of a phocomelic Creeper embryo 9 days old. C = colomba (from Gayer, 1942).

FIG. 15. Flank graft of a homozygous (prothanic) Creeper embryo onto a normal embryo. C = colomba (from Gayer, 1942).

FIG. 16. Flank graft of a normal optic vesicle. C = colomba (from Gayer, 1942).

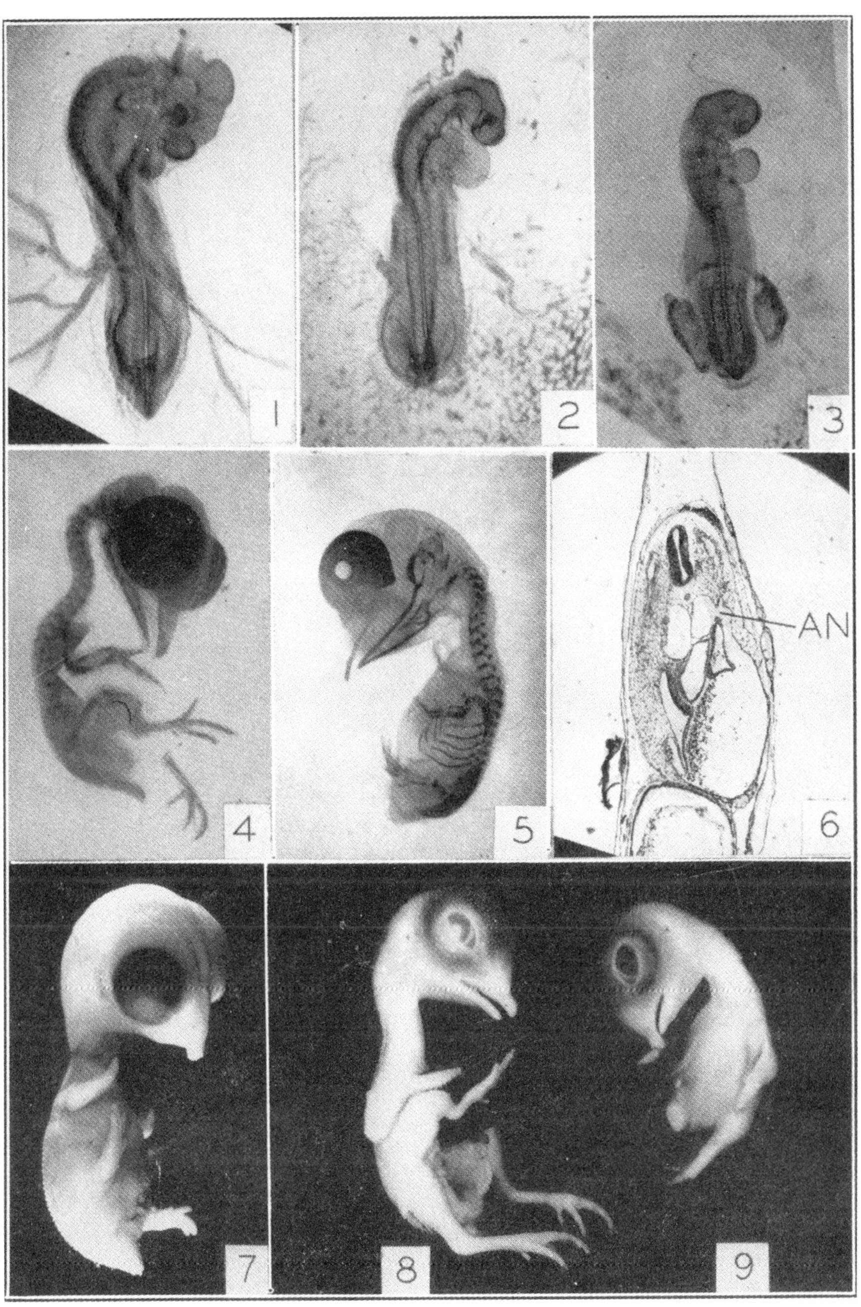

1
2
3
4
5
6
AN
7
8
9

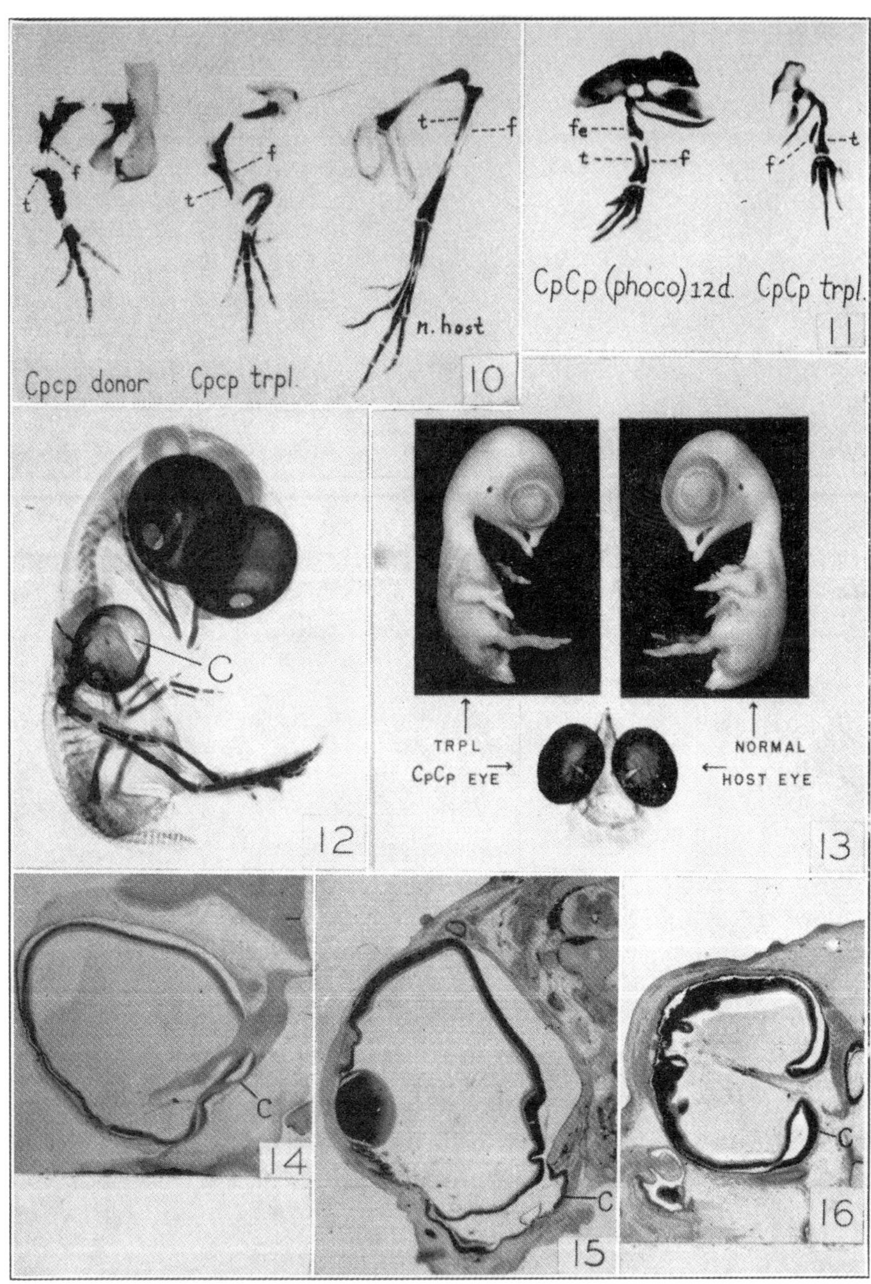
Cpcp donor
Cpcp trpl.
t----f
t----f
n. host
10
fe----
t----f
CpCp (phoco) 12d.   CpCp trpl.
11
C
12
TRPL
CpCp EYE
NORMAL
HOST EYE
13
C
14
C
15
C
16

The Evolutionary Synthesis. Perspectives on the Unification of Biology. E. Mayr and W. B. Provine, Eds. Cambridge, MA: Harvard Univ. Press, pp. 97–112 and 303–308. 1980

Darwin once expressed the opinion that embryology had provided him with the best evidence for evolution. Haeckel and his followers took up this hint and used the "law of recapitulation" to establish phyletic lineages. The result was a flowering of comparative embryology from the 1860s through the 1880s.

The basic facts of embryology, Darwin claimed in *On the Origin of Species*, all "can be explained on the view of descent with modification" (1859, p. 443). Darwin's language here is precise and crucial. Wherever possible, he claimed that phenomena could be explained by the mechanism of natural selection, not merely the looser "descent with modification." In fact, neither Darwin nor any of the great comparative embryologists could meaningfully synthesize embryology with evolution by natural selection, because the connections between hypothetical germinal material and the process of development were too vague to command the interest of serious biologists. Comparative embryology thus remained a purely descriptive science.

When *Entwicklungsmechanik* began to flourish in the late 1880s, comparative embryology went into decline. The new causal embryology, which attempted to explain development in strictly physicochemical terms soon provided important new insights into the mechanics of development. But the problem of the germinal control of development was still beyond analysis. The experimentalists who pursued the *Entwicklungsmechanik* approach had little appreciation or concern for evolution in natural populations, and by the time Mendelism arose as a science in the 1900-1910 decade, embryologists did not see their science as closely connected to the study of heredity or the mechanism of evolution. Although all the embryologists were evolutionists, very few embryologists before the 1930s, with the exception of the Oxford school (Goodrich, Huxley, de Beer), endorsed natural selection as the primary mechanism of evolution. Many prominent embryologists actively minimized the importance of natural selection, even as late as the 1940s and 1950s.

96

The role of embryology in the evolutionary synthesis raises many questions. Why did embryologists resist accepting the idea of natural selection for so long? How important were geneticists who worked on gene action and developmental genetics in bringing together the fields of genetics and embryology? Were the books of Schmalhausen (1949) and Waddington (1957) the first major works envisioning the synthesis of embryology with genetics and evolution, or did such a synthesis really require knowledge of the molecular biology of the 1950s and 1960s? What contribution, if any, did embryology make to the evolutionary synthesis of the 1930s and 1940s? The two essays in this section address all of these and related questions, particularly on the evolutionary thinking of embryologists in Germany and England.   **W.B.P.**

References

Darwin, C. 1859. *On the origin of species.* London: Murray.
Schmalhausen, I. I. 1949. *Factors of evolution.* Philadelphia: Blakiston.
Waddington, C. H. 1957. *Strategy of the genes.* London: Allen and Unwin.

Embryology and the Modern Synthesis
in Evolutionary Theory

Viktor Hamburger

Did embryology and, more specifically, experimental embryology assist in the creation of the modern synthesis during the thirties and early forties, or, on the contrary, was it a retarding element? The major works that embody the modern synthesis by Huxley (1942), Dobzhansky, Fisher, Haldane, Mayr, Simpson, and Wright hardly mention embryonic development. This omission is somewhat strange because Huxley was sufficiently interested in growth and development to write two books on this topic: one on relative growth (1932), and a text of experimental embryology with de Beer (1934). His book on evolution clearly shows the dominance of Huxley the naturalist over Huxley the embryologist. Obviously, the modern synthesis had a strong foundation in Mendelian and population genetics and its mathematical treatment, in ecology, and in field studies of speciation. It could well afford to dispense with embryology, although its implications for genetics and evolution were recognized by the founders of the modern synthesis. Conversely, the contemporary leading books on experimental embryology, by Schleip (1929), Spemann

(1936), and Weiss (1939), did not include considerations of evolution. The modern synthesis did not receive assistance from contemporary embryologists.

Could one imply that the leading experimental embryologists of the 1920s and 1930s, such as Harrison, Spemann, Lillie, Conklin, Dalcq, and Child, whose names were held in considerable esteem among most biologists, delayed the synthesis by opposition or indifference? They were all evolutionists and they all conceded the effectiveness of natural selection, at least to some extent. But many had misgivings about a key dogma of the modern synthesis; namely, the claim that natural selection is the sole explanation of all adaptations; and Lamarckist ideas were by no means dead. However, with few exceptions, such as Dalcq's somewhat later publications (1949, 1951) and some remarks in Spemann's autobiography (1943, pp. 156 ff., 272), there was little public discussion and no open opposition by this group. In fact, I believe that few embryologists in the 1920s and 1930s were aware of the emergence of a new synthesis in evolutionary theory. In Spemann's "reprint room," where his colleagues and *Doktoranden* often met during tea hour in the late twenties and early thirties, there was a lively, continuous dialogue. Spemann and his colleague, Fritz Süffert, an expert on adaptive coloration in butterflies and moths, often discussed selectionist versus Lamarckist explanations of such complex adaptations and their embryological implications, with Süffert on the selectionist side and Spemann inclined to Lamarckism. I do not remember any extension of the discussions to speciation or to the particular issues involved in the evolutionary synthesis. I can document only Harrison's awareness of the latter: "The development of modern genetics, the experimental study of the origin of mutations and the new mathematical theory of natural selection are hopeful signs of the applicability of exact methods to the study of evolutionary processes" (1937, p. 7). T. H. Morgan wrote *Embryology and Genetics* (1934), but I omit his case because a special session is devoted to him. Indeed, I believe that some leading embryologists had a retarding influence on the modern synthesis.

The lack of interest of experimental embryologists in evolutionary problems can be traced back to the founder of experimental embryology, Wilhelm Roux. Before him, for once, and only for a short time, embryology, genetics, and evolution had been united in a complete synthesis in Weismann's *The Germ-Plasm: A Theory* (1893). But this superb intellectual feat soon foundered; its foundations broke down, partly under the impact of work of the early experimental embryologists. Roux, a student of the major German prophet of evolution, Haeckel, and with impeccable credentials as a selectionist (he had extended selectionism to the explanation of adaptive structures within the organism in a book published in 1881) broke away from Haeckel in the matter of recapitulation. He founded experimental embryology or *Entwicklungsmechanik* in the 1880s as a deliberate countermove against Haeckel's categorical verdict that phylogeny is the sufficient cause of ontogeny, and that there is nothing else to explore in this matter. Roux's decisive move from ultimate or remote to proximate causes (following that of His), and the concomitant introduction of the experimental method as the indispensable tool for the analysis of proximate factors in embryonic development, started the alienation of embryological from evolutionary thinking.

By the 1920s and 1930s, experimental embryology and genetics both had accomplished a major breakthrough, experimental embryology through the achievements of Harrison and Spemann and their schools, and genetics through the Morgan school. Both fields were deeply absorbed in their own problems and took little notice of each other. The embryologists were involved in the study of epigenetic mechanisms, such as induction, gradient fields, and morphogenetic movements. Evolutionary considerations turned up rarely, as for instance in the mistaken dichotomy of "mosaic" versus "regulation" eggs.

To some extent, the evolutionists were aware of the role of embryology in evolutionary theory—to be more specific, of the fact that phenotypes, which are the target of selection, are the end result of developmental processes which in turn are the manifestation of gene activity. As Huxley says, "Any originality which this book may possess lies partly in furthering Fisher's ideas and partly by stressing the fact that a study of genes during development is as essential for an understanding of evolution as are the study of mutation and selection" (1942, p. 8). Unfortunately, his intent was not fulfilled in the writing of his book; only a short chapter deals with heterogonic growth. In a recent publication Mayr states: "The fact that fitness is determined by the phenotype is the reason for the extraordinary evolutionary importance of the developmental processes that shape the phenotype" (1970, p. 108). I assume that he was aware of this notion at a much earlier date. Of the founding fathers, only Wright was actively engaged in studies of developmental genetics. His brief résumé of 1934 goes beyond generalities. It illustrates the role of specific genes in growth processes and pattern formation and the implication of such manifestations for evolution.

I do not imply a criticism of the originators of the modern synthesis for their neglect of developmental genetics. On the contrary, I would assert that it has always been a legitimate and sound research strategy to relegate to a "black box," at least temporarily, wide areas that although pertinent would distract from the main thrust. No great discoveries or conceptual advances are possible without this expediency. Von Frisch would probably never have achieved what he did if he had allowed himself to be sidetracked by worrying over a dance center in the cerebral or thoracic ganglia of the honeybee!

Nevertheless, the modern synthesis as formulated at the time was incomplete without a chapter dealing with the effects of selection on the gene-controlled variability of developmental processes. This type of theoretical consideration was then actually in the making in the work of the Russian academician, Schmalhausen. Unfortunately, the Second World War interrupted communications between the East and the West. Schmalhausen's book, *Factors of Evolution*, which I believe offered one of the most succinct expositions of the problem and important contributions to its solution, did not become known to English readers until 1949.

Alienation between Experimental Embryology and Genetics

Before a chapter on evolutionary changes in embryonic development could be written, it was necessary to develop a new borderline field, a developmental or physiological genetics. In other words, the concepts and methods had to be created that would lead to an understanding of the role of genes and their products in embryonic development. The beginnings of such a synthesis of embryology and genetics can be traced to Boveri and Driesch at the end of the last century. But the actual analysis of gene action beginning with the study of the development of mutants and the application of the transplantation method did not come to fruition until the 1930s. It was accomplished by a younger generation, against the background of strong skepticism from at least some of the leading experimental embryologists of the older generation who were at the zenith of their accomplishments, power, and influence in the 1920s and 1930s. Some went as far as to assert a fundamental incompatibility between concepts, goals, and methods of the two fields; they saw, in principle, no chance of a meaningful amalgamation. Obviously, if this viewpoint had prevailed, there would have been no place for developmental genetics nor for a consideration of development in evolutionary thinking.

The expression of these ideas, which are presented succinctly in Lillie's essay (1927), coincides in time with the ripening of the modern synthesis. Morgan's book *Embryology and Genetics* illustrates the ambiguity of this situation. He wrote, "The story of genetics has been so interwoven with that of experimental embryology that the two can now, to some extent, be told as a single story . . . It is possible to attempt to weave them together in a single narrative" (1934, p. 9). The story goes that after the publication of the book, Morgan asked a prominent visitor what he thought of it. The visitor frankly responded that he could not find a synthesis of the two fields; whereupon Morgan, tongue in cheek, asked "What does the title say?"

The roots of the uncompromising attitude of leading experimental embryologists can be found in two major divisive issues. The first was based on the preoccupation of the geneticists with the nucleus and of the embryologists primarily with the cytoplasm. The bias of the embryologists dates back to the famous experiment of Driesch (1891) in which he isolated the first two blastomeres of the sea urchin egg, resulting in two complete larvae. The experiment demonstrated the equivalence of the blastomere nuclei, the regulative capacity of the egg, and an interaction of the two blastomeres in normal development. All three points were found to have general validity and far-reaching implications. It was argued that if every cell is in possession of the same complete genome, progressive differentiation must result from cytoplasmic differentials. The argument was reinforced by numerous impressive studies of egg structure and cytoplasmic prelocalization of organs, both in invertebrate and vertebrate eggs, and experimental proof of the importance of cytoplasmic differences for progressive differentiation. Lillie speaks of the "almost universally accepted genetic doctrine today that each cell receives the entire complex of genes. It would therefore appear to be self-contradictory to attempt to explain embryonic segregation by behavior of the genes, which are *ex hypothesi* the same in every cell" (1927, p. 13). Lillie further states:

> I do not know of any sustained attempt to apply the modern theory of the gene to the problem of embryonic segregation. As the matter stands, this is one of the most serious limitations of the theory of the gene considered as a theory of the organism. We should of course be careful to avoid the implication that in its future development the theory of the gene may not be able to advance into this unconquered territory. But I do not see any expectation that this will be possible, *even in principle*, as long as the theory of the integrity of the entire gene system in all cells is maintained. If this is a necessary part of the gene theory, the phenomena of embryonic segregation must, I think, lie beyond the range of genetics (pp. 14-15, italics added).

And Harrison wrote ten years later: "The prestige of success enjoyed by the gene theory might easily become a hindrance to the understanding of development by directing our attention solely to the genome, whereas cell movements, differentiation, and, in fact, all developmental processes are actually effected by the cytoplasm" (1937, p. 9).

Another issue formed an even more formidable obstacle to mutual understanding. A basic tenet of embryology has always been the structural and physiological unity, the individuality, of the embryo that continues through life. Lillie says: "The germ is physiologically integrated as an individual at all stages" (1927, p. 4). And the same theme pervades the entire lifework of Child, who was very influential at that period. Progressive differentiation, while creating an increase in complexity, proceeds within the confines of the individuum. This statement is not an abstraction but an expression of the epigenetic nature of development, of the internal inductive, regulative, and feedback mechanisms that integrate and synchronize developmental processes. (The embryologist has always been concerned with the individual embryo, moving from there to lower levels of organization, whereas one of the key elements of the modern synthesis was the conceptual shift from the individual to populations—another ground for alienation.)

To complicate matters, there are unitary subsystems with the same properties as the whole, which we define operationally as "morphogenetic fields." For instance, if in the tail bud stage of a urodele amphibian embryo half of the limb-forming mesodermal disc, or one half of the optic vesicle is removed, the residual group of seemingly undifferentiated cells restores the whole, and an organ of normal size and structure ensues. How would one handle such phenomena of regulation with the genetic concepts and methods available at that time? Lillie states: "Individuation is clearly an [internal] environmental relationship mediated through the cytoplasm, not through the nucleus" (1927, p. 13). Dalcq, in his review essay of Simpson and Cuénot (1951), puts the dilemma more succinctly:

> The cytoplasm of the egg is by itself an organized system along general lines, endowed with a pattern which the genetic system lacks. Moreover, it integrates the activity of numerous constituents in such a way that it is capable of regulation. Since Driesch's discovery [the experiment of regulation after blastomere isolation] the embryologists are forced to uphold more or less explicitly the notion of "configuration globale" [*Ganzheit*, whole]. This notion, so intimately tied to a pattern, is lacking in the system of concepts used by geneticists and notably in the synthetic theory. The latter is based on a particularistic, atomistic viewpoint which neglects, despite everything, this other factor which resides in the totality of the organization (p. 135; my translation).

Even though Dalcq may have been somewhat behind the times with his atomistic notion of the genome, the embryologists of that period can hardly be blamed for their failure to recognize the discrete units of the genome, which, moreover, were identical in each cell, as critical factors in the continuous flow of epigenetic development. Many contemporaries of Lillie shared his pessimistic outlook: "The progress of genetics and physiology of development can only result in a sharper definition of the two fields, and any expectation of their reunion (in a Weismannian sense) is in my opinion doomed to disappointment" (1927, p. 18).

Spemann had a more positive attitude. Coming from Boveri, he realized the close relationship of problems of localization and activation of hereditary factors. In his 1924 address to the German Society of Genetics, he tried to find common ground between the two fields. But he was no more successful than others in attaining a real synthesis. His general statement that "the activation of the genome does not occur by autonomous segregation of the hereditary factors but under far-reaching interaction between the parts, hence epigenetically" (1924, p. 78) was no more constructive than that of Wilson. But eight years later Spemann and Schotté (1932) reported the classical experiment of xenoplastic transplantation between anurans and urodeles that, though designed originally to deepen the understanding of embryonic induction, actually opened up problems of profound evolutionary significance. This aspect was taken up by Baltzer and his student E. Hadorn and their coworkers who made substantial contributions to what may be called evolutionary embryology (Baltzer, 1952).

To ensure that I have not overstated the amount of polarization, I returned to the book that we then considered as the ultimate arbiter in all such matters: Wilson's *Cell in Development and Heredity* (1925). The superb chapter, "Development and Heredity," is a succinct discussion of cytoplasmic organization with a survey of all experimental evidence for its role in development and an equally lucid presentation of nuclear organization. But, again, the synthesis is limited to the insight that all developmental processes including egg organization are controlled by genes and to the general statement that "heredity is effected by the transmission of a nuclear preformation which in the course of development finds expression in a process of cytoplasmic epigenesis" (p. 1112). If one of the most profound minds of his time with a full command of both genetics and embryology bypassed the problem of gene action in development, then obviously the embryologists of this generation were not ready to come to the aid of the architects of the new synthesis.

Beginnings of Developmental or Physiological Genetics

To overcome the impasse, it was necessary to make a fundamental conceptual shift from the antithesis nucleus versus cytoplasm to the idea of nucleocytoplasmic interactions in development. Driesch anticipated this notion as early as 1894. His hypothesis was based on the then-established premises that the nucleus is the bearer of heredity, that all nuclei in an embryo are equivalent and "totipotent," and that the egg cytoplasm has an organized structure created under the control of the oocyte nucleus. He postulated that the nucleus would effect chemical changes in its cytoplasmic environment by a "fermentative effect," in a way that would preserve its completeness or totality. "Determinative substances would not be released directly by the nucleus but originate in the cytoplasm under the control of the nucleus" (p. 88). The chemical changes in the cytoplasm, in turn, would cause the release of other specific fermentative effects in the nucleus leading to the next step in cytoplasmic differentiation, and so forth. This model, with its emphasis on the chemical nature of the interaction, on the indirect way in which genes control cytoplasmic differentiation, and on the notion of feedback between nucleus and cytoplasm, has a modern ring. But Driesch hardly made an impact on his contemporaries. His ideas were forgotten even by his friend Morgan with whom he spent several winters at the Stazione Zoologica in Naples; they certainly discussed nucleus and cytoplasm. Four decades later, Morgan (1934) presented the reciprocal feedback notion as his own novel solution to the problem of nucleocytoplasmic interaction.

A younger generation, with open minds, had to make a new beginning. Among them, Goldschmidt was the dominant figure. He became one of the founders of physiological genetics and remained its most forceful promoter. Beginning around 1915 with the study of sex determination and determination of pigment patterns in larvae of the moth *Lymantria*, he synthesized extensive experimental data and imaginative theoretical ideas in a quantitative theory of gene action, which for the first time placed the physiological role of genes as determinants of developmental processes in the center of the stage. Like Driesch who spoke of "ferments," Goldschmidt postulated that because genes have enzyme properties, their influence is of a chemical nature. A key element in his theory is the assumption that genes operate by controlling rates or velocities of developmental processes and different alleles represent quantitative differences of gene activity. It follows that they produce variations in speed of development. As in Driesch's theory, a necessary corollary of the enzymatic nature of gene action is the assumption of specific substrates located in the cytoplasm, hence the postulate of cytoplasmic organization and a chain of specific gene-cytoplasm interactions. Because he considered the entire genome as operating in an integrated fashion, some of the difficulties that Lillie and Dalcq found insuperable were overcome. From the beginning, he emphasized the importance of physiological genetics for evolutionary theory; his book, *The Material Basis of Evolution* (1940), is a synthesis of his ideas.

I am not concerned with the speculative and controversial nature of some of his novel ideas, nor with the fact that his "rate-gene" theory was soon superseded by the biochemical genetics initiated by Beadle, Ephrussi, and Tatum. The historical fact remains that he broke new ground by supplementing the contemporary genetics, which was essentially transmission genetics, with a dynamic conception of the gene as a physiological agent controlling developmental processes. By creating this link, he opened a meaningful dialogue between some embryologists and some geneticists. I and other embryologists became interested in genetics in the 1920s largely by reading Morgan's book, *The Physical Basis of Heredity* (1919), which became available to us in German in Nachtsheim's translation in 1921. And we became avid readers of Goldschmidt's books of 1920 and 1927; they opened our eyes to the challenging idea that genes as factors in development could be incorporated into our experimental-embryological thinking. In historical perspective, Boveri in his merogony experiment of 1896 had already combined experimental embryological and genetic methods and thoughts. The design of cross-fertilization of enucleated eggs with foreign sperm to elucidate nuclear-cytoplasmic compatibility had been taken up in the school of Boveri's student, Baltzer (1940) and by some others. At any rate, through Goldschmidt, Baltzer, Herbst, and others, we were immunized early against the notion of an antagonism between experimental embryology and genetics. In the late 1930s, the burgeoning field of physiological genetics had already gained foothold. At the First Growth Symposium (1939, ed. Berrill), which, like its successors, was supposed to bring us up to date on research frontiers in growth and development and brought together animal and plant embryologists, geneticists, microbiologists, biochemists, and physiologists, two of the ten presentations (by Stern and Waddington) dealt with the role of genes in development.

Before embryology could make a meaningful contribution to evolutionary theory, the basic mechanisms of gene action and the concepts derived from them had to be worked out by the endeavors of developmental geneticists. Some of those concepts proved to be useful for evolutionary theory, particularly those developed in the period before 1940, which corresponded to the formative years of the modern synthesis.

The concept of heterogonic growth (Huxley, 1932) is closely related to Goldschmidt's concept of rate genes. The significance of heterogonic growth for evolution has been discussed by Huxley (1942) and Goldschmidt (1940). If a single gene controls differential growth rates in different parts of the organism, then a mutation can bring about a multiplicity of changes while preserving relative proportions, thus "lightening the burden of natural selection." Both authors point out implications for an understanding of neoteny, metamorphosis, vestigial organs, and extinction resulting from excessive growth of specialized structures.

The concept of "norm of reaction," which is significant in this context implies that the phenotypic expression of a gene is dependent on variables in the internal and external environment during development. Each gene has a potential range of expression, the extent of the range being characteristic of each gene. Modifications of the phenotypes resulting from this developmental plasticity are important raw materials for natural selection. Closely related is the concept of sensitive periods in development: restricted phases in a developmental process or a metabolic sequence are particularly susceptible to disturbances. Each developmental process has its specific sensitive period or periods. Chemical teratogens, extreme temperature shocks, X rays, or other agents produce malformations of specific structures such as eyes or legs, or their components, when applied at the pertinent sensitive period, but at no other time. The link to genetics was Goldschmidt's discovery (1935) that malformations produced in this way often have a striking resemblance to abnormal mutants. He produced "phenocopies" of several mutants of *Drosophila* by heat shock applied at the appropriate sensitive period. The implication is that phenocopying agents and the phenocopied gene interfere with the same developmental process or metabolic sequence; as a result, sensitive periods would represent phases of determinative gene actions. Although the hope that in this way one might locate the site of primary gene action was thwarted by the consideration that a sensitive period may be anywhere along the path from primary gene effect to the phenotypic end product and not necessarily at its beginning, this point does not detract from the importance of sensitive periods for evolutionary considerations.

The concept of pleiotropic or polyphenic expressions of single genes in different organs or metabolic processes helps to simplify the complexity of the problem of genetic control of integrated developmental processes. Penetrance is the frequency (in percentages) of the phenotypic expression of a gene. Penetrance below 100 percent indicates that intrinsic or extrinsic factors set a threshold for gene manifestation. Many factors are involved in the lowering of penetrance, such as the genetic background, including modifier genes, quantitative variations in gene products, or variations in environmental factors. But again any mechanism that introduces variability in gene-controlled developmental processes gives selection a foothold.

The term "pattern genes" does not refer to a special category but to a variety of genes that control the differentiation of integrated structural patterns. The older experimental embryologists such as Dalcq found insuperable difficulties in reconciling the atomistic configuration of the genome with the holistic and regulative aspects of epigenetic development and its morphogenetic fields and gradients. The demonstration of single genes that control determinative processes resulting in complex structural patterns, such as the pigment banding pattern in the wings of moths, or the bristle pattern on the thorax of *Drosophila*, or the toe patterns in forelegs and hindlegs of guinea pigs, goes a long way toward mitigating this misconception.

The operation of pattern genes is possible only in epigenetic development with its built-in plasticity. In only a few cases was a detailed analysis of pattern-gene action possible. One of the earliest and most detailed studies was the investigation of the pigment bands on the wings of the meal moth *Ephestia kühniella* by Kühn and Henke between 1929 and 1936. The pattern consists of alternating light and dark bands of different widths, with a bilateral symmetrical arrangement on each wing. The building blocks are scales formed as flattened outgrowths of single epidermal cells. Scales are classified according to differences in shape and pigmentation. Defect and heat shock experiments were used to establish the details of the determination process of the whole pattern, in which the spreading of streams of determining agents (probably diffusible substances) plays a major role. These processes of organizing the wing surface in bands that are, of course, invisible occur in the pupal epidermis long before the actual differentiation of the scales takes place. Mutants were found that modify the "wild-type" pattern by modifying or slowing down the spreading of the determination streams, thus creating abnormal total patterns; other genes modify pigmentation. The essential point is the genetic control of supercellular invisible patterns which in turn determine the fate of the individual scale-cells (Kühn, 1936).

Later, Stern (1954) was able to carry the analysis of several pattern genes to considerably greater depth by the ingenious use of genetic mosaics on a particularly favorable and much simpler system, the bristle pattern on the thorax of *Drosophila*.

Meanwhile, a discovery of far-reaching consequences had been made in Kühn's laboratory. His student, Caspari, working with a pleiotropic gene in *Ephestia* that affects the pigmentation of eyes, skin, testis, and brain, transplanted larval mutant testis into larval coelom of wild-type individuals, and vice versa, and found pigment changes in the host eyes, or host testis, respectively. This study provided the first evidence of diffusible gene-produced substances (Caspari, 1933; Kühn, Caspari, and Plagge,

1935). Shortly thereafter, and independently, Beadle and Ephrussi (1936), using a similar transplantation technique with larval eye discs, discovered diffusible eye-pigment-determining substances in *Drosophila*. These discoveries ushered in the era of biochemical genetics. At the same time, they demonstrated a new mechanism by which primary gene effects could be amplified.

The Missing Chapter

It was clear to the founders of the modern synthesis that embryological considerations had to be incorporated in evolutionary theory, for the simple reason that the only way by which genes can produce the material for natural selection—that is, phenotypic variability—is in their capacity as controlling agents and modifiers of developmental processes. But this chapter had not been written. It would have taken a biologist with very broad interests, who would be familiar with genetics, speciation, evolution and at the same time knowledgeable in experimental embryology and the intricacies of epigenetic development to write it. The only person of this rank at the time was Schmalhausen. In the foreword to his book, *Factors of Evolution* (1949), Dobzhansky says: "The book of I. I. Schmalhausen advances the synthetic treatment of evolution starting from a broad base of comparative embryology, comparative anatomy, and the mechanics of development. It supplies, as it were, an important missing link in the modern view of evolution" (p. ix). In fact, Schmalhausen had been active in research in all these fields.

Among the younger generation, Waddington in England had independently developed ideas similar to those of Schmalhausen during the war. He was also eminently qualified to provide the missing link; he had been engaged in research in experimental embryology and genetics and written basic books in both fields. His book, *The Strategy of the Genes* (1957), in which his ideas on the relationships of genetics, experimental embryology, and evolution are synthesized, could be considered a draft of the "missing chapter."

Three issues, or trends of thought, illustrate the kind of material that I think might have become part of the missing chapter of "evolutionary biology." First, the epigenetic mechanisms that have been elucidated by experimental embryology take a great burden off the genome in its role as controlling agent of developmental processes. The activity of a single gene operating at a particular focal structure can have widespread secondary effects through the mediation of diffusible gene products, hormones, inductions, and other epigenetic or physiological mechanisms. I have mentioned pattern genes. If the segregation process in the distal part of the limb bud subdivides the mesodermal field into six instead of five units, a polydactylous phenotype emerges. This result can be accomplished by a single mutation; how the gene operates is unknown. But each digit is complete in its complex structure. Neurogenetic mechanisms provide innervation for the extra digit, without requiring direct involvement of the "polydactyly gene."

The induction of the embryonic eye by the underlying mesoderm triggers a sequence of subsequent inductions: the optic vesicle once formed induces the lens when it contacts the overlying ectoderm, and the lens, in turn, induces cornea formation. A single gene, weakening the inductive capacity of the mesoderm, or a gene producing anophthalmia, thus initiates specific, remote structural changes. For a time it was fashionable to design pedigrees of indirect gene effects illustrating such ramifications. Epigenetic mechanisms reduce the number of genes required for the production of the structural and physiological complexity of the phenotype by amplifying the primary gene effects.

Second, the remarkable regulative properties of most eggs and embryonic subsystems have intrigued and puzzled embryologists since Driesch discovered this phenomenon in 1891. The regulation of isolated blastomeres can be described as an adaptive feature, but it is doubtful whether even a harmoniously proportioned dwarf larva has a good chance of survival. However, regulation can make sense in the context of natural selection as a stabilizing agent. Schmalhausen strongly emphasized this point and elaborated on this theme. He considered the regulative capacity of undifferentiated systems (such as morphogenetic fields) as a major agency in the stabilization of the phenotype (1949, p. 221). Generally speaking, both the genome and the differentiation processes must be balanced and integrated to guarantee the survival of the embryo as a whole; any mutational or other change of the genome that disturbs the equilibrium requires a buffering device. In epigenetic development, the capacity for regulation, in the broadest sense, is one of the most effective means to accomplish this feat. The embryologist Dalcq had a somewhat bizarre idea: he envisaged "ontomutations," that is, mutational radical alterations in very early stages, as, for instance, mutants affecting egg structure or gastrulation, and considered them as the basis of the origin of higher taxa. And he argues that such drastically modified systems could survive only by the stabilizing capacity for regulation (1949). Schmalhausen and Waddington clearly realized that the stabilizing effect of regulation or other epigenetic mechanisms can serve in two different situations: in so-called stabilizing or normalizing selection that aims at retaining the status quo in a population when conditions remain constant; or as buffers against perturbations of developmental processes in so-called dynamic or directional selection that aims at shifting the phenotype in adaptation to changing conditions. Waddington speaks of canalization rather than stabilization of developmental processes.

Third, both Schmalhausen and Waddington are particularly interested in a process that Schmalhausen calls "autonomization" and Waddington, more aptly, "genetic assimilation." The evolutionist is familiar with the following situation. Suppose an environmental change occurs in a well-adapted population. It will result in the selection of extreme variants from the range of nonheritable modifications of the phenotypes, which are best adapted to the new conditions but still within the norm of reaction of the unchanged genome. If the altered condition prevails over a long period (long in evolutionary terms), mutations or recombinations that guide development in the direction of the better-adapted phenotype will occur and become the target of positive selection. In this way, in the course of time, the well-adapted phenotype will be "assimilated" in the genome, giving the impression of inheritance of acquired characters. Cornification of the foot pads in terrestrial vertebrates that are formed already in the embryo is a common example.

Schmalhausen makes an additional important point. The process of genetic assimilation makes the developmental system independent of the environmental conditions or of external stimuli that originally were required to operate continually for the purpose of selection, as long as the better-adapted variant was a nonheritable modification. The shift to the better-adapted phenotype by the creation of a new, balanced genetic system—that is, internalization of developmental determining factors—contributes significantly to the stabilization of the phenotype. Generally speaking, genetic assimilation is an effective stabilizing device in the sense that it makes the development of the better-adapted phenotype independent of environmental factors.

The relations of experimental embryology, genetics, and evolutionary theory during the first half of this century were much too complex to be treated adequately in this brief essay. For instance, I have not dealt with the problem of the embryonic and evolutionary origin of highly specialized adaptations, such as concealing coloration that many embryologists found difficult to explain solely in terms of natural selection from small-step mutations. But here we enter the borderline field of science and philosophy, or Weltanschauung, which is beyond the scope of our discussion.

I am grateful to Dr. Garland E. Allen and Dr. Jane Oppenheimer for helpful suggestions.

References

Baltzer, F. 1940. Über erbliche letale Entwicklung und Austauschbarkeit artverschiedener Kerne bei Bastarden. *Naturwissenschaften* 28:177-206.

——— 1952. Experimentelle Beiträge zur Frage der Homologie. Xenoplastische Transplantationen bei Amphibien. *Experientia* 8:285-297.

Beadle, G., and B. Ephrussi. 1936. The differentiation of eye pigments in *Drosophila* as studied by transplantation. *Genetics* 21:225-247.

Berrill, N. J., ed. 1939. *First symposium on development and growth.*

Boveri, Th. 1895. Über die Befruchtungs- und Entwicklungsfähigkeit kernloser Seeigel-Eier und über die Möglichkeit ihrer Bastardierung. *Roux's Archiv für Entwicklungsmechanik der Organismen* 2:394-443.

Caspari, E. 1933. Über die Entwicklung eines pleiotropen Gens bei der Mehlmotte, *Ephestia kühniella*. *Roux's Archiv für Entwicklungsmechanik der Organismen* 130:354-381.

Dalcq, A. 1949. *L'apport de l'Embryologie causale au Problème de l'Evolution.* Portugaliae Acta Biologica, Serie A. Goldschmidt Volume, pp. 367-400.

——— 1951. Le Problème de l'Evolution, est-il près d'être résolu? *Annales de la Société Royale zoologique de Belgique* 82:117-138.

Driesch, H. 1891. Entwicklungsmechanische Studien I: Der Wert der beiden ersten Furchungszellen in der Echinodermen-Entwicklung. *Zeitschrift für wissenschaftliche Zoologie* 53:160-178.

——— 1894. *Analytische Theorie der organischen Entwicklung.* Leipzig: Engelmann.

Goldschmidt, R. 1920. Die quantitativen Grundlagen von Vererbung and Artbildung. *Roux' Vorträge und Aufsätze 24.*

——— 1927. *Physiologische Theorie der Vererbung.* Berlin: Springer.

——— 1935. Gen und Ausseneigenschaft. *Zeitschrift für Induktive Abstammungs- und Vererbungslehre* 69:38-131.

——— 1940. *The material basis of evolution.* New Haven: Yale University Press.

Harrison, R. G. 1937. Embryology and its relations. *Science* 85:369-374.

Huxley, J. S. 1932. *Problems of relative growth.* London: Methuen.

——— 1942. *Evolution: the modern synthesis.* London: Allen and Unwin.

——— and de Beer, G. R. 1934. *The elements of experimental embryology.* Cambridge: Cambridge University Press.

Kühn, A. 1936. Versuche über die Wirkungsweise der Erbanlagen. *Naturwissenschaften* 24:1-10.

———, E. Caspari, and E. Plagge. 1935. Über hormonale Genwirkungen bei *Ephestia kühniella. Nachrichten von der Gesellschaft der Wissenschaften zu Göttingen* 2:1-30.

Lillie, F. R. 1927. The gene and the ontogenetic process. *Science* 66:361-368.

Mayr, E. 1970. *Populations, species, and evolution.* Cambridge, Massachusetts: Harvard University Press.

Morgan, T. H. 1919. *The physical basis of heredity.* Philadelphia: Lippincott.

——— 1934. *Embryology and genetics.* New York: Columbia University Press.

Roux, W. 1881. *Der Kampf der Theile im Organismus.* Leipzig: Engelmann.

Schleip, W. 1929. *Die Determination der Primitiventwicklung.* Leipzig: Akademische Verlagsanstalt.

Schmalhausen, I. I. 1949. *Factors of evolution.* Philadelphia: Blakiston.

Spemann, H. 1924. Vererbung und Entwicklungsmechanik. *Naturwissenschaften* 12:65-79.

——— 1936. *Experimentelle Beiträge zu einer Theorie der Entwicklung.* Berlin: Springer. (1938 English translation: *Embryonic development and induction.* New Haven: Yale University Press.)

——— 1943. *Forschung und Leben,* ed. F. W. Spemann. Stuttgart: Engelhorn.

——— and O. Schotté. 1932. Uber xenoplastische Transplantation als Mittel zur Analyse der embryonalen Induktion. *Naturwissenschaften* 20:463-467.

Stern, C. 1954. Two or three bristles. *American Scientist* 42:213-247.

Waddington, C. H. 1957. *The strategy of the genes.* London: Allen and Unwin.

Weismann, A. 1893. *The germ-plasm: a theory of heredity,* trans. W. N. Parker and H. R. Ronnfeldt. London: Walter Scott.

Weiss, P. 1939. *Principles of development.* New York: Holt.

Wilson, E. B. 1925. *The cell in development and heredity,* 3d ed. New York: Macmillan.

Wright, S. 1934. Genetics of abnormal growth in the guinea pig. *Cold Spring Harbor Symposium on Quantitative Biology* 2:137-147.

The Evolutionary Synthesis. Perspectives on the Unification of Biology. E. Mayr and W. B. Provine, Eds. Cambridge, MA: Harvard Univ. Press, pp. 97–112 and 303–308. 1980

Evolutionary Theory in Germany: A Comment

Viktor Hamburger

In Germany, in the twenties and thirties, the problem of adaptation was in the minds of most biologists much more than finer points of evolutionary theory. There was a widespread skepticism that straightforward selectionism could be the ultimate explanation of all complex phenomena of adaptation. Underlying the scientific discussion of these problems was also the inclination of the Germans for a metaphysical underpinning and the often subconscious need to combine their scientific thinking with a metaphysical *Weltanschauung.* Their metaphysics came, for the ones I knew, from *Naturphilosophie* of the early nineteenth century—Goethe, Carus—and from Kant and not, as in some cases in this country, from a religious, Christian background.

Let us focus on the problem of adaptation. Most scientists, at all times, are very reluctant to give away their private personal creed, in print or even in talks; that is why I consider the speech of Boveri of 1906, although it antedates the period we are interested in, as a singularly interesting document, because here he really spoke off the cuff (see chapter 3). As I understood from Baltzer, his student and biographer who knew him very well, Boveri regretted it later and never came back to this topic. But his essay on organisms as historical beings is really a unique document, and it was not repeated among my acquaintances until Spemann, shortly before his death in 1941, committed to print some similar confessions in his autobiography that I'm sure nobody has read.

The biologists I knew, including Spemann and Kühn and many other nonembryologists and nongeneticists, did not feel the need for a unitary explanation that would be valid for all adaptations. Boveri and Spemann made it very clear that they wanted to believe that different options were open to them. I think that was very important. None of them were strongly opposed to selectionism; they all conceded considerable ground to selectionism but they did not consider it a cure-all. Whenever needed they would let Lamarckian thinking slip in and some (Pauly, Spemann) would even admit psychic factors, but always in conjunction with everything else. Only Spemann had an overall psychic idea—a vitalistic view that was taken over from Pauly; Boveri never went that far.

Why this hesitation to accept selectionism as the one and only explanation? I think first and foremost it was the complexity of adaptations, such as concealing or other coloration patterns. For example, the display of the tail feathers of the peacock forms a total *Gestalt* to which each feather and each part of each feather contributes a fragment; these fragments must fit into a total pattern when all feathers come together in this display. To make things more complicated, the hormonally controlled behavior, the state of excitation, must be linked up with the display. Everything must fit, from behavior to the deposition of pigment granules by melanophores and other pigment cells during the development of the feather. Feather development poses many such complexities that challenge the imagination of the embryologist. Each feather develops from a small epidermal elevation that grows to form the shaft and its sidebranches, the barbs, and barbules. The pigment cells feed pigment granules into the barbules, and each eye spot of the completed feather, for example, with its dark center and colored rings around it, is "painted" across many barbs. But it is much more like weaving, where the dyes (granules) are arranged in a sequence of different colors in each thread (barb), each at a definite length; and the adjacent thread (barb) has the same sequence but in proportionately longer or shorter lengths, and so forth, until the deposition of black central pigment and then of the colors of the inner, then of the outer, rings is discontinued. And an eyespot is only a small fraction of the total *Gestalt*. Then the pigment-producing cells die, leaving the dead pigment pattern encased in the dead keratin of the barbs and barbules—and natural selection operates on these "dead" patterns. Even if we make allowance for epigenetic interactions between pigment-producing cells and other embryonic regulatory devices, the embryologist is faced with interpreting even minor changes of the total *Gestalt* in terms of small mutational steps. This example could be multiplied hundredfold, in protective coloration alone.

We were always confronted with arguments against selectionism, but no one made an effort, except perhaps the Kühn-Henke school, to really get at one specific example of a complex adaptive color pattern and analyze it in terms of developmental mechanisms.

Süffert, who was a colleague and friend in Freiburg, analyzed these complex adaptations structurally and behaviorally. His own position was very interesting. In Spemann's laboratory we had almost daily tea sessions in the afternoon that sometimes extended for hours, centering often through Süffert's presence on the origin of such complex patterns of adaptation. (Evolutionary problems were hardly discussed; we were not aware of population genetics.) Süffert studied adaptation in the wings of butterflies and moths and he showed how, for example, the pupa of a particular species has a pattern that, on the principle of countershading (which obliterates three-dimensionality) is effective only if the caterpillar selects for pupation the shaded side of a tree. But the hatched moth must move to the other side of the bark of the same tree to make its concealing pattern effective. The switch in behavior has evolved parallel with the change from concealing pupal to imaginal coloration. Süffert was a strict selectionist; he tried to explain to Spemann that all these extremes could be explained in principle by selectionism, but Spemann remained adamant. That dialogue was carried on over many years.

The same scientists found it difficult to accept pure selectionism for other reasons, as well. For example, the individual is always completely integrated and it seemed to Boveri, Goldschmidt, and others that a major evolutionary change would require macroevolutionary "mutations" that would lead immediately to a new organization. "It is not the evolutionary changes of organisms per se that stimulate our curiosities so powerfully, but that the changes are teleological (*zweckmässig*) measured by human standards, or more concretely: not the small modifications are important for us whereby a new species can be distinguished, but those big steps call for an explanation according to which the water animals become land animals, land animals again water animals, crawling animals become flyers, nonvisual ones become visual, and instinctive drive becomes reasoned action" (Boveri, 1906, p.16). Then Boveri asked the loaded question: "Which forces could be able to effect this?" And from there he got into psychic explanations.

It was also argued that it is hard to believe that use itself should not have any formative effects on adaptations because they can actually be seen in modifications. The living organism can individually adapt itself by developing strong muscles or keratin pads on the skin—in other words, the Lamarckian argument was revived. In summary, in the minds of many biologists who were evolutionists there were two or three very powerful phenomena that acted strongly against acceptance of selectionism as the one and only solution.

Although Spemann's ideas were not unique, he was articulate, and his way of thinking came out very well in personal discussions and also in his book. Spemann came from comparative anatomy, from Gegenbaur in Heidelberg, but also from Goethe's idealistic morphology and the principle of interconnectedness of St. Hilaire. This tradition was very much alive; Goethe was still widely read; after all, he had coined the term "morphology," so there was some legitimacy in these connections. Of course, Spemann himself, sustained by his discoveries of epigenetic mechanisms in development, always retained the holistic-interconnection picture. On the other hand he was, as he says in his autobiography, very strongly inclined toward causal analysis and he found his way into experimental embryology through Weismann, who inspired the constriction experiments whose interpretation in turn led to everything else that Spemann did. Spemann is a good example for the observation that a broad, synthetic vitalistic *Weltanschauung* is entirely compatible with a keen analytical acumen, and in his case energized it.

Spemann's only excursion into phylogeny was an article on homology (1915). I remember that Spemann got himself into deep water because he couldn't easily adjust the old concept of homology with data of experimental embryology. For example, he got into this conceptual dilemma: whether a lens that in the embryo is formed by invagination from the epidermis as a result of induction by the optic vesicle is homologous to the lens that in salamander lens regeneration originates from the upper iris. Obviously it is, but that means then that developmental mechanisms cannot be used for definition of homologies.

I did not find in Spemann a burning interest in Mendelian or Morgan-school genetics, in contrast to Baltzer who was also in Freiburg when I was (1918-1920); both were Boveri's students. Baltzer inherited this interest in genetics and actually lectured on genetics. At that time he was working on species hybrids and the hybridization method was one way of looking at evolution. I suspect that the "atomistic" aspect of the gene theory was uncongenial to Spemann although I don't remember ever having heard him say so.

The strange psycho-Lamarckian convictions that Spemann developed became stronger as he got older. Pauly's influence, which dated back to Spemann's student days, persisted throughout his life. (For more details see Hamburger, 1969.) Boveri was about fifteen years older than Spemann, and Boveri was the first one who established contact in Munich with Pauly. Pauly was a very artistic, violent-tempered biologist in Munich who developed a strong psycho-Lamarckian concept of evolution and was very intolerant of anything and anyone who contradicted or slightly disagreed with him. As a matter of fact, Boveri's *Antrittsvorlesung* of 1906 was a direct answer to Pauly's *Darwinismus und Lamarckismus* (1905). Pauly had a strange, persuasive influence on two of the most strong-minded persons of this period, Boveri and Spemann. Both were very critical of Pauly's dogmatism that went to absurd unscientific extremes. In Baltzer's book on Boveri (1962) it is told that Boveri wrote to Pauly: "I'm sure you won't be quite satisfied with my lecture," because Boveri obviously did not agree with all of Pauly's ideas. A strong letter came back from Pauly, the content of which is unknown; but it must have been hard on Boveri. He wrote back: "It isn't that bad, what I did." Nevertheless both Boveri and Spemann were attracted to Pauly by the widespread German tendency to have a unified *Weltanschauung*, a more general overview of metaphysical and scientific creeds than just having *Weltanschauung* and scientific work compartmentalized side by side. Many churchgoing scientists also have a similar compartmentalization. I believe that Spemann never quite resolved this conflict. Whereas Pauly had an all-pervasive psychic principle to explain evolutionary changes, for Spemann and Boveri it was only the last resort. They took recourse to this principle only when everything else—Lamarckism or selectionism—failed; but it was recognized as a third option.

Spemann, toward the end of his life in 1940, wrote in his autobiography a kind of confession along these lines that indicates that there were obviously conflicts in his thinking; it was not a monolithic, uniform philosophy of science. On the one hand, he said, he used psychic agents only for the specific cases that he could not explain otherwise; on the other hand, he was vitalistic on general principle. He wrote: "The encounter with Pauly was for me of decisive significance. It reinforced my basic conviction so consonant with my earlier thinking that the organism in all its living parts is *beseelt* (endowed with psychic properties), no less although in a different way than the organ from which we know this function in ourselves, the brain." Then he went on: "Today I am more firmly convinced than ever of this basic kinship of all life processes, since I have come to know through my own experimental work that the same cell group which seemed to be destined to form skin can also become brain if transplanted in early development into the region of the future brain. Hence we are standing and walking with parts of our body which we could have used for thinking if they had been developed in another position in the embryo . . . This for me is not an assumption or a supposition but an irrefutable certainty" (1943, p. 167).

Two points must be considered in deciding whether Spemann's metaphysical tendency inhibited or rather stimulated his research. The causal analytical kind of thinking in which Spemann was extremely talented was not influenced by his metaphysical thinking. On the other hand, I have pointed out that his selection of experiments from among a large number of possibilities was dictated by his holistic, organismic thinking (Hamburger, 1969). For example, very early he raised the question of the chemical nature of the organizer. He pushed it back in his mind, and only much later, in 1931, performed experiments in which organizer tissue deprived of its structure was tested for its inductive capacity; he was not successful. A year later Holtfreter proved the chemical nature of the organizer by bold, large-scale experiments. In a relatively early paper, in 1918, Spemann mentioned that completely isolating a part of the gastrula in tissue culture would be an ideal way of getting at the problem of determination, but he never performed the experiment—which indicates that the option was there but he chose not to take it. He chose instead to do other experiments in which the embryo was left intact; for example, he tried to make an organizer out of a piece of belly ectoderm by transplanting it in the organizer region. Everything that could be done with the living embryo was preferable to using dead embryos, dead tissues, and explanted, isolated tissues. In that respect, I think, his basic philosophy entered subconsciously into his scientific work.

References

Baltzer, F. 1962. *Theodor Boveri, Leben und Werk eines grossen Biologen.* Stuttgart: Wissenschaftliche Verlagsgesellschaft. (*Theodor Boveri*, trans. D. Rudnick. Berkeley: University of California Press, 1967.)

Boveri, T. 1906. *Die Organismen als historische Wesen: Festrede.* Würzburg: H. Stürtz.

Hamburger, V. 1969. Hans Spemann and the Organizer Concept. *Experientia* 25:1121-1128.

Pauly, A. 1905. *Darwinismus und Lamarckismus.* Munich: E. Reinhardt Verlag.

Spemann, H. 1915. Zur Geschichte und Kritik des Begriffs der Homologie. *Die Kultur der Gegenwart.* III. Teil, IV. Abteilung, vol. 1. Allgemeine Biologie, ed. C. Chun and W. Johannsen. Leipzig and Berlin: B. G. Teubner, pp. 63-86.

———— 1943. *Forschung und Leben*, ed. F. W. Spemann. Stuttgart: J. Engelhorn.

V. Book Reviews

Reprinted from THE QUARTERLY REVIEW OF BIOLOGY
Vol. 18, No. 3, September, 1943, pp. 263–268

EMBRYOLOGIA CHEMICA VERA IN STATU NASCENDI

Being a review of *Biochemistry and Morphogenesis*, by Joseph Needham. *Cambridge* (The University Press); *New York* (The Macmillan Company), 1942. Pp. xvi + 787. 9¾ x 6½ inches. $12.50.
By Viktor Hamburger, Washington University, Saint Louis.

Five years ago the English edition of Spemann's book, *Embryonic Induction and Development*, appeared. It marks, in a sense, the end of the "classical" period in experimental embryology. It surveys largely Spemann's own experiments, and those of his collaborators; yet it is, at the same time, the most articulate expression of the contemporary approach to embryonic development. Theoretically the epigenetic concept of animal development which it upholds has become firmly entrenched in our minds. Methodologically a marked unity of purpose was accomplished by a self-imposed limitation. Problems, experiments and terminology were kept strictly within the range of the biological level. The interactions of living systems during development were analyzed, without taking recourse to the biochemical level, and in disregard of a semi-quantitative or mathematical treatment. (It is true that the chemical nature of inductors was discovered by experimental morphologists, and not by biochemists. Spemann posed the problem as early as 1925, and most of the pioneer work was done by Holtfreter. Yet it may be said that the main emphasis of these investigations was on the morphogenetic response rather than the nature of the mediators.) The requisite of the mind for causal explanation was satisfied by establishing causal relations within this realm of biological entities. This consistent attitude has given rise to many misgivings, misunderstandings, and controversies. Its accomplishments testify to its soundness.

Meanwhile, experimental embryology has reached a new stage in its life cycle. A new category of problems has arisen which originated in a new philosophical outlook. Needham's book, though deeply rooted in the tradition of Spemann's and Harrison's ways of thinking, marshals this new period. It prepares the ground by clarifying the issues and gathering material. The author's theoretical point of view has been expressed before, in his essays *Order and Life* (1936) and in his Herbert Spencer Lecture in 1937. It is restated with clarity in the present book: "We cannot but consider the universe a series of levels of organisation and complexity, ranging from the subatomic level, through the atom, the molecule, the colloidal particle, the living nucleus and cell, to the organ and the organism, the psychological and sociological entities" (p. xiv). "Organising relations are found at the molecular level, and at the colloidal and paracrystalline level . . . just as clearly as at the anatomical level" (p. 679). "Only in the light of the conception of integrative levels can the saecular gulf between morphology and chemistry be bridged" (p. xv). It is admitted that "the regularities discovered by experimental morphology will always have their validity and will, in a sense, be unaffected by anything which either biochemistry on the one hand or psychology on the other, may discover" (p. xv). "But the important point is that although the regularities established at the level of experimental morphology are rirefragable, they will, in the absence of biochemical experimentation, remain forever meaningless" (p. xv). For Spemann, these regularities were eminently meaningful; for Needham "Meaning can only be introduced into our knowledge of the external universe by the simultaneous prosecution of research at all the levels of complexity and organisation, for only in this way can we hope to understand how one is connected with the others" (p. xv).

It is obvious that at the present time, the synthesis of biochemistry and morphogenesis is a postulate and a challenge for biochemists and experimental embryologists, rather than a fulfillment. What can be done is to collect the raw material and integrate it as far as possible. Needham is well equipped for such a task. His encyclopedic mastery of the literature is unique. Anybody who has consulted his *Chemical Embryology* knows that he is unrivalled in this respect. He draws freely from all corners of the biological universe, from experimental embryology, biochemistry, genetics, cancer research, endocrinology, general physiology. His bibliography includes approximately 6000 titles; the foreign literature is carefully covered, including publications which are not easily accessible. The list of references for the more important chapters of the book seems very nearly complete. In less important chapters where completeness is expressedly disclaimed, the author gives references for the most recent review articles.

It takes a person of Needham's constructive imagination to weld this enormous mass of data into a readable book. Fortunately, however, he did not succumb to the temptation of fitting the data into a uniform mold. Such an attempt at the present status of our knowledge would have been necessarily dogmatic and artificial. As a matter of fact, a direct biochemical attack on morphogenetic problems is in progress only with respect to the analysis of the inductor substances in Amphibian embryos, in the work of the Scandinavian school on sea urchin development, and in a few scattered investigations in developmental genetics. Most of the biochemical studies of the embryo are as yet more or less indirectly related to morphogenesis. The two main indirect avenues of approach are the chemistry of the materials which build up the embryo, and the study of metabolic and respiratory processes during development. Ultimately, they will merge, because inductors and gene products must be integral parts of metabolism. Without attempting an artificial unification, Needham is realistic enough to let the three stand singly, each on its own feet. Logically, then, the book is divided into three parts: I, The Morphogenetic Substratum (100 pages); II, The Morphogenetic Stimuli (400 pages); III, The Morphogenetic Mechanisms (200 pages). Parts I and III cover about the same ground as *Chemical Embryology* (1931); they bring this earlier book up to date by a review of the literature of the decade from 1931 to 1941. Part II is largely original. It gives extensive treatment to inductors in Amphibians, and is also a comprehensive review of the literature on embryonic induction in other vertebrates and invertebrates, with special emphasis on the agents which are the mediators of the inductive processes. In addition it includes chapters on "organisers and cancer", on "genes and organiser phenomena," on regeneration, metamorphosis, etc.

It is one of the merits of Needham's book that it stresses the theoretical issues. The following review will emphasize this aspect rather than discuss in detail the factual data, which would be an impossible task. Furthermore, it is not possible for any one reviewer to do equal justice to all parts of the book. My own interests are centered in the second part, and my review of the rest must be quite casual.

Part I. The Morphogenetic Substratum

The first chapter discusses the composition of the hen's egg and the properties of the vitelline membrane; it gives a brief survey of the effects which vitamins in the diet have on the development and hatchability of the embryo. Similar accounts are given of the chemical composition of the eggs of other vertebrates and invertebrates. The next chapter takes up a particularly interesting phase of embryonic adaptation: the dependence of different types of eggs on their environment. In the course of evolution the egg became increasingly emancipated from the environment; the process culminated in the closed-box ("cleiodic") type of egg (reptiles, birds) and reversed itself again in mammals. The remarkable adaptive changes in chemical mechanisms, such as water uptake, metabolism and excretion which are necessitated by the transition from aquatic to terrestrial habitat, are reviewed. The last chapter deals with embryonic nutrition, yolk utilization, the properties of the embryonic membranes of the chick and of the mammalian placenta, placental transmission, and the allantoic and amniotic fluids.

Part II. The Morphogenetic Stimuli

This section covers the classical field of experimental embryology, embryonic determination. It begins with a discourse on the "general concepts of causal morphology," such as "determination," "competence," "fields," etc. Their clarification is important in view of the rather confused state of terminology in this field. Unfortunately we cannot say that Needham contributes materially to the elucidation of these concepts. The term "determination" is logically traced back to Driesch. It is defined as "progressive restriction of possible fates or potencies" (p. 101). This definition is negative (is determination really a block, an inhibition?) and the vague term "potency" is not defined further. In fact, this term (without further specification) should be relegated to that most useful scrapheap of obsolete and ambiguous terms which the author has created at the end of the glossary. The term includes at least three entirely different things: self-differentiating potency, reaction potency (now called "competence") and inductive potency.

An acceptable translation of the much-used terms *ortsgemäss* and *herkunftsgemäss* is proposed: "neighborwise" for the former, and "selfwise" for the latter (p. 102). The term "morphogenetic stimulus" which is revived in the heading of this part of the book is traced back in its history to Herbst's "formative stimuli", but defined neither here nor in the glossary. Spemann's concept of the "hierarchy of organizers" is adopted. Spemann called the archenteron roof the "primary organizer," and the ones which appear later the "secondary," "tertiary," etc., organizers. (For example, the optic vesicle as lens inductor would be a secondary organizer.) It would be better to restrict the term "organizer" to the living archenteron roof, and to call the higher-grade organizers, "inductors." Among the characteristics of the archenteron roof are its regulative properties and its highly complex effects which result in the induction of a whole secondary embryo. Needham himself stresses the difference between simple induction of a structure like neural tube, and complex induction of an integrated patterned organ ("evocation" and "individuation"). The term "organizer" should be applied to the latter aspect of induction only. The term "competence" is used in preference to "reaction potency," which seems justified particularly since the useful adjective "competent" can be derived from it. "Competence" is conceived as a state of unstable equilibrium; "determination" as the stepwise transformation into more stable states of equilibrium.

The much debated "field" concept is treated briefly in this part of the book. It is also defined in terms of instability; ambiguities in its usage are pointed out. The term is widely used in the book, in the combination "individuation field" which will be discussed later. The short essay on "the liquidation of the entelechy" reduces a complex theoretical problem to its bare essentials and resolves some of the artificial and real difficulties of the age-old mechanism-vitalisms controversy.

The central, and most important part of the book, is the discussion of Amphibian morphogenesis. The author gives an exhaustive and clear account of the entire field of organizer experimentation. It is carefully done and correct in historical and factual details. The experiments on respiration and metabolism of the gastrula, including the author's own contributions, are critically considered and linked with the morphogenetic data. Of interest are the speculations on the possible combinations of the evocator substance with proteins and polysaccharides which render it inactive in the upper lip and in other tissues, and on possible mechanisms of its liberation after invagination. This part is profusely illustrated with reproductions of original photographs, drawings and graphs. As is well known, a good deal of discussion and controversy has centered around the question of the chemical nature of the medullary plate inductor (the primary evocator substance). There has been no doubt that induction is mediated chemically, ever since Holtfreter (1933) and others showed that killed tissues can induce medullary plate and since, a year later, the Cambridge and the Freiburg schools simultaneously found that cell-free extracts have inductive capacity. The question is still open whether we are dealing with a simple, definite substance which is contained in the upper lip and liberated during and after invagination. The experiments of Needham and his co-workers have led him to the conviction that there is such a substance and that it is probably a sterol. The strongest argument which,

however, has failed to convince me, is the extremely low concentration at which certain sterols are effective. Their minimal effective dose compares with that of other biologically active substances, such as vitamins, hormones and carcinogens. But in connection with this argument it should not be forgotten that other medullary plate inducing substances have not been subjected to a rigid quantitative test. At any rate, the controversial material is presented in an objective way. It is stated toward the end of the discussion that "neural inductions can be brought about by a wide variety of chemical fractions and pure or relatively pure substances. Which, if any, of these is identical with the primary evocator occurring in the dorsal blastopore lip, ... is as yet an unanswered question" (p. 186). Needham's assumption that the "primary evocator" is a sterol leads to interesting speculations as to its possible relations to sex hormones, Vitamins A and D, and carcinogens, all of which are steroid compounds. A chart (p. 262) depicts the "overlapping domains of biological specificity" of these related compounds. For instance, certain carcinogens are at the same time "neurogens" (another new term for the primary evocator substance) and so are certain oestrogens. Therefore the fields of neurogens, oestrogens, and carcinogens overlap. Such a graphic presentation is certainly striking, but I hope the chart will not find its way into the textbooks until the steroid nature of the "neurogens" is definitely proved. Similar ideas were already set forth in *Order and Life*. They are pursued at length in the chapter on organizers and cancer. The immediate link between the two is the discovery of Waddington, Needham, and others, that a number of carcinogenic substances act as medullary plate inductors. This alone would not necessarily be significant because a large number of other substances do the same. But Needham tries to find other, more general relations between organized and malignant growth. He goes back to the distinction of "evocation" and "individuation." He considers organized growth or morphogenesis as the result of two component activities: evocation, or the stimulation of competent tissue to undergo a specific but not patterned differentiation, such as nerve tissue; and individuation, or pattern induction in a previously "evocated" structure. Malignant growth he considers as evocation without individuation, or in his words, "escape from the individuation field." At present very little is gained by this formulation, because the concepts "individuation" and "individuation field" themselves are still obscure. Altogether, one has the impression that the length of this chapter is not in proportion to the few facts which link, at present, morphogenetic experiments with cancer research.

The next chapter deals with regional differentiation. Spemann discovered regional differences in the inductive capacity of the archenteron roof which led to the distinction of "head organizer" and "trunk" or "tail organizer." These and a number of experiments by

Holtfreter, Hall, and others, are reviewed, demonstrating specifically different effects not only of the organizer regions but of different adult tissues as well. How cautious one must be with the interpretation of experiments of the latter type may be seen from the results of Chuang, who found that fresh adult newt liver induces both head, trunk and tail structures, whereas mouse kidney induced only head structures. Toivenen, who implanted alcohol-coagulated liver and kidney (from different vertebrates) found just the reverse, which both he and Needham fail to point out. It will be necessary to experiment on a much larger scale using different adult structures as inductors, before the assumption that they produce specifically different effects can be considered as established.

Once the main axis of the embryo is laid down by the "primary organizer", secondary and tertiary inductors come into play which block out and determine further details of organization. When a structure has become determined it may, in turn, become an inductor for another structure. For instance, the optic vesicle, once determined by anterior archenteron roof, becomes a secondary inductor for lens, the lens becomes a tertiary inductor for cornea. Needham gives a useful tabulation of "second and lower grade organiser effects" (p. 302), but discusses in detail only a few selected cases. The account of the lens induction with all its ramifications is excellent.

At this point I should like to discuss several theoretical issues which are of importance for the evaluation of Needham's book. They are most clearly exemplified in the two chapters reviewed above. Needham is inclined to assume that the different lower grade inductors produce each a specific chemical substance. This is consistent with his stand in the matter of the primary organizer. However, the experimental evidence for this assumption is very meager and altogether not too favorable for such a hypothesis. For instance, no lens inductions have been obtained as yet from substances liberated by dead eye cups, although it must be admitted that this experiment has been made only on a small scale. On the other hand, isolated lenses were obtained in the absence of optic cups by induction from salamander liver and heart. All we can conclude from this experiment is that lens evocation is mediated by a chemical substance which is either widely distributed or which can be substituted by any number of agents. Needham argues: "If they (adult tissues) contain the primary evocator why not the secondary ones also? The disinclination to postulate a considerable number of active substances is, I believe, misguided, for there is no lack of chemical compounds isolated from living tissues for which no function has ever been found or even suggested" (p. 299). This argument of the biochemist does not carry much weight in the eyes of the morphologist. I wonder if the search for specific evocator substances promises at present the most fruitful line of attack of the induction problem. If carried to the

extreme it will certainly do more harm than good. Take for instance the case of the so-called "individuation." Needham makes ample use in his book of the distinction between "individuation" and "evocation." They are usually illustrated by the case of the induction of the central nervous system. Two components in its determination are distinguished: one, the determination of nervous structure or neural tube in general, called "evocation"; and the other the determination of its regional character, for example, forebrain, hindbrain, or spinal cord, called "individuation." Sometimes it seems that the terms "individuation" and "individuation field" are entirely restricted to the determination of the central nervous system (p. 685), which would be unfortunate because limb and ear, and other organ inductions certainly involve the same components of pattern formation. I do not wish to enter into a discussion of the merits of the term "individuation field." It seems to me that it is simply another refuge for those as yet unanalyzed components of the inductive process which make for wholeness and for pattern. Such terms are as useful, on the biological level, as the old "harmonious equipotential system." They will one day become unnecessary when our knowledge is advanced. But to invent a special type of hypothetical substance, which would be responsible for "individuation" in the same sense as the well-established evocator substance is responsible for "evocation," and even give it a name, "eidogen," certainly carries this concept too far. Take, as an example, the rôle of the host in induction processes. If trunk organizer is implanted at trunk level, it induces trunk structures; if it is implanted at head level, a secondary head results. It is legitimate to describe tentatively this host influence by stating that a "head-individuation field" exists which overrides the tendency of the transplant to induce trunk structures. It is likely that we are dealing with a complicated interplay of many factors and agents which will be difficult to analyze, but are amenable to analysis. To think of the host influence in terms of a "head-eidogen" substance means not to heed the sound advice which the author himself gives: "While it is convenient for purposes of discussion to speak of the individuation field, we must be careful that this concept is not used in such a way as to sterilize further research" (p. 278). I would suggest a more dynamic and more flexible working hypothesis for the analysis of induction. I would conceive of determination of a given structure as being brought about by the interaction of a multiplicity of agents carefully integrated in time and space. Some of these would reside within the system to be determined (they make up what is now called "competence"), others reside outside of it in the inductors. Any one of these agents is replaceable by others; no single one of these multiple factors would deserve the rank of "the specific" inductor. This is not the place to elaborate on this idea, but many puzzling experimental results are easier understood by this than by Needham's hypothesis. Among other things, gradients, which find no place in Needham's book, would fit into my scheme. The almost complete omission of gradients both in Child's particular formulation, and in the more general form of diffusion or concentration gradients of substances, is hardly justified in a book of such wide scope. One of the few exceptions is the discussion of the double gradient system in Echinoderms which is thoroughly established by the Scandinavian school. Child's theory is disposed of in a half page, and the author's position in *Chemical Embryology* is merely reiterated (p. 605).

In this second part of the book is included a chapter on "Genes and organisers." The author takes a farsighted attitude in the matter of integration of embryology and genetics. "Organisers must indeed be regarded to a large extent as the intermediary mechanisms between the gene equipment and the final form and properties of the developed animal. It is therefore important to know how genes and organisers are related in normal development" (p. 340). Several roads lead in this direction. In the first place, the experiments of Spemann and collaborators are reviewed in which it is shown, by means of heteroplastic and xenoplastic transplantations, that inductors can act across genetic boundaries. The next chapter, headed by the inadequate title "Genes and the later stages of individuation," deals with the experiments of Harrison, Twitty, and others, which give important information concerning the maintenance or adjustments of species specific growth rates of organs, when transplanted to hosts of a foreign species. The literature on the respective rôles of nucleus and cytoplasm in development is skilfully condensed. The work of Baltzer, Hadorn and coworkers on hybrids and hybrid-merogons in Urodeles is given deserved prominence. An entirely different approach to the problem "genes in development" is given by the analysis of lethal and semi-lethal genes. Again, only a small number of representative examples is selected, taken from *Drosophila*, chick and mammalian literature. The author emphasizes the important point that many phenotypic manifestations must be considered as the result of gene action on inductors; that faulty or suppressed inductor action must result in widely scattered morphological abnormalities; in other words, that many pleiotropic gene effects are secondary or indirect. Gene-produced hormones and diffusible substances are reviewed in a special chapter.

Of the other chapters in this section two deserve special mention: those on determination in insects, and in echinoderms, which give a concise and up-to-date summary. The chapter "Mosaic eggs," on the other hand is entirely inadequate and partly inaccurate. In fact, the term "mosaic eggs" should also have been buried in the grave-yard of obsolete terms. The definition *"mosaic eggs* are those in which this determination [of the main features] has taken place before fertilization and cleavage; *regulation eggs* are those in which this determination does not take place until about the time of gastrulation" (p. 131) is incorrect. Take, for instance, Echinoderm eggs as an example of the so-called "regulation" eggs, and Annelid and Mollusc eggs as an example of "mosaic" eggs. Both have definite "mosaic" features even before cleavage: a stratification of qualitatively different substances along the animal-vegetal axis. As a result, egg fragments of both types obtained by horizontal cuts before cleavage will give rise to fragmentary larvae, indicating "mosaic" development, whereas cutting along the main axis will result in "regulation" and subsequent formation of whole larvae from each half. Both types of eggs show therefore a strikingly similar degree of "determination" at this early stage. The same holds for the other representative of "regulation" eggs, the Amphibian eggs, in which the "marginal zone" (future organizer) is segregated shortly after fertilization. One of the main differences between the two "types" is in the rôle which cleavage plays in the sundering of these different substances. In Annelid and Mollusc eggs, and also in Ascidian eggs, it serves as a mechanism of segregation so that the blastomeres contain qualitatively different materials; in Echinoderms and Vertebrates, cleavage has nothing to do with the distribution of these morphogenetic agents. The author repeats a mistake which has become "classical" since Driesch made it when he interpreted his isolation experiments on sea urchins: to judge the state of determination of an egg by the outcome of isolation experiments of blastomeres.

Part III. The Morphogenetic Mechanisms

This part is a rather heterogeneous assembly of quite different topics. The main chapters deal with respiration and metabolism of the embryo. They are preceded by chapters on dissociability of fundamental morphogenetic processes and on "heterauxesis." A large amount of data is compiled to show that a number of well-defined partial processes such as growth, cell division, nuclear division, can be disengaged from each other experimentally. The yield of this information, as far as a deeper understanding of morphogenesis is concerned, is meager. One learns not much more than that there is considerable leeway in normal development. The chapter on heterogonic growth, now called "heterauxesis," is not intended to give a complete survey of this topic but is limited to the increase in amounts of chemical constituents. It culminates in what the author calls "the chemical ground plan of animal growth." He presents the growth curves for a number of chemical entities in different animals, and shows "that organisms of extremely different morphological form give identical differential growth ratios for a given chemical substance ... Not "protoplasm" only, but also the changes which it undergoes in chemical constitution during the growth of the organism, would then be identical in all animals" (p. 558). This idea is later on qualified with respect to individual specificity of protoplasm.

The chapter on respiration takes up in succession eggs and embryos of different groups. It reviews in sufficient detail recent work of Clowes and Krahl, Tyler, Bodine, Brachet, and many others; that of the author and of others on respiration of the Amphibian gastrula is taken up in the chapter on the organizer. The discussion of respiration in mammals includes that of anaerobiosis in embryonic life and the review of the work of Barcroft and collaborators. In the following part on metabolism, which was the main subject of *Chemical Embryology*, the respective chapters in that book are brought up to date. The author takes up in succession: carbohydrate metabolism, protein metabolism, nuclein metabolism, lipin and sterol metabolism, and adds a special chapter on pigment metabolism. Most of it is straight biochemistry of the embryo. The relation of the chemical changes during development to the morphological changes is yet to be established.

The last chapter of the book discusses, under the somewhat inadequate title "polarity" the submicroscopic, micellar and molecular structure of protoplasm and its possible relation to morphogenesis. In this topic, which promises to supply the most direct link between both levels of organization, our ignorance and the almost complete lack of valid information is most keenly felt. The modern literature abounds in good discussions of protoplasmic structure and its molecular and micellar basis. The review in this book is too brief to give an adequate picture. The few attempts to correlate data on fine structure with those from experimental embryology, for instance with Harrison's classical studies on axial determination in Urodele limbs,

ears, etc., or with the process of neurulation, are premature. Entirely missing in Needham's discussion is any reference to the cell surface as a possible seat of morphogenetically significant events. The author rather envisages "that embryology may become largely the study of protein fibres and their planned distortion and orientation" (p. 673).

Altogether, this is one of the most stimulating books in the current biological literature. Thanks to the wealth of material incorporated in it, it is a valuable source of information for both embryologists and biochemists. The attempt to clarify theoretical issues, however, is not too successful. It is a very readable book, written in a lucid and personal style, and much more concise than *Chemical Embryology*. It is excellently illustrated. Many of the 328 figures are photographs from original experiments; most drawings are originals. The voluminous bibliography is condensed into a relatively small space (69 pages) by a technique of rigid abbreviations of the titles of periodicals. The reader pays for this economy with a slight inconvenience; he must consult constantly the key to the abbreviations of journals, unless he is an ingenious guesser. In addition to a glossary there is a list of "terms and concepts the use of which is not recommended."

I should like to suggest two technical improvements. First, the sources of origin are omitted in the legends of almost all figures, and not even the references in the text indicate them in every instance. It is desirable that acknowledgments of the authorship of all illustrations be added. Furthermore, it would enhance considerably the usefulness of the bibliography if each reference were annotated with the page, or pages, at which the author is quoted.

Reprinted from
QUARTERLY REVIEW OF BIOLOGY
Volume 43, Number 2,   June 1968

NEW BIOLOGICAL BOOKS

The aim of this department is to give the reader brief indications of the character, the content, and the value of new books in the various fields of Biology. In addition, there will occasionally appear one longer critical review of a book of special significance. Authors and publishers of biological books should bear in mind that THE QUARTERLY REVIEW OF BIOLOGY *can notice in this department only such books as come to the office of the editor. All material for notice in this department should be addressed to Bentley Glass, Editor of* THE QUARTERLY REVIEW OF BIOLOGY, *Department of Biological Sciences, State University of New York, Stony Brook, N. Y. 11790, U.S.A.*

MALPIGHI THE MASTER

BY VIKTOR HAMBURGER

Department of Biology, Washington University, St. Louis, Missouri, 63130

A Review of

MARCELLO MALPIGHI AND THE EVOLUTION OF EMBRYOLOGY. *Volumes I-V.*
By Howard B. Adelmann. Cornell University Press, Ithaca, New York. $200.00 (complete set of 5 volumes). Vol. I: xxiv + 726 p.; Vol. II: p. 727–1013 + 12 pl.; Vol. III: p. 1016–1526 + 1 pl.; Vol. IV: p. 1527–2062 + 1 pl.; Vol. V: p. 2063–2475 + 1 pl.; all volumes ill.; subject index (Vol. V). 1966.

This is a monumental historical treatise, centered around one of the founders of scientific biology, the 17th Century Italian, Marcello Malpighi (1628–1694). Its author, the outstanding embryologist and historian of embryology, Howard Adelmann, is best known for his book on the 16th Century anatomist and embryologist, Fabricius. The present treatise, which consists of five folio-size volumes, is a monument not only to Malpighi but also to the author, who spent several decades in preparing it, to the Cornell University Press which edited it, and to the craftsmanship of the printer, the Oxford University Press in England.

At the core of the treatise are the two famous dissertations on the embryology of the chick which Malpighi sent to the Royal Society of London (of which he was a foreign member) during the year 1672. The first, "On the formation of the chick in the egg," was published in the same year, and the second, "Repeated and additional observations on the incubated egg," was printed two years later. Both together comprise only 27 printed pages and

11 plates of drawings. As Adelmann's story unfolds, it becomes clear how greatly their scientific significance transcends their modest length. Before commenting on their impact on science, I shall dwell briefly on the master, whose biography fills the first volume.

Malpighi was born in Bologna; he spent the better part of his life and his most productive years in his home town, as a member of the Medical Faculty of the University of Bologna. His stay in Pisa at the beginning of his career (1656–1659) was fateful in that it brought him into a close scientific association with an eminent scientist, Giovanni Borelli, who "more than any other man seems to have given bent and focus to Malpighi's nascent powers as an investigator" (p. 144). Borelli, twenty years his senior and the last student of Galileo, was mathematically and physically minded; he was one of the founders of a strongly mechanistic school of biological thought. For the next decade, which was crucial for Malpighi's scientific development, the two men were in continuous rapport. Although not too strongly influenced by Borelli's philosophical outlook, Malpighi profited greatly from his stimulating and critical advice. The years following Malpighi's return to Bologna were very productive, culminating in the significant epistles on the structure of the lungs. He left Bologna once more and, for several years, held the chair of Medicine at the University of Messina. But from 1666 on he remained in Bologna until near the end of his career.

Malpighi was first and foremost a passionate researcher, with an amazing breadth of interest; but he was also a practitioner of medicine of great renown. In Adelmann's words, "he was no routine practitioner; for him practicing medicine meant a perpetual search for the structural and physiological explanation of disease as the basis for sound procedure" (p. 321). He was in such high esteem that

175

Pope Innocent XII invited him in 1691 to move to Rome and attend to him as his personal physician, which Malpighi did, although reluctantly. He had been in poor health for many years and was rightly apprehensive of being uprooted. He died in Rome in 1694.

His embryological studies constitute only a small fraction of his scientific production. His range of interests was immense. As an innovator and founder of new branches of biological sciences he was unsurpassed for a century after his death. He was one of the first to make use of the microscope, which recently had been developed, and gave the first interpretation of the microstructure of the body on the basis of observation rather than speculation, thus initiating what we now call histology. His observations were based on comparative studies, and he always insisted on the interrelation of structure and function. His pioneer work on lung structure set the pace for a series of investigations on glands, liver, kidney, brain, and other organs. While studying the lungs of the frog, he made one of his major discoveries: he observed the capillary system that connects the arterial with the venous system. By demonstrating that blood vessels do not empty into open spaces, he closed the one crucial gap in Harvey's work. In the 1670's he published two volumes on the anatomy of plants which, along with the work of his English contemporary, N. Grew, established plant anatomy as a science. His monograph of the silkworm, based on minute and accurate dissections which were reproduced in beautifully executed plates, was matched only by Swammerdam's similar efforts in insect anatomy. To all his studies he brought a passionate interest and an unsurpassed gift of observation.

Let us turn next to the two dissertations on embryology. They are day-by-day descriptions of the development of the chick embryo and of the changes which he could discern with the naked eye and the primitive magnifying instruments that were at his disposal. The descriptions were accompanied by drawings which in a way embodied the substance of his observations. In the words of Adelmann, the first dissertation "was a little masterpiece, incomparably the best study of the subject that had ever been made" (p. 377). Of the drawings of the second, he says that they were "even more beautifully drawn than those illustrating the first dissertation" (p. 380). No better ones were produced for over a century. The greater clarity of the drawings in the second work is due to a simple but important technical advance; he had learned to remove the embryo and surrounding tissue from the yolk, to spread it on a glass plate and light it from below. The student of embryology who nowadays studies the so-called whole mounts of chick embryos looks at slides

prepared in exactly the same way. But he misses the excitement of seeing structures for the first time, and is instead told what he is supposed to see and draw. Malpighi depicts with astonishing precision the development of the structures which can be observed directly, that is, the heart and the blood vessels, the brain parts, eyes and other neural structures, and the somites. The excellent reproductions of the plates in Volume Two were made directly from originals still in the possession of the Royal Society.

Power of observation alone, however, cannot account for the impact which these two slender publications have made on the later development of embryology. The mark of Malpighi's genius was his unflinching dedication to unbiased direct observation and his refusal to let authorities of the past influence his judgment. All of his conclusions and deductions were based on what he had seen with his own eyes. Only one aware of the tremendous restraint imposed on the science of his age by shackles of authoritative thinking can appreciate the liberating force emanating from works like that of Malpighi. It is in this sense that he, together with his great predecessor, William Harvey, and a few others, ushered in the era of modern biological science.

In contrast to not a few of his contemporaries, he was cautious and undogmatic in the interpretation of what he observed, and aware of the limitations imposed by the senses and by his modest optical tools. He knew that it would be premature to formulate a theory of development, and he did not take a stand in the controversy between theories of preformation and epigenesis which preoccupied the minds of the late 17th and 18th centuries. "Dogmatism was evidently repugnant to Malpighi's mind" (p. 826). Nevertheless, he is widely quoted as a protagonist or founder of the preformation theory. This theory postulated the pre-existence of a miniature replica of the adult in the egg, and in its extreme form, the encapsulation of preformed future generations in the ovarian eggs. The theory denies any increase of complexity during development, which is the basic tenet of epigenetic theories. Adelmann shows convincingly that none of Malpighi's statements can be interpreted as anticipating or embracing preformationist notions as they were understood at his time or later. Indeed, anybody who has traced in Malpighi's drawings and in his descriptions the development of the heart from a simple to a coiled tube, and then to the compact four-chambered heart, or the changes of the embryonic brain, has actually witnessed epigenetic processes. "Certainly, the notion of progressive change and development is implicit in his descriptions of the central nervous system and the heart, at various stages. He nowhere makes clear how the tubular heart he describes in the embryo becomes the four-chambered heart of the adult, but surely it would be an unwarranted reflection upon his intelligence to suppose that he imagined the embryonic heart to be the adult heart in miniature" (p. 885).

Why then the imputation of a strongly preformistic view? Undoubtedly, much of the misunderstanding of his position derives from an unexplained inaccuracy in his observation or interpretation of the unincubated egg. In both dissertations he describes and depicts in the cicatrix (blastoderm) of the unincubated egg the outline of an embryonic anlage. To be sure, its features are indistinct. They are described as the head of a fetus with the first filaments (stamina) of the "carina" (i.e., primordium of the nervous system) appended to it, and enclosed in a saccule. He says: "It is therefore proper to acknowledge that the first filaments of the chick preexist in the egg." It is not stated that the adult or even the advanced embryo is preformed. Yet, the presence of any visible structure in the unincubated egg is exceptional. Most commentators point to Malpighi's remark that the observations were made "in August, when the weather was very warm" (p. 943); hence, there might have been some development before incubation. Adelmann suggests that what he saw was perhaps a primitive streak with head process. Be this as it may, the essential question is how Malpighi interpreted his observations. Adelmann argues cogently that in the absence of an explicit theoretical statement, Malpighi's views on the matter of preformation have to be extrapolated from his entire opus of which the embryological studies are only a small part; and that they have to be considered in the context of contemporary thought.

One cannot deny that Malpighi was inclined toward the idea of an early origin of embryonic structure. His bias was founded on general considerations as well as on his own research experiences. In the pioneer days of microscopy, revelations of hitherto unimagined details of fine structure opened up a new dimension in the understanding of organismic organization. Yet the enthusiasm was dampened by the frustrating awareness of the limitations that were imposed by the optical tools available at that time. "In view of the new wonders that the microscope was every day revealing, it was quite natural for the early microscopists to believe that only the inadequacy of their instruments prevented them from seeing finer and yet finer details" (p. 843). It would have been tempting for Malpighi to extrapolate from his observations and to postulate the preexistence of the complete organism in the egg. Yet he refrained carefully from such a commitment. "He was a superb observer, and it is to his

eternal credit that he never says categorically that anything is present before he can see it" (p. 885).

His own extensive investigations directed his mind even more forcefully toward the notion of early formation of structure. In 1669, he had published the monograph of the silkworm which, incidentally, presents the discovery of the excretory tubules bearing his name, and of the tracheae which he identified correctly as respiratory organs. He had also dissected pupae and had detected in them minute details of the preexisting imaginal organs. "In the silkworm, then, Malpighi believed that the parts of the moth are formed in very early stages, a conclusion to which Swammerdam also came in his work on insects, published a few months after Malpighi's *de bombyce* had appeared. Malpighi believed further that these parts arise from fluid material contained in saccules, in which the parts solidify and from which they are finally withdrawn. The saccules he saw were, of course, the imaginal discs" (p. 845). Extensive observations on plant buds and seeds further strengthened his belief in the early inception of structure.

Nevertheless, far from imagining that organs preexist in their final form, he had quite definite ideas of their coming into existence *de novo*, by processes of solidification of fluids inside of vesicles as indicated in the last sentences quoted above. A passage taken from a posthumous essay on chick development (in Adelmann's translation) indicates a very dynamic view of the developmental process, though his specific ideas are now obsolete. "Thus from a frequently repeated series of observations on incubated eggs of the hen, one may conclude that the fecund egg possesses as its principal constituent the cicatrix (i.e., blastoderm). This is nothing but a collection of fluid which is confined by an embankment (i.e., inner part of area opaca), as it were, and in which the first rudiments of the animal are contained. And so, after incubation has been initiated by the action of moderate heat . . . . the first filaments of the parts begin to become visible. In constructing them, Nature seems first to erect rising walls so to speak, whose various bendings and incurvings give rise to cavities. The spaces thus cut off are entered and filled by fluid, whereupon the carina or rudiment of the spine emerges. In these walls are formed longitudinally arranged lateral swellings which resemble spherical locules (i.e., somites) rendered visible by the gradual entrance of ichor, and from these the first framework of the vertebrae arises. But with the admission of fluid the space between the erect walls displays large vesicles over a wider area, forming the commencement of the brain. Extending from these vesicles a long canal filled with fluid initiates the framing of the appended spinal medulla. It seems therefore to be

Nature's plan to form singly all the parts from fluid as the prime material" (p. 866). The tenor of this essay and, of other similar statements gives a vivid testimony to his epigenetic way of thinking: that is, organs come into existence by complex processes of transformation.

Obviously, the adherents of the orthodox preformistic creed of encapsulation, who fancied the miniature adult in the unfertilized egg, cannot claim Malpighi as their witness and authority. He described and depicted the unfertilized blastoderm as devoid of structure and realized that whatever structure can be seen later on must have come into existence after fertilization. "There is not the slightest hint of encapsulement or emboîtement in Malpighi's thinking" (p. 885). One cannot but agree with Adelmann that "Malpighi's provisional theory is thus rather a theory of epigenesis" (p. 885).

One can hardly do justice in a few words to the contributions of Adelmann. His annotated translation of the two embryological communications with the Latin and English texts side by side is but a small part of his efforts. As is indicated in the subtitle, the work places Malpighi in a broad historical perspective and elucidates his role in the evolution of embryological ideas and concepts which are traced from ancient times to the 19th Century. In the first volume of over 700 pages, Adelmann gives a vivid picture of the intellectual atmosphere of Bologna and the University, and of the leading personalities who were Malpighi's friends or adversaries. The biography is essentially a penetrating, detailed and critical discourse covering the entire scientific production of Malpighi, which was immense by any standards. Abundant use is made of his extensive correspondence — all quoted in translation — whereby the relation of his work with contemporary science comes into relief. Not much is said of Malpighi's personality in a direct way; but the account of the personal circumstances under which his achievements were created reflects on his character and on his scientific philosophy. His was not an easy life. While he had loyal friends, he was harassed by vindictive and acrimonious rivals who fought in him the spirit of a new era. In addition, he was troubled by chronic ill health.

The second volume is dedicated entirely to the text, translation, and detailed analysis of the two embryological dissertations. It begins with a long essay tracing embryological thoughts and concepts from the Presocratics to the time of Malpighi. Other chapters explore influences on Malpighi's thinking and the reactions of his contemporaries to his publications. The extensive annotations are a substantial aid to the modern reader who is unfamiliar with 17th Century concepts and terminology.

The last three volumes deal in 28 Excursuses with the embryonic development of special structures, such as heart, brain, lungs, amnion, yolk sac, and allantois. In these chapters, Adelmann traces the currents of embryological thought about each of these structures, through the ages from the Ancients to the middle of the last century, with a focus on the contributions of Malpighi, thus setting them in historical perspective from a different angle. He follows with admirable expertness and insight, and in great detail, the gradual advances and clarifications that were made in the understanding of organ development through refinement of observation, methods and interpretation. The text is interspersed with scores of quotations and translations from the major sources. For instance, the Excursus on the heart, which is the most elaborate (200 pages), includes translations from Fernel, from Malpighi's friends Lancini and Bellini, from Maître-Jan, Haller, C. F. Wolff, Tiedemann, Pander, C. E. von Baer, Remak and others, some of them of considerable length. Although Adelmann does not claim completeness, these Excursuses are certainly the most comprehensive and authoritative contributions to the history of organogenesis that are available today.

At times, the reader who is not primarily a historian of science will find himself overwhelmed by details. He might have wished for summaries; or perhaps Adelmann could be persuaded to give us an abridged version of his work. Altogether, it is gratifying to see that at a time of haste, overflow of scientific publications, and disdain for the "classical stuff," a historical work of such scope and perfection could be produced and published in such an elegant style and attractive form. It is hoped that, apart from the specialists in the history of science for whom this work is indispensable, a few connoisseurs will take the time to enjoy it.

American Scientist, 58, pp. 321–322, 1978

The Scientists' Bookshelf

Organization and Development of the Embryo

by Ross Granville Harrison, edited by Sally Wilens. 290 pages. $15. Yale University Press, 1969.

The title of this book is also that of the Silliman Lectures which Dr. Ross G. Harrison delivered in 1949 at Yale University. Prolonged illness prevented him from preparing this, his scientific testament, for publication. The editor of the book, Miss Sally Wilens, Dr. Harrison's long-time research associate and friend, wisely chose not to reassemble the actual lectures from incomplete notes, but instead to republish five of Harrison's major articles on which the six Silliman Lectures were based. Several of them were not readily accessible before. Some of Harrison's lecture notes and comments are inserted in the text.

There are several other most welcome additions: illustrations have been provided for two articles which had been published originally without them, and in the others additional figures used in the lectures are included. Furthermore, the stage series of *Amblystoma punctatum*, with the beautiful drawings from life by Lisbeth Krause which had been circulated among Dr. Harrison's students and colleagues and have since become indispensable, is now published for the first time, with Harrison's descriptive text. The editor has contributed an informative biographical sketch, a bibliography of the Silliman Lectures, and a complete list of Harrison's publications. The design and typesetting of the book are very handsome, and the book altogether is a worthy memorial to the leading American experimental embryologist of the last generation.

The reprinted articles are general lectures rather than specialized research communications; they give the expert as well as the general reader a very good idea of Harrison's approach to fundamental problems of embryonic development and of his singularly imaginative and productive contributions to their solution.

In experimental embryology, as in other fields, progress is often enhanced by advances in methods. Significantly, two of the five reprinted articles center around methods that were inaugurated or exploited principally by Harrison. He designed the tissue culture method in 1907 as a crucial test for the nerve outgrowth theory. Both the method and the proof of the theory were of signal importance. In the lecture "On the status and significance of tissue culture" (1927) he surveyed the advances in the field of cell differentiation made during the first two decades with the aid of this method. Another chapter deals with the method of heteroplastic (i.e. interspecies) transplantation. It originated with G. Born but found in Harrison's hands ingenious applications to problems of heterogenic growth, embryonic interactions, pigmentation, etc.

The Croonian Lecture "On the origin and development of the nervous system, studied by the methods of experimental embryology" (1935) covers a field which can be said to be the creation of Harrison. Beginning with his earliest scientific investigations, he was concerned with problems of nerve outgrowth and the factors involved in neurogenesis. He formulated the major problems in this special field of embryology, and, together with students and coworkers, brought the tools of experimental embryology to bear on their solution.

The article on "Cellular differentiation and internal environment" (1939) is concerned with a recurrent theme in Harrison's thought: the cell as an elementary unit in development; the significance of its cytoplasmic fine structure, its polarity and symmetry for embryonic differentiation; and, in particular, the relation of the structural organization of the egg cell to that of the embryo. Cellular interactions during cleavage, embryonic induction, and other epigenetic mechanisms are viewed in this context.

These considerations lead directly to the ideas advanced in the chapter on the "Relations of symmetry in the developing embryo" (1936), on which the fifth Silliman Lecture is based. This includes the exposition of one of Harrison's major achievements: the solution of the difficult problem of the origin of mirror-image asymmetry in paired organs. Ingenious transplantations of early embryonic organ primordia, such as limb and inner ear rudiments, from one side of the embryo to the other, including rotations, gave the clue to an understanding of the origin of asymmetry in terms of the sequential order in which the main axes of these organs are determined. Each primordium is initially isotropic, but it acquires in very early stages a monaxial polarization by fixation of its anterior-posterior axis. The dorso-ventral axis becomes determined considerably later through the influence of factors residing in the cellular environment of the primordium. Hence, by experimental manipulation and careful choice of the appropriate stage of development, an initially left leg or ear primordium can be made to form a right organ.

In this paper and in the one on cellular differentiation, Harrison carries the discussion beyond the immediate implications of the experimental results. He was aware that they did not give the final answers. With remarkable foresight he realized (in 1936) that the ultimate solution of the symmetry and asymmetry problem must be sought on the molecular level, and he formulated quite specific ideas in this matter. He considered the cytoplasm as being in a paracrystalline state, and he postulated a repeat pattern and specific configurations of asymmetrical macromolecules in the egg and embryonic cells as the basis of symmetric and asymmetrical structure found at higher levels. In anticipation of modern ideas he implicated specifically the proteins as the building units of cellular ultrastructure. He states as a bold generalization: "In the foregoing exposition the specific form of the organism is taken as an expression of the molecular configuration of its protoplasm, in which the protein constituents play the chief part" (p. 205).

This book, written in the lucid and unpretentious style that characterizes all of Harrison's work, is a classic that should be required reading for all budding developmental biologists. It is of more than historical interest. It is a profound statement of major problems of classical experimental embryology, many of them still unresolved, and, through the perspicacity of the author, it connects the present with the past.—*Viktor Hamburger, Zoology, Washington University, St. Louis.*

Reprinted from
THE QUARTERLY REVIEW OF BIOLOGY
VOL. 45, NO. 2 JUNE 1970

NEW BIOLOGICAL BOOKS

The aim of this department is to give the reader brief indications of the character, the content, and the value of new books in the various fields of Biology. In addition, there will occasionally appear one longer critical review of a book of special significance. Authors and publishers of biological books should bear in mind that THE QUARTERLY REVIEW OF BIOLOGY can notice in this department only such books as come to the office of the editor. All material for notice in this department should be addressed to Bentley Glass, Editor of THE QUARTERLY REVIEW OF BIOLOGY, Department of Biological Sciences, State University of New York, Stony Brook, N. Y. 11790, U.S.A.

VON BAER: MAN OF MANY TALENTS

BY VIKTOR HAMBURGER

Department of Biology, Washington University, St. Louis, Missouri 63130

A Review of

KARL ERNST VON BAER, 1792–1876. *Sein Leben und sein Werk. Acta Historica Leopoldina, Number 5 (1968).*
> *By Boris Evgen'evič Raikov; translated from the Russian into German by Heinrich von Knorre. Johann Ambrosius Barth, Leipzig.* M 68 (paper). 516 p.; ill.; index of persons. 1968.

This biography is of great significance in that it enables us for the first time to assess the full stature of Karl Ernst von Baer, and to rank him among other leading biologists of the last century. His classic work *Ueber Entwicklungsgeschichte der Tiere, Beobachtung und Reflexion* (1828, 1837) established him as the founder of embryology as a modern science. But probably few biologists are aware of his other achievements: as an explorer and anticipator of modern ecology; as one of the intellectually most powerful pre-Darwinian evolutionists; as the founder of Russian anthropology, to mention only a few of his scientific activities. Several circumstances have contributed to the fact that these facets of his life have remained relatively obscure. His move, in 1834, from the Prussian University of Koenigsberg to St. Petersburg (now Leningrad) coincided with an abrupt change in his scientific work. He abandoned embryology completely and turned to other interests. From then on, the base of all his activities was the St. Petersburg Academy of Sciences which had conferred on him active membership in 1828. The records of his voyages and his manifold activities

as an academician, as well as his scientific investigations, are partly published in the Annals of the Academy and in other more or less inaccessible magazines, and partly buried unpublished in the Academy a.o. Archives. Strangely enough, neither his classic work on animal development, nor his very revealing and charmingly written autobiography, nor the three volumes of his essays have been translated into English (except for long passages on organogenesis in H. Adelmann's treatise, *M. Malpighi and the Evolution of Embryology*, Cornell Univ. Press, 1966) and, apart from a recent reprint of the *Entwicklungsgeschichte*, the German originals are hard to come by. In a word, his work is hardly accessible to the western world.

The author of the present biography, B. E. Raikov, is a renowned Russian historian of biology, who had access to all unpublished source materials, diaries and letters. He has compiled a complete annotated bibliography, which includes over 400 publications of von Baer: this by itself is a valuable contribution, considering the wide scattering of his papers in more or less obscure Russian and Western European journals. He has given us a comprehensive biography of high distinction, a well-balanced and critical evaluation of the work of von Baer, and, at the same time, he has made an important contribution to the history of some central biological ideas in the last century.

Raikov is the author of a four-volume treatise on pre-Darwinian Russian biologists-evolutionists, which indicates the direction of his major interest. He is uniquely qualified to appraise von Baer's status as an evolutionist. It is understandable that the author, not being an embryologist, does not do equally full justice to von Baer's fundamental contributions to embryology. I do not consider this as a serious shortcoming, however, because von Baer has always received full recognition for his embry-

173

ological discoveries and for creating the conceptual foundation for our modern theory of epigenesis. Raikov deals in detail with the circumstances that led von Baer into embryological pursuits: his relation to the eminent comparative anatomist and embryologist Döllinger in Würzburg, who inspired his interest in this field; to his friend Chr. Pander, who, at Döllinger's instigation had started the work on the chick embryo, the incompleteness of which set von Baer to work; and the inspiration that came from Natural Philosophy (of which more will be said later). He tells also the story of the publication of the *Entwicklungsgeschichte,* where von Baer did not show the best of judgment. Raikov's discussion of the discovery of the mammalian ovum and its implications for comparative embryology, the germ layer concept, and the fundamental insight into the nature of development as a progression from general to specific is adequate but not of special distinction.

The complete break in von Baer's scientific life, and the abandonment of the science for which he seemed predestined, has attracted the attention of everybody concerned with von Baer. Raikov weighs all motives and, in my opinion, he comes to an adequate interpretation of this fateful step. The eight years from 1819 to 1827 in which von Baer was completely absorbed in embryological investigations were scientifically his most productive years. Yet an unrestrained exertion undermined his health and sapped his nervous energies to the point of exhaustion and led to a crisis which forced him to take a drastic step for its relief. In a memorable letter written over ten years later to the noted embryologist Th. Bischoff (which is cited in full) he recalls the last year of his embryological research: ". . . I had started when the snow melted. And when I walked outdoors for the first time I found the wheat bearing grain. I then collapsed on the ground and cried. It occurred to me that literary ambitions deceive life. We believe that we must bring sacrifices to science and don't see that we bring it to our own ambitions. Science, however, bears the conditions for its development in itself and is not in need of the single individual. . . . Part of the cure was the resolution not to read anything concerning embryology in 9 years. I don't know whether I was right. . . . It now seems to me that in doing this I had drawn the best of my heart's blood" (p. 14?). And he actually kept his vow. There were other motives; not the least was the lack of recognition and even hostility on the part of his German colleagues, which hurt him deeply and made the offer to move to Russia more attractive. His reactions in this crisis reveal a good deal of his personality.

Of equal interest are the inner forces which motivated his completely new style: the life of an explorer, administrator, scientific statesman, educator. Concerning his scientific travels which consumed many of his later years, Raikov shows that they were by no means imposed on him by the new circumstances as had been implied by some, but that they were actually the fulfillment of earlier projects and dreams. For instance, plans for the study of life in the Arctic date back to his student days, and were even pursued during the Königsberg period. The new opportunities permitted him to follow a strong latent inner urge to explore the forces of nature as they operate on a larger scale in the outer world.

In his autobiography, von Baer deals only briefly with his voyages, and he never took the time to gather the data on his observations, discoveries, and ideas in a book. We find in Raikov's biography the first comprehensive account of his travels, as well as an evaluation of their scientific significance, based on the source material and diaries. One of his first expeditions, which took him to the island of Novaya Zemlya in 1837, fulfilled his desire to study conditions for life in the Arctic. Though not very extensive, the enterprise showed his basic approach and the universality of his interests. To quote Raikov: "His voyage is of considerable scientific significance. It is an outstanding example of a many-sided exploration of a limited territory in which the mutual dependency of the different factors in nature — the meteorological, geological, botanical, zoological and geographic are demonstrated. . . . We recognize in Baer already the beginnings of an ecological-morphological method of investigation as it is now generally practiced. . . . His procedure has stimulated many investigators. Baer took this position because he did not carry out his investigations as a one-sided specialist. He was a researcher with a comprehensive mind, very well trained philosophically, and in possession of a truly encyclopedic knowledge in the field of natural sciences" (p. 180).

His four expeditions to the Caspian Sea, between 1853 and 1857, were the culmination of his exploratory activities, which brought his ecological approach to full fruition. They were instigated, not just by von Baer's own interests, but by a practical problem. The Academy requested him to investigate the causes of the decline of the fishing industry which was, and still is, of considerable economic importance. The study of fishing practices and of all ichthyological aspects related to them was pursued with typical thoroughness; but, characteristically he turned the expeditions into a scientific enterprise of the first order, which completely absorbed all his energies for five years. He threw himself into this work with the same passionate intensity that had characterized his embryological period. He concerned himself with equal zest with changes in the vegetation, food chains, fish migration, animal and plant distribution, physico-chemical hydrological factors, and geology. He gained a superb insight in the inter-connections of biotic communities and their relation to environmental factors; and some of his ecological summaries have a remarkably modern ring. His stamina and personal courage were taxed severely. The travels were fraught with personal hardships and illness, and were filled with adventures and dangers at a time when ordinary travel in these remote regions was hazardous. And he was already in his sixties. Only one driven by a demonic urge to know all of nature could have carried out this enterprise. The universality of his genius is shown by a discovery in an entirely unrelated field. He observed that the western banks of rivers like the Volga, which flow in parallel with the earth's meridians, are often steep and the cliffs undermined by water, whereas the eastern banks are flat. He related this phenomenon immediately to the rotation of the earth.

The years intervening between travels were filled with an endless series of activities. Raikov portrays him as an influential and highly respected member of the Academy, as one of its most outstanding minds who effectively fostered science and science education in Russia. He initiated the systematic collections of animals, plants, fossils, human skulls. He founded and participated actively in scientific societies and almost single-handedly established anthropology in Russia. For over a decade he also held a professorship at the Medico-Surgical Academy which involved lecture courses. Along with these activities went investigations in comparative anatomy, anthropology and other fields. Yet, he never again attained the height of scientific productivity that characterized his Königsberg period. Was it the vastness of the country, the unlimited range of opportunities that dispersed and dissipated his energies?

The most significant parts of Raikov's biography seem to me to be the ones that deal with von Baer as an evolutionist. Raikov traces with perspicacity the roots of his evolutionary ideas and focuses on the intriguing interplay of idealistic philosophical thought and sober critical scientific deductions from facts and observations that characterize the originality of von Baer's views. The last chapter on his philosophical and biological creed is a masterpiece of exposition. Of great value was the discovery of unpublished texts of public lectures of a very general nature which he gave in Königsberg between 1822 and 1825. Lengthy passages from these are quoted. They enable us to trace his ideas on evolution to their beginnings. At that time he had already for-

mulated an elaborate theory of evolution clearly anticipating that of Darwin. (One remembers that the voyage of the "Beagle" was in the thirties.) The analysis of Raikov leaves no doubt that von Baer thought of transformation in terms of blood relationship. In a lecture "On the development of life on earth" (1822) he aligns himself with the Lamarckian idea of transformation as a result of environmental influences.

The origins of von Baer's evolutionism as well as of his embryological interests are traceable to very general philosophical ideas that came to him through his great teachers, Burdach and Döllinger, both of whom were adherents of the Naturphilosophie of Schelling. Its core was the idea of a transcendental cosmic force which moves the universe continuously toward higher perfection. The idea of an immanent teleological principle remained an article of faith for von Baer throughout his life. It led later to his irreconcilable rejection of selectionism. The principle of change and progression in nature which is inherent in Schelling's philosophy, directly inspired his embryological work. He was prepared philosophically for the discovery of the epigenetic mode of animal development which is one of his great contributions to biology. And in a bold extrapolation from the individual to the species he states in a lecture of 1822: "The law reigns in Nature that the different forms (i.e., species) of organic bodies always maintain themselves, while the individuals always perish. Just as in the development of the individual we always see a transformation and never an absolute beginning, so the different forms which we call species must have formed gradually, one out of the other, rather than being generated originally in their diversity" (p. 63). As Raikov comments: "Here the idea of evolution is no longer the outcome of philosophical constructs but the end result of conclusions derived from concrete observations" (p. 64). At about the same time, he made another conceptual advance of major importance. He recognized the diversity in the structural plans of animal forms and, independently of Cuvier, conceived of four major types (Vertebrates, segmented forms, Molluscs and radial forms) anticipating our concept of Phyla. He surmised that common ancestry can be expected only within each type, a view which was perfectly justified at the time. Many other, more specific examples could be quoted to show his critical, realistic, scientific approach to transformation, completely divorced from speculative philosophy. In fact, in some instances he was more cautious and critical than Darwin, whose sweeping generalizations he did not find sufficiently supported by the evidence that was then available. In other words, philosophical reflections set his mind on its course, but he did not let them interfere with his critical scientific judgment.

Yet the inner conflict between idealistic teleological creed and the sober critical scientific approach was never fully resolved. The last chapter of the book is devoted to a very thoughtful and lucid analysis of the complex mode of thought which created the deep insights in the workings of nature, and of the subtle changes of his views during his lifetime. I would accept the viewpoint of Raikov: that his mind operated on two levels without much conscious awareness, and that each level had its own inner logic (p. 402).

Von Baer emerges in this biography as a man of an intellectual stature comparable to the great minds of the century: Cuvier, Lamarck, Darwin. Yet, only in embryology was his impact commensurate to the power of his mind. Perhaps he failed where Darwin almost failed: he never published in full his version of the origin of species, nor his

typology, nor his ecological-geographic observations and reflections.

The German translation by Heinrich von Knorre was published in 1968 by the Academia Leopoldina in Leipzig where, incidentally, von Baer had published his typology in 1827. The Russian original text of the biography had appeared in the Annals of the USSR Academy of Sciences in 1961. The translation reads very well: it suggests a crisp and lucid style of the original version. The translator has made valuable additions: he has provided biographical data on Raikov, who died in 1966. For Raikov's annotated bibliography of von Baer he has provided cross references to the translation and to the autobiography of von Baer. Furthermore, he has compiled a complete list of references cited by Raikov, and an index of names referred to in the text. The latter is a small substitute for a general index, which is lacking. The book is beautifully printed, and includes 20 illustrations.

VI. Biographical and Autobiographical

Separatum EXPERIENTIA 25, 1121 (1969)
Birkhäuser Verlag, Basel (Schweiz)

Hans Spemann and the Organizer Concept

By V. Hamburger[1]

Department of Biology, Washington University, St. Louis (Mo. 63130, USA)

In his autobiography[2] (p. 171), Spemann says: 'My strongest inclination and talents are a combination of inquiry into general problems (allgemeine Fragestellung) and technical invention.' He achieved what he did by bringing these inclinations and talents to their fullest development. Those who are not familiar with the 'métier' of the experimental scientist can hardly realize how much attention to minute details and mere drudgery go into experimental work. Spemann had his full share, but no matter how painstaking the labors, the findings were never more than a few steps away from the problem. And it is true that he derived an extraordinary enjoyment and gratification from his very considerable manual skill. He could spend hours playing with glass tubing and rods over the Bunsen – or microburner, preparing instruments for his microsurgery. After his retirement, he writes in a letter to me (8 July 1937): 'At the moment my passion for playing (with instruments) celebrates orgies, and I feel transferred back to the times when I invented the glass needle technique; the same oblivion of time; the same curiosity and tension and "herzklopfende Seligkeit".' (He worked at that time on a technique of dissecting and mounting small insects.) This was certainly not a technological mind in the modern sense; it was more the spirit of the skilful craftsman of earlier times to whom the pride in the perfection of what he accomplished with his own hands was the better part of his reward. And with Spemann, the aesthetic and artistic pleasure in the design of the tools as well as of the experiments were an important element. When I prepared my first glass needles and hair loops, he looked over my shoulder and told me that the shape of the handle was not acceptable. This was not mere whimsey. His experimental success depended as much on his self-invented tools, as on his

analytical acuity. This is more true in his case than in that of many other experimental embryologists, such as Roux, Driesch and Herbst. Perhaps Harrison and Hoerstadius had similar sensitivity for their tools.

There is a third motive underlying his chosen vocation, the experimentation with the living embryo. He recognized it quite clearly. He says in his autobiography[2] (p. 203): 'I wish to influence the living [the German word 'einwirken' has an ingredient of controlling, governing] in order to participate in other life. My passion for experimentation with the developing organism is innermost related to my pedagogic inclinations; with my enjoyment in knowing people, in teaching and education.' The psycho-analytically minded may detect a faint echo of the power-over-nature motive. Maybe so. But, of course, the embryo had to be handled gently! I detect a pedagogic streak also in the style of his writings: in the great precision in expressing what he wanted to convey, never omitting a single step in the argumentation. Actually, he invested a considerable amount of energy and time in pedagogic pursuits outside the university. He was the founder and director of the Adult Education Program ('Volkshochschule') in Freiburg from 1920 to 1933, the type of cultural innovations which flourished in the Weimar Republic, until Hitler terminated them. And his close friendship and association with one of the pioneers in progressive education, Hermann Lietz,

[1] Address, given at a Symposium on the 'History of Experimental Embryology' held at Washington University, St. Louis (Mo., USA), on 29 May 1969, in commemoration of the hundredth anniversary of H. Spemann (27.6.1869–12.9.1941).
[2] H. Spemann, *Forschung und Leben* (Ed. F. W. Spemann; Engelhorn Verlag, Stuttgart 1943). All translations are performed by the present author.

the founder of one of the private progressive schools, brought him close to this movement. As an overseer of this school he attended their meetings regularly; and he derived much satisfaction from the contact with young people. All this is significant, because he did not have a strong constitution, and he had to husband his strength carefully.

Scientific work and its motives

During the 20s and 30s the Freiburg laboratory under SPEMANN, and the Osborn laboratory at Yale under HARRISON, were the hub of the experimental embryological universe.

I do not intend to retrace SPEMANN's extraordinary scientific achievements on this occasion. But as an eye-witness of the 'classical Freiburg period', first as a candidate for the Ph.D., 1920–1924, and then as an assistant and instructor, 1927–1932, I shall try to convey some of the spirit of these times. When I began to contemplate what to say on this occasion of his 100th anniversary, I re-read his publications and his autobiography; and I became interested in the inner relations between his work and his personality, and how his personal philosophy is reflected in his work – in a word: his personal style. This will be the topic of my remarks.

The strength of both SPEMANN and HARRISON was in the self-imposed concentration on a few basic themes and problems, and on a single object, the amphibian embryo. SPEMANN complained occasionally that he missed out on so much in life and science by limiting himself so severely. He enjoyed playing with other forms, particularly invertebrates, and he indulged occasionally in this pleasant pursuit at Naples and in Woods Hole. But he never permitted himself to stray far away from his chosen path.

He entered the scientific scene around the turn of the century rather inauspiciously, with a series of constriction experiments on salamander eggs which were not particularly original in the problem nor in the experimental design. The basic problem, that is, the relation of the egg structure to the organization of the fully formed organism, the century-old problem of preformation versus epigenesis, had been restated by ROUX and DRIESCH in a new conceptual frame of reference that made it accessible to an experimental approach. The preformistic concept had been dealt its death blow by DRIESCH (1889) when he showed that half of an egg of a sea-urchin could give rise to a whole larva; and the isolation of blastomeres had become fashionable in the 90s. Constriction of salamander eggs had been done more or less successfully by others. And the tremendous regulative capacity of embryonic parts had been placed in the center of the stage by DRIESCH to stay with us and haunt us to this day. What was novel in SPEMANN's experiments was their greater thoroughness, their more extensive range and the astute exploitation of some unexpected results. He

found that a number of eggs which were constricted in the 2-cell stage did not give identical twins but one embryo and a relatively unorganized spherical body which though remaining alive failed to develop axial organs. He related these differences to differences in the plane of constriction; if the latter corresponded to the median plane of the future embryo, then identical twins were obtained; but if it separated the dorsal part from the ventral part, then only the dorsal part gave an embryo, and the ventral part gave the spherical 'belly piece'. And he asked immediately: Why no axial organs in the belly piece, if the egg is as fully regulative as it was supposed to be? In the typical SPEMANN fashion, he considered the alternative explanations: either the building material is lacking or a stimulus for differentiation which, of course, would have to emanate from the dorsal half. He even thought of a chemical stimulus. In 1901[3] he says: 'The experiments give no information about the differentiation substance which is lacking in the ventral cell' (p. 256).

Upon further analysis, in 1903[4], he got the first inkling of the existence of an organizing center in the dorsal part. 'It is conceivable that at the beginning of gastrulation only certain cells in the middle of the upper blastoporal lip can be designated as an 'anlage', whereas cells lateral to them are not capable of self-differentiation and are only later incorporated in the

[3] H. SPEMANN, Wilhelm Roux Arch. EntwMech. Org. *12*, 223 (1901).
[4] H. SPEMANN, Wilhelm Roux Arch. EntwMech. Org. *16*, 606; *16*, 616 (1903).

'anlage'. And furthermore, in the same paper: 'It is not excluded that the differentiation of the medullary plate is induced by the archenteron' (p. 616). This is an almost uncanny premonition based on scanty data. But the essential point is that through these experiments, he had discovered the central theme of his life's work: to understand and analyze the factors which create the unit – organization, the individuality, of the organism, in an egg which is capable of producing more than one individual. And, perhaps more importantly, it was apparent that there is an experimental approach to this problem: namely, the analysis of the *determination of the axial organs* which identify the invidivual. The general problem was caught in the net of the experimenter.

But the problem had also its mystical undertones. Referring to his discovery that partial constriction results in double monsters with 2 heads, he states: 'It was probably the fascination with that mystery which surrounds the partially split individuality[5] – and then the enjoyment of the elegant technique... that forced me to lock myself up spring after spring, instead of roaming in the beautiful world, and to tie hair loops around slippery newt's eggs, until I had constricted a thousand and a half' (autobiography[2], p. 181).

I shall skip the period from 1904–1912 during which he was preoccupied with *lens induction*, except to say that these experiments put embryonic induction on the map as an important mechanism of progressive differentiation. This was also the occasion for the invention of the glass needle technique, without which further progress was not thinkable. This technique enabled him to do the ingenious transplantation and combination experiments on the salamander gastrula[6] which paved the way for the organizer experiment. They were done at the Kaiser-Wilhelm-Institut für Biologie in Berlin-Dahlem during the First World War. In 1919, he moved to Freiburg to occupy the chair formerly held by AUGUST WEISMANN, who had inspired his early interest in experimental embryology.[7]

The discovery of the *organizer*, the crowning experimental success in the Freiburg period, seemed, for a moment, to fulfil his major goal; to localize and analyze the factors that are responsible for the determination of the axial organs, and thus the individual. By transplantation of a small piece of the upper blastoporal lip of a salamander gastrula into a relatively indifferent region of another embryo, a second individual embryo was created which arose partly by self-differentiation of the material of the transplant and partly by complex induction on the part of the transplant, whereby host tissue was assimilated in an organized fashion. Exchange between embryos of different pigmentation, belonging to different species, made the distinction between donor and host embryo possible. In his own words, 'At the beginning of gastrulation, the individuality of the embryo is so-to-speak represented by the cells of the upper blastoporal lip which represents an organization center from which the other most important parts of the body are formed'[8]. (Notice that the concept of 'organization center' preceded that of 'organizer' by several years.) The experiments were done by HILDE PROESCHOLDT for her Ph.D. thesis. The first results were communicated in 1921[9] in a postscript to another paper, and the full publication appeared in 1924[10]. A few months later, HILDE PROESCHOLDT, who in the meantime had become Mrs. MANGOLD, died in a tragic accident at the age of 26. SPEMANN was awarded the Nobel Prize in 1935.

If later experiments did not sustain the high hopes and the enthusiasm which inspired the choice of this magic word, 'organizer', the experiment was certainly of signal importance. To the extent that the upper lip material created an integrated axial system, and in the best cases, a rather complete secondary embryo, the name 'organizer' seemed to be justified. But soon it became clear that its radius of action in *normal development* was more limited than had been assumed. When 10 years later, HOLTFRETER made the momentous discovery that dead tissues could induce highly organized axial systems, the term would seem to have been ready for retirement, since as SPEMANN himself put it 'a dead organizer is a contradiction in itself'[11]. Yet, it lived on. One of the most important conclusions from the inductions by dead tissue was the shift of emphasis from the inducing stimulus to the responding system which in these instances seemed to bear the major responsibility for the formation of the secondary embryos. What seemed imposition of organization by an outside agency was really to a major part *self-organization* within the reacting system.

In all fairness, it should be stated that SPEMANN himself recognized early the provisional nature of the organizer concept; and in most of his writings and discussions the question of the relative share of inductor and responding system was clearly on his mind.

We may then ask the heretical question: What motivated SPEMANN to coin this term with the strange holistic and psychological undertones – a term which even at that time would seem strangely out of place in a rigorous scientific vocabulary? If SPEMANN had misgiv-

---

[5] He might have said 'personality'; when in a whimsical mood, he would contemplate the compassions and competitions between the two alter egos.
[6] H. SPEMANN, Wilhelm Roux Arch. EntwMech. Org. *43*, 447 (1918).
[7] He spent the winter of 1896/97 in Arosa (Switzerland) for reasons of health. As his only scientific reading matter, he took along AUGUST WEISMANN's book on *The Germ-Plasm* (1892), through which he became acquainted with the experimental work of W. ROUX and H. DRIESCH. They inspired directly his own constriction experiments (see autobiography[2], p. 178). He attested on several occasions to his intellectual debt to A. WEISMANN.
[8] H. SPEMANN, Naturwissenschaften 7, 591 (1919).
[9] H. SPEMANN, Wilhelm Roux Arch. EntwMech. Org. *48*, 568 (1921).
[10] H. SPEMANN and H. MANGOLD, Wilhelm Roux Arch. EntwMech. Org. *100*, 599 (1924).

ings in 1932, why did he carry it over into his scientific testament, the book on 'Theory of Development'?[11] What are the roots of this strange ambiguity?

Let me get at the answer to this question in a roundabout way. The organizer experiments, and in fact all his experiments, were planned very carefully and methodically, as a sequence of steps with their inner logic, leaving very little to improvisation. The strategy of his research was well expressed in a letter to me: '... to tackle the most immediately solvable question, until, piece by piece, the whole is achieved.' Yet, progression in scientific research is not quite that simple. When one is on the road, the goal can be perceived only dimly. At the end of each breeding season, when the harvest was in, and the plans were laid for the next, there were usually more interesting possibilities than one, and choices had to be made. Nothing is more revealing than to track down the lines which the master found worthwhile pursuing, and, on the other hand, the experimental ideas which were put on ice. What intrigued him, for instance, was the regional structure of the living organizer: experiments which led to the discovery of the head- and trunk-tail organizer[12]. He placed much emphasis on the experiments which showed the awakening of inductive powers in a tissue which did not have inductive capacity before, because they led to the general concept that progressive differentiation is a chain of inductive processes: one part gets an instruction for organ formation by induction, and at the same time acquires the capacity to induce another still indifferent part to proceed along another line of differentiation[13] He was particularly gratified when an old experimental dream of his was realized. The bold idea was to exchange embryonic tissue between frog and salamander which differ widely in many structures, as for example in mouth implements. Would the inductors of the urodele accept the challenge to produce inductions in the anuran (and vice versa), and, if so, what would the induced structures be like? In the early 30s Oscar Schotté performed this difficult technical feat successfully[14], and, as expected, profound new insight in the inductive mechanism was obtained. Briefly, the experiment showed that the prospective head ectoderm of the frog embryo when transplanted to the head of a salamander can respond to the inductors of the salamanders, but it does so according to its own genetic repertory: it forms frog-type mouth implements, that is, horny jaws and suckers, which are foreign to the salamander. The inductor gives the general instruction to form mouth parts to which the reacting system answers in its own specific way. These were some of the experimental ideas that were materialized. But how about those which were not?

As early as 1924[15], Spemann contemplated an experiment which nowadays would appear to be as exciting as any. 'By implantation in the blastocoel it will be possible to test whether only the living cells are capable of induction or also a structureless "Brei" or extract.' Significantly, the matter was not taken up until 1931, when Spemann[16] found that minced and squashed organizer retains its inductive capacity; however, his experiments with dried and frozen organizers were unsuccessful. The breakthrough came with Holtfreter's spectacular success of obtaining complex inductions not only from killed organizer but from a variety of dead animal tissues[17]. These experiments established the chemical nature of inductions, and paved the way for the chemical analysis of inductive agents.

Another idea of Spemann that had great potential was not followed up either. In 1918[18] he suggested the tissue culture method as one way of testing the state of determination of early gastrula parts. Again, it was Holtfreter[19] who many years later devised the culture medium which became a powerful tool in the analysis of amphibian development.

Clearly, the living organizer always took precedence; the commitment was to the embryo as a whole. As Jane Oppenheimer[20] put it in one of her essays: 'The integrative powers of the embryo at all of its levels are so pervasive that they never permit themselves to be overlooked by those who avail themselves of the privilege of looking at the embryo as a whole. ... Spemann, for instance, who analyzed the relations between layers in terms of cellular interactions, never lost sight of the whole embryo.' And I may add: neither did Harrison. But the theoretical and philosophical paths which the two men chose to follow from there led in opposite directions.

[11] H. Spemann, *Experimentelle Beiträge zu einer Theorie der Entwicklung* (J. Springer Verlag, Berlin 1936), p. 276. The English translation appeared under the title *Embryonic Development and Induction* (Yale University Press, New Haven 1938). All quotations are translations by the present author from the German edition.
[12] H. Spemann, Wilhelm Roux Arch. EntwMech. Org. *123*, 389 (1931).
[13] H. Spemann and B. Geinitz, Wilhelm Roux Arch. EntwMech. Org. *109*, 129 (1927).
[14] H. Spemann and O. Schotté, Naturwissenschaften *20*, 463 (1932).
[15] H. Spemann, Naturwissenschaften *12*, 1093 (1924).
[16] H. Spemann, Verh. dt. zool. Ges. *129* (1931).
[17] J. Holtfreter, Naturwissenschaften *20*, 973 (1932); Wilhelm Roux Arch. EntwMech. Org. *128*, 585 (1933). The preliminary report of Holtfreter's experiments was published jointly with H. Bautzmann, H. Spemann and O. Mangold (Naturwissenschaften *20*, 971, 1932). Full credit should go to Holtfreter who at that time (spring 1932) had obtained a considerable number of complex inductions not only from killed organizer but also from dead embryonic tissues which do not induce when alive. While the other contributors reported only few cases of neural inductions, not all of them convincing, Holtfreter's embryos were reared to advanced stages and showed fully differentiated induced structures such as brain and eyes, thus establishing the inductivity of dead tissues beyond doubt.
[18] H. Spemann, Wilhelm Roux Arch. EntwMech. Org. *43*, 526 (1918).
[19] J. Holtfreter, Wilhelm Roux Arch. EntwMech. Org. *121*, 404 (1931).
[20] J. Oppenheimer, *Essays in the History of Embryology and Biology* (M.I.T. Press, Cambridge, Mass. 1967), p. 9.

The philosophical foundations

The organizer concept symbolizes, perhaps subconsciously, Spemann's deep and strange conviction that all vital phenomena, including, of course, the doings of the embryo, are the emanation of a psychic force akin to, or identical with, the workings of our mind. He has expressed this idea most succinctly in the closing paragraph of his book[21]: 'Over and again, expressions have been used which denote psychological rather than physical analogies. This is meant to signify more than a poetical metaphor. What I mean to say is that the neighborwise ["ortsgemässe"] reaction of a pluripotential embryonic part in an embryonic field, its behavior in a particular situation is not an ordinary simple or complex chemical reaction. I imply that these developmental processes, like all vital phenomena, irrespective of whether or not they may once be resolved into chemical and physical processes, resemble in the way they are connected nothing more closely than those vital phenomena of which we have the most intimate knowledge, viz. the psychic phenomena.'

The sober voice of Harrison, the wisest of all experimental embryologists, whose life-long preoccupation with the *Amblystoma* embryo brought him face to face with the same phenomena of regulation and determination, has a more modern ring. In the Harvard Tercentenary Lecture of 1936[22], the same year in which Spemann's book appeared, he says: 'In the amphibia, not only are early segmentation and gastrulation stages capable of regulation but also later embryos. This quality of "wholeness" in the parts of the organism, particularly the embryo, had led to much speculation and even to a system of philosophy. It is the capital problem of embryology [here Spemann would have ended the sentence; but Harrison continues] to find the physico-chemical basis for it.' And further: 'In endeavoring to reach a physico-chemical description of life, one is baffled rather by the bewildering array of possibilities than by such dearth of material as would warrant recourse to so-called vitalistic forces. At least I am unwilling to accept the defeatism of the vitalist, so long as means of investigation by experiment are available.' I would disagree with Harrison in only one point: If Driesch's vitalism may have sprung from defeatism, that of Spemann had entirely different roots; it was in the line of Goethe and the German natural philosophy.

The origin of Spemann's particular type of psychological vitalism can be traced back readily to his friend and mentor August Pauly, a Munich zoologist and theorist, a bearded sage, with a fiery spirit. His neo-Lamarckistic ideas on cell psyche had a lasting influence on Spemann, from his impressionable student days on. The central issue in Pauly's doctrine was the pervasive phenomenon of adaptation in organisms. He rejected violently selectionism as its explanation. Incidentally, he shared this view with many leading biologists of the turn of the century, as is well told in an illuminating article by Garland Allen on T. H. Morgan[23], who himself was not converted until 1910. Here, then is actually an element of defeatism. But Pauly built up from there a special brand of neo-Lamarckism which endowed all cells with psychic powers of inventiveness, enabling them to meet new challenges with innovations the way we invent new tools to deal with new demands. The idea was elaborated in great detail in his book[24]. Although Spemann admitted the great shortcomings of Pauly's arguments and was profoundly critical of his dogmatism, it remained his innermost conviction that psychic properties are all-pervasive and not limited to the mind of man and higher animals[25].

If Spemann's natural philosophy sounds anachronistic today, we realize how deeply the advances of biology in recent decades have influenced our theoretical thoughts about vital phenomena. I think that, indeed, some of them are now less enigmatic than they were 30 or 40 years ago. But there is little cause for a particular pride or complacency. The holistic view is still with us and, I think, rightly so, though stripped of its metaphysical undertones. And do we really have a better understanding today of the regulative capacities of the embryo, of morphogenetic fields and self-organization? Or, for that matter, of the plasticity of our brain and our mind? Do we really believe that our efforts to reduce the biological phenomena to physical and chemical processes will answer all questions?

And we had better realize that the scientific approach altogether opens only a small window to the universe. We cannot expect our intellect to fathom all depths. In Spemann's words: 'All really fundamental questions are of an elementary, simple kind, and they do not come from pure reason but from the totality of the human personality'[26]. Spemann's was a very complex personality. The sharp analytical mind lived together with the artist's approach to life and work, and with the pedagogue's inclination to mould human beings --and to influence embryos. If his personal creed was the mainspring of his inner strength and achievements, that much the better for experimental embryology.

[21] H. Spemann, *Experimentelle Beiträge zu einer Theorie der Entwicklung* (J. Springer Verlag, Berlin 1936), p. 278.
[22] R. G. Harrison, Trans. Conn. Acad. Arts Sci. *36*, 281 (1945).
[23] G. E. Allen, J. Hist. Biol. *1*, 103 (1968).
[24] A. Pauly, *Darwinismus und Lamarckismus* (E. Reinhardt Verlag, München 1905).
[25] See [2], p. 161. It is of great interest that the great cell biologist Th. Boveri, Spemann's teacher and himself a friend of A. Pauly, shared the scepticism of the latter concerning the all-powerful role of natural selection in evolution and accepted the essential elements of Pauly's psycho-lamarckistic doctrine of cell psyche. See his 'Rektoratsrede', *Die Organismen als Historische Wesen* (Würzburg, Universitäts-Druckerei 1906) and F. Baltzer's biography *Theodor Boveri* (Wiss. Verlagsanstalt, Stuttgart 1962), p. 166.
[26] See [2], p. 151.

Reprinted from Trends in NeuroSciences – *September, 1985*   Vol. 8, No. 9, pp. 385-7

Hans Spemann, Nobel Laureate 1935

V. Hamburger

Fifty years ago, in October 1935, Hans Spemann received the Nobel Prize in Physiology or Medicine for 'his discovery of the organizer effect in embryonic development'. The organizer experiment by Spemann and Hilde Mangold was published in 1924 under the title 'On Induction of Embryonic Primordia by Implantation of Organizers from a Different Species'[1]. It was the crowning event of Spemann's 25-year search for the developmental mechanisms that create the pattern of the major organ systems in vertebrate embryos. The actual experiment was performed on early amphibian embryos by H. Mangold as the project for her PhD thesis. Her tragic death in an accident shortly after the publication of the paper deprived her of her share in the world-wide recognition of the experiment as one of the most important in embryology[2].

Spemann's early work followed the tradition of the founder of experimental embryology, the German anatomist and embryologist, W. Roux. He had made the breakthrough from descriptive to causal-analytical embryology and had introduced the experimental approach for the analysis of organ determination. One of the most significant discoveries in the early days was the finding of H. Driesch that if the first two blastomeres (cleavage cells) of the sea urchin egg are separated by shaking them apart, each cell regulates and forms a whole pluteus larva[3]. The egg structure thus proved to be extraordinarily plastic and far from the mosaic of predetermined organ primordia, as had been envisaged by the school of preformationists. The challenge, then, was to elucidate the developmental mechanisms by which pluripotency is transformed into a pattern of organs. Spemann addressed this fundamental problem by repeating the experiment of Driesch on amphibian embryos in the hope of finding clues that would yield deeper insights than the sea urchin experiment had given.

The separation of the two blastomeres of the salamander egg was accomplished by constriction with a fine hair. In many instances, Spemann obtained two complete embryos, demonstrating that the amphibian egg has the same capacity for regulation as the sea urchin egg. In other instances, however, only one cell gave rise to a complete embryo; the other formed a hollow sphere which, though viable, failed to undergo further differentiation. Here, for the first time, Spemann showed his remarkable capacity for astute interpretation of data and for asking the right questions. He surmised that in the case of twin formation the first cleavage plane had coincided with the median plane of the future embryo. Each lateral half then had reconstituted the whole embryo. In the second instance, however, the constriction had separated the dorsal from the ventral half of the future embryo. Hence, the undifferentiated sphere was called the 'belly piece'. This meant that the dorsal axial organs can develop in the absence of the ventral half and even reconstitute the ventral viscera (intestine, heart), but not *vice versa*. (In the embryo, the axial organs are represented by the neural tube, the precursor of the CNS; the notochord, a skeletal rod, beneath it; and two rows of segmental somites, the precursors of trunk musculature, vertebrae and dermis. Neural tube and notochord are in the midline, flanked by the somites.)

Thus the random variation in the

Fig. 1. *Hans Spemann.*

relation of the first cleavage plane to the bilateral egg structure had given Spemann the hoped-for clue. He stated: 'My experiment gives no information concerning the kind of "differentiation substance" which is lacking in the ventral blastomere. One can think of an unorganized substance which is necessary either for the triggering or the formation of the (axial) organs, or else it could be an embryonic material which has the capacity to differentiate to form the respective organs, or perhaps to incite other cells to differentiate'[4]. This was an uncanny prescience of the organizer action.

The constriction experiment gave one further clue. When it was extended to older stages, it was found that duplications can be obtained only up to the early gastrula stage. Constrictions after gastrulation, in the neurula stage, resulted in the formation of two half-embryos. This meant that at the neurula stage the capacity for regulation had been lost and the irreversible determination of the axial organs had occurred. [Gastrulation is a sequence of morphogenetic movements of cell assemblies during which the ventral half of the blastula, a hollow sphere, invaginates and the dorsal half expands to cover the invaginated parts. Invagination begins at a dorsally located sickle-shaped cleft, the blastopore (Fig. 2) which later extends laterally and ventrally. The region above the blastopore is called the dorsal lip. The invaginated material represents the precursors of the mesodermal axial organs and of the endodermal intestine: the outer layer, the ectoderm, gives rise to the neural plate and epidermis.]

Further progress required a method by which the state of differentiation of different parts of the early gastrula could be tested. A suitable method of microsurgery on amphibian embryos had been worked out by Spemann in the context of another project. He had become interested in the development of the eye and had focused on the origin of the crystalline lens. The lens develops as an ectodermal invagination at the point where the optic vesicle, an outpocketing of the fore-brain, establishes contact with the ectoderm. Spemann suspected that the optic vesicle provides a stimulus for lens formation. To test for a causal relation, the eye anlage had to be removed before it contacted the ectoderm. For this purpose, Spemann fashioned very fine pointed glass needles by drawing out a thin glass rod over a small gas flame. With this instrument he extirpated the eye anlage. As expected, the lens failed to develop. When an optic vesicle was transplanted beneath the flank ecto-derm, a lens differentiated from this foreign material. These experiments marked the seminal discovery of embryonic induction which proved to be a major mechanism of progressive differentiation, and also an important component of organizer action. Ob-viously, an inductor is more than a trigger, it has an 'instructive' com-ponent. The definitive publication on lens induction appeared in 1912[5].

In 1914 Spemann became the Director of the Division of Experi-mental Embryology at the Kaiser-Wilhelm Institute for Biology in Berlin-Dahlem. His first priority was to analyse the state of determination of gastrula parts. Using the glass needle technique, he exchanged a small piece of ectoderm representing prospective neural plate with another piece of ectoderm representing prospective flank epidermis. Both transplants developed according to their new environment. This meant that they had not been programmed as yet and that the new environment had incor-porated them by a process later to be known as 'assimilative induction'. A very different result was obtained when the upper blastoporal lip was transplanted to the flank of another embryo. It invaginated, as it would have done at its original site, and subsequently it self-differentiated to notochord and somites. Several of these mesodermal structures were accompanied by neural tubes; but since their origin could not be ascertained, not much attention was paid to them.

The import of the observation that the region directly above the blasto-pore was already capable of self-differentiation whereas more distant regions were still dependent on induc-tive influences by their environment, did not escape Spemann's attention. He conceived the idea that a stream of determination spreads forward from the upper blastoporal lip, and he

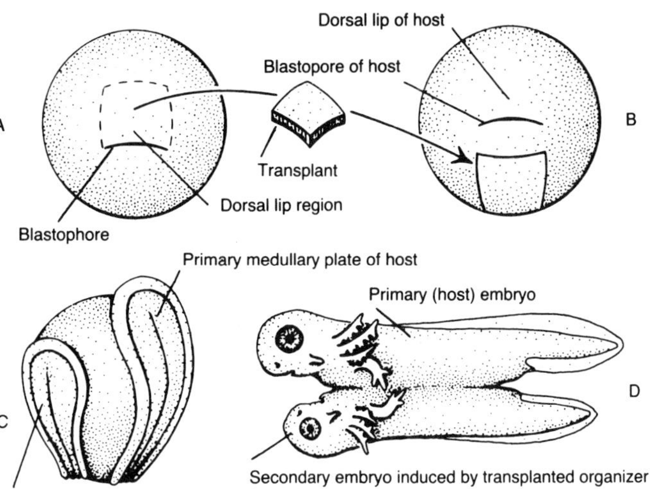

Fig. 2. *The organizer experiment (drawings by H. Holtfreter).*

designated the latter as the 'deter-mination or organization center' for the axial organs[6]. There was then no direct evidence for the spreading of an inductive agent; but conceptually this was the decisive step on the road to the organizer.

Further progress was impeded by the inability to distinguish between donor and host tissue. What was needed was a reliable cell marker. This problem was solved by using the method of heteroplastic transplanta-tion, that is, transplantation between two different species. Such experi-ments had been done before in amphibians but not at very early stages. The German fauna obliged by providing Spemann with the unpig-mented eggs of the salamander spe-cies, *Triturus cristatus* and the brown-ish eggs of *Triturus taeniatus*. Exten-sive tests of the method by exchanging different ectodermal regions of the early gastrula were successful; but for reasons that are not obvious, hetero-plastic transplantations of the upper blastoporal lip were not undertaken when the experiments were done in 1917. The publication of the results was delayed until 1921[7], due to Spemann's move to the University of Freiburg as the Director of the Zoological Institute, in 1919. In a postscript of this publication, dated May 1921, Spemann reported that his student, H. Mangold, had trans-planted an unpigmented blastoporal lip to a pigmented gastrula and the experiment had given a spectacular result: a well-formed, pigmented sec-

ondary embryo had developed on the flank of the host embryo (Fig. 2). Of course, Spemann realized the impor-tance of this finding. On the basis of this single case, he coined the term 'organizer': 'Such a piece of the organization center can be called an 'organizer'. It creates in the indifferent material to which it is transplanted an organization field of a definite direc-tion and extent."

The results of the experiments of H. Mangold were published in 1924[1]. Observations of the living embryos had shown that the transplants in-vaginated as they would have done *in situ*. They then proceeded to self-differentiate to notochord and som-ites. But since the transplants repres-ented only the central part of the fairly large chorda-mesoderm anlage, a process of regulation set in by which the missing parts were restored. This was achieved by recruiting uncom-mitted host cells at the margins of the transplants. This was another instance of 'assimilative induction'. As a result, notochord and somites were composed of a mixture of donor and host cells; pigmented host cells were interspersed in the otherwise unpigmented noto-chord and many somites were com-posed of a mosaic of pigmented and unpigmented cells. Other somites were either entirely unpigmented or entirely pigmented. Thus, the regula-tion of the mesodermal axial organs had occurred in complete disregard of the boundary between host and donor tissue. But the most striking feature of the secondary embryos was that their neural tubes were composed entirely of host tissue. This meant that the chorda-mesoderm, that is, the organizer, had induced the neural plate, the precursor of the neural tube, in the overlying ectoderm. Neural induction has become the hallmark of organizer action.

The origin of the secondary embryos was thus explained in terms of a combination of self-differentiation, regulation, assimilative induction and induction by contact. But the most impressive and unique feature of the organizer experiment was the integration of these components in the formation of a harmonious whole (Fig. 3). Considering that a well-proportioned secondary embryo was created by the upper blastoporal lip transplant which represented a small fragment of the chorda-mesoderm anlage, its designation as the 'organizer' seemed to be entirely appropriate. Thus Spemann's goal to come to an understanding of the embryonic origin of vertebrate organization, typefied by the axial organs, had been achieved beyond expectation. Singleness of purpose, ingenious analytical acuity and mastery of the craft of microsurgery were the keys to his success.

The magic of the term 'organizer' has undoubtedly contributed to the fame of the Spemann–H. Mangold discovery that extended far beyond the boundaries of experimental embryology. To this day, the organizer experiment stands as a symbol of the triumphs of classical experimental embryology, many of whose other achievements are now largely forgotten. But the term has also invited criticism and misunderstandings. It is a profound misinterpretation of the organizer to consider it as the general manager or master regulator of early vertebrate development. Its field of

action is limited in normal development essentially to the dorsal axial organs, though in the transplantation experiment it can extend its activity to the creation of viscera and heart by secondary inductions. Misgivings have been expressed by some who have insinuated that the term 'organizer' reflects Spemann's belief in a vitalistic force that guides and regulates embryonic development. This claim is not justified. It is true that vitalistic ideas had a place in his personal beliefs, his 'Weltanschauung'[8]. But the term 'organizer' was conceived strictly in a causal–analytical frame of reference. Its operational definition as a piece of the upper blastoporal lip of the early gastrula which, when transplanted causes the formation of a secondary embryo by a combination of self-differentiation, regulation, assimilative induction and induction by contact, is unassailable. All these processes have been well established as basic components of animal development. But, of course, it is a truism that concepts like induction and regulation have an explanatory value only in the framework of classical experimental embryology. While some progress has been made in the analysis of the chemical nature of induction, other phenomena, and in particular, regulation, have remained enigmatic to this day. But the fact that this and other supracellular processes, revealed by the experimental embryologists are hardly accessible to a reductionist approach at the present state of cellular and molecular biology is no reason to dismiss them as intractable. Like other 'mysteries' of this kind, they are a challenge to the future.

To the developmental neurobiologist, the most intriguing aspect of the organizer is its crucial role in the induction of the neural plate, the

precursor of the CNS. But more than that, the PNS which derives mostly from the neural crest is likewise the product of the primary neural induction, since the anlage of the neural crest can be traced back to the margin of the neural plate, the future neural folds. No embryonic cell, no matter what its clonal origin, can become a neuron, unless it becomes exposed to an extrinsic stimulus. In normal development the stimulus is provided by the chorda-mesoderm.

I may add a historical footnote. Experimental neuroembryology started as a modest offshoot of experimental embryology. Its founder was R. Harrison. He is best known as the originator of the tissue culture method by which the controversial issue of the origin of the axon was settled in favor of the outgrowth theory[9]. At the time of the organizer experiment he had turned to other issues. His student, S. R. Detwiler continued the tradition; he wrote the first textbook of neuroembryology, in which Harrison's and his own accomplishments are summarized[10]. As fate would have it, experimental embryologists nowadays try to redefine the old problems, while developmental neurobiology flourishes as never before.

Selected references
1 Spemann, H. and Mangold, H. (1924) *Wilhelm Roux Arch. mikrosk. Anat. Entwicklungsmech.* 100, 599–638. English transl. in Willier, B. H. and Oppenheimer, J. (1974) *Foundations of Experimental Embryology.* pp. 38–50, Haffner Press, New York
2 Hamburger, V. (1984) *J. History Biol.* 17, 1–11
3 Driesch, H. (1891) *Z. Wiss. Zool.* 53, 160–183. English transl. in Willier, B. H. and Oppenheimer, J. (1974) *Foundations of Experimental Embryology,* pp. 144–184, Haffner Press, New York
4 Spemann, H. (1901) *Wilhelm Roux' Arch. Entwicklungsmech. Org.* 12, 224–264
5 Spemann, H. (1912) *Zool. Jhrb. Abt. Allg. Zool. Physiol. Tiere* 32, 1–98
6 Spemann, H. (1918) *Wilhelm Roux Arch. Entwicklungsmech. Org.* 43, 448–555
7 Spemann, H. (1921) *Wilhelm Roux Arch. Entwicklungsmech. Org.* 48, 533–570
8 Hamburger, V. (1969) *Experientia* 25, 1121–1125
9 Hamburger, V. (1981) *Trends Neurosci* 4, 151–155
10 Detwiler, S. R. (1936) *Neuroembryology: An Experimental Study,* Macmillan, New York
11 Holtfreter, J. and Hamburger, V. (1955) in *Analysis of Development,* (Willier, B. H., Weiss, P. A. and Hamburger, V., eds), pp. 230–296, W. B. Saunders, Philadelphia

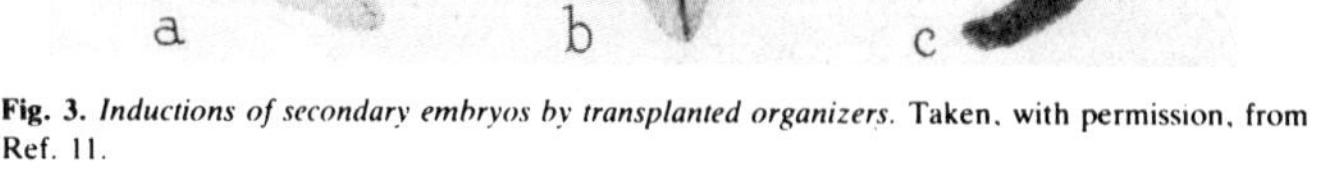

Fig. 3. *Inductions of secondary embryos by transplanted organizers.* Taken, with permission, from Ref. 11.

V. Hamburger is at the Department of Biology, Washington University, St Louis, MO 63130, USA.

Reprinted with permission, from the *International Journal of Developmental Neuroscience*

Preface to The S. Kuffler Lecture
in the *International Journal of Developmental Neuroscience*

Although the *International Journal of Developmental Neuroscience* publishes original research papers, from time to time we feel it is appropriate to publish historical essays by eminent developmental neuroscientists. In commoration of the 11th anniversary of the founding of the International Society for Developmental Neuroscience, we are pleased to publish the following personal account of the ontogeny of the field of developmental neuroscience by Dr. Viktor Hamburger, who was one of the key players. Dr. Hamburger's perspective is unique in providing an eyewitness account of the beginnings of experimental neuroembryology and its evolution into developmental neuroscience. In acknowledgement of this important role, Dr. Hamburger was awarded the first honorary membership in the International Society for Developmental Neuroscience in July 1983. Dr. Hamburger is Mallinckrodt Professor Emeritus and former chairman of the Department of Zoology, Washington University, St. Louis. The following essay was delivered by Dr. Hamburger as the Steven J. Kuffler lecture in the Department of Neurobiology, Harvard University, May 8, 1989.

Jean M. Lauder, President
International Society for
Developmental Neuroscience

The S. Kuffler Lecture
The Rise of Experimental Neuroembryology
A Personal Reassessment

Viktor Hamburger

The Germans have a penchant for condensing their collective wisdom in proverbs: "All is well that ends well," "Pride comes before the downfall," and "that which wants to become a hook, bends early." At age 14 I was exploring the countryside around the small German town in which I grew up. I collected plants, animals, fossils, and brought home salamander and frog eggs and watched them develop and metamorphose. There was never any doubt in my mind that I would become a naturalist. Years later, I made the conscious choice to study Zoology. I felt that the Natural Sciences brought me closer to their subject matter than the Cultural Sciences, as the Humanities were then called. Animals and plants could be ordered in a rational taxonomic system, and frog embryos displayed a reassuring regularity. There was hope to understand some of the rules that govern these phenomena. Surely, others would find *history* more dramatic. No doubt, the lives of Caesar, Napoleon, or George Washington could be more fascinating than those of my tadpoles. But to unravel the mysteries of their genius seemed to me beyond my ken.

I discovered my calling at age 19, in a graduate seminar at the Zoological Institute in Heidelberg. Its Director, the distinguished experimental embryologist Curt Herbst, admitted me, a second-year student, probably on the recommendation of an older friend. At any rate, this was a refreshing sign of informality that prevailed then at German Universities. We read and discussed some publications of Wilhelm Roux, who was the founder of Experimental Embryology. Despite his long-winded and somewhat pompous style I detected my "elective affinity" to his causal–analytical way of thinking and I looked forward to experimenting on living embryos.

When I moved to Freiburg, at the foot of the Black Forest, hiking and skiing were much on my mind, but I had also found out that the Director of the Zoological Institute, Hans Spemann, was a distinguished experimental embryologist. He had discovered embryonic induction, in the case of lens induction by the optic vesicle, and he had designed a method of transplantation on early amphibian embryos in the pursuit of problems of determination of organ primordia.

Two fellow students arrived in Freiburg at the same time, in 1920: Johannes Holtfreter, who became Spemann's most innovative student and my life-long friend, and Hilde Proescholdt, who later became Mrs. Mangold. She was fortunate in having the organizer experiment assigned to her as her Ph.D. thesis. The organizer experiment consisted of the transplantation of the so-called dorsal lip of the blastopore of the gastrula (an early stage of the amphibian embryo) to the flank of another gastrula, where it induced a secondary neural plate (the precursor of the nervous system) and eventually the formation of an entire secondary embryo. I think the designation "organizer" for the tiny piece of embryonic tissues that accomplishes this feat is entirely justified. Some detractors of Spemann have implied that this term has vitalistic implications. This is not the case. The formation of the secondary embryo can be fully understood in terms of then well-established mechanisms, such as assimilative and contact inductions, selfdifferentiation, etc.

I still remember the great excitement in the laboratory when Hilde Proescholdt showed us late in the Spring of 1921 her first success, a well-formed induced secondary embryo. And we were deeply saddened when she died in an accident in 1924, shortly after the publication of the famous organizer paper by her and Spemann (1924), that earned Spemann the Nobel Prize in 1935.

Spemann's approach to the causal analysis of development has influenced me profoundly. His emphasis was first and foremost on fundamental problems of a general nature. The specific experiment served the solution of a general problem. When he had discovered embryonic induction by demonstrating that the optic vesicle causes lens formation in the overlying ectoderm, he commented: "In the final analysis, what counts is not the eye of the frog, but a universal law of development" (Autobiography, 1943, p. 200).

Spemann was a master in the critical assessment of his experiments. When one is limited essentially to two methods, extirpation and transplantation of embryonic parts, one has to exploit the experimental results to the limits. All possible implications were carefully considered. I'll give you one example from his earliest experiments (Spemann, 1901). He was impressed by a finding of his contemporary, Hans Driesch, who had isolated the two cleavage cells of the tiny sea urchin egg and had made the startling discovery that each half regulated to form a whole larva. Spemann wondered whether vertebrate eggs were endowed with the same regulative capacity. He constricted the 2-cell stage of a salamander egg with a baby's hair and obtained likewise identical twins. But it did not escape his attention that in many cases only one cell had formed an embryo; the other had formed an undifferentiated spherical vesicle which he called the "belly piece." He chose this name because he assumed—correctly as it turned out—that in these cases the constriction had separated the dorsal from the ventral half of the future embryo, rather than left from right. He then argued that the dorsal region must contain a substance or a structure that is indispensable for the formation of the dorsal axial organs, but was missing in the ventral half. The single-minded pursuit of this agent led to the discovery of the organization center in 1921, and of the organizer in 1924. Another part of Spemann's success was his proficiency in designing the tools for his microsurgery: glass needles, hair loops, and micropipettes. He was rather pedantic in this matter and insisted that our instruments were not just adequate but perfect or even elegant.

During the 10 years which I spent in his laboratory, first as a Ph.D. candidate and later as a junior faculty member, I got to know Spemann very well. He was much more personable and less austere than he is often portrayed. It is true that the atmosphere at German Universities was much more formal than in this country, but when he came home one summer from a visit to the Marine Laboratory at Woods Hole, he told us how much he had enjoyed its relaxed atmosphere, when the young students greeted him with "Hey, Doc." And he had a keen sense of humor. When a colleague pointed out in a review article that an American anatomist, Warren Lewis, had actually transplanted the upper blastoporal lip of frog gastrulae, in 1907, and that he deserved priority for the organizer experiment, Spemann wrote: "One can make a discovery without intending it, but not without noticing it." And when after his favorite lecture on birds a student asked him whether it is true that birds are lighter than air, he said "Yes, that is why they have to grasp the branches so tightly."

One of the many visitors in the Freiburg laboratory in the 1920s was Ross G. Harrison from Yale, who was a friend of Spemann. His visits became important to me, because I was working on a Ph.D. thesis that dealt with embryonic correlations involving the nervous system. Harrison was *the* expert in neurogenesis, whereas Spemann's interest in this topic ended with neural induction.

Harrison is rightly credited with the establishment of experimental neuroembryology as a branch of experimental embryology in its own right. He is best known for the design of the tissue culture method which settled the long-standing controversy over the origin of the axon. While the German anatomist Wilhelm His and the Spanish anatomist S. Ramón y Cajal had presented strong evidence that the axon is an outgrowth of the embryonic neuron, the majority of their colleagues insisted that protoplasmic fibers, or plasmodesms, which were bridges between non-nervous cells, created, or at least contributed to, the formation of axons. Harrison realized that the histological methods available at that time could not decide the issue. He stated, "The really crucial experiment remained to be done, and that was to test the power of the nerve centers to form nerve fibers within some foreign medium which could not by any possibility be suspected of contributing organized protoplasm to them" (1910, p. 790). He found the suitable medium in clotted blood plasm of the frog in which he grew pieces of spinal cord of very early frog embryos. In 1907 he was the first to observe the outgrowth and elongation of the axon from individual embryonic nerve cells, and the spreading of filopodia from the growth cone (Harrison, 1970a). The old dispute was thus settled. Interestingly enough, Harrison never used the *in vitro* culture method again.

He designed another model experiment that has become an indispensable stand-by in neuroembryology: the limb transplantation which he used for the analysis of nerve pattern formation. He observed that limb primordia of frog embryos transplanted to the flank would be innervated by foreign nerves which formed a perfectly normal nerve pattern; and he concluded that the tissues of the growing limb provide the cues that guide the growth cones to their targets (Harrison, 1907b).

Harrison then turned to other problems and left neuroembryology to the younger generation, that is, mostly his own students. One of the most active among them was Sam Detwiler. Harrison had assigned to him the experiment of shifting the limb primordia of salamander embryos to the flank, to see whether they would perform coordinated movements. They did so, as long as they were connected with the brachial plexus. The histological examination of the material revealed a novel phenomenon: the brachial dorsal root ganglia (DRG) which had been deprived of their targets were hypoplastic and the thoracic ganglia which were overloaded were hyperplastic. Strangely enough, the motor centers seemed to be unaffected (Detwiler, 1920). It should be mentioned that in salamanders the motor neurons are not condensed in columns, but scattered in the spinal cord and thus difficult to identify.

A few years later, I encountered this problem in the reverse: a German experimental embryologist had asserted that in frog embryos the nerve supply has an influence on the normal development of the limb. But the issue was controversial, and Spemann suggested that I repeat his experiments. The results of my Ph.D. thesis were somewhat ambiguous. I then decided to perform the crucial experiment of producing nerveless legs by extirpating the lumbar spinal cord in early tail bud stages of frog embryos, that is prior to nerve outgrowth. The legs developed in a perfectly normal fashion in every respect, though the muscles atrophied later. The result also excluded functional activity as a factor in muscle or joint differentiation. This was my first original contribution (Hamburger, 1928).

Early on, I felt the need for fitting my data into a broader frame of reference. In 1927 I wrote my first review article in which I identified the 3 major issues in the developmental correlations between the limbs and their innervation:

1) the role of nerve supply in limb development, an issue which I considered as settled by my experiment;
2) the determination of nerve patterns, for which Harrison had given a preliminary answer;
3) the role of the targets in the differentiation of their nerve centers.

The last topic was disposed of in less than half a page, because there were only two experiments to be considered: the one by Detwiler which I have mentioned, and one by Miss Shorey on chick embryos published in 1909. She had destroyed the wing bud by electrocautery and had found that both the brachial DRG and the brachial lateral motor columns (lmc) were hypoplastic, in contrast to Detwiler's claim that the motor centers were not affected. It seemed that the motor systems of salamander and chick embryos reacted differently to target deprivation, unless Detwiler had overlooked a deficiency in the motor system of his embryos.

I knew Detwiler from his visit to Freiburg, but I knew nothing of Ms. Shorey; her name had disappeared from the literature. I found out a few years later that she had been a student of Dr. Frank Lillie of the University of Chicago, whose classic book "The Development of the Chick" had put the chick embryo on the map for research and teaching. As fate would have it, in 1932 the Rockefeller Foundation gave me a one-year fellowship to be spent in Dr. Lillie's laboratory, with the assignment of trying out Spemann's technique of microsurgery on chick embryos. When Hitler came to power in the following spring, the Rockefeller Foundation continued to support me and the one year extended to over half a century.

Dr. Lillie reminded me that Ms. Shorey had done her experiment at his suggestion, and, being aware of my interest in neuroembryology, he suggested that I repeat her experiment with my more refined technique, and try to settle the discrepancy of her and Detwiler's results with respect to the motor system. Within a few months I had mastered the technique of extirpating and transplanting limb buds (Hamburger, 1934, 1939). As

Figure 1. Nine-day chick embryo with wing transplant behind normal wing. (from Physiol Zool 1939; 12)

examples I present to you first, an ordinary transplant to the flank, and then a more fanciful, chimerical case: this is Dick and Chuck; a chick embryo with a duck leg and a duck embryo with a chick leg. It was certainly a relief to get results within days rather than weeks and months and not to be frustrated by high mortality. Since then, I have never touched an amphibian embryo again, except in the classroom, and the chick embryo has moved to a privileged position in experimental embryology.

The wing extirpation experiments confirmed Ms. Shorey: both DRG and lmc were hypoplastic. To remove any doubt that the deficiencies might be due to a trauma inflicted by the operation, I used my large material of limb transplantations and showed that the DRG and lmc that contributed to the innervation of the transplants were hyperplastic (Hamburger, 1939).

I realized that the target effects on their nerve centers raised a fundamental issue and that its analysis could be carried much further than Shorey and Detwiler had done. Beginner's luck played also into my hands. My extirpations were not very accurate and I obtained a wide range of muscle deficiencies, but the skin loss was quite invariable. When I quantified the data, I found a close correlation between the wide range of muscle

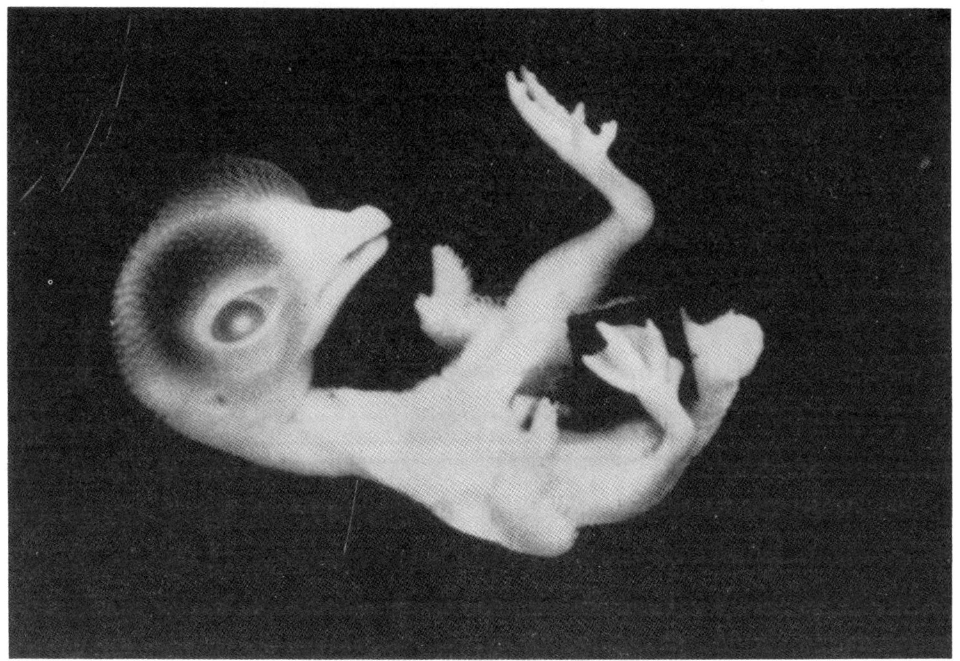

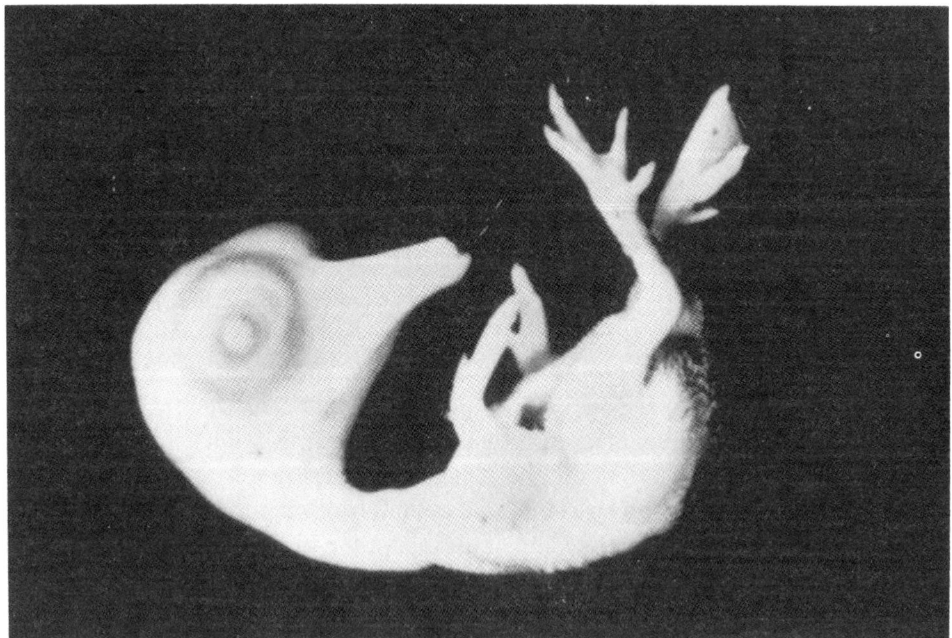

Figure 2. "Dick and Chuck." A duck embryo with a transplanted chick leg, and a chick embryo with a transplanted duck leg.

deficiencies and a corresponding range in the hypoplasia in the motor columns and the same correspondence of skin deficiency and DRG hypoplasia. This suggested a direct and specific effect of each target on its own center. I stated in the publication of 1934 "that every structure within the growing limb, muscles as well as sense organs send stimuli to the nervous system. Each part of the peripheral field controls directly its own nerve center, that is, the limb muscles affect the lateral motor centers and the sensory fields control the ganglia" (p. 470), and I implied "that the nerve fibers themselves serve as mediators between the two links of the correlation" (p. 475).

The paradigm of specific agents produced by the different targets, that are transported retrogradely to their respective centers, and regulate the differentiation of these centers in a quantitative way has remained the foundation of all subsequent investigations in this field, and led eventually to the discovery of NGF.

In 1934, one particular problem had remained unresolved: the mode of action of the hypothetical agent. In accordance with my experimental embryological background in Spemann's laboratory, I formulated an induction or recruitment hypothesis. I postulated a pool of undifferentiated precursor cells in the nerve centers. I assumed further that early differentiating neurons would send their axons to the target area and that these pioneer fibers would explore the extent of the target area. The pioneer fibers would then send signals to their perikarya, which in turn would recruit as many un-differentiated cells as were required to saturate the target area. Both hypo- and hyper-plasia could be explained in this way, but unfortunately, my brain child was short-lived.

I have been taken to task for making a wrong guess in this important issue. But my critics have failed to give me credit for establishing the basic paradigm which had to be there before I could have been right or wrong about a specific aspect of the paradigm. The underlying notion of a specific agent that is transported retrogradely from the target to the center has never been challenged.

Perhaps my wrong guess was a blessing in disguise. I suspect that it motivated a young Italian investigator, Rita Levi-Montalcini, and her mentor, Giuseppe Levi, to repeat my experiment, when they got hold of my publication of 1934 during the war. For reasons that I shall explain later they focused on the DRG. They repeated my limb bud extirpations and counted sensory neurons in limb ganglia at regular intervals after the operation and found a gradual decline in the numbers of fully differentiated neurons. They concluded that the hypothetical target-produced agent does not influence progres-sive differentiation, but is necessary for the survival of neurons (Levi-Montalcini and Levi, 1943).

When I became acquainted with these investigations after the war, I was amazed and puzzled. The thought that such a morbid event as massive cell death should be part of progressive differentiation was alien to an experimental embryologist. I was eager to pursue the matter further, and I was very pleased when Rita accepted my invitation to join me in St. Louis, in 1947. We repeated the wing extirpation experiments. Our publication two years later advanced the analysis on 3 points: We saw for the first time degenerating cells in large numbers in brachial DRG subsequent to wing extirpation. Second, Rita observed massive degeneration in cervical and thoracic DRG, which had not been affected by the operation. This was the discovery of naturally occurring neuronal death. Finally, all data taken together suggested the *competition hypothesis.* We postulated that the target area produces a trophic maintenance factor which is in short supply and that even under normal conditions a considerable percentage of neurons would lose out in the competition for the factor.

I do not have to recount the story of the discovery of the Nerve Growth Factor. I participated actively in the adventure and in the two publications of 1951 and 1953, which dealt with the startling effect of mouse sarcoma on the enlargement of sensory and sympathetic ganglia. In the meantime, the biochemist Stan Cohen had joined us. When the analysis had moved altogether to the biochemical level in the mid-1950s, I realized that I could be of no further help, and I turned to other interests.

It is intriguing to contemplate that all this might not have happened, were it not for Rita's predilection for the sensory system which had prevailed over my predilection for the motor system. Those who might attribute our preferences to the gender difference would be on the wrong track. Our differences were actually due to early imprinting. Mine occurred in my first wing extirpation experiment, which focused on the discrepancy between Shorey's and Detwiler's experiments with respect to the motor system. But how about Rita's partiality to the DRG? I found out that the very first research project which her teacher, Guiseppe Levi, had assigned to her was the not very inspiring task to count sensory neurons in newborn mice to find out whether significant differences exist between ganglia of individuals of the same litter and of different litters (Levi and Sacerdote, 1934). I assume that you are not particularly interested in the results, but this study established the permanent bond between Rita and the sensory ganglia. It was only natural that when she repeated my experiment in Italy, she turned again to the ganglia; and when we repeated the wing extirpation once more in St. Louis, the old habit had become routine. This was the first of quite a few fortuities that has given the story of the discovery of NGF its special flavor.

Another fortuity brought about my return to the motor system. One day, in the laboratory of a pre-medical embryology course, in which the 10-mm pig embryo was obligatory, I was struck by a peculiar feature in the cross sections of the spinal cord. Large numbers of mitotic cells were clustered in the dorsal part, but there were none in the ventral region. When I checked on my slides of chick embryos, an interesting picture emerged. In early stages all mitotic activity was concentrated in the ventral, that is, the motor region; in later stages it had subsided there and shifted to the dorsal region which is the source of internuncial and secondary sensory neurons. Hence the motor system matures several days in advance of the sensory system (Hamburger, 1948). This observation rang a bell. I remembered a classic book in my library, the "Physiology of the Embryo" by a famous German physiologist, William Preyer, which had been published in 1885. In it he had reported an observation which he, in his own words, considered to be "one of the most important facts in the whole area of the physiology of the embryo." He had seen motility in very early chick embryos, 4 or 5 days before he could obtain responses to stimulation. He spoke of "impulsive" movements; I referred to them later as pre-reflexogenic *spontaneous* movements. Obviously, I had found the structural basis of this phenomenon.

Embryonic motility had been a lively but contentious issue in the 1920s and 1930s.

A dispute had developed between two camps that were poles apart. The behaviorists who dominated the scene and had concentrated on mammalian fetuses had predictably come to the conclusion that behavior starts with local reflexes which become gradually integrated into more complex behavior patterns. At the other pole stood a lone fighter, the ingenious neuroembryologist, G.E. Coghill, who had chosen the salamander embryo. A long series of investigations had led to the conclusion that all behavior is totally integrated from the first bending of the head to the complex performances of swimming, walking, and feeding, and that local reflexes originate by emancipation, or, as he called it, "individuation" from the total pattern. His position was strengthened by the careful analysis of the structural basis for his behavior study. He found that each step in behavioral development was preceded by the formation of the appropriate synapses in the central nervous system. The only error he made was that he thought that his model could be generalized to all vertebrates (Coghill, 1929). I met Coghill one summer in Woods Hole and I remember lengthy discussions with him, long before I became involved in this topic. What I admired most was his broad philosophical outlook which is well recorded in his biography by his friend Judson Herrick.

At any rate, the two extreme positions were irreconcilable; an impasse had been reached, and the field was almost deserted, when I began to take an interest in it. Obviously, spontaneous motility had no place in either one of the two schemes. I wondered whether the chick embryo represented a special case, or, on the contrary, whether its behavior might give a clue to the resolution of the conflict. I was also encouraged to enter this field by the idea that the experimental method might help to break the deadlock; so far, observation and stimulation experiments had been the only approach. The project which I launched in the late 1950s kept me and a team of very able young collaborators busy for a decade. We found that spontaneous motility is the typical mode of behavior of the chick embryo not just in early stages but throughout the incubation period, and that evoked responses play a minor role, if any. A glance through a window in the shell will convince anybody that the chick embryo does not conform at all to Coghill's notion of total integration. On the contrary, its motility is entirely uncoordinated in the sense that at any moment any combination of parts of the embryo can be in motion, while other parts are immobile. The embryo presents a rather bizarre picture of convulsive-like, aimless fidgeting. Another characteristic is that during the first half of incubation, motility is periodic in the sense that activity phases alternate with inactivity phases (Hamburger, 1963). Clearly, the chick embryo had nothing in common with the salamander embryo.

Next, we decided to eliminate all sensory input in a critical deafferentation experiment. It gave clear evidence that we were dealing with non-reflexogenic, intrinsically activated motility. The posterior body and legs were deafferented by the removal of several thoracic spinal cord segments and the removal of the posterior neural crest which produces the lumbar DRG. We found that leg motility was the same as in controls with isolated lumbar spinal cords (Hamburger, Wenger, Oppenheim, 1968). Then we tried to find out how the motility is generated. Extracellular recordings from the ventral part of the spinal cord gave polyneuronal burst patterns which were precisely synchronous with the observed motility patterns throughout the incubation period. The burst patterns persisted even when the embryo was curarized. Thus we had conclusive evidence that the motility was autonomously generated by action potentials in the spinal cord (Hamburger, 1970).

But that leaves an interesting question unresolved: How does the chick embryo manage to escape from the shell? Certainly, it would have to abandon its aimless fidgeting. Ron Oppenheim and I spent several months figuring this out. We found that about 3 days before hatching the embryo starts a new highly coordinated behavior pattern which culminates in powerful backthrusts of the head and beak that break the shell (Hamburger and Oppenheim, 1967).

Rat fetuses showed a behavior pattern similar to that of the chick, and we concluded that this pattern is characteristic of all birds and mammals, including the human fetus. In fact, all modern studies of mammalian and human fetal activity are based on this paradigm. However, we get little credit from most developmental psychobiologists. For them, life begins with the rat fetus.

Altogether, we had obtained a clear overall picture. We did not have the tools to carry the analysis any further, and I returned to my interest in trophic interactions. But rather than turning to this theme, I should like to turn the clock back and reflect on some other major events that contributed to the coming of age of experimental neurogenesis. When I started in Chicago in the 1930s, the number of embryologists interested in the nervous system was small, perhaps 20 or 30. We knew each other and met at the annual meetings of the Anatomists and Zoologists, but there was little meeting of the minds. Our outlook was limited and so were our tools. This is reflected in the first book on neuroembryology by Detwiler, published in 1936. It is remarkable how few of the experiments that filled the 200 pages ever came to fruition; most of them are now forgotten. Among the survivors were Detwiler's and my experiments on the dependence of nerve centers on their targets which I have already discussed. Another topic that got new attention was the old problem of the mechanism of axon guidance. Detwiler reviewed tissue culture experiments of Paul Weiss which had moved the problem on a new track. At that time the idea prevailed that targets attract the appropriate nerve fibers by action at a distance. Chemotropism had been advocated by R. y Cajal as far back as the 1890s. The Dutch anatomist Ariens Kappers had promoted a theory of galvanotaxis under the fancy name of neurobiotaxis, according to which electric currents conducted in longitudinal fiber tracts would elicit directional outgrowth of axons and dendrites in nearby neurons. It was the merit of P. Weiss to have discredited the notion of action at a distance and drawn attention to the microenvironment of the growth cone. Weiss was at that time in the early 1930s a guest in Harrison's laboratory; he had adopted the otherwise rather neglected tissue culture method. He explanted brain fragments and ganglia of chick embryos and showed that he could control the direction of outgrowth of axons by structuring the substrate. This was done in a rather crude fashion by stroking with a brush or needle (Weiss, 1934). But the basic concept of *contact guidance*, as he called it, was sound. By extending this concept to guidance by submicroscopic, micellar fibrous elements, he paved the way to modern molecular approaches. Actually, Wilhelm His, Harrison, and others before him had demonstrated that axons require a solid substrate for elongation, and they had coined the term *stereotropism*. This notion and the refined version of Weiss had in common the emphasis on the physical properties of the substrate. But it was difficult to imagine how contact guidance alone could account for the highly selective pathways and synaptic contacts of axons. Weiss conceded that chemical signals might provide some additional information, but in general he greatly overrated mechanical and underrated chemical factors.

The crucial shift of emphasis from contact guidance to *chemoaffinity* was achieved by his student, Roger Sperry, in the 1940s. It is of interest to trace the steps that led to Sperry's defection from his mentor. To do this, I have to mention another of Weiss' strong convictions, this time in the matter of specificity of synapse formation. He shared with most of his contemporaries the belief that nerves distribute themselves randomly to skin and muscles, and that they become specified secondarily by the targets. Opinions were divided as to the mechanism of specification. Behaviorists who were then at the top claimed that functional activity or conditioning plays the decisive role. Weiss rejected this idea and developed the notion that specifications are established very early in development, in prefunctional stages, by developmental mechanisms akin to induction. He assumed that in the case of the neuromuscular synapse the muscle primordia acquire specific labels very early and then confer them on the nerves which make contact with them. His term *myotypic specification* conveys this idea (Weiss, 1936).

On the basis of some experiments of his own, Sperry had become suspicious of the notion that neurons are initially unspecified and distributed indiscriminately to their targets. In a radical departure from these ideas, he asserted that neurons acquire a high degree of cytochemical specificity in very early stages of their differentiation, independently of their targets; and furthermore, that they establish synaptic connections only with other neurons and muscles that have a matching specificity. The *chemoaffinity theory* was tested in the early 1940s on the retino–tectal system of amphibians. The switch from the neuromuscular to the retino–tectal system turned out to be most fortunate. Its major asset was the precise projection of the retinal map on the tectal map which offered many possibilities of experimental manipulation. You will recall Sperry's classical experiments of the 180° eye rotation in salamanders and frogs (Sperry, 1944). They were predicated on the earlier finding that the optic nerve can regenerate and visual perception can be restored. The maladapted behavior that the frog displayed after eye rotation was not just blurred vision. Rather, the visual perception was also inverted precisely by 180°. The animals responded as if they saw everything upside-down and reversed front to back. The obvious explanation was that regeneration had occurred in an orderly fashion and that the synaptic connections had been reestablished on the basis of highly selective affinities between retinal fibers and tectum cells. It was a stroke of genius to have solved one of the most subtle problems of neurogenesis by the combination of simple surgery with a simple behavior test.

A great variety of experiments, mostly on other sensory systems, corroborated the theory. In earlier publications Sperry still made some concessions to the notion of imprinting of axon specificity by the target cells, but in a 1965 review the last reservations were withdrawn. In the meantime, the chemoaffinity theory had been extended to selective pathfinding of axons. Sperry's experiments in collaboration with D. Attardi (1963), in which parts of the retina had been removed, had shown that retinal fibers make correct choices at branching points and altogether make no errors in reaching the target.

Of course, Sperry did not know of Spemann's precept mentioned earlier that "in the final analysis what counts is not the eye of the frog but a universal law of development," but he acted accordingly and generalized, "One must assume that the cells and fibers of the brain and cord carry some kind of individual identification tags, presumably cytochemical in nature, by which they are distinguished one from another, and further, that the growing fibers are extremely particular, when it comes to establishing synaptic connections, each axon linking with certain neurons to which it becomes selectively attached by specific chemical affinities" (1963, pp. 703–704).

A major virtue of the retino–tectal system was that the mechanisms by which the retinal map was projected on the tectal map could be explored by the experimental manipulation of both the retinal and the tectal system. A further refinement of the analysis was achieved when Gaze and Keating in England introduced recordings from the tectum in the 1960s. It is hardly surprising that the abundance and perhaps super-abundance of experiments that were done in this field sometimes blurred the picture and revealed that the situation was actually much more complex than anybody could have anticipated. This was due in part to the formidable plasticity and regulative capacity of the embryos of the lowly amphibians. I suspect that a lot of trouble could have been avoided if one had consulted the chick embryo. At any rate, Sperry's inspired experimentation set the stage for the ongoing search for the cues that guide growth cones to their destination.

In the meantime, NGF had begun its meteoric rise. However, its counterpart, neuronal death, suffered from gross neglect. Apart from the work of two English investigators, M. Prestige and A. Hughes, who had made a detailed analysis of neuronal death in amphibian embryos in the 1960s (Hughes, 1968), there was little interest in this subject, and the latency period lasted for two decades. But in the 1970s it became clear that neuronal death is a widespread and almost universal phenomenon, and that a loss of 50% or more is quite common. How can one explain this strange phenomenon? There are two ways of looking at it—they are not alternatives, but complementary. Taking again the limb innervation as an example, one has to realize that the lmc and the limb bud differentiate entirely independently of each other, both spatially and chronologically. Hence an overproduction of motor neurons would have to be built in as a safety factor. In other words, overproduction is a precondition for what has been called "*systems matching*" (Gaze and Keating, 1972). But why an excess of 50%, if an additional 15–20% would suffice to guarantee the saturation of all muscles? At this point, the previously mentioned *competition hypothesis*, which refers to competition for a trophic maintenance agent, comes to mind. If this agent is in very short supply, then a high degree of mortality is to be expected. But perhaps the last word on this matter has not yet been said.

In this connection it occurred to me that in spite of the vast expansion of NGF research that revealed ever new facets of NGF activities, a simple but important question had not been addressed: Is NGF identical with the hypothetical trophic agent produced by limb tissues, for which the DRG neurons compete? At that time there was no way of a direct identification of NGF in embryonic tissues and we had to use an indirect approach. First (in collaboration with Judy Brunso-Bechtold, 1979) we showed that NGF is transported selectively to DRG during the critical period. This was done by implanting pellets containing radioactive NGF into the limb. Only sensory nerves and lumbar DRG showed radioactivity. Next (in collaboration with Judy Brunso-Bechtold and Joe Yip, 1981) we were successful in rescuing the majority of sensory neurons that were destined to die, by daily injections of NGF into the yolk sac. Finally, (in collaboration with Joe Yip, 1984) we subjected our assumption to the most severe test; wing bud extirpation was combined with daily injections of NGF. Again, the majority of sensory neurons that would have died did survive. These findings, taken together, give strong support to the idea that NGF is indeed the normally occurring trophic maintenance factor for sensory neurons which we had postulated in 1949.

When I was still actively engaged in research, I witnessed the gradual transformation of experimental neuroembryology to developmental neurobiology. This sounds innocuous, but actually it was a fundamental shift in perspective and conceptualization. Both the term "experimental" and the "embryo" were dropped in the process. Not that experimentation *per se* disappeared, but the old-style *causal–analytical* experiment, the legacy of general experimental embryology, lost its relevance. It had served us well in our search for single or even multiple causes in morphogenesis and differentiation. When the analysis moved several orders of magnitude down from the supracellular to the molecular level, complexity increased to the same degree. Nowadays, every event is seen as the result of an intricately interwoven web of signals, growth factors, ultrastructural and molecular interactions, and the findings are stated in a new language. The only claim that the old guard can make is that it did identify the half dozen or so major issues in neurogenesis; that it struggled with them, and since we were not encumbered by a bewildering amount of details, we came up with relatively simple, first-approximation, answers which then served as guideposts for more penetrating analysis.

The loss of the "embryo" when the term "neuroembryology" was dropped is a more serious matter. To me it is symbolized in the indispensable role that the tissue culture method plays in modern research. Most of us, including its creator, R. Harrison, did not find it suitable for our approach. We found great satisfaction in handling living embryos, and no doubt, the present generation finds equal satisfaction in handling molecules. But all this should not obscure the fact that we all have a central concern in common. Whatever we explored with transplantations, and what you now explore with infinitely more sophisticated technology, is actually played out in the living, developing embryo. It is a reassuring thought that the embryo which already has outlived the experimental neuroembryologists will still be around when their molecular successors have, likewise, become part of history.

References

Attardi D, Sperry RW: Preferential selection of central pathways by regenerating optic fibers. Exp. Neurol. 1963; 7: 46.

Brunso-Bechtold J, Hamburger V: Retrograde transport of nerve growth factor in chick embryo. Proc. Natl. Acad. Sci. (USA). 1979; 76: 1494.

Coghill GE: *Anatomy and the Problem of Behaviour.* Cambridge: Cambridge University Press, 1929.

Detwiler SR: On the hyperplasia of nerve centers resulting from excessive peripheral loading. Proc. Natl. Acad. Sci. 1920; *6*: 96.

Detwiler SR: *Neuroembryology.* New York: The Macmillan Company, 1936.

Gaze RM, Keating MJ: The visual system and "neuronal specificity." Nature. 1972; *237*: 375.

Hamburger V: Entwicklungsphysiologische Beziehungen zwischen den Extremitaeten der Amphibien und ihrer Innervation. Naturwiss. 1927; *15*: 657, 677.

Hamburger V: Die Entwicklung experimentell erzeugter nervenloser und schwach innervierter Extremitaeten von Anuren. Roux'. Arch. f. Entw. mech. 1928; *114*: 272.

Hamburger V: The effects of wing bud extirpation on the development of the central nervous system in chick embryos. J. exp. Zool. 1934; *68*: 449.

Hamburger V: Motor and sensory hyperplasia following limb bud transplantations in chick embryos. Physiol. Zool. 1939; *12*: 268.

Hamburger V: The mitotic patterns in the spinal cord of the chick embryo and their relation to histogenetic processes. J. Comp. Neur. 1948; *88*: 221.

Hamburger V: Some aspects of the embryology of behavior. Quart. Rev. Biol. 1963; *38*: 342.

Hamburger V: Development of embryonic motility. The Neurosciences Second Study Program (F.O. Schmitt, ed.). New York: Rockefeller Univ. Press, 1970.

Hamburger V, Brunso-Bechtold JK, Yip J: Neuronal death in the spinal ganglia of the chick embryo and its reduction by nerve growth factor. J. Neurosci. 1981; *1*: 60–71.

Hamburger V, Levi-Montalcini R: Proliferation, differentiation and degeneration in the spinal ganglia of the chick embryo under normal and experimental conditions. J. exp. Zool. 1949; *111*: 457.

Hamburger V, Oppenheim R: Prehatching motility and hatching behavior in the chick. J. exp. Zool. 1967; *166*: 171.

Hamburger V, Wenger E, Oppenheim R: Motility in the chick embryo in the absence of sensory input. J. exp. Zool. *162*: 133–160.

Hamburger V, Yip J: Reduction of experimentally induced neuronal death in spinal ganglia of the chick embryo by nerve growth factor. J. Neurosci. 1984; *4*: 767.

Harrison RG: Observations on the living developing nerve fiber. Anat. Rec. 1907a; *1*: 116.

Harrison RG: Experiments in transplanting limbs and their bearing on the problem of the development of nerves. J. exp. Zool. 1907b; *4*: 239.

Harrison RG: The outgrowth of the nerve fiber as a mode of protoplasmic movement. J. exp. Zool. 1910; *9*: 787.

Hughes AFW: *Aspects of Neural Ontogeny.* London: Logos Press, Academic Press, 1968.

Levi-Montalcini R, Hamburger V: Selective growth stimulating effects of mouse sarcoma on the sensory and sympathetic nervous system of the chick embryo. J. exp. Zool. 1951; *116*: 321.

Levi-Montalcini R, Hamburger V: A diffusible agent of mouse sarcoma, producing hyperneurotization of viscera in the chick embryo. J. exp. Zool. 1953; *123*: 233.

Levi-Montalcini R, Levi G: Récherches quantitatives sur la marche du processus de différentiation des neurones dans les ganglions spinaux de l'embryon du poulet. Arch. de. Biol. 1943; *54*: 189.

Levi-Montalcini R, Sacerdote E: Ricerche quantitative sul sistema nervoso di mus musculus. Monitore Zool. Ital. 1934; *45*: 162.

Lillie, FR: *The Development of the Chick.* New York: Henry Holt, 1909.

Preyer W: *Die Specielle Physiologie des Embryo.* Leipzig: Grieben's Verlag, 1883.

Shorey M: The effect of the destruction of peripheral areas on the differentiation of the neuroblasts. J. exp. Zool. 1909; *7*: 25.

Spemann H: Entwicklungsphysiologische Studien am Tritonei I. Roux'. Arch. f. Entw. mech. 1901; *15*: 448.

Spemann H: Die Erzeugung tierischer Chimaeren durch heteroplastische Transplantation zwischen *Triton cristatus* und *taeniatus.* Roux'. Arch. f. Entw. mech. 1921; *48*: 533–570.

Spemann H: *Forschung und Leben.* Stuttgart: J. Engelhorn, 1943.

Spemann H, Mangold H: Ueber Induktion von Embryonalanlangen durch Implantation artfremder Organisatoren. Roux'. arch. f. Entw. mech. 1924; *100*: 599.

Sperry WR: Optic nerve regeneration with return of vision. J. Neurophysiol. 1944; *7*: 57.

Sperry WR: Chemoaffinity in the orderly growth of nerve fiber patterns and connections. Proc. Natl. Acad. Sci. 1963; *50*: 703.

Sperry WR: Embryogenesis of behavioral nets. in: *Organogenesis* (ed. DeHaan and Ursprung), 161, 1965.

Weiss, P: *In vitro* experiments on the factors determining the course of the outgrowing nerve fiber. J. exp. Zool. 1934; *68*: 393.

Weiss, P: Selectivity controlling the central-peripheral relations in the nervous system. Biol. Rev. 1936; *11*: 494.

VII. Miscellaneous

AM. ZOOLOGIST, 2:119-215 (1962).

AN EMBRYOLOGIST VISITS JAPAN

V. HAMBURGER, *Washington University, St. Louis, Mo.*

It was my good fortune to be invited to Japan in spring, when the famous cherry and peach blossoms cover the countryside and invade even the serene temple gardens which symbolize the eternal life of nature and of the spirit and banish all other bright-colored flowers. Perhaps they remind you there of the evanescence of life. But this is not the mood of the blossom-viewing parties and picnics which have been popular for ages, and immortalized by poets and painters. As the 17th century poet Basho says in one of his famous short poems or *haiku* which are the classical form of Japanese lyrical poetry:

> Under the trees
> A flurry of cherry petals
> On soup and fish salad

Throngs of people still make their pilgrimage to the famous orchards and gardens; whole family clans and villages are under way, with baskets of soup and salad and sushi; they spread their mats under the trees and exhilarated by sake wine indulge in the delights of blossom viewing and chatting. On such an occasion, the overpopulation of the country was impressed on me very tangibly, when, on a balmy April day, crowds from near and far away converged on the celebrated 1000-year-old cherry groves on the mountain sides of Yoshino, and closed in on our small car, on the narrow village street, like an enormous wave which our driver found difficult to navigate.

Of course, my scientific preoccupations interfered somewhat with such pleasures. I was in Japan on a six-week visit at the invitation of the Japanese Society for the Promotion of Science. It was suggested that I make a round of visits to embryological laboratories, and give a series of lectures and informal discussions. My trip took me to Tokyo, Nagoya, Kyoto, Kobe, and to

FIG. 1. Buddhist temple near Kamakura.

Fukuoka, on the Southern Island of Kyushu. Wherever I went, I experienced the proverbial Japanese hospitality, which is much more than a traditional courtesy. It is a genuine concern and respect for the guest and for his special interests; it is the art of accommodating and helping him, in a reserved and yet warm-hearted way.

Early relations to American zoology

My first lecture was on a cold day in April at the Zoology Laboratory of the University of Tokyo. The laboratory building reminded me of similar survivors of the austere architecture of the turn of the century, in old European universities, where the inner light, rather than the sunlight, was supposed to illuminate the lecture halls. It was a small lecture room, and a little coal stove near the speaker's lectern created a pleasant temperature gradient which reached me and part of the audience. The only other Western faces in the room were the photographs of two scientists, on opposing walls. The one looked familiar and turned out to be Charles Otis Whitman, the long-time director of the Chicago Zoological Laboratory and founder of the Woods Hole Marine Biological Laboratory; and the other was E. S. Morse, another unique figure in American 19th Century zoology. Dr. T. Fujii, the present Chairman, a physiological embryologist with a broad range of interests and, like other

Address of Vice President and Chairman of Section F, AAAS, delivered at the New York Meeting, December 29, 1960.

(119)

Japanese colleagues, with a philosophical bent, introduced these two men as the founders of academic zoology in Japan.[1]

The University of Tokyo was inaugurated in 1877 in the wake of the Meiji reformation which opened the country to the Western spirit and Western educational ideas after centuries of complete seclusion. A few months after the opening of the University, an adventurous zoologist, E. S. Morse, a student of Louis Agassiz, arrived in Japan to hunt for his favorites, the brachiopods. His colorful and exuberant personality apparently impressed the authorities of Tokyo University so deeply that he was asked immediately to become the head of the newly established Zoology Department. During the same summer, he began his collecting activities on the island of Enoshima, south of Tokyo. He installed himself with considerable difficulties in a fisherman's hut near the seashore which he proudly referred to as "the first marine laboratory on the coasts of the Pacific." Much later, it was moved to Misaki, where it flourishes now as the largest marine biological station of Japan; just as the small marine laboratory created by his teacher Agassiz on Penikese Island became the leading laboratory of his country, after its transfer to Woods Hole. His large collections of marine animals became the foundation of the Tokyo Museum of Natural History. His lectures at the University, as well as his public lectures before large audiences brought Darwin's ideas to Japan. He founded the Tokyo Biological Society from which the Japanese Zoological Society descends. He was a remarkable personality with a wide range of interests. He represents the collector type of a scientist, in the best sense of the word. He collected not only brachiopods, mollusc shells, and marine animals in general, but also notes and sketches on every detail of Japanese life, which are published in his amusing two-volume diary, called "Japan Day by Day."[2]

He made one of the first systematic collections of Japanese pottery, and monographed them with the same taxonomic and artistic skill as his brachiopods. They are now in the possession of the Boston Museum of Fine Arts.[3] He also wrote the first book on Japanese houses[4] which was a classic. After a two-year tenure he was succeeded by his friend C. O. Whitman, who served for another two years. While Morse was the naturalist, Whitman brought to Japan the laboratory methods and microscopy. But the most highly valued gift of those two men to Japan was their dedication to scientific pursuit which kindled the spirit of independent research in many of their students. A surprising number of them became professors of zoology and productive scientists who firmly established indigenous Japanese zoology within a generation. However, close relations to this country continued. A number of young Japanese went to Johns Hopkins to study under W. K. Brooks, who probably attracted them not only by his scientific prominence but also by his interest in the more philosophical problems of life. These men, among them Mitsukuri and Watase, brought embryology to Japan. However, experimental embryology was transplanted from Columbia University, through Yatsu, of *Cerebratulus* fame, who was a student of E. B. Wilson. Since then, this field has flourished in Japan. Most experimental embryologists of the present generation have been inspired by Dr. Yô K. Okada, the dean of Japanese Zoology, himself a student of Watase, hence the scientific grandson of Brooks. His congenial, unpretentious personality, breadth of interests, and sense of humor reminded me of Dr. Harrison.

---

[1] Dr. Fujii very kindly prepared for me a "Short History of the Zoological Institute, University of Tokyo" from which I took some of the data presented in the following pages. See also F. R. Lillie "Charles Otis Whitman" Journ. Morphol. *22:* XIX ff, 1911; Dorothy G. Wayman, "Edward Sylvester Morse"; Harvard Univ. Press, 1942.

[2] Houghton Mifflin Co., Boston and New York, 1917.

[3] Catalogue of Morse Collection of Japanese Pottery. Boston Museum of Fine Arts, Cambridge, 1901.

[4] Japanese Homes and Their Surroundings. Boston, 1886. His arduous "collecting" zeal embraced such esoteric subjects as prehistoric shell mounds, tiles, the arrow release methods in native cultures, —and latrines of the East (published in *Am. Architect,* 1893).

Impressions of university life

I shall not give you an account of the numerous informal and, to me, very informative round table discussions over innumerable cups of tea, with colleagues and graduate students. At each place, a special meeting was arranged for graduate students who presented their own research; each one had to give a short talk, sometimes illustrated by his slides. In Tokyo, this turned out to be a lengthy session, since there are a few dozen graduate students, working on a great variety of topics, in the fields of endocrinology, cytology, physiology, embryology. Among them is the youngest son of the Emperor, Prince Masahito, more familiarly known as Prince Yoshi, who works in physiological embryology. Like other members of his family, he enjoys the liberation from strict court rules and moves about as a free citizen. He joins the staff and graduate students at luncheon each noon and shares their tea and shop talk. Most students have an adequate command of English and no trouble listening to lectures. Their attitude toward the professors has more of the respectful deference which I had been used to, in Germany; but there are signs of relaxation which should not be surprising at a time when the country, and particularly the young generation, is going through a period of profound changes in social attitudes and a radical re-evaluation of their traditional ties.

In the laboratories of the larger universities there is little of the hustle and bustle of undergraduate students; the atmosphere is more that of a graduate and research institute. In places like Tokyo and Nagoya, the first two college years, called "Junior" years, are taught in separate departments at the Faculties of Cultural Education. The Ph.D. is usually not granted for a single thesis but for evidence of sustained research ability, and predoctoral publications are quite common.

Practically every graduate (or "postgraduate") student is the recipient of a government scholarship which however is hardly sufficient for his support and is usually supplemented by a job, mostly tutoring. Students do not assist in courses, which I consider a shortcoming in their training.

But I found them all to be strongly research-minded, well trained in modern fields of biochemical and molecular biology, and remarkably well acquainted with recent literature. However, there is perhaps the danger of early specialization.

Notes on experimental embryology

Time does not permit me to give even a cursory account of the active and intense research in the field of experimental and physiological embryology that is being pursued in a number of laboratories. I became most intimately acquainted with the investigations of the different groups at Nagoya University, where I spent two pleasant weeks. The contributions of Dr. T. Yamada to our knowledge of the biochemical mechanism of embryonic induction are well known in this country. At Kyoto University, Dr. T. Okada, is also involved in immuno-embryological work and tissue dissociation studies. Dr. T. Sato at Nagoya, well known for his work on lens regeneration in amphibians, now studies this phenomenon in certain teleost species (Cobitids). This is remarkable from the evolutionary viewpoint; two distantly related small groups seem to have acquired this unusual talent quite independently of each other, while closely related species have lost it or were never in its possession. Dr. T. Yamamoto has large cultures of mutant strains of the famous Japanese medaka (*Oryzias latipes*), in which he can produce functional sex reversal in both directions, by oral administration of sex hormones.

At Konan University in Kobe, Dr. Takaya, a particularly skillful experimentalist, works on a number of induction problems; so do Dr. and Mrs. I. Kawakami, in the modern, imposing Biology Laboratory at Fukuoka. Dr. Ichikawa and his very active group in Kyoto University have worked for years on the endocrine control of insect development, using largely Lepidoptera as their experimental material. They have available a diapausing and a nondiapausing strain of swallow-tails which differ in pupal pigmentation and other characters, and are used for cell-physiological, cytochemical, and biochemical investigations.

The Zoology Department of Tokyo University includes a large group of physiological embryologists under Dr. T. Fujii. In earlier years, Dr. Fujii has contributed to the biochemical analysis of inductivity in amphibians; more recently, he has become much interested in the role of trace elements in cellular physiology. Of 32 Ph.D. candidates, eight are engaged in embryological experimental work, among them Prince Masahito. Dr. K. Takewaki represents the field of endocrinology; Dr. H. Kinoshita works in cellular physiology, and Dr. J. Ishida in the field of physiological chemistry. They all have their research associates and graduate students.

Unfortunately, time was too short for a visit to the Marine Biological Station of Misaki which is affiliated with Tokyo University. However, I met a member of the station, Dr. H. Kanatani, who acquainted me with his interesting experiments of producing bipolar forms in planaria.

In this very brief account, a reference to the leading role of Japanese Zoologists in the analysis of the fertilization process should not be omitted. I had the welcome opportunity to meet two prominent workers in this field, Dr. Sugiyama, whom I visited at the beautifully located Sugashima Marine Station which is affiliated with Nagoya University, and Dr. Katsuma Dan, who is well known to many of his colleagues in this country. He has his laboratory in a large modern building at the Metropolitan University of Tokyo which was established after the war.

In general, I was much impressed by the strong research spirit which prevails everywhere, despite the very serious handicaps of inadequate equipment and entirely inadequate financial support of research. This is mitigated by the great ingenuity and technical skill with which the researchers build their own equipment. Modern, well-equipped laboratories are growing up here and there, but some of the best work is being done under conditions which most of us would consider as aggravating, to say the least. I acquired the greatest respect for those whose research spirit is undaunted by lack of heat on cold winter days, by unreliable supply of electricity, lack of secretarial help, and inadequate library facilities; not to mention the low living standards and housing difficulties which seem to prevail everywhere. The top salaries were equivalent to about $180.00. Even if you double or treble this sum, to take account of the greater purchasing power of the yen, you are faced with a living standard below that of lower ranks in this country. Needless to say that a private telephone is a luxury that few can afford, that there are no parking problems on the campus, and that the official car of the Faculty, which is available to the Dean, is perhaps an inducement to accept this office.

Pearl Island

A trip to the widely advertised pearl culture island of the late Mr. Mikimoto took me to one of the centers of tourist traffic. Pearl Island itself, near the town of Toba, south of Nagoya, is merely a display and demonstration place with all kinds of exhibits; the extensive pearl cultures are being done in the secluded Bay of Ago, further south. The pearl oysters, mostly *Pinctada martensii*, are reared to the age of two or three years, then inoculated and placed in wire-mesh cages which are suspended from huge rafters and submerged for a period of five to seven years, well protected, and surrounded by a rich food supply. The embryologist is intrigued by the superb skill with which the girls perform the implantation of the foreign body. The oyster is mounted on a stand, pried open, and a very small bead of limestone is placed in a kind of scooping spoon, at the tip of a holder, together with a tiny strip of outer mantle tissue, and both are implanted in the ovary. All this is done with incredible speed and dexterity which is matched only by the girls who recover and sort out the finished products according to size, smoothness, luster and color. I was told that less than 20% of the pearls are of commercial value. Incidentally, the limestone for the beads is supposed to be imported from somewhere in the Mississippi Valley. Pearl culturing seems to be the only instance where the transplantation experiment became a lucrative business. As might be expected, the promoter, Mr. Mikimoto, took all the credit, and earned millions,

FIG. 2. Laboratory of the Emperor on the Palace Grounds in Tokyo.

whereas the zoologist who did the research and development, Dr. Nishikawa, was kept in the background. However, he got even with Mr. Mikimoto by marrying his daughter!

A visit at the laboratory of the Emperor

I had the privilege of meeting our most prominent colleague in Japan, the Emperor Hirohito, in a very informal way, in his biological laboratory on the Palace grounds in Tokyo. His life-long interest in biology is well known; but I did not realize that his preoccupation with marine zoology is on a highly professional level and very productive and far from an amateurish hobby.

My visit was arranged by my old friend Dr. Tadao Sato, with whom I had spent several years in the laboratory of Dr. Spemann in Freiburg, when he started his work on lens regeneration. He had been a schoolmate of the Emperor for six years, and after his return from Europe had been his tutor and assistant at his laboratory for several years. He accompanied me on this visit and was a most helpful guide and interpreter.

One morning, we met at one of the side gates of the Palace Grounds. This is a very large area of 250 acres, in the heart of Tokyo, surrounded by a wide moat and a massive stone wall with impressive watch towers at the corners. The palace grounds are really a large park with old trees, gardens, including a famous collection of bonsai or dwarf trees, residences, administration buildings and shrines. The public is admitted only twice a year, on New Year's Day and on the Emperor's birthday, and

then only to a restricted area. On the long ride through the park we could see very little of the residential area. Not only the Imperial family, but all Japanese protect their privacy and their homes with high walls and fences. When we arrived at the laboratory we were welcomed by one of the chamberlains and by an assistant in a white coat, and we settled down immediately to the inevitable cup of tea.

We had arrived more than an hour before the Emperor was expected, and I had ample time and unrestricted access to all parts of the building, of which I took full advantage. The laboratory is a modest two-story building with a greenhouse. On the first floor there are a few medium-sized laboratories for the Emperor and for the two or three assistants and a technician, and a small working library. The rest of the building houses a large collection of marine animals and plants, with emphasis on invertebrates. The collection is made up entirely of specimens which were collected by the Emperor and his staff, on dredging trips in Sagami Bay, south of Tokyo. In Hayama, a well-known resort place, he has a villa to which is attached a small marine laboratory, the home base of his yacht which is fully equipped for dredging. The flora and fauna of the Bay seem to be almost as rich, as colorful, and as diversified as that of the Gulf of Naples. The Emperor and his staff have made systematic collecting trips in the Bay over a period of more than 30 years; these dredgings have yielded by far the most comprehensive and, in fact, unequalled knowledge of marine life any-

FIG. 3. Yacht of the Emperor, used for collecting trips.

where along the Japanese coast. The material is at first carefully inspected and recorded on board and in the laboratory and then turned over to specialists for expert taxonomic description. Practically all material is published. The special interest of the Emperor is reflected in several monographs published by the Imperial Household and beautifully illustrated with color plates: two volumes on Opisthobranchia[5] and one on ascidians[6], with Japanese and English text. The collection covers all phyla and a number of rare groups. It is well organized and expertly displayed, with many labels written by the Emperor himself. I was particularly impressed by the showcases on the first floor, in which the large number of newly discovered species are on display. My estimate of 200 was confirmed by a complete annotated list which was kindly prepared for me by Dr. Tomyiama, and by the Laboratory staff. The list includes representatives of practically all major phyla. The opisthobranch Mollusca are represented by no less than 72 new species, a remarkable "catch" in a relatively small area. Many of them are bizarre forms with interesting, typically Japanese, color designs. There are 23 new ascidian species and 30 new algae. Among the curiosities is the largest known sessile ctenophore, *Lyrocteis imperatoris*, Komai,[7] a remarkable lyra-shaped creature; and a new, rare pterobranch (*Atubaria heterolopha*, Sato).[8]

The large array of mollusc shells with

FIG. 4. The Emperor on a collecting trip.

their fascinating color patterns suggests that the esthetic beauty of marine life adds to the Emperor's enjoyment of his avocation.

The meeting took place in the private laboratory, where the Emperor spends regularly two or three afternoons of the week. It is as unpretentious as his personality: a large working table near the window, with a modest microscope and a dissecting scope, and a camera for microphotography; a few book shelves, drawers for microscope slides, and a case for notebooks. The Hydrozoa were the subject of his investigations. The notebooks are filled with drawings of specimens and microscopic preparations, done by an obviously talented artist-zoologist.

The meeting was entirely unceremonious and not different from other discussions with colleagues. No time was lost with formalities. I was introduced by Dr. Sato. We pulled the laboratory chairs and a stool to the middle of the room, and I began to report briefly on my visits to laboratories and my impressions of Japanese zoology. Dr. Sato translated, but I had the impression that the Emperor followed my English well. He then inquired about my field of specialization and the work in our labora-

---

[5] a. Opisthobranchia of Sagami Bay collected by His Majesty the Emperor of Japan. By Dr. Kikutaro Baba (edited by Dr. H. Hattori, Chief, Biol. Lab., Imp. Household), Tokyo 1949, 194 pp., 50 pls., text-figs.

   b. Opisthobranchia of Sagami Bay collected by His Majesty the Emperor of Japan. Supplement. *ibid.* 1955, 59 pp., 20 pls., text-figs.

[6] Ascidians of Sagami Bay collected by His Majesty the Emperor of Japan. By Dr. Takasi Tokioka (edited by Dr. H. Hattori, Chief, Biol. Lab., Imp. Household), Tokyo 1953, 315 pp., 79 pls., 1 map, text-figs.

[7] T. Komai, 1942, The structure and development of the sessile ctenophore *Lyrocteis imperatoris* Komai. Mem. Coll. Sci., Kyoto Imp. Univ., Ser. B, Vol. 17, No. 1, 1-36.

[8] T. Komai, 1949, Internal structure of the pterobranch *Atubaria heterolopha* Sato, with an appendix on the homology of the 'notochord.' Proc. Jap. Acad., Vol. 25, No. 7, 19-24.

tory, and I launched into a short version of one of my more general lectures on neuro-embryology and showed some photographs. He interrupted occasionally and his questions showed that he had a quick grasp of the problems which are quite unrelated to his own field. Three quarters of an hour passed very quickly. At the end he expressed a sincere interest in the exchange program of scientists as one valuable device to create a better mutual understanding.

Other colleagues who have known him longer and much better confirmed my impression: that he is a remarkable personality, with a serious devotion to his scientific work to which he brings a very active mind, sustained interest, and, as I was told, a remarkably good memory. Considering the handicaps imposed by his official position, including the lack of personal contact with the stream of living science, his achievements are remarkable, indeed.

Conclusion

I should like to endorse most emphatically the idea of visits of individual scientists to foreign countries, as it is put into practice by the Japanese Society for the Promotion of Science. Such individual visits accomplish a much more intense and thorough scientific and personal communication and mutual appreciation than larger meetings can do. This is particularly important in the case of Japan. A good deal of first-rate research is being done, but part of it is published in Japanese or, though written in English, published in local Japanese Journals which are hardly accessible to Western scientists. More frequent visits of Japanese biologists to this country, sponsored by American biological societies or agencies, would be of definite benefit to our own biological sciences.

GOETHE'S *ZUR FARBENLEHRE:*
An excerpt and plate reproduced from the first edition

**Published in celebration of
the gift to Washington University
of the Gert von Gontard Collection**

**Friends of the Libraries
of Washington University
St. Louis, 1981**

414

 hile Goethe is considered to be Germany's greatest literary genius, his scientific studies have been appreciated by few. Yet he himself valued them highly, and for us they provide an insight into his philosophy of Nature which in turn aids us in a deeper understanding of his literary creations. His universal mind embraced the literature of the world, and it embraced with the same universality all realms of natural science: geology, comparative anatomy, botany, meteorology, and the physics and physiology of colors. His two-volume treatise *Zur Farbenlehre (Theory of Colors)* which appeared in 1810, when he was 61 years old, is his most comprehensive scientific achievement, the result of decades of observations and experiments. His subject is viewed under the broadest possible perspective: the physical basis of colors, their perception by the eye, the way they are experienced subjectively, their aesthetic effects and their use in painting. Goethe was an excellent observer of natural phenomena; to him, the eye was the noblest of the senses. But what makes this work unique in the scientific literature is the intuitive and symbolic (in contrast to inductive) interpretation of the phenomena.

A basic tenet of his theory is the assertion that *white light* is uniform, homogeneous. It is what he called an Urphaenomen, a primordial or ultimate phenomenon that can be comprehended intuitively, but not explained in terms of more basic units. This view is in direct opposition to Newton's theory of colors which had asserted 140 years earlier that white light is a composite of the spectral or rainbow colors. To Goethe this claim seems absurd: how can red, green and other colors be contained in white when our eye tells us that this is not so? Nevertheless, Newton's theory had been accepted universally, even in Goethe's time, and his own theory rejected, much to his distress. Yet fundamental errors of a great mind can be more provocative than the undisputed discoveries of lesser minds. In this instance, Goethe's position in his theoretical confrontation with Newton is of abiding interest, since it challenges the foundations of modern science. Before we address this issue, we shall present a few specific examples which illustrate Goethe's way of experiencing and interpreting colors.

Polarity played an important role in his mind: attraction and repulsion; separation and union; light and darkness. In his view, colors result from mixing of light and dark. All colors are darker than white and lighter than black. He says: "Closest to the light, a color appears which we call yellow; another appears next to the darkness which we call blue. When these in their purest state are so mixed that they are exactly equal, they produce a third color, called green." Hence, yellow and blue are the polar colors. In addition to polarization, he develops the principle of *intensification.* Intensified blue changes to violet, intensified yellow leads to orange, and pure purple-red is the most intense of all colors. Hence, Goethe's color circle moves from purple through violet, blue, green, yellow, orange to red.

Colors can also be viewed in a different way: turbid media, such as smoke or particles in the atmosphere can create colors. Sunlight, when seen at sunset through the turbid atmosphere, appears yellow or orange; dark mountains in haze appear blue. Of course, he knew that intermediates of white and black are shades of gray; but since all bodies are to some extent turbid, they appear colored under appropriate conditions of illumination.

While the physical aspects of Goethe's theory did not stand the test of time, it is generally acknowledged that his most valuable and insightful contributions deal with the *subjective sensations* of the eye. He was one of the first to study positive and negative after-images. To the latter category belongs the observation that if one fixates a red body on a white sheet and then removes the body, a green after-image is seen. This and similar observations led him to a very thoughtful consideration of complementary colors. He noticed that red poppies have a blue margin in twilight, and he assumed that red "calls up" or "summons" the opposite or polar color, blue.

As an artist, he paid special attention to the *aesthetic* effects of colors. He has this to say of yellow: "It excites a warm and agreeable impression. Hence, in painting it belongs to the illumined and emphatic side. If we look at a landscape through a yellow glass, particularly on a gray winter day, the eye is gladdened, the heart expands and the soul is cheered. An immediate warmth seems to envelop us." "Blue gives us a feeling of coldness and reminds us of shadows. Rooms papered with blue paper appear spacious but empty and cold." "The effect of purest red is that of seriousness and dignity but also of gracefulness and charm." "A well-illuminated landscape seen through purple glass appears awe-inspiring. It conjures up the color that heaven and earth may assume on the day of the Last Judgement."

I alluded to a profound schism between Goethe's view of Nature and modern science. To bring this into focus, we return to white light. The modern scientist would reject Goethe's claim that since the eye does not perceive colors in white it must be uniform; that the sensation of white requires a stimulus equivalent to white. He would point out that complex external stimuli may well be perceived as uniform, and that Newton and his followers have proven precisely that for colors. He would reprove Goethe for failing to distinguish between the objective external stimuli and their subjective impression. Modern science is built on a strict separation of the objective world and the subjective observer. Goethe was fully aware of this, but he refused to make this distinction because it was incompatible with his deepest intuitive insight into Nature and the nature of Man. He could not comprehend Man outside of Nature, and in his scientific endeavors, particularly in his *Theory of Colors,* he tried to reconcile what appeared to him an unnatural, unacceptable dualism. In the Introduction to his work he says: "The eye owes its existence to the light. Out of lower animals' subsidiary organs, light procreates for itself a sense organ which as it were is akin to itself; thus the eye forms itself on the light for the light, so that the inner light corresponds to the outer light." And he recreates the words of an old mystic:

> If the eye were not akin to the sun (sonnenhaft)
> How could we perceive light?
> If God's own power lived not in us,
> How could we delight in the Divine?

Obviously, this view is "unscientific." It is an effort to combine the scientific approach with poetic symbols and imagery, to separate and then reunite.

Perhaps his attitude becomes more clear in his position toward a very important aspect of science: the need to quantify and to measure. To the scientist, the transformation of qualities like red and blue into measurable quantities—in this instance, wavelengths—is an indispensable step in the direction toward the understanding of natural phenomena in abstract terms, and ultimately in general laws. To Goethe, quantifying, measuring and abstracting impoverishes the phenomena and deprives them of their sensuality and their essential nature. The moment of contemplation of the red poppy with its blue fringe captures its eternal meaning; the act of contemplation becomes the revelation of a general law or divine truth. The specific object reveals intuitively the idea of which it is a manifestation. Both are inseparable.

He goes a step further: he fully realizes the potentials of modern science, its beneficial promises, but also the demonic Faustian forces moving it relentlessly along a never-ending path in search of answers which inevitably become more abstract and incomprehensible. (What would the claim of the physicist that mass and energy are two aspects of the same have meant to Goethe?) The thought that this search is unchecked by humane and ethical considerations, that it is a path of ultimate non-responsibility frightened him. In a prophetic way the unleashing of the atom and the gene are anticipated.

For himself, he deferred to a limit, a boundary self-imposed by reverence for the divine and fear of the demonic. For him, there are questions— mysteries if you will—that should be left untouched in a spirit of humility. From this perspective, the *Theory of Colors* is more than a scientific treatise. At the threshold of modern science it is a last titanic effort to stem the tide and to chart for science a course in which human measure is respected and the micro- and macrocosmos remain united.

Viktor Hamburger
Edward Mallinckrodt
Distinguished University
Professor Emeritus of Biology,
Washington University in St. Louis

Wenn nun die objectiven Versuche gewöhnlich nur
mit dem leuchtenden Sonnenbilde gemacht wurden, so
ist ein objectiver Versuch mit einem dunklen Bilde bis;
her fast gar nicht vorgekommen. Wir haben hierzu aber
auch eine bequeme Vorrichtung angegeben. Jenes gro;
ße Wafferprisma nehmlich stelle man in die Sonne und
klebe auf die äußere oder innere Seite eine runde Pap;
penscheibe; so wird die farbige Erscheinung abermals
an den Rändern vorgehen, nach jenem bekannten Ge;
setz entspringen, die Ränder werden erscheinen, sich in
jener Maße verbreitern und in der Mitte der Purpur
entstehen. Man kann neben das Rund ein Viereck in
beliebiger Richtung hinzufügen und sich von dem oben
mehrmals angegebenen und ausgesprochenen von neuem
überzeugen.

Objective experiments have been usually made with the sun's
image: an objective experiment with a dark object has hitherto scarcely been thought
of. We have, however, prepared a convenient contrivance for this also. Let the large
water-prism before alluded to be placed in the sun, and let a round pasteboard disk be
fastened either inside or outside. The coloured appearance will again take place at the
outline, beginning according to the usual law; the edges will appear, they will spread in
the same proportion, and when they meet, red will appear in the centre. An intercepting
square may be added near the round disk, and placed in any direction *ad libitum,*
and the spectator can again convince himself of what has been before so often described.

German text:
Goethe, Johann Wolfgang von. *Zur Farbenlehre.*
Tübingen: J. G. Cotta, 1810. Vol. 1, pp. 125-26.

English text:
Goethe, Johann Wolfgang von. *Theory of Colours.*
Translated from the German with notes by Charles
Lock Eastlake. Cambridge, Mass. and London:
The M.I.T. Press, 1970. p. 137.

Permissions

Birkhäuser Boston would like to thank the original publishers of the papers of Viktor Hamburger for granting permission to reprint specific papers in this collection.

I.   Developmental Neurobiology—Reviews

[3]   English translation by the author, from *Naturwissenschaften* **15**, © 1927 by Springer-Verlag Heidelberg.

[28]   Reprinted from *Annals N.Y. Acad. of Sci.* **55**, © 1952 by Annals of New York Academy of Science.

[32]   Reprinted from *Biochemistry of the Nervous System*, edited by H. Wallsch, © 1955 by Academic Press.

[41]   Reprinted from *Journal of Cellular and Comparative Physiology* **60**, © 1962 by Alan R. Liss Inc.

[66]   Reprinted from *Neurosciences Research Program Bulletin* **15** (suppl.), © 1976 by MIT Press.

[71]   Reprinted from *Scripta Varia* **45**, © 1980 by Pontificia Academia Scientiarum and Elsevier Science Publishers BV.

[75]   Reprinted from *Neurosci. Comment.* **1**, © 1982 by the Journal of Neuroscience.

[76]   Reprinted from *Medicine, Science and Society*, edited by K.J. Isselbacher, © 1984 by Churchill and Livingstone.

II.   Development of Motility and Behavior

[43]   Reprinted from *Quart. Rev. of Biol.* **38**, © 1963 by Stony Brook Foundation, Inc.

[49]   Reprinted from *27th Symp. of the Soc. of Develpm. Biol.: Developm. Biol. Suppl.* **2**, © 1986 by Academic Press.

[54]   Reprinted from *Neuroscience Second Study Program*, 1970, **15**, pp 141–151 by copyright permission of the Rockefeller University Press.

[58]   Reprinted from *Studies in the Development of Behavior and the Nervous System*, edited by Gilbert Gottlieb, © 1973 by Academic Press.

III.   History of Neurogenesis

[61]   Reprinted from *Perspect. Biol. Med.* **18**, © 1975 by The University of Chicago Press.

[70]   Reprinted from *Perspect. Biol. Med.* **23**, © 1980 by The University of Chicago Press.

[74]   Reprinted from *Trends in Neuroscience* **4**, © 1981 by Elsevier Publications Cambridge.

[81]   Reprinted from *J. Neurosci.* **8**, © 1988 by the Journal of Neuroscience.

IV.   Developmental Genetics and Evolution

[16]   Reprinted from *Biol. Symposia* **6**, © 1942 by R.R. Bowker.

[72]   Reprinted from *The Evolutionary Synthesis*: Perspectives on the Unification of Biology, edited by E. Mayr and W.B. Provine, © 1980 by the President and Fellows of Harvard College. Reprinted by permission of Harvard University Press.

V.   Book Reviews

[A1]   Reprinted from *Quart. Rev. of Biol.* **18**, © 1943 by Stony Brook Foundation.

[A6]   Reprinted from *Quart. Rev. of Biol.* **43**, © 1968 by Stony Brook Foundation.

[A7]   Reprinted from *Am. Scientist* **58**, © 1970 by Sigma Xi, Scientific Research Society.

[A8]   Reprinted from *Quart. Rev. of Biol.* **45**, © 1970 by Stony Brook Foundation.

VI.   Biographical and Autobiographical

[51]   Reprinted from *Experientia* **25**, © 1969 by Birkhäuser Verlag AG.

[79]   Reprinted from *Trends in Neuroscience* **8**, © 1985 by Elsevier Publications Cambridge.

[83]   Reprinted from *International Journal of Developmental Neuroscience* **8**, 121–131 © 1990 by Pergamon Press.

VII.   Miscellaneous

[42]   Reprinted from *Am. Zool.* **9**, © 1962 by American Society of Zoologists.

[A9]   Reprinted from Goethe's "Zur Farbenlehre" issued by Friends of the Libraries of Washington University, Introduction © 1981 by Viktor Hamburger.